THE GEORGE FISHER BAKER
NON-RESIDENT LECTURESHIP IN CHEMISTRY
AT CORNELL UNIVERSITY

PRINCIPLES OF POLYMER CHEMISTRY

BY PAUL J. FLORY

PRINCIPLES OF
POLYMER CHEMISTRY

By Paul J. Flory

Professor of Chemistry, Stanford University

Cornell University Press

ITHACA AND LONDON

International Standard Book Number 0-8014-0134-8

PRINTED IN THE UNITED STATES OF AMERICA BY THE

GEORGE BANTA COMPANY, INC., MENASHA, WISCONSIN

Preface

IT WAS the author's privilege to hold the George Fisher Baker Non-resident Lectureship in Chemistry at Cornell University during the spring of 1948. This book had its inception at that time. The uncompleted manuscript was laid aside shortly thereafter, and the work was not resumed until 1951. During the intervening years much new material became available from the ever-increasing literature, and major revision of manuscript and outlines consequently was necessary. The momentum of extensive investigations on the kinetics of addition polymerization, undertaken in various laboratories at the close of the war, carried this phase of the subject to an advanced stage of development. The material on dilute polymer solutions, appearing in the later chapters, had to be completely recast and expanded in the face of recent theories and experiments. The book consequently goes considerably beyond the Baker lectures both in scope and in the inclusion of new developments.

The field of polymer science has grown very large indeed, and it would scarcely be possible in a single volume to do justice to all the excellent researches in various branches of the subject. Even with a less ambitious objective, selection of material to be included is difficult and an element of arbitrariness is unavoidable. The author has been guided in his choice of material by a primary concern with principles. Out of the vast research effort carried on by many investigators over the past twenty years, and especially during the last decade, certain reasonably well-defined generalizations have emerged. It was felt that the time had come when these should be brought together in a book. In accordance with this objective, experimental results have been introduced primarily for illustrative purposes and to develop the basis for these principles. Descriptions of the properties of specific polymers and extensive cataloging of accumulated data, except as they may serve the foregoing objective, lie outside the intended scope

v

of this project. Some of the more important experimental methods are summarized, but there has been no attempt to offer detailed descriptions of apparatus or procedures.

Even in the selection of that which may be regarded as having achieved the status of a principle, the author admittedly exercised arbitrary judgment, and to a degree that may evoke censure. In particular, he may be criticized for myopic preoccupation with his own work. While granting that the allegation may be well founded, he would nevertheless call attention to the desirability of a unified point of view in a book of this sort, and he could scarcely have hoped to maintain a proper perspective from the outlook of someone else. However this may be, it is undeniably true that some important subject matter has been omitted. The subject of mechanical properties has been slighted, dynamic behavior having been omitted altogether. Originally a chapter on these topics was contemplated, but before Chapter XIV was completed the book had considerably exceeded its projected size and the author's patience. He therefore sought a measure of comfort in the dubiously tenable position that the subject of dynamic properties, being in the process of rapid growth on the one hand while deficient in theoretical interpretations on the other, should perhaps be postponed for some other volume, or possibly for a revision of this one—preferably by another author.

There are two introductory chapters. The first presents an interpretation of early developments which proved rewarding to the author, though it may seem an unnecessary embellishment to some readers. The second chapter, written primarily for the newcomer, is intended to set forth essential definitions and elementary concepts. The next three chapters deal with polymerization and copolymerization reaction mechanisms and kinetics. Chapters VI through IX cover the broad field of polymer constitution, including structure, molecular weight determination, and molecular distribution. The final five chapters treat molecular configuration and associated properties of polymers and their solutions. Familiarity with the material ordinarily included in courses in organic chemistry, physical chemistry, physics, and calculus are prerequisites, and some knowledge of thermodynamics and statistical mechanics is assumed in the later chapters. Explanation of standard topics belonging to these fields has not, in general, been attempted; the consultation of textbooks by the reader to the extent required seemed a preferable solution. No previous knowledge of polymers has been assumed, and the various chapters are addressed primarily to the beginner. At the same time it is hoped that some of the subject matter may prove useful to the experienced investigator as well. Certain derivations, possibly of less interest to the general reader, have been included in Appendixes to a number of the chapters. In addition to the usual indexes, a glossary of the more widely used symbols has been included.

The author takes pleasure in acknowledging the generous assistance of many of his colleagues. He is especially indebted to Prof. P. Debye, who tendered the invitation to the Baker Lectureship; to Drs. T. G. Fox, Jr., L. Mandelkern, W. R. Krigbaum, and A. R. Shultz, whose expertly conducted investigations while officially collaborators of the author are repeatedly referred to in this book; to Prof. R. M. Fuoss for many valuable criticisms of the manuscript and to Prof. H. A. Scheraga and Dr. L. Mandelkern for reading and criticizing portions of the manuscript; to Dr. Helen Bedon, Mr. T. E. Dumitru, and Mr. A. T. McIntyre for proofreading and assistance with the indexing.

Permissions granted by the following journals and publishers are gratefully acknowledged: Academic Press, Inc., Annual Reviews, Inc., *Canadian Journal of Research, Chemische Berichte, Die Makromolekulare Chemie, India Rubber World, Industrial and Engineering Chemistry*, Interscience Publishers, Inc., *Journal of the American Chemical Society, Journal of Applied Physics, Journal of Chemical Physics, Journal of the Chemical Society* (London), *Journal of Colloid Science, Journal of Polymer Science, Journal of Research of the National Bureau of Standards, Transactions of the Faraday Society*, Williams and Wilkins Company, and *Zeitschrift für physikalische Chemie*.

P. J. F.

Ithaca, New York
September 3, 1953

Contents

CHAPTER IV

POLYMERIZATION OF UNSATURATED MONOMERS
BY FREE RADICAL MECHANISMS   106

CHAPTER V

COPOLYMERIZATION, EMULSION POLYMERIZATION AND IONIC POLYMERIZATION

CHAPTER VI

THE STRUCTURE OF VINYL POLYMERS

CHAPTER VII

DETERMINATION OF MOLECULAR WEIGHTS 266

CHAPTER VIII

MOLECULAR WEIGHT DISTRIBUTIONS IN
LINEAR POLYMERS   317

CHAPTER IX

MOLECULAR WEIGHT DISTRIBUTIONS IN NONLINEAR
POLYMERS AND THE THEORY OF GELATION 347

CHAPTER X

CONFIGURATION OF POLYMER CHAINS 399

Chapter XI

RUBBER ELASTICITY   432

CHAPTER XII

STATISTICAL THERMODYNAMICS OF
POLYMER SOLUTIONS   495

CHAPTER XIII

PHASE EQUILIBRIA IN POLYMER SYSTEMS

Chapter XIV

CONFIGURATIONAL AND FRICTIONAL PROPERTIES OF THE POLYMER MOLECULE IN DILUTE SOLUTION

PRINCIPLES OF POLYMER CHEMISTRY

CHAPTER I

Historical Introduction

THE hypothesis that high polymers are composed of covalent structures many times greater in extent than those occurring in simple compounds, and that this feature alone accounts for the characteristic
properties which set them apart from other forms of matter, is in large
measure responsible for the rapid advances in the chemistry and
physics of these substances witnessed in recent years. This elementary concept did not gain widespread acceptance before 1930, and
vestiges of contrary views remained for more than a decade thereafter.
The older belief that colloidal aggregates, formed from smaller molecules through the action of intermolecular forces of mysterious origin,
are responsible for the properties peculiar to high polymers is repudi-
ated by this hypothesis. Such characteristic properties as high viscosity,
long-range elasticity, and high strength are direct consequences of the
size and constitution of the covalent structures of high polymers.
Intermolecular forces profoundly influence the properties of high
polymers, just as they do those of monomeric compounds, but they are
not primarily responsible for the characteristics which distinguish
polymers from their molecularly simple analogs. As a corollary of the
prevailing viewpoint, the forces binding atoms of high polymers to one
another, i.e., the covalent bonds, may be considered entirely equivalent to those which occur in analogous monomeric substances; intermolecular forces likewise are of a similar nature.

The implications of the foregoing concept have profoundly influenced
modern trends in polymer research. If polymers owe their differences
from other compounds to the extent and arrangement of their "primary
valence" structures, the problem of understanding them is twofold.
It is necessary in the first place to provide appropriate means, both
experimental and theoretical, for elucidating their macromolecular
structures and for subjecting them to quantitative characterization.
Secondly, suitable relationships must be established to express the de-

pendence of physical and chemical properties on the structures so evaluated. It may appear incredible that significant investigations from this obvious point of view were not undertaken in so manifestly important a field before about 1930, and that noteworthy advances have occurred principally since 1940. The reasons for the delay in the evolution of a rational approach to the study of high polymers would be difficult to explain adequately in few words. An insight into them and a perspective with respect to the course of more recent investigations may be gained, however, through a survey of the circumstances leading up to the eventual acceptance in the early 1930's of the primary valence, or "molecular," viewpoint concerning the constitution of polymeric substances.*

The earliest reported studies of polymeric substances which may be regarded as significant from the point of view expressed above were carried out by two essentially independent groups of investigators. On the one hand, there were those concerned with the physical and chemical constitution of the natural polymers—starch, plant fibrous material (cellulose), proteins, and rubber. The other group consisted of the synthetic organic chemists of the latter half of the nineteenth century who, though not primarily interested in polymeric substances, inadvertently came upon synthetic polymers incidental to the pursuit of other objectives. Neither group of investigators appears to have been aware of the significance with reference to the naturally occurring polymers of these occasionally reported syntheses of polymeric products. The conclusions which were reached from results obtained in each of these two fields will be examined below.

1. EARLY INVESTIGATIONS ON NATURALLY OCCURRING POLYMERS

The tendency to attribute the unique properties of the naturally occurring high polymers to a different type of organization of matter is almost as old as modern chemistry. Thomas Graham[1] in 1861 called attention particularly to the slow, or even negligible, rates of diffusion of certain polymers in solution, and to their inability to pass through semipermeable membranes. He coined the descriptive term colloids, signifying "gluelike," for these substances. Other chemical species, which in most cases can be obtained in macroscopically crystalline

---

* The term molecule has been purposely avoided in the above discourse for the reason that it cannot be applied appropriately to a most important class of polymers, namely, those composed of indefinitely large network structures. Even in the case of linear polymers, which are composed of finite (although in many cases exceedingly large) molecules, the continuity of structure may deserve greater emphasis than the individuality of the molecules. This is particularly true insofar as properties of the bulk polymers are concerned.

form, were classified as "crystalloids." The distinction is valid and the inclusion within a single class of the diverse materials now known to be polymeric is commendable.

Later colloid chemists[2] (Wo. Ostwald in particular) advanced the concept of the colloidal *state* of matter. Virtually any substance, according to this view, may be obtained in the colloidal state, just as it may occur under various conditions as a gas, a liquid, or a solid. If a colloidal solution is defined as one in which the dispersed particles comprise many chemical molecules, this may be an acceptable assertion; surely it is conceivable that any molecular substance may, under suitable conditions, aggregate to particles which are nevertheless small, i.e., of colloidal dimensions. The implied converse of the concept of colloid particles as the manifestation of a state of matter does not follow, however. That is to say, many colloidal substances (as originally defined by Graham) are known which do not revert to "crystalloids" without chemical change. Thus, the individual molecules of cellulose and of high molecular weight polystyrene are typical colloids according to the intent of Graham's definition, but they cannot be disaggregated by any process corresponding to a physical change of state. A "crystalloidal solution" of such substances is therefore unattainable. Hence the concept of the colloidal state as a purely physical state of organization is inapplicable to the very substances for which the term colloid was originally chosen. For many years investigators were seldom concerned with, or aware of, the distinction between a colloidal particle composed of numerous molecules of ordinary size held together by intermolecular "secondary valence" forces of one sort or another and a polymer *molecule* made up of atoms held together exclusively by covalent bonds.

As a result of this extension and alteration of the term, other forms of matter having little in common with macromolecular substances eventually came to be designated also as colloids. Gold sols, soap solutions, and colloidal solutions of tannic acid resemble solutions of high polymers to the extent that the dispersed particles are large. The internal organization of these particles is so unlike that of high polymers, however, that their designation as colloids led eventually to unfortunate misinterpretations.

Somewhat in advance of the widespread usage of the colloidal connotation of Ostwald and others, investigators in a number of instances seem to have favored the view that cellulose, starch, rubber, etc., are polymeric, the term being used in much the same sense as it is used today. The idea that proteins and carbohydrates are polymeric goes back at least to Hlasiwetz and Habermann,[3] who, in 1871, considered that these substances include a host of isomeric and polymeric species differing from one another with respect to the degree of *molecular* condensation. It is interesting to note further that Hlasiwetz and Habermann differentiated "soluble and unorganized" members of these substances, e.g., dextrin and albumin, from "insoluble organized" members such as cellulose and keratin. This distinction may be looked upon as a precursor of present-day differentiation between non-crystalline and crystalline polymers.

Prior to the work of Raoult,[4] who developed (1882–1885) the cryoscopic method for determining molecular weights of dissolved substances, and to van't Hoff's[5] formulation (1886–1888) of the solution laws, no method was available for quantitatively determining the molecular weights of substances in solution. The vapor density method obviously could not be applied to any but very low polymers. No means was at hand for determining the state of polymerization even in instances where polymerization was suspected.

To answer the question as to whether starch and the dextrins are isomeric or polymeric forms of the simple sugars, Musculus and Meyer[6] (1881) measured their diffusion rates. They concluded that dextrin molecules must be much larger than those of the sugars. This evidence apparently was not convincing, for Brown and Morris[7] in 1888 hailed Raoult's method as a means for settling the same controversy. In the following year[8] they arrived at a value of about 30,000 for the molecular weight of an "amylodextrin," formed in the hydrolytic degradation of starch, from measurements of the freezing point depressions of its aqueous solutions. These authors, in common with other investigators of the same period, were concerned over the possible inapplicability of Raoult's method to colloidal solutions. On the basis of other measurements on solutions known to be of a colloidal nature, Brown and Morris concluded that the method should be applicable.

Lintner and Düll[9] (1893) assigned the formula $(C_{12}H_{20}O_{19})_{54}$ to amylodextrin. The degree of polymerization was based on a cryoscopic molecular weight of 17,500. Rodewald and Kattein[10] (1900), from osmotic pressure measurements carried out on starch iodide solutions, arrived at the somewhat higher molecular weights, 39,700 and 36,700, for starch. A value of 10,000 was obtained in 1900 by Nastukoff[11] for the molecular weight of cellulose nitrate from the boiling point elevations for its solutions in nitrobenzene. He was aware of the importance of extrapolating his measurements to zero concentration, a requisite which other investigators as much as forty years later chose to disregard.

Contemporaneously with the investigations of Brown and Morris[7,8] on the molecular complexity of starch and its degradation products (dextrins), Gladstone and Hibbert[12] applied Raoult's method to rubber, obtaining values from 6000 to "at least 12,000 . . . if the method holds good" for the molecular weights of various specimens. In view of its noncrystallinity, sensitivity to heat, slow solution rate, and high molecular weight, they concluded that caoutchouc belongs to the class of colloids and that the "colloidal molecule" contains a very large number of atoms. Their molecular weight values are now known to have been much too low; this was due, in part at least, to failure of the authors to extrapolate their measurements to zero concentration. Nevertheless, these values indicated molecules containing hundreds of isoprene units, a size which seems to have been too great to be accepted by investigators of that day.

It is apparent from the above citations that a few early investigators were led by their experiments to favor the view that cellulose, starch, and rubber are composed of very large molecules. If the results of their molecular weight determinations had been accredited, the concept of giant molecular structures might have been established long before the 1930's. Actually, they were not accepted for several reasons, some of which have been alluded to. In the first place, these naturally occurring materials were among those originally classified as colloids by Graham. With the advent of the notion that typical colloids such as these are substances distinguished merely by an exceptional tendency to occur in that condition referred to as the colloidal state, the applicability of physicochemical laws to colloids was questioned.[13] Negligible freezing point depressions and boiling point elevations were observed for various colloidal solutions. In a solvent in which the substance could be molecularly dispersed, the behavior was "normal."[14] One or the other of two conclusions was indicated: either the solution laws are inherently inapplicable to colloidal solutions or the negligible freezing point depressions afford further evidence of the large size of the colloidal aggregate. The former alternative came to be preferred; the behavior of colloidal solutions was regarded as anomalous in many respects, and it was popularly agreed that laws which hold for ordinary solutions often do not apply to colloids. Furthermore, the Raoult method was new, and apparent deviations occurring for other substances (e.g., solutions of electrolytes) seemed to testify to its fallibility.

A second factor of importance in this connection is found in the emphasis which was placed in the 1890's and early 1900's on secondary association of molecules. Coordination complexes, the concept of "partial valences," and van der Waals forces attracted wide attention and at the same time tended to overshadow the literal valency interpretation of molecular constitution which had evolved from Kekulé and his successors. The publication by Ramsay and Shields[15] in 1893 of their well-known expression for the temperature coefficient of the molar surface energy and their observation that associated liquids exhibit abnormally large values provided a method for detecting association and for estimating its extent. Subsequent investigators quickly took advantage of this means for gaining insight into the nature of the liquid state. Molecular association unfortunately was referred to as polymerization. So great was the interest in associative polymerization that the significance of the term as applied to chemical compounds, in accordance with its definition as originally set forth by Berzelius,[16] was obscured from about 1895 to 1920. Index references to polymers and polymerization during this period almost invariably refer to physically associated substances.

In addition to the confusion in terminology, association of molecules became a favorite subject for investigation by theoretical and physical chemists. It was even suggested[17] that most physical and chemical changes are merely various manifestations of tendencies toward polymerization, in the physical sense mentioned above. That many authors of this period did not clearly distinguish large covalent structures, on the one hand, from aggregates formed through the operation of secondary forces between smaller molecules, on the other, is not surprising. The language of authors on the issue was ambiguous;* usually the distinction was ignored altogether. The emphasis placed on the not infrequent occurrence of coordination compounds and of association in liquids sustained the view that polymeric substances are similarly composed of *physical* aggregates of smaller molecules.

Mention also should be made of the reluctance of chemists to give consideration to molecules of a size commensurate with the indications of the results obtained from physical methods of measuring molecular weights. The gap between molecules of ordinary size and polymers hundreds (actually thousands) of times as large was too great to be bridged in a single leap. Organic chemists were motivated by the desire to devise concise formulas and to isolate pure substances, the term pure being synonymous with "homo-molecular" and invariably implying a formula of convenient size. Hence the quest for *the* cellulose molecule and *the* rubber molecule continued. We shall elaborate this point in the discussion of synthetic polymers in the next section.

The constitution of rubber was essentially unknown prior to the in-

* See, for example, S. V. Lebedev and B. K. Merezhkovskii, *J. Russ. Phys. Chem. Soc.*, **45**, 1249 (1913).

vestigations of Harries[18] beginning in 1904. Faraday deduced the empirical formula C_5H_8 in 1826, and isoprene was obtained by destructive distillation of rubber by Williams in 1860. Although the C_5H_8 unit was thus indicated to be a constituent of the rubber structure, its status therein remained a mystery. A relationship to the terpenes was suspected, and formulas such as $(C_5H_8)_x$ or $(C_{10}H_{16})_x$ were frequently applied to rubber, the latter formula denoting a "polymerized terpene."

The levulinic aldehyde and acid obtained by Harries[18] on hydrolyzing the ozonide of rubber demonstrated recurrence of the structure

$$\begin{array}{c} CH_3 \\ | \\ =CH—CH_2—CH_2—C= \end{array}$$

in "the rubber molecule." The structural unit is preferably expressed in terms of the sequence of atoms corresponding to an isoprene molecule as follows:

$$\begin{array}{c} CH_3 \\ | \\ —CH_2—C=CH—CH_2— \end{array}$$

These units obviously might be combined either in cyclic structures or in open chains. Choice of the latter alternative would have posed the problem of accounting for end groups, whose presence seemed to be precluded by chemical evidence indicating exactly one double bond per unit and a composition accurately complying with the empirical formula C_5H_8. These difficulties would not enter if the chain were extremely long, but this possibility did not appeal to most chemists for reasons mentioned above. Furthermore, Harries' measurements indicated molecular weights of only a few hundred for the ozonide of rubber. He concluded,[18] therefore, that the rubber *molecule* is the cyclic dimer (dimethylcyclooctadiene), many of these molecules being combined through the action of "partial valences" into much larger aggregates. These ideas were expressed in the formula

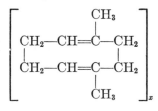

Later (1914) Harries[19] agreed to a larger ring of five, and eventually of seven, isoprene units.

Pickles[20] objected to the associated dimethylcyclooctadiene formula of Harries on the grounds that the parent molecule cannot be distilled from the assumed aggregate at high temperatures, and that saturation of the double bonds with bromine does not destroy the colloidal character of the substance. He thought that the "chain" must contain at least eight units, probably many more, but he clung to the view that the ends of the chain were joined to form a ring. A few years later Caspari[21] (1914) reported osmotic pressure measurements on dilute rubber solutions which indicated a molecular weight of about 100,000 when extrapolated to infinite dilution. Puzzled by the marked effect of concentration on the apparent molecular weight, he rejected the osmotic estimate of the molecular weight entirely.

Seeligmann, Torrilhon, and Falconnet,[22] in a treatise on rubber published in France in 1896, recorded interesting early observations on the *sol* and *gel* components now known to be present in most specimens of undegraded raw rubber. They referred to them as the two "isomeric hydrocarbides" of rubber, one being the "adhesive principle" and the other the "nervous principle." It was observed that the latter refused to dissolve on repeated treatment with fresh portions of solvent and that the approximate percentage of this constituent was roughly the same when different solvents were used to extract the soluble portion. The authors called attention also to the enormous swelling capacity of the "nervous hydrocarbide." Vulcanization was attributed to a reaction of sulfur with the adhesive principle whereby the adhesive characteristics are suppressed or eliminated.

We recognize today a remarkable measure of truth in these early interpretations. Across the channel, however, Weber[23] vigorously denounced the view that the insoluble constituent is in any way related to vulcanization, saying, in part: "The extent to which the fact of the existence of an insoluble constituent in India rubber has inspired the imagination of our French colleagues is truly astonishing."

Approximate conformity of the composition of cellulose and starch with the formula $C_6H_{10}O_5$ was deduced very early, but often the literal exactness of this formula was obscured by failure to obtain the polymer in the anhydrous condition.* In the first decade of the present century the $C_6H_{10}O_5$ formula was shown to represent the composition of

---

* The history of the achievement of present-day understanding of the constitution of cellulose has been splendidly presented by C. B. Purves in Chapter II, A, of *Cellulose and Cellulose Derivatives*, ed. by Emil Ott (Interscience Publishers, New York, 1943), pp. 29–53. The author takes pleasure in acknowledging indebtedness to this source.

cellulose precisely, and the demonstration shortly thereafter that the end product of cellulose degradation, glucose, may be obtained in near-quantitative yield indicated a unique relationship of the structural units to this monosaccharide.[24,25] The nature of the structural units occurring in cellulose and in starch was further clarified by the identification of the trimethylglucose obtained on hydrolyzing methylated cellulose[26] and methylated starch.[27]

Such observations did not, however, resolve the conflict between the ring and the chain formulas. The ring formula was overwhelmingly preferred since, as in the case of rubber, it avoided all difficulty of accounting for the apparent total absence of end groups, which in the case of cellulose would have been easily detected by their reducing character had they been present in sufficient quantity. The opinion commonly held at that time was expressed by Heuser[28] in 1922 as follows: "According to the most recent investigations the principle of the chain formula must be abandoned." He concluded that cellulose probably consists of a cyclic dimer of anhydrocellobiose. Colloidal character was attributed to partial valences or other residual affinities which supposedly caused the relatively simple cyclic molecules to associate.

We shall not attempt to review here the early history of ideas relating to proteins, the other major class of natural high polymers. Most proteins are believed to consist of a single molecular species, or of a narrow range of similar molecular species. Pioneers in this field were not confused, therefore, by the variability in properties which would have arisen from wide differences in average molecular weights of different preparations, such as occur in other natural and synthetic polymers. Although more recent work has shown that many proteins previously considered to be "pure" actually are composed of two or more quite different components, no major error was committed in considering the proteins to consist of chemical individuals. The complex arrays of amino acid constituents (and iron in the case of hemoglobin) pointed to high molecular weights. The values so estimated were not so high as we now know them to be, nor were they always accepted, but recognition of the polymeric nature of proteins is as old as the peptide theory. Emil Fischer[29] considered his eighteen-membered polypeptide to be similar in molecular weight to most natural proteins, although he recognized that different amino acids were involved and that their order along the chain did not correspond to that occurring in proteins. His polypeptide theory has received abundant confirmation, but his estimate of the length of the chain was too small by one to four orders of magnitude, depending on the protein.

2. EARLY ENCOUNTERS WITH CONDENSATION POLYMERS

The possibility of the existence of indefinitely large covalent structures is implicit in the basic concepts of structural organic chemistry. While no stress was placed on this point by early organic chemists, occasionally in the course of their investigations they inadvertently came upon reactions which led to polymeric materials. Untold numbers of polymers have been cast aside as unwanted tars or undistillable residues, the constitution of which seldom attracted interest. These were the by-products which persistently frustrated organic chemists in their quest for pure compounds in high yields. Nevertheless, a few synthetic polymeric substances did attract the attention of nineteenth century investigators. The earliest recorded instances in which the constitution of a synthetic chain polymer was assigned with some degree of correctness actually antedate Kekulé's benzene ring (1865) by a few years.

In 1860–1863 Lourenço[30] reported the synthesis of polyethylene glycols of the formula* $HO—(C_2H_4O)_n—H$ by condensing ethylene glycol in the presence of an ethylene dihalide.† Individual members of the series up to $n=6$ were isolated by distillation. Noting that the viscosity increased with n, Lourenço reasoned that the highly viscous undistillable products obtained under more drastic conditions must be of correspondingly greater complexity, i.e., n could be much greater than 6. He noted that as n increases toward infinity in his formula, the chemical composition approaches that of ethylene oxide. His remarkably accurate comprehension of the nature of the polymeric products he obtained is attested by the following conclusion:

We have demonstrated that the percentage composition and even the chemical reactions [of the hydroxyl groups] are insufficient to determine the chemical molecule of certain compounds, and that there should exist organic and inorganic bodies having the same apparent percentage composition, presenting the same reactions, and having however entirely different degrees of condensation.

Lourenço[32] also prepared poly-(ethylene succinate)

$$[—OCH_2CH_2OCOCH_2CH_2CO—]_x$$

* Lourenço[30] expressed his formulas in accordance with the conventions of the period as follows

$$\left.\begin{array}{c} n(C^2H^4) \\ H^2 \end{array}\right\} O^{n+1}$$

Although this was not an explicit representation of the structure, Lourenço clearly comprehended the complexity of his products.

† Wurtz[31] (1859) prepared the lower polyethylene glycols by condensing ethylene oxide.

by heating ethylene glycol and succinic acid, but paid little attention to its constitution. Presumably he included it as a member of the class of "highly condensed" materials. Thirty-one years later Vorländer[33] assigned to it the cyclic formula (sixteen-membered ring) composed of two units. In the same year Roithner[34] tentatively suggested a cyclic tetrameric formula for polymers prepared from ethylene oxide, in spite of cryoscopic and ebullioscopic measurements which indicated an approximate empirical formula $(C_2H_4O)_{30}$.

Only four years after Kekulé proposed the ring structure for benzene, and before the configuration of salicylic acid was established, Kraut[35] concluded on the basis of sound chemical evidence that the products obtained on heating acetylsalicylic acid possess chain structures formed through intermolecular esterification. He assigned the dimeric and tetrameric formulas

$$HOC_6H_4COOC_6H_4COOH$$

and

$$HOC_6H_4COOC_6H_4COOC_6H_4COOC_6H_4COOH$$

to two products thus obtained. To the "salicylide," $C_{14}H_8O_4$, obtained by Gerhardt[36] (1853) through the action of phosphorous oxychloride on sodium salicylate, Kraut[35] assigned the analogous octameric chain formula which he considered to be the end product of salicylic acid condensation. This curious notion of condensation in discrete steps from one intermediate to another of twice the degree of polymerization, no other species being formed, has been advanced repeatedly, appearing even within recent years. Kraut, like investigators for many years thereafter, disregarded the fact that the octamer, and probably his "trisalicylsalicylic acid" as well, were mixtures of polymeric species. Nevertheless, recognition of the essential nature of chain polymeric structures was clearly expressed in his work.*

Polymers isomeric with the polysalicylides were prepared from m- and p-hydroxybenzoic acids in 1882 by Schiff[39] and independently by Klepl[40] in the same year. To his various products Schiff, like Kraut, assigned dimeric, tetrameric, and octameric chain formulas, which probably are of limited significance since they were based principally on carbon and hydrogen analysis. Klepl obtained in addition to lower condensates a product, $C_7H_4O_2$, which he concluded must be of "high molecular weight." Piutti[41] (1883), working in Schiff's labora-

---

* Researches on the polysalicylides have been continued intermittently over the period extending from 1853 to the present. Cyclic dimers and tetramers have been reported in addition to the linear polymers.[37,38]

tory, prepared an analogous polymer from *m*-aminobenzoic acid. It probably represents the first synthetic polyamide.

The preparation of polymers in this early phase of synthetic organic chemistry seems not to have been uncommon. It must not be concluded from the above citations, however, that their polymeric nature usually was comprehended. In the vast majority of instances this was not the case.

Märcker[42] reported in 1862 on the formation of a product in the pyrolysis of "salicylide" which melted at 103°C and possessed the empirical formula C_6H_4O. These observations were confirmed by Kraut,[35] and the product was given the name phenylene oxide. Unquestionably it was polymeric but, like numerous other polymers encountered by accident, it has never been investigated as such.

In 1863 Husemann[43] prepared an intermediate, to which he assigned the formula C_2H_4S, by the action of sodium sulfide on ethylene bromide. From it he obtained the cyclic dimer, dithiane, by distillation. Mansfeld[44] (1886) reinvestigated the intermediate and concluded that it is a polymer. As a reminder of the significance of the term polymer at that time it is to be noted, however, that Mansfeld suggested the cyclic trimeric formula for the intermediate, which is now known to be a linear polymer. Other polymers prepared similarly by Husemann[45] (1863) include methylene sulfide ($-CH_2-S-)_x$ and methylene trithiocarbonate ($-CH_2-S-CS-S-)_x$. Neither was recognized as a polymer, and neither has since been investigated from this standpoint.

Recognition of the basic nature of condensation polymers did not advance perceptibly from the early work of Lourenço and Kraut up to 1910 or 1920, or even somewhat later. In fact, evidence of retrogression could be cited. Some further examples serve to illustrate the state of affairs prevailing at the turn of the century.

Birnbaum and Lurie[46] in 1881 assigned the cyclic formula

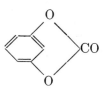

to the condensation product obtained from resorcinol and phosgene in the presence of pyridine. Einhorn,[47] who repeated this work and extended the reaction to hydroquinone, asserted in 1898 that products prepared in this manner are polymeric. He chose formulas such as

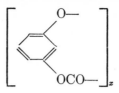

which are somewhat indefinite since they do not specify whether the structure consists of a ring (a large one, perhaps) or an open chain. In 1902 Bischoff and von Hedenström[48] prepared oxalates and succinates in addition to the carbonates of the dihydric phenols. Their formulas, such as the following:

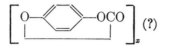

were scarcely less ambiguous. (Formulas of this nature were not uncommon at that time and even later.) Doubtless the authors found the monomeric anhydride formulas convenient for the purpose of representing the empirical composition and the nature of the functional group. In consideration of the fact that properties of such compounds are at variance with those to be expected from the monomeric formulas, some sort of polymerization was invoked to explain the discrepancy. The use of the chain formula would entail a commitment in regard to the nature of the end groups, and some authors may have chosen the cyclic formula merely to avoid the end group enigma. In other instances, however, authors seem to have attached real significance to the ring formula, polymeric character supposedly being superimposed on the basic structure through the agency of partial valences or association forces. The question mark occurring in the Bischoff and von Hedenström formula may be taken as an indication of the uncertainty in the minds of authors of that period.

In 1893 Bischoff and Walden[49] showed that the "isomer" of glycolide previously prepared[50] by heating the sodium salt of chloroacetic acid actually is a polymer from which the cyclic dimer, glycolide

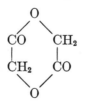

can be distilled. The dimer was observed to revert to the polymer on heating or in the presence of zinc chloride. Fichter and Beisswenger[51] (1903) observed the similar reversible polymerization of δ-valerolactone, likewise a six-membered ring. They assigned the formula $(C_5H_8O_2)_n$ to the polymer; it was estimated from ebullioscopic measurements that $n = 5$ to 7, values which doubtless are much too low. While they did not subscribe specifically to a cyclic structure for this polymer, the empirical formula given, in conjunction with the low value of n, clearly is incompatible with a chain structure. On the other hand, Blaise and Marcilly[52] (1904) specified chain formulas for the low molecular weight polyesters obtained from hydroxypivalic acid, and Blaise[53] (1906) pointed out that the anhydrides of the dibasic acids of the series $HOOC(CH_2)_nCOOH$ wherein $n = 4$ to 8 must be chain polymers.*

The polypeptides prepared from glycine and its derivatives are of particular interest. Curtius[56] observed that ethyl glycinate decomposes to the cyclic dimeric anhydride of glycine (i.e., diketopiperazine) and a biuret base. Schwarzschild[57] (1903) concluded that this latter substance is the linear heptamer, ethyl hexaglycylglycinate, but Curtius[58] (1904) claimed it to be the linear tetramer. Similar synthetic peptides were obtained by Balbiano[59] (1901) and by Maillard[60] (1914) on heating glycine in glycerol. The latter investigator preferred a cyclic octameric ring formula for the highest condensation products.

Leuchs[61] and co-workers (1906–1908) prepared various polypeptides by decomposing the N-carboxyanhydrides of α-amino acids, but they did not suggest chain formulas for their products. Instead, they assumed that a three-membered ring structure was formed as indicated by the following equation:

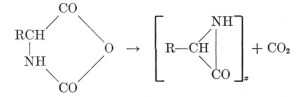

In order to account for the nonvolatility, infusibility, and limited solubility, Leuchs postulated polymerization of the "ground type" cyclic compound, as indicated by the subscript x in his formula given above. It is now well established that linear polypeptides are produced on decarboxylation of the N-carboxyanhydrides of α-amino acids, and under favorable conditions the chain length may be fairly large. Leuchs favored the view that strained rings, i.e., those of other than five or six

* Voerman[54] had contended that the anhydrides obtained from all dibasic acids of this series are monomeric. See also Staudinger and Ott[55] on the polymeric anhydride of dimethyl malonic acid.

members,* tend inherently to polymerize by "secondary valences" of some sort. In support of this view, Leuchs called attention to the polymeric nature of the condensation products synthesized about the same time by von Braun[62] from ε-aminocaproic acid and from ζ-amino-heptanoic acid. Leuchs assumed that these products likewise consisted of monomeric lactams, and commented that his three-membered lactams were as difficult to prepare in the unpolymerized form as the seven- and eight-membered homologs obtained by von Braun. The latter actually did not go so far in assigning definite structures to his products; he expressed their structures by the noncommittal formulas

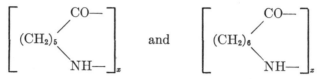

These are now known to have been linear polyamides, probably of high molecular weight.

Most linear condensation polymers occur in the crystalline condition and their melting points are fairly sharp, particularly if they are carefully determined and the molecular weights are high. They may be purified by recrystallization and their melting points are reproducible. These characteristics doubtless caused many investigators to mistake them for the desired cyclic compounds. Of the various ring structures of improbable size reported in the literature of this period, many appear to have been linear polymers,[63] some perhaps being of moderately high degrees of polymerization. A few of these are listed below:

Reported formula	Melting point, °C	Reference and year
C_6H_4S (the major product from which	295°	(64) 1897

was isolated in small yield)

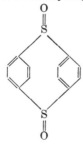

305°   (65) 1910

* The earlier suggestion of Sachse that larger rings are nonplanar, and therefore strainless, was not revived by Mohr until ten years later.

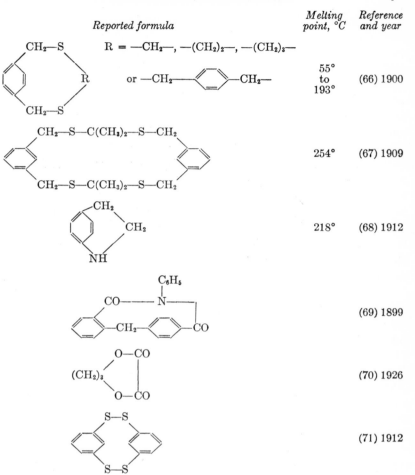

Reported formula	Melting point, °C	Reference and year
	55° to 193°	(66) 1900
	254°	(67) 1909
	218°	(68) 1912
		(69) 1899
		(70) 1926
		(71) 1912

None of these, nor many other polymers for which evidence is to be found in the chemical literature, have been investigated as polymeric substances.

In summary, it seems but a short step from the conclusions of Lourenço or of Kraut, reached at the virtual beginning of structural organic chemistry, to the consideration of much longer chain molecules and to the realization that such products necessarily would consist of mixtures of molecular species difficult to separate. Had the implications of these observations been pursued, synthetic polymer chemistry conceivably might have evolved much earlier. But chemistry did not, for one reason or another, undergo development in this direction.

Uppermost in the minds of chemists seems to have been the desire to prepare or to isolate pure substances, a pure substance being defined as one which consists exclusively (within standard limits of purity) of a single molecular component. By the turn of the century this objective had crystallized to a discipline which dominated synthetic organic chemistry. To be eligible for acceptance in the chemical kingdom, a newly created substance or a material of natural origin had to be separated in such a state that it could be characterized by a molecular formula. The investigator was obliged to adduce elementary analyses to confirm the composition, and to supplement these with molecular weight determinations for the purpose of showing that the substance was neither more nor less complex than the formula proposed. Otherwise the fruits of his labors would not be elevated to an honored place in the immortal pages of the chemical compendiums. The successes of synthetic organic chemistry in creating the hundreds of thousands of different combinations and permutations of atoms must not be discounted. In magnitude of creative achievement, they are scarcely surpassed in any other field of science.

While this discipline was strikingly successful, it also tended to narrow the outlook of contemporary researchers. They came to believe that every definable substance could be classified in terms of a single definite molecule capable of being represented by a concise formula. This point of view was not held by organic chemists alone, but was probably in the minds of physical chemists as well. Theoretical chemists focused their attention on *the molecule* as the entity from which stems all of the observed physical and chemical properties of the substance in question. They particularly preferred to frame laws for ideal molecules which could be approximated by spheres only a little larger than atomic dimensions. A full understanding of the mechanics of the molecule would suffice to explain everything. There is nothing wrong with this viewpoint in itself; it merely overlooks the existence of the large and most important class of the naturally occurring substances, and of numerous synthetics as well, which cannot be reduced to the notation of a characteristic molecular formula. The idea of a molecule as a primary entity is of limited value in the interpretation of such substances. (Metals and many inorganic materials might be included in this category also.)

While the possibility of forming giant chemical structures composed of linear chains, branched chains, or networks was implicit in the foundations of structural chemistry, chemists generally hesitated to discard the molecule as the prime goal of their researches.

3. VINYL POLYMERS

Observations on the polymerization of readily polymerizable vinyl monomers such as styrene, vinyl chloride, and butadiene date back approximately to the first recorded isolation of the monomer in each case. Simon[72] reported in 1839 the conversion of styrene to a gelatinous mass, and Berthelot[73] applied the term polymerization to the process in 1866. Bouchardat[74] polymerized isoprene to a rubberlike substance. Depolymerization of a vinyl polymer to its monomer (and other products as well) by heating at elevated temperatures was frequently noted.[73,75] Lemoine[76] thought that these transformations of styrene could be likened to a reversible dissociation, a commonly held view. While the terms polymerization and depolymerization were quite generally applied in this sense, the constitution of the polymers was almost completely unknown.

Poly-(methacrylic acid) was prepared in 1880 by Fittig and Engelhorn.[77] Mjöen[78] separated the polymer by precipitation and attempted to determine its molecular weight by cryoscopic and ebullioscopic methods. He decided that his product, which he regarded as a colloid, was an octamer but reached no conclusions as to its constitution other than that it was an octabasic acid of the formula $C_{24}H_{40}(COOH)_8$.

In 1910 Stobbe and Posnjak[79] drew the conclusion that polystyrene is a "colloidal body" after noting that boiling point elevations of its solutions are negligible. They proposed cyclic formulas composed of four, five, or possibly more structural units, e.g.

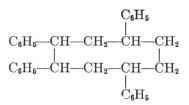

Butadiene was polymerized (1911–1913) by Lebedev[80] and by Harries,[81] both of whom at first assigned the cyclooctadiene structure

$$\left[\begin{array}{c} CH_2-CH=CH-CH_2 \\ | \qquad\qquad\qquad | \\ CH_2-CH=CH-CH_2 \end{array} \right]_x$$

to polybutadiene on the basis of the identification of succinic acid in the products obtained from the ozonide. The cyclic unit was assumed to be associated by "partial valences" or other forces, in complete analogy with the structure assumed for natural rubber. Lebedev[82,83] subsequently proposed chain structures such as

$$(-CH_2-CH=CH-CH_2-CH_2-CH=CH-CH_2-)_z$$

for polybutadiene and for rubber. While Lebedev[82,83] recognized that the polymers were of high molecular weight, he erroneously concluded[82] that the polymer is formed by a gradual passage of the monomer through successive stages of aggregation and that the entire process resembles intermolecular association.

In spite of the proposals of large primary valence structures for rubber by Pickles[20] and somewhat ambiguously for polybutadiene by Lebedev,[82] prevailing opinion favored rings of moderate size for vinyl and diene polymers. Structures similar to those widely accepted for cellulose and rubber were generally assumed.

4. RISE OF THE MACROMOLECULAR HYPOTHESIS

In an important paper published in 1920 Staudinger[84] deplored the prevailing tendency to formulate polymeric substances as association compounds held together by "partial valences." He specifically proposed the chain formulas

$$-CH_2-\underset{\underset{C_6H_5}{|}}{CH}-CH_2-\underset{\underset{C_6H_5}{|}}{CH}-, \text{ etc.}$$

$$-CH_2-O-CH_2-O-CH_2-O-, \text{ etc.}$$

for polystyrene and polyoxymethylene (paraformaldehyde), which are the ones accepted at the present time. He also advocated the long chain formula for rubber. The colloidal properties of these substances were attributed entirely to the sizes of their primary valence molecules, which he guessed might contain of the order of a hundred units. Staudinger disposed of the end group problem with the utmost facility by simply suggesting that no end groups are needed to saturate terminal valences of the long chains. At that time it seemed not implausible that free radicals at the ends of very long chains would be unreactive owing to the size of the molecule. He called attention also to the frequent erroneous assignment of relatively simple cyclic structures to substances which actually are chain polymers, mentioning as examples dimethylmalonic anhydride,[55] adipic anhydride, and glycolide.

Staudinger relentlessly championed the molecular, or primary valence, viewpoint in the years which followed. He supported his original contentions with the observation that hydrogenation of rubber, as well as its conversion to other derivatives, does not destroy its colloidal properties.[85] In contrast to association colloids, high polymers (or *macromolecules* as he chose to call them[86]) exhibit colloidal properties in all solvents in which they dissolve.[87] Polyoxymethylenes were extensively investigated,[88] leaving no real basis for doubt as to their linear polymeric nature.

The views of Staudinger were not widely accepted at once, most investigators tenaciously adhering to the finite certainty offered by ring formulas in preference to the vagaries of chains of undefined length.*,[89] Molecular weight measurements[91] which yielded moderately low values, now known to have been seriously in error,[92] seemed to support this opposition. Ring formulas composed of one or several $C_6H_{10}O_5$ units were advocated for cellulose and starch.[93,94,95,96,97] X-ray diffraction indicated unit cells for crystalline rubber and cellulose similar in size to those of simple substances,[98,99] from which it was argued that the molecules must likewise be small. The fallacy in the assumption that the molecule could be no larger than the unit cell had been pointed out earlier,[100,101] but it remained for Sponsler and Dore[102] to show in 1926 that the results of X-ray diffraction by cellulose fibers are consistent with a chain formula composed of an indefinitely large number of units. The structural units occupy a role analogous to that of the molecule of a monomeric substance in its unit cell; the cellulose molecule continues from one unit cell to the next through the crystal lattice. This interpretation nullified one of the final arguments mustered in support of the associated ring theory; it was soon extended to other linear polymers showing characteristic X-ray fiber patterns.

Meyer and Mark, who were among the foremost of the early advocates of long chain, covalent structures for polymers, retained the association hypothesis to the extent that they regarded the crystallites to be discrete units or "micelles" formed by aggregation of polymer molecules.[103,104] They proceeded in 1928 to estimate the size of the crystallite from the breadths of the X-ray diffraction spots. In this way, lengths of 50 to 150 units were deduced for the cellulose[103] and rubber[104] micelles. Micelle length was considered to be identical with molecular length, molecular weights of the order of 5000 being estimated in this manner. The much higher values, 150,000 to 400,000, which they obtained from careful osmotic pressure measurements on dilute rubber solutions were attributed to solvation, and later[105] were cited as evidence that the micelle observed by X-rays persists in solution.

Staudinger's opposing view[106] that the size of the crystallite bears no relationship to the size of the polymer molecule has been largely sub-

---

* Duclaux[90] declared in 1923 that it is useless to seek explanations for the properties of rubber on the basis of its molecular structure inasmuch as rubber should be regarded as a physical state rather than as a chemical compound.

stantiated.* Thus, neither the dimensions of the unit cell nor those of the entire crystallite are related (directly at least) to the polymer chain length. A polymer molecule may pass through the many unit cells reaching from one end of a crystallite to the other, then meander through an amorphous region and into another crystallite, etc. Extending the covalent structure idea one step further, Staudinger[107] in 1929 differentiated linear from nonlinear, or *network*, polymers. He attributed the characteristic infusibility and insolubility of the latter to the formation of network structures of great extent. Meyer and Mark,[104] a year earlier, had proposed that the properties of vulcanized rubber are to be accounted for by the formation of covalent cross-linkages, a view subsequently confirmed.

In 1929 Carothers[108,109] embarked on a series of brilliant investigations which were singularly successful in establishing the molecular viewpoint and in dispelling the attitude of mysticism then prevailing in the field. The object of these researches, clearly expressed at the outset,[108] was to prepare polymeric molecules of definitive structures through the use of established reactions of organic chemistry, and further to investigate how the properties of these substances depend on constitution. The major contributions of Carothers and his collaborators lay in the field of condensation polymers—the polyesters, the polyamides, and so forth. In a sense, Carothers extended the synthetic approach emphasized by Emil Fischer (and intimated much earlier in the work of Lourenço), discarding, however, the unnecessary and severely encumbering insistence on pure chemical individuals under which Fischer and his colleagues labored.

Scarcely had the covalent chain concept of the structure of high polymers found root when theoretical chemists began to invade the field. In 1930 Kuhn[110] published the first application of the methods of statistics to a polymer problem; he derived formulas expressing the molecular weight distribution in degraded cellulose on the assumption that splitting of interunit bonds occurs at random.

The statistical approach has since assumed a dominant role in the treatment of the constitution, reactions, and physical properties of polymeric substances. The complexities of high polymers are far too great for a direct mechanistic deduction of properties from the detailed structures of the constituent molecules; even the constitution of

---

* The views of Staudinger, on the one hand, and Meyer and Mark, on the other, actually were not so far apart as the tenor of their polemics might indicate. Their agreement on the essential macromolecular nature of high polymers appears to have contributed materially to the rejection of the simple cyclic formulas widely accepted up to that time.

polymers often is too complex for an exact description. The very complexities which make the task of rational interpretation of polymers and their properties appear formidable actually provide an ideal situation for the application of statistical procedures.

The average size and shape of a long chain molecule endowed with the ability to assume all sorts of configurations through rotations about its valence bonds have held a fascination for theoretically minded chemists and physicists for many years. In 1934 Guth and Mark[111] and Kuhn[112] independently discussed the problem, arriving at similar solutions. These theoretical investigations furnished the background for an attack on such problems as the high viscosities[112] exhibited by dilute solutions of high polymers, double refraction of flow,[112] and rubber elasticity.[111,113]

The quantitative treatment of polymer constitution and behavior, which is a necessary counterpart of the theoretical approach, could not take place until methods for quantitatively characterizing polymer structures had been established. The various properties of polymers which depend on molecular weight obviously could not be satisfactorily elucidated prior to the establishment of means for assigning correct molecular weight values. Associated therewith is the related problem of characterizing molecular weight distributions. Staudinger deserves credit for emphasizing that the molecular weight of a linear polymer may be related directly to the viscosity of its dilute solutions.[114] Solution viscosity is a readily measured quantity, and such measurements are widely used technically for this reason. The relationship between polymer solution viscosity and the molecular weight of the polymer is therefore of considerable utility for both pure and applied investigations. Staudinger erroneously concluded, however, that a direct proportionality exists between the quantity now called the *intrinsic viscosity* and the molecular weight, i.e., that

$$\left(\frac{\eta_r - 1}{c} \right)_{c \to 0} = KM$$

where η_r is the relative viscosity of a polymer solution of concentration c, K is a constant characteristic of a given polymer series, and M is the molecular weight. Widespread use of Staudinger's formula led to the assignment of molecular weights which were generally too low, sometimes by factors of ten or more. This situation was not fully rectified until the mid-1940's.* At present a rather impressive num-

---

* Empirically established relationships between molecular chain lengths in various polymer series and their intrinsic viscosities have been summarized by Goldfinger, Hohenstein, and Mark.[115]

ber of different polymer series have been subjected to quantitative molecular weight measurements, and reliable relationships between solution viscosity and M have been established in many cases.

Not until the 1940's did suitable methods, both experimental and theoretical, become available for reducing the constitution of polymers, including the nonlinear, network-forming types, to tractable quantitative terms. Since such means are a prerequisite to the quantitative treatment of polymer properties in relation to constitution, advances in this direction necessarily were delayed.

REFERENCES

1. Thomas Graham, *Trans. Roy. Soc.* (London), **151**, 183 (1861).
2. See, for example, Wolfgang Ostwald, *Kolloid Z.*, **1**, 331 (1907); *Z. Chem. Ind. Kolloide*, **3**, 28 (1908); *An Introduction to Theoretical and Applied Colloid Chemistry*, 2d Amer. ed., trans. by M. H. Fischer (John Wiley and Sons, New York, 1922). See also various texts by other authors, for example, A. W. Thomas, *Colloid Chemistry* (McGraw-Hill, New York, 1934), pp. 1–8.
3. H. Hlasiwetz and J. Habermann, *Ann. Chem. Pharm.*, **159**, 304 (1871).
4. F. M. Raoult, *Compt. rend.*, **95**, 1030 (1882); *Ann. chim. phys.*, [6], **2**, 66 (1884); *Compt. rend.*, **101**, 1056 (1885).
5. J. H. van't Hoff, *Z. physik. Chem.*, **1**, 481 (1887); *Phil. Mag.*, [5], **26**, 81 (1888).
6. F. Musculus and A. Meyer, *Bull. soc. chim. France*, [2], **35**, 370 (1881).
7. H. T. Brown and G. H. Morris, *J. Chem. Soc.*, **53**, 610 (1888).
8. H. T. Brown and G. H. Morris, *J. Chem. Soc.*, **55**, 465 (1889).
9. C. J. Lintner and G. Düll, *Ber.*, **26**, 2533 (1893).
10. H. Rodewald and A. Kattein, *Z. physik. Chem.*, **33**, 579 (1900).
11. A. Nastukoff, *Ber.*, **33**, 2237 (1900).
12. J. H. Gladstone and W. Hibbert, *J. Chem. Soc.*, **53**, 688 (1888); *Phil. Mag.*, [5], **28**, 38 (1889).
13. See for example, E. Paterno, *Z. physik. Chem.*, **4**, 457 (1889); *Gazz. chim. ital.*, **19**, 195 (1889).
14. A. Sabanseff, *J. Russ. Phys. Chem. Soc.*, **22**, 102 (1890).
15. W. Ramsay and J. Shields, *J. Chem. Soc.*, **63**, 1089 (1893). *Trans. Roy. Soc.* (London), **A184**, 647 (1893).
16. J. J. Berzelius, *Jahresbericht*, **12**, 63 (1833).
17. V. Harcourt, *J. Chem. Soc.*, **71**, 591 (1897).
18. C. Harries, *Ber.*, **37**, 2708 (1904); *ibid.*, **38**, 1195, 3985 (1905).
19. C. Harries, *Ann.*, **406**, 173 (1914); C. Harries and F. Evers, *Chem. Abstracts*, **16**, 3232 (1922).
20. S. S. Pickles, *J. Chem. Soc.*, **97**, 1085 (1910).
21. W. A. Casperi, *J. Chem. Soc.*, **105**, 2139 (1914).
22. T. Seeligmann, G. L. Torrilhon, and H. Falconnet, *Le caoutchouc et la gutta percha* (Paris, 1896).

23. C. O. Weber, *The Chemistry of India Rubber* (Charles Griffin and Co., London, 1902), p. 48 ff.

24. R. Willstätter and L. Zechmeister, *Ber.*, **46**, 2401 (1913).

25. J. C. Irvine and C. W. Souter, *J. Chem. Soc.*, **117**, 1489 (1920); J. C. Irvine and E. L. Hirst, *ibid.*, **121**, 1585 (1922).

26. J. C. Irvine and E. L. Hirst, *J. Chem. Soc.*, **123**, 518 (1923).

27. J. C. Irvine and J. MacDonald, *J. Chem. Soc.*, **129**, 1502 (1926).

28. E. Heuser, *Textbook of Cellulose Chemistry*, trans. from the 2d German ed. (1922) by C. J. West and G. J. Esselen (McGraw-Hill, New York, 1924), p. 183.

29. E. Fischer, *J. Am. Chem. Soc.*, **36**, 1170 (1914).

30. A.-V. Lourenço, *Compt. rend.*, **51**, 365 (1860); *Ann. chim. phys.*, [3], **67**, 273 (1863).

31. A. Wurtz, *Compt. rend.*, **49**, 813 (1859); **50**, 1195 (1860).

32. A.-V. Lourenço, *Ann. chim. phys.*, [3], **67**, 293 (1863).

33. D. Vorländer, *Ann.*, **280**, 167 (1894).

34. E. Roithner, *Monatsh.*, **15**, 665 (1894).

35. K. Kraut, *Ann.*, **150**, 1 (1869).

36. C. Gerhardt, *Ann.*, **87**, 159 (1853).

37. R. Anschütz, *Ber.*, **25**, 3506 (1892); *Ann.*, **273**, 79 (1893); *Ber.*, **52**, 1875 (1919); *ibid.*, **55**, 680 (1922); *Ann.*, **439**, 1 (1924). G. Schroeter, *Ber.*, **52**, 2224 (1919).

38. L. Anschütz and R. Neher, *J. prakt. Chem.*, **159**, 264 (1941).

39. H. Schiff, *Ber.*, **15**, 2588 (1882).

40. A. Klepl, *J. prakt. Chem.*, [2], **25**, 525 (1882); *ibid.*, **28**, 193 (1883).

41. A. Piutti, *Ber.*, **16**, 1319 (1883).

42. C. Märcker, *Ann.*, **124**, 249 (1862).

43. A. Husemann, *Ann.*, **126**, 280 (1863).

44. W. Mansfeld, *Ber.*, **19**, 696 (1886).

45. A. Husemann, *Ann.*, **126**, 292 (1863).

46. K. Birnbaum and G. Lurie, *Ber.*, **14**, 1753 (1881).

47. A. Einhorn, *Ann.*, **300**, 135 (1898).

48. C. A. Bischoff and A. von Hedenström, *Ber.*, **35**, 3435 (1902).

49. C. A. Bischoff and P. Walden, *Ber.*, **26**, 262 (1893); *Ann.*, **279**, 45 (1894).

50. T. H. Norton and J. Tcherniak, *Bull. soc. chim. France*, [2], **30**, 102 (1878).

51. F. Fichter and A. Beisswenger, *Ber.*, **36**, 1200 (1903).

52. E. E. Blaise and L. Marcilly, *Bull. soc. chim. France*, [3], **31**, 308 (1904).

53. E. E. Blaise, *Bull. soc. chim. France*, [3], **35**, 665 (1906).

54. G. L. Voerman, *Rec. trav. chim.*, **23**, 265 (1904).

55. H. Staudinger and E. Ott, *Ber.*, **41**, 2208 (1908).

56. T. Curtius, *Ber.*, **16**, 753 (1883).

57. M. Schwarzschild, *J. Chem. Soc.*, Abstracts, **84(I)**, 780 (1903).

58. T. Curtius, *Ber.*, **37**, 1284 (1904).

59. L. Balbiano, *Ber.*, **34**, 1501 (1901).

60. L.-C. Maillard, *Ann. chim.*, [9], **1**, 519 (1914).

61. H. Leuchs, *Ber.*, **39**, 857 (1906); H. Leuchs and W. Manasse, *ibid.*, **40**, 3235 (1907); H. Leuchs and W. Geiger, *Ber.*, **41**, 1721 (1908).

62. J. von Braun, *Ber.*, **40**, 1834 (1907). See also A. Manasse, *ibid.*, **35**, 1367 (1902).

63. W. H. Carothers, *Chem. Rev.*, **8**, 353 (1931).

64. P. Genvresse, *Bull. soc. chim. France*, [3], **17**, 599 (1897).

65. T. P. Hilditch, *J. Chem. Soc.*, **97**, 2579 (1910).

66. A. Kötz and O. Sevin, *J. prakt. Chem.*, [2], **64**, 518 (1901); *Ber.*, **33**, 730 (1900).

67. W. Autenrieth and F. Beuttel, *Ber.*, **42**, 4349 (1909).

68. J. von Braun and W. Gawrilow, *Ber.*, **45**, 1274 (1912).

69. H. Limpricht, *Ann.*, **309**, 120 (1899).

70. M. D. Tilitcheev, *Chem. Abstracts*, **21**, 3358 (1927).

71. T. Zincke and O. Krüger, *Ber.*, **45**, 3468 (1912).

72. E. Simon, *Ann.*, **31**, 265 (1839).

73. M. Berthelot, *Bull. soc. chim. France*, [2], **6**, 294 (1866).

74. G. Bouchardat, *Compt. rend.*, **89**, 1117 (1879).

75. G. Wagner, *Ber.*, **11**, 1260 (1878).

76. G. Lemoine, *Compt. rend.*, **125**, 530 (1897); **129**, 719 (1899).

77. R. Fittig and F. Engelhorn, *Ann.*, **200**, 65 (1880).

78. J. A. Mjöen, *Ber.*, **30**, 1227 (1897).

79. H. Stobbe and G. Posnjak, *Ann.*, **371**, 259 (1910).

80. S. V. Lebedev and N. A. Skavronskaya, *J. Russ. Phys. Chem. Soc.*, **43**, 1124 (1911).

81. C. Harries, *Ann.*, **395**, 211 (1913).

82. S. V. Lebedev and B. K. Merezhkovskii, *J. Russ. Phys. Chem. Soc.*, **45**, 1249 (1913).

83. S. V. Lebedev. *J. Russ. Phys. Chem. Soc.*, **45**, 1296 (1913); *Chem. Abstracts* **9**, 798 (1915).

84. H. Staudinger, *Ber.*, **53**, 1073 (1920).

85. H. Staudinger and J. Fritschi, *Helv. Chim. Acta*, **5**, 785 (1922).

86. H. Staudinger, *Ber.*, **57**, 1203 (1924); *Kautschuk*, 63 (1927).

87. H. Staudinger, *Ber.*, **59**, 3019 (1926); H. Staudinger, K. Frey, and W. Starck, *ibid.*, **60**, 1782 (1927).

88. H. Staudinger and M. Lüthy, *Helv. Chim. Acta*, **8**, 41 (1925); H. Staudinger, *ibid.*, **8**, 67 (1925); H. Staudinger, H. Johner, R. Signer, G. Mie, and J. Hengstenberg, *Z. physik. Chem.*, **126**, 434 (1927).

89. R. Pummerer and P. A. Burkard, *Ber.*, **55**, 3458 (1922).

90. J. Duclaux, *Rev. gen. colloides*, **1**, 33 (1923).

91. R. Pummerer, H. Nielsen, and W. Gündel, *Ber.*, **60**, 2167 (1927).

92. G. Gee, *Trans. Faraday Soc.*, **38**, 109 (1942).

93. P. Karrer, *Polymere Kohlenhydrate* (Akademische Verlagsgesellschaft m.b.H., Leipzig, 1925)

94. K. Hess, W. Weltzein, and E. Messmer, *Ann.*, **435**, 1 (1924).

95. M. Bergmann, *Ber.*, **59**, 2973 (1926).

96. H. Pringsheim, *Naturwissenschaften*, **13**, 1084 (1925).

97. R. O. Herzog, *Naturwissenschaften*, **12**, 955 (1924); *J. Phys. Chem.*, **30**, 457 (1926).

98. E. Ott, *Physik. Z.*, **27**, 174 (1926).

99. E. Ott, *Naturwissenschaften*, **14**, 320 (1926). See also E. A. Hauser and H. Mark, *Kolloidchem. Beihefte*, **22**, 94 (1926).

100. M. Polanyi, *Naturwissenschaften*, **9**, 288 (1921).

101. R. O. Herzog, *Physik. Z.*, **27**, 378 (1926).

102. O. L. Sponsler, *J. Gen Physiol.*, **9**, 677 (1926); O. L. Sponsler and W. H. Dore, *Colloid Symposium Monograph* (Chemical Catalog Co., New York, 1926), IV, 174.

103. K. H. Meyer and H. Mark, *Ber.*, **61**, 593 (1928).

104. K. H. Meyer and H. Mark, *Ber.*, **61**, 1939 (1928); K. H. Meyer, *Naturwissenschaften*, **16**, 781 (1928).

105. K. H. Meyer, *Naturwissenschaften*, **17**, 255 (1929). See also H. Mark, *Ber.*, **59**, 2997 (1926).

106. H. Staudinger, *Ber.*, **61**, 2427 (1928).

107. H. Staudinger, *Angew. Chem.*, **42**, 67 (1929).

108. W. H. Carothers, *J. Am. Chem. Soc.*, **51**, 2548 (1929).

109. *Collected Papers of Wallace Hume Carothers on High Polymeric Substances*, ed. by H. Mark and G. S. Whitby (Interscience Publishers, New York, 1940).

110. W. Kuhn, *Ber.*, **63**, 1503 (1930).

111. E. Guth and H. Mark, *Monatsh.*, **65**, 93 (1934).

112. W. Kuhn, *Kolloid Z.*, **68**, 2 (1934).

113. W. Kuhn, *Kolloid Z.*, **76**, 258 (1936).

114. H. Staudinger and W. Heuer, *Ber.*, **63**, 222 (1930); H. Staudinger and R. Nodzu, *ibid.*, **63**, 721 (1930); H. Staudinger, *Die hochmolekularen organischen Verbindungen* (Verlag von Julius Springer, Berlin, 1932).

115. G. Goldfinger, W. P. Hohenstein, and H. Mark, *J. Polymer Sci.*, **2**, 503 (1947).

CHAPTER II

Types of Polymeric Substances:

Definitions and Classifications

1. PRIMARY DEFINITIONS

THE constitution of a polymeric substance is customarily described in terms of its *structural units*. These may be defined in the most general terms as groups having a valence of two or more; the terminal units coming at the ends of polymer chains represent minor exceptions in that they possess a valence of only one. The structural units are connected to one another in the polymer molecule, or polymeric structure, by covalent bonds. Although the structures of polymers vary widely, nearly all of those of interest may be expressed as combinations of a limited number of different structural units; in many cases a single type of structural unit suffices for the representation of the entire polymer molecule. This feature, namely, the generation of the entire structure through repetition of one or a few elementary units, is the basic characteristic of polymeric substances, as is indeed implied by the etymology of the term *polymer* (i.e., "many member").

The structural units may be connected together in any conceivable pattern. In the simplest of all polymers, the *linear polymers*, the structural units are connected one to another in linear sequence. Such a polymer may be represented by the type formula

$$A'—A—A—A— \cdots —A'' \quad \text{or} \quad A'(—A—)_{x-2}—A''$$

where the principal structural unit is represented by A, and x is the *degree of polymerization*, or number of structural units in the molecule. The structural units, other than the terminal units, A' and A'', must necessarily be bivalent in this case. The group A' may or may not be identical with A'', but neither may be identical with A. In many cases the terminal units have the same composition as the principal units, but their structures must differ inasmuch as the former are monovalent.

Alternatively, the structural units of the polymer may be connected together in such a manner as to form *nonlinear*, or *branched*, structures of one sort or another. Some, at least, of the structural units must then possess a valency greater than two. A typical nonlinear polymer structure in which the branching units represented by Y are trivalent may be indicated as follows:

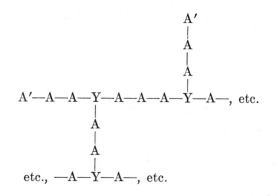

The branching units may, of course, have a valency exceeding three. Highly ramified molecular species can be formed through further propagation of the branched structure. It may, in fact, interconnect with itself to yield a *network* structure. A *planar network* analogous to that of graphite might conceivably occur, or a regular three-dimensional *space network* analogous to diamond could be imagined. The random nature of polymerization processes, however, renders unlikely the assemblage of units in a structure which conforms to a regular pattern. The *space-network* structures frequently encountered among nonlinear polymers are highly irregular labyrinths. The structure may abound with "closed circuits" leading from a given junction via a sequence of linear chains and polyfunctional junctions back to the given junction, but these circuitous paths will be extremely varied and in general will involve many chains and many polyfunctional junctions. Any analogy to the closely connected spatial array occurring in crystalline substances is therefore likely to be misleading.

The structural units represent residues from the monomeric compound(s) employed in the preparation of the polymer. Usually, there is a direct correspondence between the monomer(s) and the structural unit(s). Several illustrative examples of units occurring in linear polymers are listed on the following page:

Polymer	Monomer(s)	Structural unit(s)
Polystyrene	$CH_2{=}CH$ $\quad\vert$ C_6H_5	$-CH_2-CH-$ $\qquad\vert$ $\qquad C_6H_5$
cis-1,4-Polyisoprene (natural rubber)	CH_3 $\quad\vert$ $CH_2{=}C-CH{=}CH_2$	CH_3 $\qquad\vert$ $-CH_2-C{=}CH-CH_2-$
Poly-(sebacic anhydride)	$HOCO(CH_2)_8COOH$	$-OCO(CH_2)_8CO-$
Poly-(hexamethylene adipamide) (nylon)	$H_2N(CH_2)_6NH_2$ $+HOCO(CH_2)_4COOH$	$-NH(CH_2)_6NH-$ $-CO(CH_2)_4CO-$

In each of the first two examples the structural unit and the monomer from which it is derived (hypothetically, in the case of natural rubber) possess identical atoms occupying similar relative positions; only a rearrangement of electrons is involved in the conversion of the monomer to the structural unit. In the latter two examples the elements of water are eliminated in converting monomers to structural units. In the last example there are two types of monomeric reactants, and these give rise to two supplementary structural units which alternate one after the other along the polymer chain. It is convenient in cases such as this to define the pair of alternating structural units as the *repeating unit* of the chain. Thus, the repeating unit

$$-NH(CH_2)_6NH-CO(CH_2)_4CO-$$

of poly-(hexamethylene adipamide) consists of two *structural* units. In polymers containing a single structural unit, the structural unit ordinarily is also the repeating unit.

The polymers listed above, and all other linear polymers as well, are formed from monomers which enter into two, and only two, linkages with other structural units. This statement corresponds to the previous remark that the structural units of linear polymers necessarily are bivalent. The interlinking capacity of a monomer ordinarily is apparent from its structure; it is clearly prescribed by the presence of two condensable functional groups in each monomer in the third and fourth examples above. The ability of the extra electron pair of the ethylenic linkage to enter into the formation of two bonds endows styrene with the same interlinking capacity. In accordance with the *functionality* concept introduced by Carothers,[1,2] all monomers which when polymerized may join with two, and only two, other monomers are termed *bifunctional*. Similarly, a *bifunctional unit* is one which is attached to two other units. It follows that linear polymers are composed exclusively (aside from terminal units) of bifunctional units.*

* The designation of vinyl monomers (e.g., styrene) as bifunctional in accord-

Nonlinear polymers are obtained from monomers at least some of which possess a functionality exceeding two. In other words, nonlinear polymers may be defined as those containing units some of which are *polyfunctional*, this term being reserved for functionalities exceeding two. Thus, the condensation of glycerol, a trifunctional reactant, with phthalic acid, a bifunctional reactant, yields a nonlinear polymer comprising structures such as

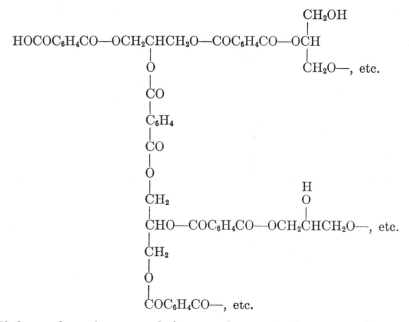

If the condensation proceeds far enough, a network structure is easily formed. Similarly, copolymerization of a little divinyl adipate with vinyl acetate[3] yields a nonlinear polymer in which chains of bifunctional vinyl acetate units are bridged, or *cross-linked*, by the *tetrafunctional* divinyl adipate units, as indicated on the following page, where the divinyl adipate unit is enclosed between vertical dashed lines:

---

ance with the above definition conflicts with the conventional terminology of organic chemistry, according to which the ethylenic linkage represents a single functional group. This digression from precedent is justified by the universal applicability to virtually all polymers of the functionality definition[1] as applied to polymers and polymerization.

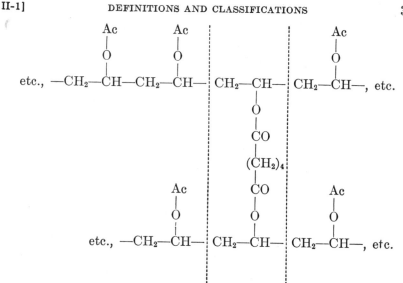

Again network structures may be formed.

An example of a nonlinear polymer derived by cross-linking an initially linear polymer is afforded by vulcanized natural rubber. In the usual vulcanization procedure involving the use of sulfur and accelerators, various types of cross-linkages may be introduced between occasional units (about one in a hundred) of the polyisoprene chains. Some of these bonds are indicated to be of the following type:[4,5]

In complete analogy with the preceding example, the portion of the structure shown between vertical dashed lines, consisting of two isoprene residues and connecting sulfur atoms, represents a single tetrafunctional unit; four otherwise independent portions of the polymer structure are bonded directly to this unit. The choice of the portion of the structure of a nonlinear polymer to be designated as the unit of higher functionality obviously is somewhat arbitrary. The tetrafunctional unit in this case might be designated as two trifunctional isoprene units bonded to each other by sulfur. Since these trifunctional isoprene units occur in pairs, the quantitative treatment of such structures is greatly simplified if the pair and its sulfur cross-link are considered as a single unit—a tetrafunctional unit.

Structural units of functionality exceeding four may occur in non-linear polymers. The terminology set forth above is easily extended to include them. It should be noted further that various polyfunctional units having differing functionalities may occur in the same structure, in the same way that bifunctional units coexist with the polyfunctional units.

With regard to the formal distinction between linear and nonlinear polymers based on the absence or presence, respectively, of polyfunctional units, it is to be noted that harsh application of this definition would relegate the term linear polymer to the status of an ideal which may be approached but never realized. The merest trace of a spurious polyfunctional unit would disqualify an otherwise linear polymer. In practice, polymers prepared by reactions which on chemical grounds should proceed overwhelmingly through bifunctional interlinking, branching of any sort being limited to unknown side reactions, are regarded as linear polymers unless their properties give definite evidence for the occurrence of nonlinearity. Thus, the polyesters prepared by condensing glycols with dibasic acids, the polyamides from diamines and dibasic acids, and the various vinyl polymers obtained from unsaturated monomers are ordinarily classified as linear polymers. Certain vinyl polymers,[6,7] e.g., poly-(vinyl acetate), may possess occasional branches—at perhaps one unit in a thousand— which result from free radical side reactions during polymerization. A very small degree of branching will vastly modify the pattern of the molecule if the molecular weight is very large. If such branching occurs in polystyrene, its extent is so small as to have no discernible effect on the properties of the polymer. Polystryene is therefore ordinarily included in the linear classification. On the other hand, the polymerization of diolefins is generally capable of yielding partly insoluble products, from which it is inferred that they are nonlinear polymers.

Even among nonlinear polymers, many of the materials of interest are composed of a preponderance of bifunctional units with only a minority of polyfunctional units. This applies to vulcanized rubber where no more than about 1 or 2 percent of the isoprene units are cross-linked. It also applies to the amylopectin fraction of starch which consists of chains composed of an average of about 20 glucose units, these chains being connected to one another by trifunctional units to yield an irregular branched array; to wool consisting of polypeptide chains cross-linked by the S—S linkages of cystine (i.e., the diamino dibasic acid, cystine, is combined in wool as a tetrafunctional unit); and to synthetic diolefin polymers in which both double bonds of an occasional monomer participate in the polymerization, thereby yielding a tetrafunctional unit. Thus, many nonlinear polymers consist principally of linear sequences of units which are interrupted only at rather wide intervals by units of higher functionality. It must not be concluded, however, that the relatively small fraction of polyfunctional units are of little consequence. They may bring about profound changes in certain physical properties, just as they are capable of altering the structural plan of the polymer from that of a finite linear chain to the "infinite" space network type of structure.

In certain highly branched polymers the structural unit concept is necessarily less explicit. Examples are the glyceryl phthalates (see above), phenol-formaldehyde resins (Table I, p. 42), the urea-formaldehyde resins (Table I), and polymers of drying oils in which pairs of unsaturated acid radicals of the triglyceride monomer combine in an irregular fashion. In these examples a given monomer residue may enter the polymer as a mono-, di-, tri-, or sometimes higher functional unit, and, furthermore, nonequivalent reactive groups are present (e.g., the primary and the secondary hydroxyls of glycerol or the *ortho* and *para* positions of phenol). The architectural plan cannot be so conveniently expressed since the component structural units are manifold in variety. Types of linkages can be specified, however, and the complex array of structural possibilities need not be a serious deterrent to quantitative interpretation.

The degree of polymerization represented by x, which expresses the number of structural units in a given polymer molecule, becomes at once ambiguous when applied to an actual specimen composed of many polymer molecules, owing to the fact that all of them are not characterized by the same value of x. High polymers (with the *possible* exception of certain naturally occurring polymers) invariably consist of molecular species varying in degree of polymerization, usually over a considerable range. The form of this *molecular size distribution*, or *molecular weight distribution* (the molecular weight can be taken to be proportional to x), assumes importance in various problems dealing with polymer constitution and physical properties. It is sufficient to mention here that the hazards of ambiguity can be avoided by referring to the *average* degree of polymerization, provided that the manner of averaging is specified. Unless stated otherwise, the term *average degree of polymerization* refers to the ordinary average (*number average*, see Chap. VII) obtained by dividing the total number of structural units by the total number of molecules.   Similarly, the *average molecular weight* (more accurately, the *number average molecular weight*) represents the mass of the sample divided by the number of moles it contains.   Where no confusion is likely to arise, the qualifying adjective *average* may be omitted.   Implicit in all discussions of polymer phenomena, however, must be the recognition of their inherent molecular heterogeneity.

Throughout the preceding discussion the term *polymer* has been used in the general sense of a substance which is polymeric in nature, irrespective of whether it contains a single structural unit or a plurality of different units.   This term and its derivatives sometimes are applied in the restricted sense of a polymeric substance containing a single repeating unit (disregarding terminal units).   Polymeric substances containing two or more structural units combined more or less in random sequence are then distinguished by the term *copolymer*.   A linear copolymer composed of two bifunctional units A and B, for example, may be represented as follows:

$$-A-B-B-A-A-A-B-A-B-B-A-, \text{ etc.}$$

The structural units may occur in random sequence, there may be a preponderance of long sequences of like units, or the two types may tend to alternate.   In any event the substance is included within the meaning of the term *linear copolymer*.   If the units alternate with perfect regularity, however, the substance is preferably regarded as a *polymer* of the repeating unit —AB—.   Polyamides formed from diamines and dibasic acids afford examples of this type.   Although these substances might be looked upon as *copolymers* of the diamine and the dibasic acid *structural units*, it is preferable to classify them as *polymers* in which the *repeating unit* consists of one residue from each of the two reactants.

Typical copolymers are prepared by copolymerizing styrene with methyl methacrylate, or butadiene with acrylonitrile, for example. Copolyamides composed of two repeating units are obtained by condensing two diamines with a single diabasic acid, or by condensing a single diamine with two dibasic acids.   A higher multiplicity of reactants may of course be used, in which case the variety of repeating units is correspondingly increased.   The term copolymerization implies that the mixed monomers combine mutually in the same polymer chains.   If the monomers were to polymerize separately to form molecularly distinct species, each containing a different unit, the product would be a mixture of polymers and not a copolymer.

It should be remarked that vinyl polymers such as polystyrene which possess asymmetric carbon atoms ordinarily will consist of mixtures of *d* and *l* structural units. Rigorous application of the foregoing definition of terms would place polystyrene in the class of copolymers. Similarly, the polymer of butadiene containing no less than four distinct structural units, the *cis* and *trans* configuration of the 1,4 unit

$$-CH_2-CH=CH-CH_2-$$

and the *d* and *l* configurations of the 1,2 unit

$$-CH_2-CH-$$
$$\quad\quad\quad |$$
$$\quad\quad CH=CH_2$$

actually is a coplymer, although ordinarily neither polystyrene nor polybutadiene is so regarded. Each of these substances is formed by *polymerization* of a single monomer rather than by *copolymerization* of a plurality of monomers. On this basis there is some justification for custom in applying the restricted term polymer in such instances.

2. CLASSIFICATION OF POLYMERS

In 1929 Carothers[8] proposed a generally useful differentiation between two broad classes of polymers: *condensation polymers*, in which the molecular formula of the structural unit (or units) lacks certain atoms present in the monomer from which it is formed, or to which it may be degraded by chemical means, and *addition polymers*, in which the molecular formula of the structural unit (or units) is identical with that of the monomer from which the polymer is derived. Condensation polymers may be formed from monomers bearing two or more reactive groups of such a character that they may condense intermolecularly with the elimination of a by-product, often water. The polyamides and polyesters referred to above afford prime examples of condensation polymers. The formation of a polyester from a suitable hydroxy acid takes place as follows:

$$x \ \ HO-R-COOH \ \rightarrow \ HO[-R-COO-]_{x-1}R-COOH$$
$$+ (x-1)H_2O$$

The molecular formula of the condensation polymer is not an integral multiple of the formula of the monomer molecule owing to the elimination of a by-product, which in this case is water. The most important class of addition polymers consists of those derived from unsaturated monomers, such as the vinyl compounds

$$CH_2=CH \ \rightarrow \ -CH_2-CH-CH_2-CH-, \ etc.$$
$$\quad | \quad\quad\quad\quad\quad | \quad\quad\quad |$$
$$\quad X \quad\quad\quad\quad\quad X \quad\quad\quad X$$

where X may be phenyl, halogen, acetoxy, etc. The molecular formula of the polymer is x times the molecular formula of the monomer. This terminology may be applied also to the processes by which the polymers are formed. Thus the former of these two reactions is a *condensation polymerization*, and the latter is an *addition polymerization*.

Whether or not the structural unit differs in composition from the monomer from which it is derived is of no particular significance. The principal justification for the differentiation between condensation and addition polymers (and polymerizations) lies in the marked contrast between the processes by which they are formed. Condensation polymerizations proceed by stepwise intermolecular condensation of functional groups; addition polymerizations usually proceed by a chain mechanism involving active centers of one sort or another. Consequently, the two types of polymerization processes differ so drastically that their separate consideration is imperative. Differences in chain structure occurring in the two types of polymers afford further justification for the distinction. The structural units of condensation polymers usually are joined by interunit functional groups (e.g., ester or amide) of one sort or another; most addition polymers do not possess functional groups *within* the polymer chain, although they may be present as lateral substituents. Linear condensation polymers usually, though not always, conform to the type formula

$$\text{—R—X—R—X—, etc.}$$

where R is a divalent radical and X is a functional group (e.g., —OCO—, —O—, —NHCO—, —S—S—) polar in nature and susceptible to cleavage by reagents such as water or alcohol. The presence of polar groups at regularly spaced intervals within the chains of linear condensation polymers endows these polymers with chemical and physical properties not ordinarily found among addition polymers, whose chain skeletons usually consist exclusively of carbon atoms.

Since Carothers[8] introduced the distinction between condensation and addition polymers based on the composition relationship between monomer and structural unit, a number of examples of polymerization processes have been found which formally resemble the condensation type but which proceed without evolution of a by-product. For example, a glycol and a diisocyanate react intermolecularly as follows:

$$x \ \text{HO—R—OH} + x \ \text{OCN—R'—NCO} \ \rightarrow$$

$$\text{HO}\left[\text{—R—O}\overset{\overset{\text{O}}{\|}}{\text{C}}\text{NH—R'—NH}\overset{\overset{\text{O}}{\|}}{\text{C}}\text{O—}\right]_{x-1}\text{R—O}\overset{\overset{\text{O}}{\|}}{\text{C}}\text{NH—R'NCO}$$

The process proceeds through the reaction of pairs of functional groups which combine to yield the urethane interunit linkage. From the standpoint of both the mechanism and the structure type produced, inclusion of this example with the condensation class clearly is desirable. Later in this chapter other examples will be cited of polymers formed by processes which must be regarded as addition polymerizations, but which possess within the polymer chain recurrent functional groups susceptible to hydrolysis. This situation arises most frequently where a cyclic compound consisting of one or more structural units may be converted to a polymer which is nominally identical with one obtained by intermolecular condensation of a bifunctional monomer; e.g., lactide may be converted to a linear polymer

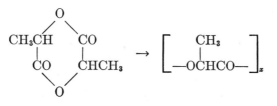

which may also be prepared by the elimination of water from lactic acid.

As these examples indicate, the original Carothers distinction between condensation and addition polymers (and polymerizations), if applied quite literally, oftentimes fails to serve the desired purpose. We shall in the following pages include various ambiguous polymerizations and polymers in one or the other type according to their similarities to representative members of that type, at the sacrifice of clearcut definitions. In general, a *polymerization process* which proceeds by a reaction between pairs of functional groups with the formation of a type of interunit functional group not present in the monomer(s) will be regarded as a condensation polymerization. If the net process involves merely the scission of a bond in the monomer and the re-formation of similar connections with other monomers (as in the opening of the double bond of an unsaturated monomer or the opening of the ring of a cyclic monomer), and if no by-product is evolved, it will be referred to as an addition polymerization. Mention has already been made of the not infrequent possibility of forming equivalent polymers alternatively by different processes one of which is a condensation and the other an addition. The confusion which would arise if two such polymers were to be relegated to separate classifications can be avoided by including among condensation polymers those which are formed by the conversion of cyclic compounds to acyclic structural units of the same composition, notwithstanding the fact that the *process* is one of addition polymerization. It is thus appropriate to broaden the definition of condensation polymers to include not only those products which have been formed by a condensation polymerization process as defined above, but to include also those polymers which on chemical degradation (e.g., by hydrolysis) yield monomeric end products differing in composition from the structural units.

As previously indicated, both condensation and addition polymers may be prepared from monomers of functionality exceeding two, with resulting formation of nonlinear polymers. Hence the distinction between linear and nonlinear polymers subdivides both the condensation and the addition polymers, and four types of polymers are at once differentiable: linear condensation, nonlinear condensation, linear addition, and nonlinear addition. The distinction between linear and nonlinear polymers is clearly warranted not only by the marked differences in their structural patterns but also by the sharp divergence of their properties.

3. CONDENSATION POLYMERS AND CONDENSATION POLYMERIZATION[9]

Representative condensation polymers are listed in Table I. The list is by no means exhaustive, but it serves to indicate the variety of condensation reactions which may be employed in the synthesis of polymers. Cellulose and proteins, although their syntheses have not been accomplished by condensation polymerization in the laboratory, nevertheless are included within the definition of condensation polymers on the ground that they can be degraded, hydrolytically, to monomers differing from the structural units by the addition of the elements of a molecule of water. This is denoted by the direction of the arrows in the table, indicating depolymerization.

From the examples given in Table I it is apparent that almost any condensation reaction can be utilized for the production of polymers. The primary requisite is a monomer, or pair of monomers, bearing two or more condensable groups or centers of reactivity. (A monomer such as formaldehyde possesses a single functional group but is to be regarded as bifunctional since it enters into bond formation with two other monomers.) The commonplace nature of the reactions involved is significant. The knowledge available from the established organic chemistry of these reactions facilitates specification of optimum conditions for polymerization and prescribes with virtual certainty the structure of the product obtained in most cases. High polymers of accurately definable over-all structure may be synthesized by condensation polymerization.

A generalized kinetic treatment of the array of processes occurring in condensation polymerization might appear hopelessly complex. In the polyesterification of a hydroxy acid, for example, the first step is intermolecular esterification between two monomers, with the production of a dimer

$$HORCOOH + HORCOOH \rightarrow HORCOORCOOH + H_2O$$

This step may be followed by reaction of the dimer with another monomer to form a trimer, or the dimer may react with another dimer to form a tetramer, and so on. These species in turn may react with monomers, dimers, and so on. All of the reactions

$$x\text{-mer} + y\text{-mer} \rightarrow (x + y)\text{-mer}$$

in which x and y assume every possible combination of positive integral values, are to be reckoned with.

If it were necessary to assign a separate rate constant, k_{xy}, to each of the above reactions occurring in condensation polymerization, kinetic analysis would be extremely difficult if not impossible. However, these various steps need not be differentiated. All involve the same process, e.g., esterification, and, as will be shown in Chapter III, the rate constants for the various steps do not differ discernibly. Consequently, the entire polymerization process can be regarded as a reaction between functional groups, e.g., OH and COOH. The numerous molecular species and the manifold steps in which they are involved can be disregarded. This point of view having been adopted, the chemical reaction mechanism of polyintermolecular condensation is no more complex than the condensation of analogous monofunctional compounds. In polyesterification, for example, the rate of esterification is similar to the rate of reaction of ethyl alcohol with acetic acid under similar conditions, and it is subject to acid catalysis in a parallel manner (see Chap. III). Similarly, the polyamidation reaction parallels monoamidation in its rate, temperature coefficient, and reaction order. The essential differences lie only in the functionality of the reactants and the nature of the products produced.

The average degree of polymerization, and hence the average molecular weight, of a linear condensation polymer depends on the degree of completion achieved in the condensation reaction. To obtain products of the usually desired high molecular weight (about 10,000 or more, or a degree of polymerization x exceeding 50), it is imperative that the condensation process be an efficient one. It must be reasonably free of side reactions, particularly those which consume functional groups without producing intermolecular linkages; the reacting

TABLE I.—REPRESENTATIVE CONDENSATION POLYMERS AND THEIR PROPERTIES

Type	Interunit linkage	Examples	Approximate melting point °C	Characteristics
Polyester	$\overset{O}{\underset{\|}{-C-O-}}$	$HO(CH_2)_9COOH \rightarrow HO[-(CH_2)_9COO-]_xH$	76	Hard, waxlike, crystalline; fibers may be cold drawn above M.W. 9000[2,10]
		$HO(CH_2)_2O(CH_2)_2OH + HOOC(CH_2)_4COOH$ $\rightarrow HO[-(CH_2)_2O(CH_2)_2OCO(CH_2)_4COO-]_xH$	Amorphous	Viscous liquid to glasslike; no fibers[11]
		$HO(CH_2)_{10}OH + HOOC(CH_2)_4COOH$ $\rightarrow HO[-(CH_2)_{10}OCO(CH_2)_4COO-]_xH$	80	Hard, waxlike; fibers may be cold drawn above M.W. 10,000; good strength[12]
		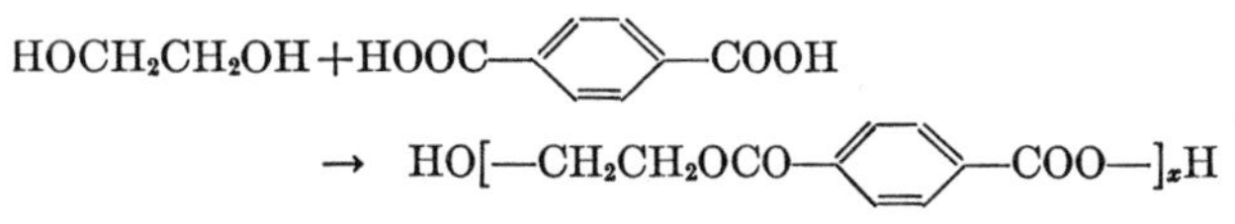 $HOCH_2CH_2OH + HOOC\!\!-\!\!\langle\bigcirc\rangle\!\!-\!\!COOH$ $\rightarrow HO[-CH_2CH_2OCO-\langle\bigcirc\rangle-COO-]_xH$	267	Hard, crystalline; fibers may be cold drawn above M.W. 12,000; excellent strength[13]
		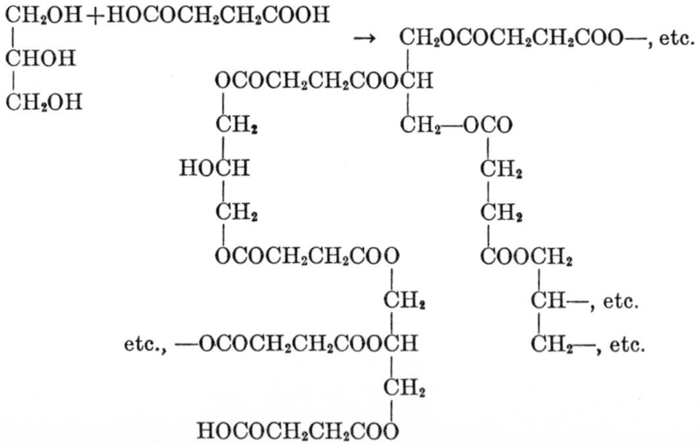 $CH_2OH + HOCOCH_2CH_2COOH \rightarrow CH_2OCOCH_2CH_2COO-$, etc. $-CHOH$ $-CH_2OH$	Amorphous	Except at rather low extents of reaction, polymers are insoluble and infusible; no fiber properties; moderate strength

etc., $-OCOCH_2CH_2COOCH$

$HOCOCH_2CH_2COO$

TABLE I.—*Continued*

Type	Interunit linkage	Examples	Approximate melting point °C	Characteristics
Poly-anhydride	$\overset{O}{\underset{}{}}\ \overset{O}{\underset{}{}}$ —C—O—C—	$HOOC(CH_2)_8COOH \rightarrow HO[-CO(CH_2)_8COO-]_xH$	83	Similar to crystalline polyesters; fibers may be cold drawn at high molecular weight; hydrolyzed by water[14]
Polysulfides	—S—S—	$ClCH_2CH_2Cl+NaS \rightarrow Cl[-CH_2CH_2S-]_xNa$	Crystalline[15] ca. 130[15]	Rubberlike; becomes crystalline only on stretching[15]
	$\begin{array}{c}S\ \ S\\ -S-S-\end{array}$	$HSCH_2CH_2SH+[O] \rightarrow HS[-CH_2CH_2SS-]_xCH_2CH_2SH$		
		$ClCH_2CH_2Cl+Na_2S_4 \rightarrow Cl[-CH_2CH_2S-S-]_xNa$		
Polyacetals	$\begin{array}{c}H\\ -O-C-O-\\ R\end{array}$	$HO(CH_2)_{10}OH+CH_2(OBu)_2 \rightarrow HO[-(CH_2)_{10}OCH_2O-]_x(CH_2)_{10}OH$	ca. 60	Fiber-forming at high molecular weight[16]
Polyamides	$\overset{O}{\underset{}{}}$ —C—NH—	$NH_2(CH_2)_5COOH \rightarrow H[-NH(CH_2)_5CO-]_xOH$	225	Fibers may be cold drawn above about M.W. 8000; high strength[17]
		$NH_2(CH_2)_6NH_2+HOOC(CH_2)_4COOH \rightarrow H[-NH(CH_2)_6NHCO(CH_2)_4CO-]_xOH$	260	Similar to the above[17]
Silk fibroin	$\overset{O}{\underset{}{}}$ —C—NH—	$NH_2CH_2COOH+NH_2CHRCOOH \rightarrow H[-NHCH_2CONHCHRCO-]_xOH$	Infusible without decomposition	Crystalline; good strength
Cellulose	C—O—C	$C_6H_{12}O_6 \leftarrow -[C_6H_{10}O_4]-O-[C_6H_{10}O_4]-O-, \text{ etc.}$	Infusible without decomposition	Crystalline; excellent strength when highly oriented

TABLE I.—*Continued*

Type	Interunit linkage	Examples	Approximate melting point °C	Characteristics
Phenol-aldehyde	—CHR—	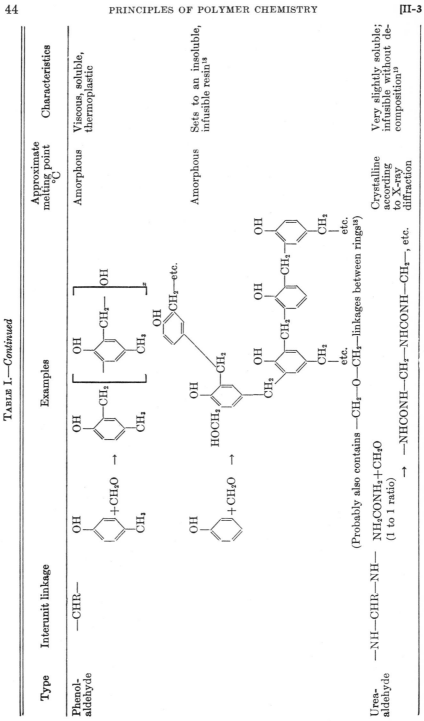	Amorphous	Viscous, soluble, thermoplastic
			Amorphous	Sets to an insoluble, infusible resin[18]
Urea-aldehyde	—NH—CHR—NH—	NH₂CONH₂+CH₂O (1 to 1 ratio) → —NHCONH—CH₂—NHCONH—CH₂—, etc.	Crystalline according to X-ray diffraction	Very slightly soluble; infusible without decomposition[19]

Table I.—Continued

Type	Interunit linkage	Examples	Approximate melting point °C	Characteristics
Urea-aldehyde	—CHR—N—CHR— $\vert$ CHR	$NH_2CONH_2 + CH_2O$ (excess CH_2O) → —NHCONH—CH₂—NCONH—CH₂—, etc. with —CH₂— and —NHCONH—, etc.	Amorphous	Infusible, insoluble, hard thermoset resins[20]
Poly-siloxanes	$\begin{array}{c} R \\ \vert \\ -Si-O- \\ \vert \\ R \end{array}$ and $\begin{array}{c} R \\ \vert \\ -Si-O- \\ \vert \\ R \end{array}$ $R-Si-R$	$\begin{array}{c} CH_3 \\ \vert \\ HO-Si-OH \\ \vert \\ CH_3 \end{array}$ → $HO\left[\begin{array}{c} CH_3 \\ \vert \\ Si-O \\ \vert \\ CH_3 \end{array}\right]_x H$ and the condensation products, etc.	Amorphous	Oils, very viscous liquids, to elastic rubbery materials, depending on x, which may vary up to 10,000[21,22]
			Amorphous	Oils, greases, to hard thermoset composition, depending on M.W. and proportion of trifunctional unit[21]
Poly-urethane	$\begin{array}{c} O \\ \parallel \\ -O-C-NH- \end{array}$	$HO(CH_2)_6OH + OCN(CH_2)_6NCO$ → [—O(CH₂)₆OCONH(CH₂)₆NHCO—]$_x$	ca. 150°	Fiber-forming[23]
Polyurea	$\begin{array}{c} O \\ \parallel \\ -NH-C-NH- \end{array}$	$NH_2(CH_2)_6NH_2 + OCN(CH_2)_6NCO$ → [—NH(CH₂)₆NHCONH(CH₂)₆NHCO—]$_x$	270°	Fiber-forming[23]

monomers must be pure; in polymerizations of the A——A+B——B type (e.g., a coreaction of a diamine with a dibasic acid) the reactants must be used in very nearly the precise stoichiometrically equivalent proportions; and the reaction between the functional groups involved must be one which may be carried very nearly to completion. These requirements are identical with those for obtaining a pure product in good yield in synthetic organic chemistry. Here, however, their stringency is far greater. Amidation and esterification, when conducted under suitable conditions, fulfill the above requirements in good measure. Under optimum conditions, average molecular weights exceeding 25,000 may be attained.

Bifunctional condensation, according to the very nature of the process, necessarily leads to products of finite molecular weight. In view of the impossibility of forcing the condensation reaction literally to completion, there will always be some few unreacted functional groups. These mark the ends of the linear molecules, which therefore are finite in length.

Nonlinear condensation polymers, on the other hand, are not restricted to growth in only two directions. It is at least conceivable that some of the molecules formed from reactants of functionality greater than two may be indefinitely large. This can be seen from the structure indicated in Table I for the product from glycerol and succinic acid. As the polymer molecule increases in size, its functionality increases, in contrast to the linear polymers which retain only two terminal functional groups per molecule. Although some of these functional groups of the nonlinear polymer may remain unreacted, others will combine, thus continuing the structure.

To look at the situation in another way, consider the condensation of two moles of glycerol with three moles of succinic acid. Each ester group formed decreases by one the number of molecules present, excluding from consideration reactions between two functional groups on the same polymer molecule. Thus, if five moles of ester groups were formed intermolecularly, all of the units would be combined into one molecule. But there were initially six moles each of hydroxyl and carboxyl groups. Hence, if five-sixths, or 83.4 percent, of the functional groups were esterified intermolecularly, all units would be combined into a single molecule; any further reaction, incidentally, would of necessity be intramolecular. It would be impossible for the intermolecular reaction to reach this stage without producing structures which assume macroscopic dimensions. Macroscopic structures involving a portion only of the total polymerizing mixture might appear at earlier stages in the condensation; the foregoing argument merely fixes a stoichiometric upper limit.

This calculation indicates that condensations of monomers having a functionality greater than two should be expected to yield indefinitely large polymer structures at sufficiently advanced stages of the process. These indefinitely large or "macro" structures will extend throughout the volume of the polymerized material. Ordinarily their sizes, which will depend on the size of the polymerized sample as well as on the extent of reaction, can conveniently be expressed in grams. On the molecular weight scale they may be considered as essentially infinite in size, hence the term *infinite network*.

Among the physical characteristics of these nonlinear condensation polymerizations, the occurrence of a sharp *gel point* is of foremost significance. At the gel point, which occurs at a well-defined stage in the course of the polymerization, the condensate transforms suddenly from a viscous liquid to an elastic gel. Prior to the gel point, all of the polymer is soluble in suitable solvents, and it is fusible also. Beyond the gel point, it is no longer fusible to a liquid nor is it entirely soluble in solvents. Linear polymers, on the other hand, remain soluble in suitable solvents and fusible to liquids as well (unless the melting point is above the temperature of thermal decomposition), regardless of the extent of condensation.

Gelation and the attendant insolubility mentioned above are encountered in all of the nonlinear polymerizations listed in Table I and in many others likewise. Naturally these characteristics have been attributed to the restraining effects of three-dimensional, or space, network structures of infinite size within the polymer. This is the feature which distinguishes most nonlinear from linear polymers.

It is possible, of course, to avoid gelation in nonlinear polymerizations by limiting the extent of condensation or by using proportions of reactants far from the amounts required stoichiometrically. For example, a mixture of four molecules of glycerol and three of succinic acid will not gel, regardless of the extent of esterification. But the product so obtained bears little resemblance to a linear condensation polymer; its average molecular weight is low and its physical properties are inferior. Thus, although a system containing reactants of higher functionality does not necessarily undergo gelation, the properties of the product may nevertheless be of such a nature as to set it apart from linear polymers.

The physical properties of the various condensation polymers included in Table I are briefly summarized in the last two columns. Two

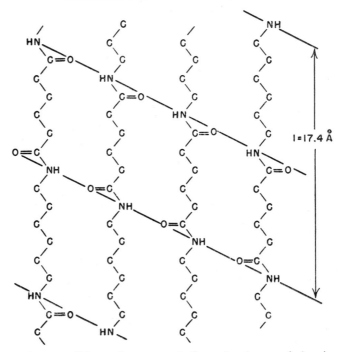

FIG. 1.—Schematic representation of polymer chains in crystalline poly-(hexamethylene adipamide). Layer structure resulting from association of polar groups is indicated by transverse parallel lines. (From Baker and Fuller, *J. Am. Chem. Soc.*, **64**, 2399 [1942].)

characteristic features of the chain structures of linear condensation polymers have an important bearing on the properties of these substances: first, the structural units generally are not subject to isomerism of any sort (e.g., they have no asymmetric atoms) and hence they repeat with perfect regularity along the chain; and, secondly, polar linkages recur *within* the polymer chain.   Both features favor the occurrence of *crystallization*, whereby portions of the polymer chains align themselves in parallel array (Fig. 1), their polar linkages being coordinated in layers transverse to the chain axes.   So-called crystalline polymers actually are semicrystalline only, for amorphous portions invariably persist which fill the intervening space between the *crystallites*, or crystalline regions.   The latter generally are shorter in length than the polymer molecule; a given chain may traverse one crystallite, pursue an irregular path through an amorphous region, then re-enter another crystallite, and so on.   The morphology of a semicrystalline polymer is indicated in Fig. 2.

FIG. 2.—Morphology of a semicrystalline polymer.

Crystalline polymers usually are opaque or translucent, although the individual crystallites are far too small to scatter visible light individually.　In the process of crystallization many crystallites ordinarily grow radially from a common nucleus, and the resulting "spherulitic" aggregate made up of many crystallites may be large enough to be seen in the polarizing microscope.　These aggregates are responsible for the opacity of crystalline polymers.

With proper precautions characteristic sharp melting points may be obtained for crystalline polymers by observing the disappearance of opacity on very slow heating, a method not unlike that applied to simple organic compounds.　One distinction must be borne in mind. however; polymers generally melt over a range of temperature preceding the final disappearance of opacity.　This final disappearance is sharply definable and the term melting point as it applies to polymers refers to the temperature at which all crystallinity vanishes (at equilibrium).　A rough approximation to the melting temperature often may be obtained from observation of the temperature at which the polymer softens, although the results so obtained are by no means reliable.

The incidence of crystallinity among linear condensation polymers to be expected on the basis of the above-cited structural features is borne out by the examples shown in Table I (pp. 42–45).　The noncrystallinity of polyesters derived from diethylene glycol is very nearly a unique exception to the rule.　The strong polarity and hydrogen-bonding capacity of the amide linkage occurring in the polyamides is responsible for their high melting points, which are a full 100 to 150°C above those for analogous polyesters.

Copolymers tend to be less crystalline than the parent polymers, as would be expected from the greater irregularity of their chains. In general, the melting point is depressed by copolymerization and the degree of crystallinity is much reduced, although not necessarily eliminated altogether. Highly nonlinear condensation polymers usually are noncrystalline owing to the pronounced irregularity of their structures. The crystallinity characteristic of the bifunctional repeating units persists, however, if the proportion of polyfunctional units is very small, as for example in the "co-nonlinear" polymers which may be obtained from decamethylene glycol, adipic acid, and a very small proportion of tricarballylic acid.

Crystallinity has a marked effect on physical properties. Depending on its molecular weight, an amorphous polymer may be a syrup (1 to 100 poises), a very viscous liquid which may be stirred with difficulty (10^2 to 10^4 poises) or a semirubbery plastic (10^5 poises or more). If the polymer is appreciably crystalline, however, it no longer displays fluidity; if highly crystalline, it may be very hard. Highly crystalline polyesters and polyamides are superficially waxlike but hard and tough enough to be machined. Those containing the terephthaloyl radical (—CO—⟨benzene ring⟩—CO—) are extremely hard. Variations in hardness and other related mechanical properties probably depend at least as much on the amount, size, and arrangement of the crystallites occurring in a given specimen as on the particular type of structural unit. If the molecular weight is sufficiently high, the polymer in the form of fibers or thin sheets may be oriented by stretching ("cold drawing") from three to five times its initial length under suitable conditions. In the process of stretching, an arrangement is created in which crystallites are aligned, with their chain axes approximately parallel to the fiber axis, and the strength of the sample is thereby greatly enhanced. This capacity for development of high strength through orientation invariably is found in crystalline condensation polymers provided their molecular weights are large enough; 10,000 to 15,000 usually is sufficient. The marked influence of crystallinity is evinced by the total loss of all vestiges of strength when the polymer is heated above its melting point. In some cases a polymer yielding fibers of high strength may even be poured as a liquid (a very viscous one) above its melting point.

The present discussion of physical structure and properties is intended to serve merely as a basis for appraising the characteristics of various polymers here surveyed. The nature of the semicrystalline state in polymers and its influence on their physical properties will be dealt with in greater detail in a later chapter.

4. ADDITION POLYMERS FROM UNSATURATED MONOMERS

Addition polymerizations of unsaturated monomers leading to the formation of products of high molecular weight invariable proceed by chain reaction mechanisms. Primary activation of a monomer M (or a pair of monomers) is followed by the addition of other monomers in rapid succession

$$M \xrightarrow{\text{activation}} M^* \xrightarrow{+M} M_2{}^* \xrightarrow{+M} M_3{}^* \rightarrow, \text{etc.}$$

until the growing chain eventually is deactivated, with the net result that a polymer molecule, M_x, has been formed from x monomers. Each growth step consists of a reaction with monomer. Combination of two growing chains results in deactivation. Unlike condensation polymerization where polymer-polymer combinations occur in profusion, reaction between two growing polymer molecules enters as a deterrent to the development of very large molecules in addition polymerization, and not as a growth reaction. The center of activation, located at the growing end of the molecule, may be a free radical, a carbonium ion (cationic polymerization), or a carbanion (anionic polymerization). The majority of polymerizations of unsaturated compounds (referred to as vinyl polymerizations) which have been investigated in detail proceed by a free radical mechanism. Free radical polymerizations usually are induced by radicals released by a decomposing peroxide, commonly referred to as the catalyst. The initiating free radicals may also be generated photochemically or thermally, however. Examples of addition polymers formed from unsaturated monomers are listed in Table II.

Regardless of the particular chain propagation mechanism involved, the entire synthesis of an individual polymer molecule from unreacted monomers occurs within a brief interval of time—often within a few seconds or less—whereas the over-all conversion of monomer to a good yield of polymer may require hours (see Chap. IV). Thus, at any instant during the polymerization process the reaction mixture consists almost entirely of unchanged monomer and of high polymer. Material at intervening stages of growth is virtually absent; the portion consisting of actively growing chains is so small as to be immeasurable by ordinary chemical methods. As the percentage conversion to polymer increases, the average degree of polymerization of the polymerized portion remains approximately the same, or at least does not change commensurately with the extent of conversion to polymer.

The synthesis of a given polymer molecule through condensation polymerization, on the other hand, is accomplished by a series of independent condensations, which ordinarily occur at intervals scattered

TABLE II.—REPRESENTATIVE ADDITION POLYMERS FORMED FROM UNSATURATED MONOMERS

Monomer	Type of polymerization	Structural unit	Approximate melting (T_m) or softening (T_g) temperature in °C[a]	Properties
Ethylene	Free radical; at pressures up to 2000 atmospheres at 100 to 200°C using traces of oxygen or aqueous persulfate as catalyst[24]	—CH₂—CH₂—	$T_m = 110$ to 130	Waxlike, highly crystalline, very strong when oriented. Insoluble in all solvents at temperatures much below T_m
Vinyl chloride	Free radical; polymerization in bulk or emulsion rapid in presence of peroxides; susceptible to photochemical polymerization	—CH₂—CH— / Cl	$T_g = 75$	Largely amorphous, except when highly oriented by stretching. Hard. Soluble in ketones and esters
Vinyl acetate	Free radical polymerization similar to the above	—CH₂—CH— / OCOCH₃	$T_g = 30$	Amorphous, even when stretched. Soluble in various solvents
Styrene	Free radical polymerization similar to the above. Also susceptible to rapid cationic polymerization induced by AlCl₃ at −80°C and to anionic polymerization using alkali metals or their hydrides	—CH₂—CH— / C₆H₅	$T_g = 100$	Amorphous, even when stretched. Hard. Soluble in aromatic hydrocarbons, higher ketones, and esters
Methyl acrylate	Free radical polymerization similar to the above	—CH₂—CH— / COOCH₃	$T_g = 0$	Amorphous, even when stretched. Soft, rubbery if molecular weight is high. Readily soluble
Methyl methacrylate	Free radical polymerization similar to the above. Also susceptible to rapid anionic polymerization[25] induced by RMgX or Na in liquid NH₃	CH₃ / —CH₂—C— / COOCH₃	$T_g = 90$	Amorphous, even when stretched. Hard. Soluble in aromatic hydrocarbons, esters, dioxane, etc.
Acrylonitrile	Free radical polymerization similar to the above, but thermal polymerization is difficult.[26]	—CH₂—CH— / CN	Infusible up to 250	Imperfectly crystalline; shows oriented crystallinity in drafted fibers. Good strength. Insoluble in common solvents. Soluble in dimethylformamide

Monomer	Polymerization	Structure	Temperature[a]	Properties
Methacrylonitrile	Free radical polymerization similar to acrylonitrile. Also undergoes vigorous anionic polymerization[25] in presence of RMgX or Na in liquid NH₃	CH₃ | —CH₂—C— | CN	T_g ca. 120(?)	Amorphous. Readily soluble in various solvents[27]
Vinyl isobutyl ether	Free radical polymerization is slow and yields only very low polymers. Vigorous cationic polymerization induced by BF₃-ether complex at temperatures down to −80°C[28]	—CH₂—CH— | OC₄H₉	$T_g = -20$	Amorphous, rubbery if molecular weight is high. When prepared by controlled reaction at low temperatures, polymer is semicrystalline[28]
Vinylidine chloride	Free radical polymerization similar to vinyl chloride	Cl | —CH₂—C— | Cl	$T_m = 210$	Highly crystalline, very strong when oriented by stretching. Difficultly soluble below melting point
Isobutylene	Not susceptible to free radical or anionic polymerization, but cationic polymerization induced by BF₃, AlCl₃, etc., is extremely rapid even at −100°C	CH₃ | —CH₂—C— | CH₃	$T_g = -70$	Amorphous, except when highly oriented by stretching. Soft, rubbery. Molecular weights may range up to 10⁷. Soluble in hydrocarbons
Butadiene	Undergoes free radical, anionic, and cationic polymerization	—CH₂—CH=CH—CH₂— and —CH₂—CH— | CH=CH₂	$T_g = -88$	Amorphous, except for polymers prepared by free radical polymerization at 20°C or lower. Latter crystallize on cooling or stretching. Soft rubbery. Generally contains gel fraction insoluble even in hydrocarbon solvents
Tetrafluoroethylene	Free radical polymerization in presence of H₂O₂ in H₂O at 60°C and ca. 50 atmospheres[29]	—CF₂—CF₂—	$T_m = 327$	Very tough and resistant to chemicals and all solvents. Rubbery but nonplastic above T_m[29]
Allyl acetate	Slow free radical polymerization at 80°C in presence of large amounts of peroxide (benzoyl)[30]	—CH₂—CH— | CH₂OCOCH₃		Amorphous. Polymeric oils of low molecular weights from 1000 to 2000[30]

ᵃ Softening temperature T_g refers to the temperature at which the polymer changes from a glassy or brittle condition to a "liquid" or "rubbery" one. The values reported for T_g are highly dependent on the method of determination, those given in the table being approximate only.

over the period during which the polymerization is carried out, and not during a single comparatively brief interval as is the case in addition polymerization. Interruption of a condensation polymerization at an early stage of the process yields a polymer of low average molecular weight. Unless the average degree of polymerization is very low—less than about ten units—the polymerizing mixture will contain a negligible amount of monomer (less than 1 percent). As the reaction is continued, the low polymers first formed condense further, with the result that the average molecular weight continues to increase. Thus, the monomer disappears almost completely during the initial phase of the (linear) condensation polymerization process, but in order to attain a high average molecular weight it is necessary to continue the polymerization until the reaction approaches closely to completion. In vinyl addition polymerization high polymer makes its appearance at the outset; the duration of the process is determined by the yield of polymer desired and not by the molecular weight required. Addition polymer molecules ordinarily do not respond to further polymerization by interreaction with one another.

The characteristics of vinyl polymerizations outlined above follow naturally from the fact that they are chain reactions in the kinetic sense of the term. They may proceed quite rapidly under proper conditions, and polymers having molecular weights exceeding 10^5 usually are obtained with no difficulty. In some instances molecular weights may exceed 10^6 or even 10^7. Polymerizations of unsaturated compounds which proceed by an ionic mechanism are especially rapid; the reaction may be violently so, even at very low temperatures (e.g., even at $-100°C$ in the case of isobutylene polymerization catalyzed by BF_3). Condensation polymerizations generally require a much greater length of time, particularly if a satisfactorily high molecular weight is to be reached. Molecular weights much above 25,000 are rare for linear condensation polymers.

Vinyl addition polymerizations, being chain reactions, are more complex than condensation polymerizations; and they are generally more likely to include minor side reactions leading to branching or crosslinking. The structures of the polymeric products are consequently more difficult to establish with certainty.

Nonlinear addition polymers are readily obtained by copolymerizing a divinyl compound (e.g., divinylbenzene) with the vinyl monomer (e.g., styrene), as already mentioned. Products so obtained exhibit the insolubility and other characteristics of space-network structures and are entirely analogous structurally to the space-network polymers produced by the condensation of polyfunctional compounds. Owing to

the length of the chains in vinyl polymerizations, as little as 0.01 percent of the divinyl compound may be sufficient to bring about gelation,
with the formation of a space-network polymer insoluble in liquids
which would readily dissolve the polymer were it not for the network
structure.   A copolymer obtained from styrene and a fraction of a
percentage of divinylbenzene, for example, swells to many times its
initial volume when placed in benzene, but ultimately it reaches a
swelling limit if sufficient solvent is present.   No appreciable portion
of the polymer dissolves in the excess solvent not absorbed by the
swollen gel.

Polymers of dienes such as butadiene frequently contain a substantial
portion of gel which will not dissolve in a good solvent, though it may
swell to a volume 20 to 100 or more times that of the polymer itself.
This gel, which may comprise up to 90 percent or more of the polymer,
consists of a space-network structure formed as a result of a very few
cross-linkages provided by occasional (perhaps 1 in 1000 or less) diene
units both double bonds of which have entered into the polymerization
(see Chap. VI).

Substituted ethylenes in which substituents occur on both carbon
atoms (with the exception of the fluoroethylenes) usually are not prone
to polymerize, although some of them, such as the maleates and fumar-
ates, copolymerize readily with other monomers.   The further fact
that, with rare exceptions, the monomers unite through the addition
of the substituted carbon atom of one unit to the unsubstituted carbon
atom of the next permits representation of nearly all vinyl addition
polymers by the general structural formula

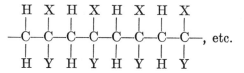

where X and Y represent any of a variety of substituents, including
hydrogen.   X and Y may be the same or different (see Table II).   No
assignment of configurations to the asymmetric atoms occurring when
X and Y are different is intended.   The repeating unit is identical
with the structural unit, which contains two chain atoms.   On this
basis the vinyl addition polymers as a group possess structural similarity.   Wider variations in the structure of the repeating unit obviously are possible among condensation polymers.   Addition polymers
formed from 1,3-dienes differ in structure from those of monovinyl
compounds, the principal structural unit consisting of four atoms, as
previously pointed out.

Referring to Table II, it will be noted that crystallinity is less prevalent among the addition polymers obtained from unsaturated monomers than among the linear condensation polymers. This is in large part due to the asymmetry of alternate chain atoms when the substituents X and Y in the chain formula given above are not the same. Such polymers are in reality copolymers of d and l structural units. If Y = H and X is not too large, as for example in polyvinyl chloride, or if the interactions between pairs of the substituents X belonging to different molecules are large, as in polyacrylonitrile or polyvinyl alcohol[31] (not included in Table II), crystallinity nevertheless may occur in spite of the d,l heterogeneity in the chain structure. The resulting crystalline regions are, of course, imperfectly ordered. When X is a large substituent and Y = H, no evidence for crystallinity is discernible.

When X = Y, as in polyethylene, poly-(tetrafluoroethylene), polyisobutylene, and poly-(vinylidene chloride), the polymers are highly crystalline products with sharply definable melting points (except for polyisobutylene, which crystallizes readily on stretching but with difficulty on cooling). Oriented specimens of high strength may be obtained, exactly as in the crystalline condensation polymers.

For every amorphous polymer there exists a narrow temperature region in which it changes from a viscous or rubbery condition at temperatures above this region, to a hard and relatively brittle one below it. This transformation is equivalent to the solidification of a liquid to a glass; it is not a phase transition. Thus, above about 100°C polystyrene is a rubbery material if its molecular weight is high, or a very viscous liquid if it is low; it quickly becomes a hard glass below this temperature in either case (although if the molecular weight is very low the critical temperature for the transition may be lowered somewhat). Not only do hardness and brittleness undergo rapid changes in the vicinity of the glass temperature, T_g, but other properties such as the thermal expansion coefficient, the heat capacity, and the dielectric constant (in the case of a polar polymer) also change markedly over an interval of a few degrees. T_g is regarded variously as the brittle temperature, the critical temperature for the glassy state, or the second-order transition temperature, although as mentioned above no phase transition is involved and reference to a transition may therefore be misleading.

Approximate values of T_g are included in Table II for the purpose of indicating the temperature region in which the polymer characteristically changes from a hard, more or less brittle glass to a rubbery or viscous polymer within which motions of portions of the chains, usually called segments, are comparatively unhampered by the interactions between neighboring chains. It will be noted that structural units having higher van der Waals interactions generally confer a higher value of T_g on the polymer.

The strengths of polymers in the glassy state exceed those of amorphous polymers at temperatures above T_g, but they are generally much inferior to the strengths attainable in oriented crystalline polymeric substances.

5. POLYMERIZATION OF CYCLIC COMPOUNDS

In Section 3 of this chapter it was mentioned that polymers obtained by intermolecular condensation of bifunctional monomers may often be prepared alternatively by an addition polymerization of a cyclic compound having the same composition as the structural unit. Typical examples are shown in Table III. The processes indicated are appropriately regarded as addition polymerizations. Each of these polymers may also be prepared through the condensation of suitable bifunctional monomers. The dimethylsiloxane polymer, for example, may be prepared, as indicated in Table I (p. 45), through the condensation of dimethyl dihydroxysilane formed by hydrolysis of the dichlorosilane

$$\underset{\underset{CH_3}{|}}{\overset{\overset{CH_3}{|}}{HO-Si-OH}} \rightarrow \left[\underset{\underset{CH_3}{|}}{\overset{\overset{CH_3}{|}}{-Si-O-}} \right]_x + H_2O$$

The chemical and physical properties of the polymers obtained by these alternate methods are identical, except insofar as they are affected by differences in molecular weight. In order to avoid the confusion which would result if classification of the products were to be based on the method of synthesis actually employed in each case, it has been proposed that the substance be referred to as a condensation polymer in such instances, irrespective of whether a condensation or an addition polymerization process was used in its preparation. The cyclic compound is after all a condensation product of one or more bifunctional compounds, and in this sense the linear polymer obtained from the cyclic intermediate can be regarded as the polymeric derivative of the bifunctional monomer(s). Furthermore, each of the polymers listed in Table III may be degraded to bifunctional monomers differing in composition from the structural unit, although such degradation of polyethylene oxide and the polythioether may be difficult. Apart from the demands of any particular definition, it is clearly desirable to include all of these substances among the condensation

TABLE III.—ADDITION POLYMERIZATION OF RING COMPOUNDS

Monomer type	Example	Reaction conditions	Properties of polymer
Lactone	$O(CH_2)_5CO \rightarrow [-O(CH_2)_5CO-]_x$	150°C, preferably in presence of K_2CO_3[32]	Crystalline, m.p. 51–53°C
Lactam	$NH(CH_2)_5CO \rightarrow [-NH(CH_2)_5CO-]_x$	200°C or higher in presence of small amounts of water[33]	Crystalline, m.p. 225°C
Cyclic anhydride	$CO(CH_2)_8CO \rightarrow [-CO(CH_2)_8COO-]_x$	Heating and/or presence of traces of water[14]	Crystalline, m.p. 82°C
Cyclic thioether	$(CH_2)_3{=}S \rightarrow [-(CH_2)_3S-]_x$	In dry hexane containing HCl at room temperature[34]	Crystalline, m.p. 80–100°C
Cyclic siloxane	$\left[-\underset{CH_3}{\overset{CH_3}{Si}}-O- \right]_x$	In anhydrous medium containing H_2SO_4[35]	Amorphous (See Table I p. 45)
Ethylene oxide	$CH_2CH_2 \rightarrow (-CH_2CH_2O-)_x$	Heating (with caution!) in presence of acidic or alkaline catalysts	Crystalline, m.p. 66°C

polymers, for they contain recurrent functional groups within their polymer chains.

The polymerization of a ring compound usually proceeds by an interchange reaction, induced either catalytically[16,35] or by the presence of small amounts of end-group-producing substances.[36,37] For example, the polymerization of lactide, the cyclic dimer of lactic acid, is accelerated by small amounts of water. The water undoubtedly hydrolyzes the lactide to lactyllactic acid, which may then react with other lactide molecules by ester interchange.[37] This reaction scheme can be represented as follows:

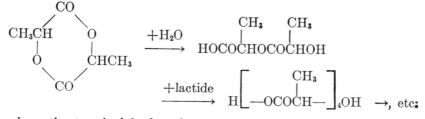

where the terminal hydroxyl group of the polymer chain adds the lactide in a process which amounts merely to an ester interchange. Similarly, water accelerates the polymerization of ϵ-caprolactam,[38] presumably via a similar mechanism

$$CO(CH_2)_5NH + H_2O \rightarrow HOCO(CH_2)_5NH_2$$

$$\xrightarrow{+lactam} HOCO(CH_2)_5NHCO(CH_2)_5NH_2$$

$$\downarrow$$
$$etc.$$

Ethylene oxide polymerization may be initiated similarly by substances (alcohols, amines, mercaptans) capable of generating a hydroxyl group through reaction with the monomer.[39] In the presence of strongly acidic or basic catalysts, successive addition of ethylene oxide molecules proceeds rapidly in the following manner:

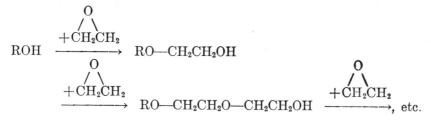

The polymerization of N-carboxyanhydrides of α-amino acids to polypeptides[40]

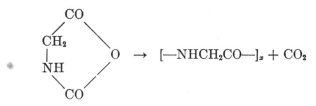

doubtless proceeds according to a similar mechanism. A small amount of water, alcohol, or an amine (or an amino acid) accelerates the polymerization.[41,42] These substances provide functional groups capable of opening the ring of the monomer through an interchange reaction such as the following:

The free amino group of the amino ester may then react analogously with another molecule of the monomer, etc. The kinetics of the polymerization are in harmony with a mechanism of this sort.[42] The final polypeptide may contain up to 300 or more structural units.[42] While the polymerization of N-carboxyanhydrides is closely analogous to the addition polymerizations of ethylene oxide and of other cyclic substances, definition unfortunately classifies it as a condensation polymerization inasmuch as carbon dioxide is eliminated in the process.

The stepwise polymerizations of cyclic compounds appear to resemble vinyl-type addition polymerizations in that they proceed exclusively (according to the mechanisms given) by addition of the monomer (or cyclic dimer) to chain molecules. If the rate constant governing the successive additions of monomer molecules is much more rapid than the first step (i.e., than the addition of monomer to the initiating substance), then the over-all polymerization may assume the character of a chain reaction. Usually, however, the initiating step and subsequent monomer additions are closely similar in nature and may be expected to proceed at comparable rates; kinetically, these are not therefore chain reactions. It will be observed that the number of polymer molecules is fixed by the number of initiating species (e.g., ROH) initially introduced. Hence the average molecular weight will depend on the ratio of monomers polymerized to the number of initiator molecules used.

Ring-to-chain polymerizations which proceed according to the above stepwise mechanisms are intermediate in character between condensation and vinyl addition polymerization.   Molecular growth occurs exclusively by addition of the cyclic reactant, and at an intermediate stage of the process the mixture should consist of polymer and unreacted monomer, the yield of polymer advancing with continuation of the process as it does in addition polymerization.   Like condensation polymerizations, however, the various polymer molecules undergo more or less simultaneous growth throughout the polymerization process, and the average molecular weight of the polymer also increases as the process progresses.

Polymerization of cyclic compounds may also occur by ionic mechanisms under the influence of strong acids or bases and in the absence of water and alcohols.   Thus, in the presence of a strong acid or electron acceptor (BF_3), ethylene oxide may polymerize violently.   The mechanism may be the following, where the electron acceptor is represented by the hydrogen ion:

$$H^+ + \overset{\displaystyle O}{\overset{\displaystyle /\ \backslash}{CH_2CH_2}} \rightarrow HOCH_2CH_2^+$$

$$\xrightarrow{\quad +\overset{\displaystyle O}{\overset{\displaystyle /\ \backslash}{CH_2CH_2}} \quad} HOCH_2CH_2OCH_2CH_2^+ \rightarrow, \text{ etc.}$$

A corresponding anionic mechanism in the presence of a strong base (or electron donor) is plausible.   Other cyclic compounds may be susceptible to polymerization by similar ionic mechanisms.   Inasmuch as the growth step must be extremely rapid, a chain reaction is indicated and classification with vinyl-type addition polymerizations should be appropriate in such cases.

Polymerizations of cyclic compounds, particularly those involving six- or seven-membered rings, frequently are reversible (see Chap. III).

6. ANOMALOUS CONDENSATION POLYMERIZATIONS

The difficulties of devising a basis for conveniently classifying various polymerizations in an appropriate manner have been discussed earlier in this chapter and several borderline examples which offer particular difficulty have been mentioned.   One of these, the polymerization of the N-carboxyanhydrides, falls within the definition of a condensation polymerization, proceeds by a mechanism resembling a vinyl addition polymerization, and yields a product which possesses the structure of a typical condensation polymer.   Definitions have been so adjusted as to permit the polymerization of a diisocyanate with a glycol or a diamine to be included as a condensation type. A superficially similar polymerization is exemplified by the condensation of a dithiol with a diolefin (preferably nonconjugated), e.g.[43]

$$HS—R—SH + CH_2{=}CH—R'—CH{=}CH_2 \rightarrow$$
$$—S—R—S—CH_2CH_2—R'—CH_2CH_2—, \text{ etc.}$$

which takes place readily in the presence of a source of free radicals. Consistency demands that this process also be regarded as a condensation polymerization. This is unfortunate, for the mechanism of the reaction bears little resemblance to the stepwise coupling of functional groups which characterizes other condensation polymerizations. It proceeds by a free radical chain mechanism;[44] a mercaptyl radical adds to the α-carbon of the ethylenic linkage, whereupon the β-carbon removes a hydrogen atom from a mercaptan molecule, and so forth. This chain mechanism differs from that involved in vinyl polymerizations, since the two ends of the diene react quite independently of one another. The process appears to be in a class by itself; the polymeric product, however, qualifies for the condensation polymer class in fulfillment of definition.

Two further examples illustrate the limitations of the differentiation between condensation and addition polymerizations. One of these is the polymerization of a benzyl halide, e.g.

$$C_6H_5CH_2Cl \rightarrow [—C_6H_4CH_2—]_x + HCl$$

a well-known reaction which occurs readily in the presence of a small amount of a Friedel-Crafts catalyst. (Both *ortho* and *para* positions are subject to attack in this example; hence the polymer is nonlinear.) Although the process is a condensation, a chain reaction appears to be involved.[45] The other example consists in the condensation of a dihalide with a metal such as sodium or magnesium, e.g.,[46]

$$Br(CH_2)_{10}Br + Na \rightarrow [—(CH_2)_{10}—]_x + NaBr$$

In principle, polymers equivalent to those obtained from vinyl and divinyl monomers may be synthesized by this method. The product in the above example possesses the same chain structure as polyethylene. The polymerization process, notwithstanding the likelihood of a metal alkyl intermediate, should conform satisfactorily to stepwise condensation. However, the product, and those obtained by Friedel-Crafts condensation as well, lack the recurrent functional groups which generally characterize condensation polymers.

The occurrence of the various atypical condensation and addition polymerizations mentioned above does not necessarily undermine the usefulness of the distinctions which have been drawn between the two. They emphasize merely that such generalizations as may be made concerning one or the other class of polymerizations (or polymers) will be subject to exceptions. These exceptions to the classification scheme set forth above actually are not so common as the emphasis placed on them, possibly excessive, may seem to indicate.

7. INORGANIC POLYMERS

The capacity to form long chain molecules is by no means limited to organic substances. In the organosilicon polymers previously discussed, the main chain consists of alternating silicon and oxygen atoms, the carbon atom occurring only in the substituent groups. Representative inorganic substances known to possess polymer chain-like structures are listed in Table IV. The physical properties of these substances resemble those of typical organic polymers. Phosphonitrilic chloride and amorphous sulfur display rubberlike elasticity. The former exhibits a stress-strain curve similar to that of vulcanized

TABLE IV

INORGANIC CHAIN POLYMERS[a]

Compound	Method of preparation	Structural unit	Physical properties
Phosphonitrilic chloride	PCl_5+NH_4Cl at 150°C, or by heating cyclic trimer:	$\begin{array}{c} Cl \\ \mid \\ -P=N- \\ \mid \\ Cl \end{array}$	Rubberlike, crystallizes on stretching.[47] Swells in benzene
Amorphous sulfur	Melting rhombic sulfur and heating at about 150°C	$-S-$	Rubberlike, crystallizes on stretching[48]
Sulfur trioxide	Rubbery form of SO_3 precipitates from solution of SO_3 in SO_2[49]	$\begin{array}{c} O \\ \parallel \\ -S-O- \\ \parallel \\ O \end{array}$	Low melting fibrous form melts at 31.5°C
Silicon disulfide	$Al_2S_3+SiO_2$ at 1200°C		Fibrous crystals. Sublimes without melting. Attacked by water[50]

[a] The polymeric nature of the structures of various crystalline inorganic substances has been discussed at length by K. H. Meyer.[51]

rubber.[47]   Both crystallize on stretching, and their crystal structures have been investigated by X-ray diffraction.[47,48]   Silicon disulfide melts too high for possible rubberlike properties to be observed, but its tensile strength has been reported to be very high,[50] as would be expected from its crystalline character.   In consideration of their structures and properties, there are good reasons for including these inorganic compounds within the scope of a general discussion of polymeric substances.

According to the opening remarks of this chapter, polymers may be described as substances consisting of many structural units connected by valence bonds in *any* conceivable pattern.   If these statements were construed as a definition (which was not intended), a host of additional inorganic substances could be included as polymers.   Most of the mineral silicates, the complex phosphates, and the borate and silicate glasses would qualify as polymers, as would also many crystalline substances in which the atoms are connected in a continuous array by covalent bonds.[51]   In the latter category would be included substances such as graphite, boron nitride, and black phosphorus in which the atoms are connected in planar sheets, the sheets being held together by van der Waals forces.   Talc, mica, and various clays are similarly constituted except that adhesion between layers is afforded by ions.   These substances might be regarded as "sheet" polymers.   Diamond, various silicates, and numerous other substances should then be included among three-dimensional net polymers.   As a matter of fact, common ionic crystals such as those of sodium chloride would seem to be about equally eligible for inclusion among the high polymers; the only substantial difference from the crystalline substances just mentioned occurs in the ionic nature of the bonds.   This obviously carries the boundaries of definition much too far.

There is sound justification for including certain of the silicates and complex phosphates among representative polymeric substances, but not for including structures which are regularly bonded in two or three dimensions.   They have little in common with the conventional nonlinear, or network, polymers.   The structures of the latter conform to no set pattern, being controlled largely by the laws of chance.   Circuitous connections in polymer networks may be manifold, but the number of units included in a closed circuit varies widely and generally is fairly large.   This situation is in sharp contrast to the six-membered rings which repeat with perfect regularity throughout the structures of graphite and diamond.   The physical properties of network polymers bear little resemblance to those of standard crystalline substances in which the valence structure is regularly propagated in two or three dimensions.*   It is neither appropriate nor otherwise desirable, therefore, to include such substances within the meaning of the term polymer.

The question of whether or not silicate glasses in which the constituents are connected to one another in a somewhat irregular fashion should be considered to be high polymers is debatable.   The transiency of these connections casts doubt on the desirability of doing so, as does also their approximate spatial regularity.

8. CONCLUDING REMARKS ON THE STRUCTURE AND PROPERTIES OF HIGH POLYMERS

In view of the diversity of properties displayed by different high polymers, some being viscous liquids, others rubbery, and still others very hard and tough, the question arises as to how these most apparent physical properties relate to structure and to composition.   It is to be noted at once that no one of these attributes should be considered as an invariant characteristic of polymers of a given structural unit. Starting with virtually any structural unit, or units, polymers conceivably may be constructed which are oils, rubbers, or fibers, depending on the temperature and on the molecular chain length, or, more generally, on the manner in which the units are connected.   Thus it is no more fitting to ascribe rubberlike character to the polymers obtained from given monomers than to state that compounds containing certain elements are gaseous without specifying the molecular constitution, temperature, and pressure.   On the other hand, the temperature at which the polymer becomes brittle is specific for the given structural unit (assuming that the molecular weight is high).   Similarly, the melting point of a crystalline polymer relates directly to the symmetry and interaction forces of its repeating unit.

Other properties such as solubility, viscosity (above T_g and T_m), modulus of elasticity, and strength are highly dependent on the polymer architecture, or pattern of the interconnections between units.   In other words, these properties vary to a marked degree with changes in the molecular weight and in the degree of cross-linking.

It is instructive to consider what steps may be taken to ascertain qualitatively the structural type and physical state of a given polymeric substance.   To this end one should first of all determine whether or not the substance is soluble without decomposition in any solvent,

* If the covalent bonds connect elements of the structure along one dimension only, as in silicon disulfide, it may be desirable to consider the substance as a polymer.   This will certainly be appropriate if the valence structure supersedes whatever crystalline arrangement prevails, i.e., if the substance can be melted without seriously disrupting the continuity of the interunit connections.

and whether or not it will, on heating, soften to a liquid which displays measurable fluidity. A positive result from either test assures that the substance is of the linear, or non-network, type. Negative results do not necessarily prove the presence of a network structure: the melting point may exceed the decomposition temperature, in which case the solubility is likely to be negligible. A high degree of swelling (several-fold or more on a volume basis) in the best candidate for a good solvent would indicate a loose network structure such as occurs in vulcanized rubber. If the polymer qualifies either as a linear or as a loose network structure on the basis of these tests, the temperature at which it hardens (or softens) should be determined. Whether crystallinity or embrittlement is involved at this temperature can usually be deduced from the transparency (opacity indicating crystallinity) or, with greater certainty, from X-ray diffraction.

If the polymer is hard, insoluble, and infusible without decomposition, and if it refuses to swell greatly in any solvent, it may be assumed either that it is highly crystalline, with a melting point above its decomposition temperature, or that it possesses a closely interconnected network structure (e.g., as in a highly reacted glyceryl phthalate or a phenol-formaldehyde polymer). Differentiation between these possibilities is feasible on the basis of X-ray diffraction.

From the results of tests of this nature, the type of structure occurring in a given polymer usually can be deduced. Following these qualitative observations, chemical composition and structure determinations would be logical next steps. If the polymer is soluble, measurement of the average molecular weight and determination of the molecular weight distribution enter as first objectives in the quantitative elucidation of constitution. If it is insoluble owing to a network structure, ordinary physicochemical methods obviously cannot be applied. Certain information on the density of cross-linking within the network structure can be secured from equilibrium swelling measurements, however. Quantitative physical methods applicable to polymers will be dealt with in later chapters.

REFERENCES

1. W. H. Carothers, *Trans. Faraday Soc.*, **32**, 39 (1936).
2. W. H. Carothers, *Chem. Revs.*, **8**, 353 (1931).
3. C. Walling, *J. Am. Chem. Soc.*, **67**, 441 (1945).
4. See chapter entitled "Vulcanization" by E. H. Farmer in *Advances in Colloid Science* (Interscience Publishers, New York), II (1946), 299.
5. G. F. Bloomfield and R. F. Naylor, *Proceedings of the XIth International Congress of Pure and Applied Chemistry*, Vol. II, "Organic Chemistry, Biochemistry," p. 7 (1951).

6. G. V. Schulz, *Z. physik. Chem.*, **B44**, 227 (1939); P. J. Flory, *J. Am. Chem. Soc.*, **69**, 2893 (1947).

7. O. L. Wheeler, S. L. Ernst, and R. N. Crozier, *J. Polymer Sci.*, **8**, 409 (1952).

8. W. H. Carothers, *J. Am. Chem. Soc.*, **51**, 2548 (1929).

9. The reader is referred to the *Collected Papers of Wallace Hume Carothers on High Polymeric Substances*, ed. by H. Mark and G. S. Whitby (Interscience Publishers, New York, 1940), Part One of which consists of 28 papers on condensation polymerization, among which are included many of the most important ones on this subject. See also Ref. 2.

10. W. H. Carothers and F. J. Van Natta, *J. Am. Chem. Soc.*, **55**, 4714 (1933).

11. P. J. Flory, *J. Am. Chem. Soc.*, **61**, 3334 (1939).

12. R. D. Evans, H. R. Mighton, and P. J. Flory, *J. Am. Chem. Soc.*, **72**, 2018 (1950); W. H. Carothers and J. A. Arvin, *ibid.*, **51**, 2560 (1929).

13. J. R. Whinfield, *Nature*, **158**, 930 (1946); Brit. Pat. 578,079; H. J. Kolb and E. F. Izard, *J. Applied Phys.*, **20**, 564 (1949).

14. J. W. Hill and W. H. Carothers, *J. Am. Chem. Soc.*, **54**, 1569 (1932).

15. J. C. Patrick, *Trans. Faraday Soc.*, **32**, 347 (1936); J. C. Patrick and S. M. Martin, Jr., *Ind. Eng. Chem.*, **28**, 1144 (1936).

16. J. W. Hill and W. H. Carothers, *J. Am. Chem. Soc.*, **57**, 925 (1935).

17. W. H. Carothers, U.S. Patents, 2,071,253 (1937) and 2,130,948 (1938); see also D. D. Coffman, G. J. Berchet, W. R. Peterson, and E. W. Spanagel, *J. Polymer Sci.*, **2**, 306 (1947).

18. N. J. L. Megson, *Trans. Faraday Soc.*, **32**, 336 (1936); T. S. Carswell, *Phenoplasts* (Interscience Publishers, New York, 1947), Chap. II.

19. A. E. Dixon, *J. Chem. Soc.*, **113**, 238 (1918); K. H. Meyer, *Trans. Faraday Soc.*, **32**, 407 (1936); G. Smets and A. Borzee, *J. Polymer Sci.*, **8**, 371 (1952).

20. T. S. Hodgins and A. G. Hovey, *Ind. Eng. Chem.*, **30**, 1021 (1938).

21. E. G. Rochow, *Chemistry of the Silicones* (John Wiley and Sons, New York, 1946).

22. D. W. Scott, *J. Am. Chem. Soc.*, **68**, 1877 (1946); A. J. Barry, *J. Applied Phys.*, **17**, 1020 (1946).

23. W. E. Catlin, U.S. Patent 2,284,637 (1942); W. E. Hanford, U.S. Patent 2,292,443 (1942); P. Schlack, U.S. Patent 2,343,808 (1944).

24. British Patent 471,590; R. E. Brooks, M. D. Peterson, and A. G. Weber, U.S. Patent 2,388,225 (1945).

25. R. G. Beaman, *J. Am. Chem. Soc.*, **70**, 3115 (1948).

26. W. Kern and H. Fernow, *J. prakt. Chem.*, **160**, 281 (1942).

27. E. Mertens and M. Fonteyn, *Bull soc. chim. Belges*, **45**, 438 (1936).

28. C. E. Schildknecht, S. T. Gross, H. R. Davidson, J. M. Lambert, and A. O. Zoss, *Ind. Eng. Chem.*, **40**, 2104 (1948).

29. W. E. Hanford and R. M. Joyce, *J. Am. Chem. Soc.*, **68**, 2082 (1946).

30. P. D. Bartlett and R. Altschul, *J. Am. Chem. Soc.*, **67**, 812, 816 (1945).

31. C. W. Bunn and H. S. Peiser, *Nature*, **159**, 161 (1947); C. W. Bunn, *ibid.*, **161**, 929 (1948).

32. F. J. Van Natta, J. W. Hill, and W. H. Carothers, *J. Am. Chem. Soc.*, **56**, 455 (1934).

33. W. E. Hanford, U.S. Patent 2,241,322 (1941); J. R. Schaefgen and P. J. Flory, *J. Am. Chem. Soc.*, **70**, 2709 (1948).

34. R. W. Bost and M. W. Conn, *Ind. Eng. Chem.*, **25**, 526 (1933).

35. W. I. Patnode and D. F. Wilcock, *J. Am. Chem. Soc.*, **68**, 358 (1946); M. J. Hunter, E. L. Warrick, J. F. Hyde, and C. C. Currie, *ibid.*, **68**, 2284 (1946); D. F. Wilcock, *ibid.*, **69**, 477 (1947).

36. W. H. Carothers, G. L. Dorough, and F. J. Van Natta, *J. Am. Chem. Soc.*, **54**, 761 (1932).

37. P. J. Flory, *J. Am. Chem. Soc.*, **64**, 2205 (1942).

38. P. Schlack, U.S. Patent 2,241,321 (1941); W. E. Hanford, U.S. Patent 2,241,322 (1941); R. M. Joyce and D. M. Ritter, U.S. Patent 2,251,519 (1941).

39. H. Hibbert and S. Z. Perry, *Can. J. Research*, **8**, 102 (1933); P. J. Flory *J. Am. Chem. Soc.*, **62**, 1561 (1940).

40. H. Leuchs, *Ber.*, **39**, 857 (1906); H. Leuchs and W. Geiger, *ibid.*, **41**, 1721 (1908).

41. F. Wessely, *Z. physiol. Chem.*, **146**, 72 (1925); F. Sigmund and F. Wessely, *ibid.*, **157**, 91 (1926); W. E. Hanby, S. G. Waley, and J. Watson, *J. Chem. Soc.*, **1950**, 3009.

42. S. G. Waley and J. Watson, *Proc. Roy. Soc.* (London), **A199**, 499 (1949).

43. C. S. Marvel and R. R. Chambers, *J. Am. Chem. Soc.*, **70**, 993 (1948); C. S. Marvel and P. H. Aldrich, *ibid.*, **72**, 1978 (1950); C. S. Marvel and G. Nowlin, *ibid.*, **72**, 5026 (1950); C. S. Marvel and A. H. Markhart, Jr., *J. Polymer Sci.*, **6**, 711 (1951).

44. F. R. Mayo and C. Walling, *Chem. Revs.*, **27**, 387 (1940); W. E. Vaughan and F. F. Rust, *J. Org. Chem.*, **7**, 473 (1942).

45. Unpublished results.

46. W. H. Carothers, J. W. Hill, J. E. Kirby, and R. A. Jacobson, *J. Am. Chem. Soc.*, **52**, 5279 (1930).

47. K. H. Meyer, W. Lotmar, and G. W. Pankow, *Helv. Chim. Acta*, **19**, 930 (1936).

48. J. J. Trillat and H. Forestier, *Bull. soc. chim. France*, **51**, 248 (1932); K. H. Meyer and Y. Go, *Helv. Chim. Acta*, **17**, 1081 (1934).

49. H. Gerding and N. F. Moerman, *Z. physik. Chem.*, **B35**, 216 (1937); *Naturwissenschaften*, **25**, 251 (1937).

50. E. Zintl and K. Loosen, *Z. physik. Chem.*, **A174**, 301 (1935).

51. K. H. Meyer, *Natural and Synthetic High Polymers* (Interscience Publishers, New York, 1942), Chap. B, p. 51.

CHAPTER III

Molecular Size and Chemical Reactivity; Principles of Condensation Polymerization

IN THE period immediately following the establishment of the concept of high polymers as substances of very high molecular weight, the chemical reactivity of polymers usually was conceded to be rather low. This conclusion appears to have been based primarily on the intuitive feeling that such large structures, being sluggish in their movements in general, should undergo chemical transformation at a correspondingly low rate. Although certain experimental evidence seemed, superficially at least, to confirm this view, its support derived mainly from theoretical arguments. It was held that the collision rate for a "particle" so large must be small owing to its low kinetic theory velocity, and that the high viscosity of the liquid medium consisting of polymer molecules would further suppress that rate. Shielding of the reactive group within the coiling chain of its own molecule was often invoked in favor of a low steric factor for polymeric reactants. In the face of the aggregate of these presumed obstacles, it is indeed remarkable that chemical reactions of high polymers proceed at a measurable rate, or, for that matter, that high polymer molecules can be produced at all by chemical reactions.

It is most fortunate for the development of polymer science that these imagined complications have turned out to be almost wholly illusory. As will be brought out in the course of this chapter, the influence of molecular size and complexity on chemical reactivity may be disregarded in very nearly all polymer reactions. If this were not the case, application of the principles of reaction kinetics to polymerization and polymer degradation reactions would be difficult, and might be so complicated as to be fruitless. Not only would polymer reaction kinetics suffer, theories of polymer constitution (i.e., molecular weight distribution and the constitution of network polymers formed through non-linear polymerization) would be undermined as well, for they are based directly on the reactivity characteristics of the various (polymeric) intermediates involved in the synthesis of the final product. An understanding of the constitution of high polymers is a prerequisite to the interpretation of their various physical properties. Hence the principles governing polymer reactions are of vital importance, directly or indirectly, to the greater portion of the present-day interpretation of polymer phenomena.

1. CHEMICAL REACTIVITY IN HOMOLOGOUS SERIES OF MONOMERIC COMPOUNDS

Rate-of-reaction studies on homologous series were among the earliest investigations in the field of reaction kinetics.[1]   The results of these investigations show that the velocity constant, measured under comparable conditions, for the reaction of various members of a given series approaches an asymptotic limit as the chain length increases. Certain of these results on esterification, saponification, and etherification are shown in Table V.   The sets of rate constants given in the four columns refer to the following four reactions:

(A) $H(CH_2)_n COOH + C_2H_5OH \xrightarrow[(HCl)]{} H(CH_2)_n COOC_2H_5 + H_2O$

Carried out in large excess of ethanol containing HCl (Bhide and Sudborough[2]).

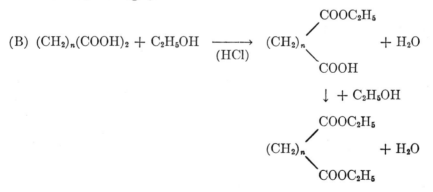

(B) $(CH_2)_n(COOH)_2 + C_2H_5OH \xrightarrow[(HCl)]{} (CH_2)_n \underset{\diagdown COOH}{\overset{\diagup COOC_2H_5}{}} + H_2O$

$\downarrow + C_2H_5OH$

$(CH_2)_n \underset{\diagdown COOC_2H_5}{\overset{\diagup COOC_2H_5}{}} + H_2O$

Carried out as in (A) (Bhide and Sudborough[2]).   First and second stages are not differentiated.   Rates were measured as equivalents of carboxyl consumed per unit of time.

(C) $H(CH_2)_n COOC_2H_5 + KOH \rightarrow H(CH_2)_n COOK + C_2H_5OH$

Saponification in 85 percent ethanol at 50°C (Evans, Gordon, and Watson[3]).

(D) $H(CH_2)_nI + NaOCH_2C_6H_5 \rightarrow H(CH_2)_nOCH_2C_6H_5 + NaI$

Reaction conducted in ethanol at 30°C (Haywood[4]).

In every case significant change in the velocity constant with chain length is confined to the region of low n.   As the chain length increases, the velocity constant rapidly approaches an asymptotic value.   The not uncommon impression that larger molecules in a homologous series

TABLE V.—VELOCITY CONSTANTS[a] FOR HOMOLOGOUS SERIES

Chain length n	$k_A \times 10^4$ at 25°C (Esterification of monobasic acids)[2]	$k_B \times 10^4$ at 25°C (Esterification of dibasic acids)[2]	k_C at 50°C (Saponification of esters)[3]	$k_D \times 10^4$ at 30°C (Etherification)[4]
1	22.1		38.7	26.6
2	15.3	6.0	24.7	2.37
3	7.5	8.7	12.2	0.923
4	7.4$_5$	8.4	13.3	.669
5	7.4$_2$	7.8	14.5	
6		7.3	12.7	
7			13.3	.668
8	7.5			.667$_5$
9	7.4$_7$			
Higher	7.6[b]			.690[c]

[a] All velocity constants are expressed in (gram equivalents/liter)$^{-1}$ sec.$^{-1}$.
[b] Average ± 0.2 for $n = 11$, 13, 15, and 17.   [c] Determined for $n = 16$.

react more slowly may arise from the diluting effect of the large chain, an effect which must be taken into account when the reaction is carried out by directly mixing the reactants without adjusting to corresponding molar concentrations.   Sometimes limited solubility of higher members of the series may be responsible for an apparently slow reaction rate.   Comparison on the basis of rate constants for the homogeneous reaction of successive members of a homologous series invariably leads to the above-stated conclusion.

The kinetics of reactions of bifunctional compounds are especially significant in relation to polymer reactions.   The esterification rate constants shown in the third column (B) of Table V for the homologous series of dibasic acids do not differ greatly from those for the monobasic acids.   These differences vanish as the length of chain separating the carboxyl groups increases.

Of perhaps greater importance is the fact that the same rate constant is applicable to each of the consecutive steps in the conversion of the dibasic acid to a diester:

$$(CH_2)_n(COOH)_2 + C_2H_5OH \xrightarrow[(H^+)]{} (CH_2)_n \overset{\displaystyle COOC_2H_5}{\underset{\displaystyle COOH}{<}} + H_2O$$

$$(H^+) \quad \downarrow +C_2H_5OH$$

$$(CH_2)_n(COOC_2H_5)_2 + H_2O$$

It is convenient to express the rate constants in units of gram equivalents of the reacting functional group (COOH in this case) per liter, rather than in moles of the reactant per liter as is customary; this scheme has been adopted in Table V. Then the rate of the first and second steps may be written, respectively

$$d[\text{monoester}]/dt = k_1[COOH]''[H^+] \tag{1}$$

$$d[\text{diester}]/dt = k_2[COOH]'[H^+] \tag{2}$$

where $[COOH]''$ and $[COOH]'$ represent the concentrations of carboxyl groups belonging to unreacted acid and to monoester, respectively.* The alcohol is considered to be present in very large excess, as in the above-quoted experiments; hence second-order rate expressions are used. If the reactivity of one of the carboxyl groups in the dibasic acid is unaltered by esterification of the other, then k_1 and k_2 will be equal, and we may express the total rate of formation of ester groups as

$$d[\text{ester groups}]/dt = k[COOH][H^+] \tag{3}$$

where $[COOH]$ represents the total carboxyl group concentration. Eq. (3) is applicable to the esterifications of all polymethylene dibasic acids for which n is unity or greater. No variation in the value of k is apparent throughout the course of the esterification, and separate esterification rate measurements on the dibasic acid and on the monoester yield identical rate constants within experimental error. Only in the case of oxalic acid is there a significant difference between the rates of the first and second steps.[5]

Results deduced from the aqueous hydrolysis rates of ethylene glycol

* The value of k_1 will be half that ordinarily obtained when molar rather than equivalent concentrations are used. Rate constants will be expressed universally according to the convention stated above.

TABLE VI.—HYDROLYSIS RATES AT 25°C FOR THE ESTERS OF GLYCOL, GLYCEROL, AND SUCCINIC ACID[a]

Ester	Acid hydrolysis[6]		Base hydrolysis[7]		
	$k_1 \times 10^4$	$k_2 \times 10^4$	k_1	k_2	k_3
Glycol di- and monoacetates	0.793	0.786	0.272	0.272	
Glycerol, tri-, di- and mono-acetates (mixtures of isomers)			.280	.297	0.287
Diethyl and monoethyl succinates	.192	.202	$\sim$.25	.027	

[a] Rate constants are expressed in (gram equivalents/liter)$^{-1}$ sec.$^{-1}$.

diacetate (k_1) and of ethylene glycol monoacetate (k_2) are given in the first row of Table VI. The values of k_1 and k_2 are identical within experimental error for basic hydrolysis[7] and very nearly so for acid hydrolysis[6] as well. Similarly, the ester linkages of the tri-, di-, and monoacetates of glycerol undergo base-catalyzed hydrolysis at rates which are about equal. Although the di- and monoacetates consisted of mixtures of isomers,[7] these results nevertheless indicate that reaction of one functional group in the molecule does not affect appreciably the reactivity of the neighboring group.

It must not be concluded that the reactivity of a hydroxyl or ester group is unaffected by the presence of a polar group such as hydroxyl or ester on the adjacent carbon atom; acid hydrolysis of either the mono- or diacetate of ethylene glycol is several times more rapid than the hydrolysis under comparable conditions of ethyl acetate, for example. In the case of glycerol, reactivity at the secondary hydroxyl group is somewhat lower than at the neighboring primary positions. The results mentioned above demonstrate merely that ordinarily the reactivity of a given functional group in a polyfunctional molecule can be assigned a definite value which does not change during the course of the reaction. This reactivity may be markedly influenced by neighboring substituents, but the influence of one functional group on the other is not sensibly altered by reaction of one of them in any of the examples discussed thus far. Furthermore, the effect of one functional group on the other rapidly diminishes with the distance of separation in the molecule.

The acid-catalyzed hydrolysis of diethyl succinate[6] conforms to the above generalizations, as shown in the last row of Table VI, but the base-catalyzed hydrolysis presents a quite different situation.[7] The rate constant for the base hydrolysis of diethyl succinate approximates the normal value for aliphatic esters, but the rate for the mono-ester is one order of magnitude smaller. This observation is at-tributable to the electrostatic repulsion between the carboxylate ion (in basic solution) of the monoester and the reacting hydroxyl ion

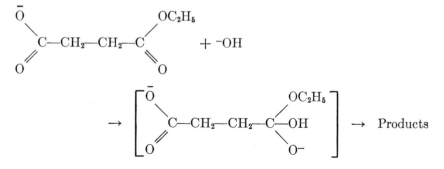

None of the other reactions so far discussed involve interaction between a pair of charged species. This is but another instance of the electro-static effect shown by Kirkwood and Westheimer[8] to be responsible for the disparity between the first and second ionization constants of di-basic acids,[9] for the effect of the carboxylate ion on the basicity of an α-amino acid,[9] and for the difference in reactivity of ionic compounds compared with analogous nonionic species in acid- or base-catalyzed reactions.[10]

Approximate ratios of the velocity constants for the hydrolysis in basic media of di- and monoesters of homologous dibasic acids are given in the second column of Table VII.[11] The logarithm of this ratio, given in the third column, may be compared with the ΔpK values taken from the first and second ionization constants K_1 and K_2 of the acids.[9] Here K_1 and K_2 are expressed using the conventional molar units. Hence it is necessary to divide by the symmetry factor 4, since either of the two carboxyl groups in the molecule may yield a hydrogen ion in the first ionization, and in the reversal of the second equilibrium either of the two carboxylate ions may combine with the hydrogen ion.[9] Except at the lowest values of n, $\log(k_1/k_2)$ correlates reasonably well with ΔpK; both approach zero gradually as n increases, as should be expected for an electrostatic effect.

The electrostatic effect of an ionic substituent on the reactivity of a functional group subject to attack by another ion persists over a much greater length of intervening chain than is found for the influ-ence of an unionized substituent (compare column B of Table V). This follows of course from Coulomb's law. From the standpoint of polymer reactions, it is important to bear in mind that when charged

TABLE VII.—COMPARATIVE SAPONIFICATION RATES AT 25°C OF CARBOETHOXY
GROUPS IN DI- AND MONOETHYL ESTERS OF DIBASIC ACIDS[a]

| n | HOCO(CH$_2$)$_n$COOH | | |
	k_1/k_2	log (k_1/k_2)	$\Delta pK = \log (K_1/4K_2)$
0	8000	3.9	2.36
1	40	1.60	2.26
2	3.5	0.55	0.84
3	3	.47	.47
6	1.5	.18	.28
7	1.4$_5$	.16	.26
8	1.4	.14$_5$	

[a] Data given in the second and third column are from Skrabel and Singer.[11] The
pK values for the dissociations of the acids are taken from the tables published by
Westheimer and Shookhoff.[9]

groups are involved the reactivity of a functional group may not be
entirely independent of the status of other groups in the molecule.
The range over which these electrostatic effects are transmitted is small,
however, in comparison with usual polymer dimensions.

2. THEORY OF THE REACTIVITY OF LARGE MOLECULES

The results cited above show that the rate of reaction of a functional group does not depend on the size of the molecule to which it is
attached, at least within the range covered by investigations on monomeric systems. Extrapolation of these conclusions to polymers, including those of very high molecular weight, might seem precarious.
This much, however, can be concluded without hazardous assumptions,
namely, that the *intrinsic reactivity* of a functional group remains independent of molecular size, except, of course, when the molecule is
quite small. Lengthening of the polymer chain bearing a reactive
group at one end can be considered the equivalent, insofar as intrinsic
reactivity is concerned, of introducing a substituent at the other end
of the polymer molecule. It follows from the results cited above that
such changes as may occur in the intrinsic reactivity of a polymer
molecule when its size is increased progressively should be confined to
the very early stages of its development, vanishing by the time the
chain length reaches about ten atoms. If a terminal ionic group is
involved, its influence may persist somewhat farther. Thereafter,
further increase in chain length should exert no effect on the intrinsic
reactivity.

As pointed out at the beginning of this chapter, arguments favoring
a decrease in reactivity with molecular size usually have arisen from
considerations of the mechanics of interaction of two functional groups

attached to large molecules, rather than from any inherent abnormal character of the functional groups themselves. It is well established that large molecules diffuse slowly, but the collision rate of the functional group must not be confused with the diffusion rate for the molecule as a whole. The mobility of the terminal functional group is much greater than would be indicated by the macroscopic viscosity (which may reach many thousand poises). While the range of diffusion of a terminal group within an interval of time which is small compared to the period required for displacement of the molecule as a whole will be limited by its attachment to the polymer molecule, the group may nevertheless diffuse readily over a considerable region through rearrangements in the configurations of nearby segments of the chain. Its actual oscillations against immediate neighbors may occur at a frequency comparable to, or at any rate not much less than, that prevailing in simple liquids. Thus the actual collision frequency will bear little relationship to the mobility of the molecule as a whole or to the macroscopic viscosity.

As Rabinowitch and Wood[12] have demonstrated, a pair of neighboring molecules, or functional groups, in the liquid state may collide repeatedly before diffusing apart. The lower the diffusion rate, the greater the prolongation of this series of encounters between functional groups, but it will be proportionately longer before the functional group diffuses to a new position in which it is again an immediate neighbor of another functional group. Another series of collisions will ensue, and this pair will be separated by diffusion of one of them, unless, of course, that exceedingly rare event, chemical reaction, has occurred during one of the collisions. Thus, decreased mobility, due to large molecular size and/or to high viscosity, will alter the time distribution of collisions experienced by a given functional group, but it should not affect to any great extent the collision rate averaged over a period of time that is long compared with the interval required for diffusion from one partner to the next.

Even if this average collision frequency is diminished somewhat by a possible reduction in internal mobility as the polymerization progresses, the duration of the collided state will be prolonged proportionately. Hence the concentration of pairs of functional groups sufficiently close together to permit the condensation reaction to occur (if the necessary activation energy is available) is independent of mobility. This conclusion can be established more rigorously by following Eyring's[13] theory of reaction rates, according to which the velocity constant is given by

$$k = K^*(kT/h) \tag{4}$$

where k is Boltzmann's constant, T is the absolute temperature, h is Planck's constant, and K^* is the "equilibrium" constant for the transitory activated complex. For a bimolecular reaction

$$K^* = (F_{ab}^*/F_a F_b) \exp(-E_0/kT) \tag{5}$$

where F_a, F_b, and F_{ab}^* are the partition functions for the two reactants and the activated complex, respectively, and E_0 is the energy of the activated complex at absolute zero. An increase in the complexity of one or both of the reactants modifies F_{ab}^* and the product $F_a F_b$. Unless this increase in complexity involves alteration of the molecule in the immediate vicinity of the functional group, F_{ab}^* and the product $F_a F_b$ will be modified by very nearly identical factors for the added degrees of freedom, and no change in the equilibrium constant K^* should be observed. In other words, mobility within the liquid will not affect the equilibrium:

$$\text{reactants} \; \underset{\longleftarrow}{\overset{\longrightarrow}{}} \; \text{activated complex}$$

Since the rate of reaction is proportional to the concentration of the activated complex, it will not be affected by the mobility of the molecules, the diffusion rate, or the viscosity.

Exceptions to these conclusions will occur when the molecular mobility is extremely low, as, for example, when the molecular weights of the reactants are large and the viscosity is extremely high, or when the reaction rate constant is exceptionally great and the mobility is low.[12] If the reaction is too fast, or the mobility too low, to allow maintenance of the equilibrium concentration of pairs of reactants adjacent to one another in the liquid, then diffusion will become the rate-controlling step. Condensation polymerizations generally proceed by reactions of moderate rate such that no more than about one bimolecular collision in 10^{13} between reactants is fruitful. Within the interval of time for this number of collisions, considerable diffusion of the molecule, and especially of its terminal functional group, may occur. The concentration of reactant pairs should be easily maintained substantially at equilibrium even at the highest molecular weights, and viscosities, attainable (bearing in mind that the mobility of a terminal functional group will be much greater than the macroscopic mobility of the system indicated by its viscosity).

In addition polymerizations which proceed exclusively by addition of monomer to the end of a growing chain, the mobility of the monomer, being affected only to a comparatively small degree by surrounding polymer molecules, should be adequate to maintain an equilibrium concentration of monomers in the vicinity of active centers. Even in the relatively rapid chain growth processes occurring in vinyl polymerizations (one reaction per 10^9 collisions; see Chap. IV), diffusion would not be expected to be rate-controlling. The chain termination step, involving interaction between two active centers attached to large polymer molecules, occurs at a relatively large proportion of the collisions (one in about 10^5) between two such centers. Here it is not improbable that diffusion may under some circumstances become rate-controlling, particularly when the viscosity of the medium is very large. (See Chap. IV for further discussion of this subject.)

Of the various arguments for low reactivity in polymers, the contention that the terminal group of a very large polymer molecule will be shielded by the coilings of the chain, and hence that its reaction rate will be reduced, remains to be disposed of. Such an effect may be real in very dilute solutions where sufficient space is available to permit the irregularly coiled polymer molecules to exist more or less independently of one another. In concentrated systems the polymer chains intertwine extensively in a most irregular fashion. In choosing its environment, a given functional group shows no preference whatever for units of its own chain and in general will be surrounded by units belonging to various other molecules. These chain units act like so much diluent and are appropriately taken into account by writing the rate expression in terms of concentrations of the *functional groups* participating in the reaction *per unit volume*, and not in terms of molecules or mole fractions of molecules.

We may conclude that no valid theoretical justification for a generally abnormal chemical reactivity in polymer systems has been found, and that the reaction rates observed should therefore be expected to be independent of the molecular weights of the polymeric species participating. Exceptions to these generalizations may be anticipated only among processes for which the specific rate constant is very great while at the same time, owing to the sizes of the reactant molecules and/or to the viscosity of the medium, the diffusion rate is low. Diffusion may then become rate-determining, with the result that the process in question proceeds more slowly than standard kinetic extrapolation would predict. Shielding of functional groups attached to large polymer molecules may be expected to reduce the rate of reaction only in solutions which are very dilute.

Unqualified application of these principles to actual polymer reactions presupposes a homogeneous reaction. Occasionally limitations on miscibility in systems containing polymer molecules impose difficulties in bringing the reacting species together in the same phase—difficulties not encountered with the analogous monofunctional reaction.

3. KINETICS OF CONDENSATION POLYMERIZATION

The progress of polyester-forming reactions between glycols and dibasic acids is easily followed by titrating the unreacted carboxyl groups in samples removed from the reaction mixture.   Simple esterification reactions are known to be acid-catalyzed.   In the absence of an added strong acid, a second molecule of the acid undergoing esterification functions as catalyst.[14]   The rate of the polyesterification process should therefore be written

$$- d[COOH]/dt = k[COOH]^2[OH] \qquad (6)$$

where concentrations are expressed as equivalents of the functional groups in accordance with the convention established earlier in this chapter.   By so expressing the rate, the complications of writing a separate rate equation for the reaction of each molecular species with every other are avoided.   In adopting this procedure we assume, however, that the velocity constant k is the same for *all* functional groups and that it does not change as the average molecular weight increases. The primary purpose of kinetic experiments on condensation polymerizations has been to test this hypothesis.

If the hydroxyl and carboxyl group concentrations are equal, both being given by c, Eq. (6) may be replaced by the standard integrated expression for a third-order reaction:

$$2kt = 1/c^2 - \text{Const.} \qquad (7)$$

If the process is uniformly of the third order without change in velocity constant throughout, the integration constant will be $1/c_0^2$ where c_0 is the initial concentration of the functional groups.   It is convenient here, and for many other purposes as well, to introduce a parameter called the *extent of reaction* and designated by p, which represents the fraction of the functional groups initially present that have undergone reaction at time t.   Then $c = (1-p)c_0$, and Eq. (7) may be replaced by

$$2c_0^2kt = 1/(1 - p)^2 - \text{Const.} \qquad (8)$$

In the polyesterification process p is directly calculated from the carboxyl group titer.   Results[15] for the polyesterification reaction between diethylene glycol and adipic acid at 166° and 202°C are plotted in Fig. 3 in accordance with the third-order equation (8).   For comparison purposes, the course of the non-polymer-forming reaction of diethylene glycol with the monobasic acid, caproic, is also shown.   Eq. (8) is not obeyed from zero to 80 percent esterification $[1/(1-p)^2 = 1$ to $25]$, as is shown by the curvature over this region.   From 80 to 93 percent esterification the reaction appears to be third order.   The non-polymer-forming esterification of diethylene glycol with caproic acid (and other simple esterifications as well[15]) follows a similar course. The curves for mono- and polyfunctional esterifications are superimposable by merely adjusting all time values for each by a constant factor. The lower rate of the, DE-C esterification is largely due to the lower value of c_0 for this mixture.

Failure of these reactions to follow the third-order law over the earlier portion of the reaction seems to be due to the pronounced changes in the characteristics of the medium with the disappearance of so many hydroxyl and carboxyl groups[15]—changes to which an ion-catalyzed reaction should be sensitive. Whatever the exact cause of this behavior may be, it is typical of simple esterifications as well as of those which bring about polymer formation and is not therefore in any way related to the sizes of the reacting molecules. The similarity in the courses followed by mono- and by polyesterifications[15] provides direct evidence for nondependence of reactivity on molecular size.

In a linear polyesterification involving bifunctional monomers exclusively, the hydroxyl and carboxyl groups being present in equal numbers, the number of unreacted carboxyl groups must equal the number of molecules present in the system, provided that no side reactions occur; the number of molecules in moles per unit volume is equal, therefore, to $c_0(1-p)$. If each residue from a glycol and from a di-

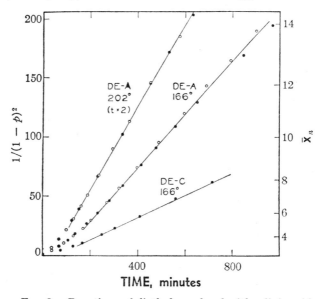

FIG. 3.—Reactions of diethylene glycol with adipic acid (DE-A) and of diethylene glycol with caproic acid (DE-C). Time values at 202°C have been multiplied by two.[15]

basic acid is regarded as a structural unit, so that the number of structural units equals the total number of bifunctional monomers initially employed, the average number of structural units per molecule, or the number average *degree of polymerization* will be given by

$$\bar{x}_n = \frac{\text{No. of units}}{\text{No. of molecules}} = 1/(1 - p) \qquad (9)$$

(The *repeating unit* of the chain in this case consists of two *structural units* as here defined.) The bar included in the symbol $\bar{x}_n$ denotes an average over the number x of units in a molecule, and the subscript n specifies the *number* average rather than any other average which might be taken (cf. Chap. VII). The number average molecular weight (defined as the total weight divided by the total number of molecules) is given by

$$\overline{M}_n = M_0/(1 - p) \qquad (10)$$

where M_0 is the mean molecular weight of a structural unit.

The average degree of polymerization is shown along the right-hand ordinate of Fig. 3. These results show, therefore, that the reactivity is unaffected by molecular size up to $\bar{x}_n = 14$. By substitution of Eq. (10) in (8), $\bar{x}_n$ is found to be approximately proportional to the square root of t (for a third-order condensation), except during the early stages of the reaction. Although formation of low polymers proceeds rapidly, the rate of increase of molecular weight with time diminishes as the third-order reaction proceeds, and the attainment of very high molecular weights requires a wholly unreasonable time of reaction. This is evident from the $\bar{x}_n$ scale of Fig. 3. The decrease in the rate of advancement of the degree of polymerization in direct polyesterifications is a consequence of the third-order nature of the esterification reaction and should not be attributed to low reactivity of large molecules, as was suggested by early investigators in this field.

Greater success in extending kinetic measurements to higher degrees of polymerization has been achieved with polyesterifications catalyzed by a small amount of a strong acid catalyst.[15] The catalyst concentration being constant throughout the process, the second-order rate expression

$$- dc/dt = k'c^2$$

may be used, where the concentration of the catalyst is included in the second-order rate constant k'. Then

$$c_0 k't = 1/(1 - p) - \text{Const.} \qquad (11)$$

and $\bar{x}_n$ increases linearly with the time of reaction, a much more favor-

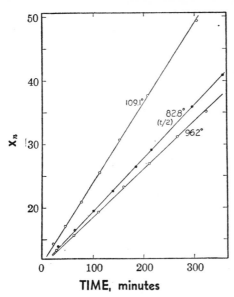

FIG. 4.—Reaction of decamethylene glycol with adipic acid at the temperatures indicated, catalyzed by 0.10 equivalent percent of p-toluenesulfonic acid. The time scale for the results at 82.8°C is to be multiplied by two.[15]

able situation for obtaining high average molecular weight than that encountered in the "uncatalyzed" third-order reaction. Results for the condensation of decamethylene glycol with adipic acid in the presence of p-toluenesulfonic acid at three temperatures are shown in Fig. 4. The process is quite accurately second order throughout the range investigated. Results[15,16] reported on acid-catalyzed polyesterifications of other glycols and dibasic acids show that the reaction continues on a second-order course at least up to a degree of polymerization of 90, which corresponds to average molecular weights of about 10,000. In spite of the manifold increase in molecular size and the concurrent increase in the viscosity of the medium by a factor exceeding 2000, there is no indication of a diminution in the velocity constant.

The course of nonlinear polyesterifications involving glycerol in reaction with phthalic anhydride, succinic acid, or another dibasic acid may be followed up to the gel point by titration of carboxyl groups. After gelation has occurred, this method cannot be used owing to partial insolubility. Kienle and co-workers[17] have pursued the course of the reaction through the gel point by measuring the rate of water evolution. Exact kinetic interpretation of their results is complicated by the difference in reactivity of primary and secondary hydroxyl groups in glycerol, as well as by the peculiarities of the esterification process up to 80 percent conversion as previously discussed. A most significant feature of these studies is the fact that water evolution continues smoothly through the gel point and beyond, with no suggestion of abnormality arising from the formation of a macroscopic network structure at the gel point and the consequent acquisition of infinite macroscopic viscosity.

Satisfactory kinetic data on condensation polymerizations other than polyesterification are meager. Polyamidation, however, has been found[18] to be second order with respect to the functional groups throughout its course, in conformity with monofunctional amidations.

4. KINETICS OF DEGRADATION OF CONDENSATION POLYMERS

4a. Hydrolysis of Cellulose and Cellulose Derivatives.—When dissolved in strong acids, cellulose is degraded through hydrolytic splitting of the β-glucoside linkages between the structural units of the cellulose chain:

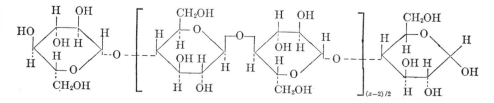

The process can be followed viscometrically, polarimetrically, or by chemical determination of the aldose end group produced for each bond split. The latter two methods are equivalent in that they afford direct measures of the number of bonds hydrolyzed. The former method involves the deduction of the degree of polymerization of the partially hydrolyzed polymer from the viscosity of its solution, making use of a previously established empirical relationship between solution viscosity and molecular weight. The degree of polymerization having been determined in this manner, the extent of condensation p can be obtained from Eq. (9). If c_0 is the total concentration of glucose units in the system, $c_0(1-p)$ gives the number of polymer molecules. The increase in this quantity with the time of reaction represents the rate of hydrolysis of interunit bonds.

Freudenberg, Kuhn, and co-workers[19] have applied the polarimetric and chemical end group methods to the rate of hydrolysis in 51 percent sulfuric acid at 18°C of cellulose and of the lower polysaccharides for which $x = 2, 3$, and 4 in the formula given above. The first-order velocity constant for the splitting of interunit linkages in cellulose increased over the range from 10 to 100 percent hydrolysis, corresponding to a decrease in $\bar{x}_n$ from 10 to 1. The linkages remaining near completion of the degradation apparently hydrolyze with greater ease, on an average, than those initially present. Hydrolysis of cellobiose ($x = 2$) proceeded in strict accordance with the first-order expression, whereas the rate constants for bond splitting in the tri- and tetrasaccharides increased with the degree of hydrolysis. Values of the ini-

TABLE VIII.—RATES OF HYDROLYSIS OF CELLULOSE AND ITS LOWER MOLECULAR
HOMOLOGS IN 51 PERCENT SULFURIC ACID AT 18°C[19]

Polysaccharide	Schematic structure	Initial rate constants min.$^{-1} \times 10^4$
Cellobiose	a	1.07
Cellotriose	a　b	0.64
Cellotetrose	a　c　b	.51
Cellulose	a　c　b，$x-3$	.305

tial rate constants, expressed as the fraction of the interunit linkages disappearing per minute, are given in Table VIII.[19]

Freudenberg, Kuhn and their co-workers[19] showed that both the velocity constants and the courses followed by the hydrolyses of these various polymers can be accounted for by postulating that one or the other or both of the terminal linkages, a and b of Table VIII, in these various species hydrolyze more rapidly than the internal c linkages. All of the latter can be assumed to hydrolyze at the same rate. If, for example, one of the two terminal linkages, a or b, in an x-mer reacts at a rate equal to cellobiose, 1.07×10^{-4}, and the rate for each of the other $x - 2$ linkages corresponds to the initial average rate, 0.305×10^{-4}, of hydrolysis of the bonds in cellulose ($x \to \infty$), then the calculated initial rates for cellotriose and cellotetrose, 0.69×10^{-4} and 0.56×10^{-4}, respectively, are in fair agreement with the observed values. Furthermore, the course of the hydrolysis of cellulose is in good agreement with Kuhn's mathematical analysis[20] of the consecutive reaction problem precipitated by the postulate mentioned above. The results can be explained equally well perhaps by assuming that the enhanced reactivity is shared jointly by both terminal linkages a and b. In any case, the results with cellulose indicate that the reactivity of internal c linkages is about constant, independent of molecular weight over the range investigated, i.e., up to $\bar{x}_n = 10$. The lesser reactivity of the internal linkages may reasonably be ascribed to the short-range substitution effect previously discussed.

Wolfrom, Sowden, and Lassettre[21] measured the rate of hydrolysis of methylated cellulose in fuming hydrochloric acid at 0°C by continuous mercaptalation of the aldose group formed for each bond ruptured.

The degree of mercaptalation increased linearly with time for small extents of degradation corresponding to changes in the degree of polymerization from 150 to 50. Evidently no appreciable change in reactivity accompanies this threefold change in molecular weight.

The degradation of cellulose of high molecular weight is best observed viscometrically. The change in the number of end groups over the initial portion of the degradation is too small to be measured accurately by chemical or polarimetric methods. The degradation of celluloses varying from 130 to 1500 units in length have been investigated viscometrically by Husemann, Schulz, and others.[22] Small degrees of degradation of the celluloses of highest molecular weight yielded products which on fractionation (by fractional precipitation) were found to cover a narrower distribution of molecular weights than would be expected for random splitting of chains. Their results suggested the occurrence of easily hydrolyzable bonds at intervals of about 500 units along the chain, giving rise to an excessive proportion of molecular species of this length in the early stages of hydrolysis.* From a degree of polymerization of about 500 units downward (but short of the advanced degradation stage where the anomalies discovered by Freudenberg and co-workers are manifested) the hydrolysis of linkages is of the first order.

From the various results discussed above it is clear that the internal linkages between the units of cellulose and its derivatives are equivalent in their reactivity toward hydrolysis regardless of their location in the chain and independent of the length of the chain, at least up to a degree of polymerization $\bar{x}_n$ of 500. At higher molecular weights, weaker linkages may be present, owing to specific differences in the chemical character of occasional units of the chain. There is no evidence for an effect of molecular weight per se on reactivity up to $\bar{x}_n = 1500$ (molecular weight $= 250,000$).

4b. Hydrolysis of Polyamides.—Hydrolysis of the polypeptide of glycine

$$NH_2CH_2CO[—NHCH_2CO—]_{x-2}—NHCH_2COO^-$$

in alkaline solution[24] follows a pattern similar to that observed for cellulose. In view of the short length of the structural unit and the negative charge at one end of the chain (in alkaline solution), the rate

---

* Husemann and Schulz[22] suggested that these more readily hydrolyzable linkages occur at xylose units spaced at regular intervals of about 500 units along the chain. Schulz[23] has obtained evidence to indicate that these linkages, however they may differ from the normal β-glucoside linkage, are also more sensitive to oxidative degradation in ammoniacal cupric hydroxide (cuprammonium solutions).

of alkaline hydrolysis of the terminal interunit bond nearest the carboxylate ion would be expected to be less than that for internal bonds. Kuhn, Freudenberg, and co-workers[24] have shown that the course of the hydrolysis of a synthetic polyglycine having $\bar{x}_n = 8$ is accurately described by assigning a mean rate for the two terminal bonds (*a* and *b* of the schematic representation used in Table VIII) equal to the observed rate of hydrolysis of the peptide bonds in diglycylglycine ($x = 3$) and a rate about three times as great for all internal bonds (*c* bonds in Table VIII). This scheme conforms with the ratio of the rates of hydrolysis of internal and terminal bonds in triglycylglycine ($x = 4$).[24]

The protein collagen undergoes hydrolytic degradation to gelatin in a manner which indicates the presence of a small minority of comparatively easily hydrolyzable bonds. Scatchard, Oncley, Williams, and Brown[25] concluded that these are regularly spaced at intervals of about 1200 units in the collagen molecule.

No counterpart for such regularities is found in synthetic polymers. The hydrolysis of poly-ϵ-caproamide

$$NH_3^+(CH_2)_5CO[-NH(CH_2)_5CO-]_{x-2}-NH(C I_2)_5COOH$$

in 40 percent sulfuric acid at 50°C proceeds uniformly as a first-order splitting of equivalent bonds over the entire range investigated from $\bar{x}_n = 220$ to 6.[26]

4c. Alcoholysis of Polyesters.—The partial degradation of polymeric decamethylene adipate by limited amounts of a higher alcohol or glycol according to the reaction

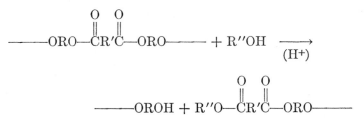

has been investigated[27] by measuring the viscosity of the mixture as a function of time. [R and R' represent divalent groups, e.g. $(CH_2)_n$.] By means of an empirical correlation between melt viscosity and molecular weight, it is possible to compute from the viscosity the extent to which the above degradation has proceeded. In order to avoid simultaneous polymerization by the condensation of terminal hydroxyl and carboxyl groups, the polymers used in these experiments were prepared by reacting a small excess of the glycol with the dibasic acid until the condensation was substantially complete. The end groups of polymers so obtained consist almost exclusively of hydroxyl groups; hence the possibility of further condensation is virtually eliminated. In carrying out the degradation reaction, only a small amount of the alcohol R″OH was used in relation to the total ester groups. The process should be of the first order, therefore, with respect to the added alcohol. That this is true is shown in Fig. 5. Over the range covered by these results, from a degree of polymerization of about 40 to 15, the mean reactivity of interunit bonds shows no change.

4d. Interchange Reactions in Condensation Polymers.—The ready occurrence of the interchange reaction between an alcohol and a poly ester suggests at once that polyester molecules bearing termina hydroxyl groups should be capable of reacting with one another in the following manner:

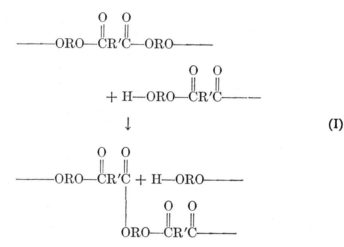

(I)

A similar interchange might occur between a carboxy acid end group and an ester, but the rate of this reaction is known to be very much slower than the alcoholysis reaction given above. In these interchange reactions between polymer molecules there is no net change in the number of interunit linkages, and the number of molecules on either side of the equation is the same. The number average molecular weight is therefore unaffected by the interchange. Such processes can, however, bring about changes in the molecular weight distribution. For example, two average molecules may react to produce one much longer and another correspondingly shorter than the average. Conversely, if a mixture is composed of two species, one very long and

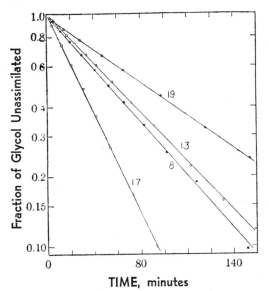

Fig. 5.—The partial degradation of decamethylene adipate polyester with small percentages of decamethylene glycol (experiments 8, 13, and 17), or with lauryl alcohol (experiment 19), at 109°C, catalyzed with 0.1 equivalent percent of *p*-toluenesulfonic acid. The fraction of added glycol, or alcohol, unassimilated has been calculated indirectly from melt viscosity measurements and is plotted on the logarithmic ordinate scale.[27]

the other short, they may conceivably interchange with each other to produce polymers intermediate in size.

A completely satisfactory account of the role and importance of these interchange processes in condensation polymers would necessitate detailed consideration of the subject of molecular weight distribution, which is covered in later chapters. For the purpose of the present discussion it is sufficient to note that regardless of the initial distribution of molecular weights at the commencement of interchange, ultimately the interchange processes will lead to a state of (dynamic) equilibrium in which the concentration of each species remains constant. These concentrations define the equilibrium molecular weight distribution[28] for a given (fixed) extent of polymerization, or number of interunit linkages. This distribution happens to be identical with that which is obtained directly from random intermolecular condensation without simultaneous, or subsequent, occurrence of interchange reactions. Although interchange may occur freely, ordinarily it will have no effect on the molecular weight distribution of a condensation polymer. Analogous interchange processes may, of course, occur in nonlinear polymers, but again without altering the molecular distribution or structure in most cases.

The occurrence of interchange reactions between linear polyesters may be demonstrated by mixing two polymers, one of low and another of high average molecular weight, and observing the viscosity of the mixture under conditions permitting ester interchange. Such experiments have been carried out on mixtures of two decamethylene adipate polyesters,[28] each having hydroxyl end groups so as to prevent simultaneous further condensation. At 109°C in the presence of p-toluenesulfonic acid, the viscosity decreased rapidly owing to reshuffling of the molecular weight distribution, finally approaching a limiting value near that calculated for the equilibrium distribution consistent with the fixed number of molecules in the given mass of polymer.

Similar interchange processes may occur in other condensation polymers under suitable conditions. There is ample evidence in the literature on polyamides,[29] and on monomeric amides as well, for the occurrence of the amine-amide interchange reaction (II)

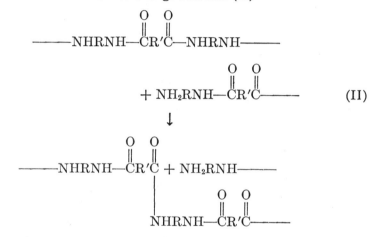

which is analogous to the polyester interchange (I) given above. Another type of interchange, namely, interchange between two amide linkages

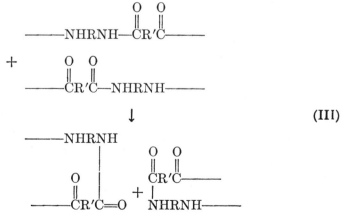

also is possible in polyamides. When two polyamides formed from different monomers (e.g., one formed from triglycol diamine and adipic acid and the other from hexamethylene-diamine and adipic acid) are heated together for about 30 minutes at 285°C, they combine to give a product differing from both initial polymers.[29] This product is not merely a mixture of molecules of the two original components, as is shown by their inseparability by the action of a selective solvent. Evidently interchange occurs by either, or both, of these processes (II and III) to the extent that few of the initial polymer molecules remain intact. The product differs from the copolymer of the same composition prepared directly from the monomers because interchange was not continued until a state of equilibrium was reached in which the units would be combined in random arrangement.

The counterpart of reaction (III) in polyesters, i.e., ester-ester interchange, probably does not take place to any appreciable degree. Polymeric polysulfide rubbers ("Thiokol" type)

$$-CH_2CH_2-S-S-CH_2CH_2-S-S-, \text{ etc.}$$

give evidence for this type of interchange, as is to be expected from the comparatively low strength of the S—S bond. These polymers are rubberlike and can be vulcanized (i.e., cross-linked) to yield network structures. The stress produced by stretching is observed to diminish gradually when the stretched sample is heated at fixed length. This relaxation phenomenon appears to be due to interchange reactions which provide the means whereby the network may alter its structure in favor of an arrangement in conformity with the applied deformation.[30]

Polyanhydrides are susceptible to interchange reactions with carboxyl groups in analogy to (I) and (II). Polymeric dimethylsiloxanes readily interchange in the presence of sulfuric acid[31] by a mechanism which may be presumed to involve cations. In some polymers the interunit linkage is too stable to enter readily into interchange reactions. Such an example is polyethylene oxide

$$HO[-CH_2CH_2O]_{x-1}-CH_2CH_2OH$$

The concept of mobile reshuffling of sections of high polymer molecules at temperatures far below their decomposition points, and of the consequent transitory existence of the individual molecules, is interesting in itself. As pointed out above, the process is of importance in altering the distribution of an artificial mixture of two polymers from the same monomers but differing in average molecular weight, and in cross-blending the molecules of two polymers prepared from different monomers and subsequently heated together. It should also be mentioned that interchange reactions may be employed in producing high molecular weight polyesters from a substantial stoichiometric excess of ethylene glycol over dibasic acid.[32,33] The initial products are of low molecular weight, owing to the excess of glycol, but further heating in vacuum removes excess glycol and the molecular weight may be advanced. Now the amount of free glycol remaining at completion of the esterification of 5 mole percent excess of glycol over dibasic acid (without loss of glycol) is extremely small—about 0.25 percent of the initial amount of glycol. However, as this is removed by vaporization, interchange between hydroxyl terminal groups and the ester linkages nearest the ends of chains will tend to restore the equilibrium concentration of free glycol. Hence more glycol can be removed and the average degree of polymerization gradually increases.

5. MOLECULAR WEIGHTS OF LINEAR CONDENSATION POLYMERS

The connection between average degree of polymerization, or average molecular weight, and the extent of reaction has been pointed out in the discussion of the kinetics of condensation polymerizations. For a strictly linear, or bifunctional, condensation involving coreaction of A groups with B groups present in precisely equivalent amounts, the number of molecules at a given stage of the process necessarily is equal to the concentration of unreacted A groups and also to the concentration of unreacted B groups. The average degree of polymerization (number average, to be specific) and the average molecular weight are reciprocally related to the number of molecules; this relationship is expressed by Eqs. (9) and (10). These statements apply equally to the condensation of a pure A——B monomer (e.g., an amino acid or hydroxy acid)

$$A——B \rightarrow A——BA——BA——, \text{ etc.} \qquad \text{(type i)}$$

and to the condensation of equivalent quantities of A——A with B——B (e.g., glycol plus dibasic acid)

$$A——A + B——B \rightarrow A——AB——BA——, \text{ etc.} \qquad \text{(type ii)}$$

The requirement that A and B shall be present in equivalent quantities is automatically taken into account in the type (i) polymerization if the monomer is pure and no side reaction occurs. In type (ii) it is also necessary to specify that the reactants are present in equivalent proportions if Eqs. (9) and (10) are to apply.

The problem of achieving the highest possible molecular weight in a linear condensation polymerization resolves itself into the problem of reducing the number of end groups to the lowest possible value. Just as a chemical reaction cannot be carried to absolute completion, the molecular weight reached in a *bifunctional* condensation must always remain finite. Maximum molecular weight will be attained under conditions favoring the most efficient reaction of the functional groups. Although virtually any condensation reaction is suitable in principle for the formation of condensation polymers, only those which can be carried very nearly to completion, substantially free from undesirable side reactions, are eligible for the production of condensation polymers of high molecular weight. "Undesirable side reactions" means, essentially, any process leading to the creation of monofunctional units; for example, the consumption of one functional group by some reaction other than condensation with a group of opposite type would be such a process. Monofunctional units, whether introduced as impurities or formed by side reactions, contribute terminal chain units incapable of condensing with other molecules, which therefore limit the ultimate molecular weight attainable. An excess of either reactant in a type (ii) condensation acts similarly; loss of a small proportion of one ingredient during the polymerization may produce the same situation, which, however, can be overcome by replacing quantitatively the component lost.

The depression of the molecular weight brought about by nonequivalence of reactants, by monofunctional ingredients, or by unbalance in the stoichiometric proportions may be expressed quantitatively as follows. Suppose that a small amount of a reactant designated by B—|—B is added either to a pure A——B monomer or to an equimolar mixture of A——A and B——B; B—|—B may or may not be identical with B——B. Let

N_A = total number of A groups initially present

N_B = total number of B groups initially present

$r = N_A/N_B$

p = fraction of A groups which have reacted at a given stage of the reaction.

The total number of units is

$$(N_A + N_B)/2 = N_A(1 + 1/r)/2$$

The total number of ends of chains can be expressed as

$$2N_A(1 - p) + (N_B - N_A) = N_A[2(1 - p) + (1 - r)/r]$$

and this must equal twice the total number of molecules. The number average degree of polymerization is given therefore by

$$\bar{x}_n = \frac{\text{Number of units}}{\text{Number of molecules}} = \frac{1 + r}{2r(1 - p) + 1 - r} \qquad (12)$$

which reduces to Eq. (9) when $r = 1$. At completion of the reaction $(p = 1)$

$$\bar{x}_n = (1 + r)/(1 - r) \tag{13}$$

$$= \frac{(\text{Moles bifunctional units exclusive of B—|—B})}{(\text{Moles of B—|—B})} + 1 \tag{13'}$$

The same equations are applicable to polymers containing small amounts of a monofunctional reactant B—|—, provided that r is redefined as follows:

$$r = N_A/(N_A + 2N_{B-|-})$$

Eq. (13') requires no revision. Other special cases can be similarly treated, as, for example, when two or more types of added substances are employed simultaneously.

Equations (13) and (13') emphasize the marked effect on the ultimately attainable molecular weight of a small amount of a monofunctional impurity, or of a small excess of one bifunctional reactant. One mole percent of an extraneous unit in the system limits the average degree of polymerization to about 100 units, for example. In a type (ii) condensation, loss of a fraction of a percent of one ingredient through volatilization from the reaction mixture, or a comparable loss of functional groups through side reactions occurring in either type of condensation, may appreciably limit the molecular weight which can be reached.

Molecular weights of synthetic condensation polymers usually are deduced from chemical end group determinations and the stoichiometry of the ingredients present. If the amount of one functional group is determined analytically and the ratio r is known, $\bar{x}_n$ may be computed from Eq. (12). Often r is maintained as near to unity as possible, and the average molecular weight is calculated from analytical determination of one end group assuming r is exactly equal to unity, in which case Eq. (9) applies. A given small deviation of r from unity (or any other assumed value) introduces an error in $\bar{x}_n$ which, though small at low degrees of polymerization, increases approximately as the square of $\bar{x}_n$. The *percentage* error increases directly with the degree of polymerization and is given approximately by

$$(\text{Percentage error in } r) \times \bar{x}_n/2$$

Thus, at $\bar{x}_n = 200$, an unaccounted loss of only 0.1 percent of one or the other of the bifunctional reactants A——A or B——B introduces a 10 percent error in the calculated molecular weight.

Bifunctional condensing systems sometimes may contain minute amounts of reactants of higher functionality; or, through the occurrence of side reactions of one sort or another, a few of the monomers may enter the polymer as units of higher functionality.* These complications are more difficult to deal with in a general fashion. Usually, however, they may be taken into account through the use of the nonlinear polymerization theory discussed in Chapter IX. As a rule, appreciable deviations toward higher functionality are accompanied by gelation, which is easily recognized by the loss of fluidity of the polymer or by its incomplete solubility.

Amidation is particularly well adapted to use as a polymer-forming condensation reaction. The reaction is rapid above 180° to 200°C, it is remarkably free from side reactions, no catalysts are required (indeed, none are known), and the process is of the second order so that the molecular weight increases directly as the time of reaction. Molecular weights of 20,000 to 30,000 are attainable with no great difficulty under favorable conditions. This is not true of particular polyamide reactants susceptible to side reactions, as, for example, in the reaction of a diamine with glutaric acid wherein the inherent instability of the glutaric amide unit leads to decomposition.

Esterification of hydroxyl groups with carboxy acid groups, or through ester interchange with a carbalkoxy group, requires catalysis of one sort or another if the reaction is to be carried well toward completion. Acid catalysts usually cause side reactions (dehydration or etherification) of the hydroxyl group at elevated temperatures and hence are of limited utility. Small amounts of basic substances, such as sodium alkoxides preferably used in conjunction with magnesium, are effective in the ester interchange process at high temperatures.[35] Rapid and efficient polyesterification may be achieved in favorable cases by coreacting the glycol with a dibasic acid chloride.[36] With pure decamethylene glycol and pure terephthaloyl chloride, for example, molecular weights exceeding 35,000 are obtainable by this method within a few hours.

* The suggestion of Staudinger and co-workers[34] that polyesters formed from bifunctional reactants are nonlinear owing to the occurrence of occasional ortho-ester linkages

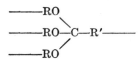

is altogether unlikely since these polyesters are soluble and they exhibit finite viscosities in the molten state. If linkages of the above type occurred even to a small extent, gelation should be observed, at least at advanced stages of the polymerization.

The condensation of dimethylsilanediols (produced by hydrolyzing dichlorodimethylsilane) proceeds efficiently at an advantageous rate. Polymeric dimethylsiloxanes having molecular weights of the order of one million have been reported,[37] but these very high polymers apparently were prepared through rearrangement of one of the lower cyclic polymers, perhaps cyclooctamethyltetrasiloxane, carefully purified beforehand, rather than by condensation of the silanediol.

Oftentimes a condensation polymer having an average molecular weight less than the maximum attainable is desired. The molecular weight may, of course, be controlled by discontinuing the condensation at the desired stage, but polymers so prepared are subject to further change in molecular weight on heating. In order to avoid this susceptibility to further polymerization, it is common practice to achieve molecular weight "stabilization" by employing a small excess of one bifunctional reactant A——A or B——B, or by adding a small amount of a monofunctional reactant. The ultimate molecular weight is then limited to an extent depending on the proportion of either "stabilizer," as discussed above.

6. RING FORMATION VS. CHAIN POLYMERIZATION

Polyfunctionality of the reactants is not sufficient in itself to assure formation of polymer; the reaction may proceed intramolecularly with the formation of cyclic products. For example, hydroxy acids when heated yield either lactone or linear polymer (or both),

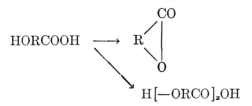

the direction of the reaction depending on the particular hydroxy acid and, to a lesser extent, on the reaction conditions. α-Hydroxy acids such as lactic acid condense to give both the dimeric cyclic ester, lactide, and a linear polymer:

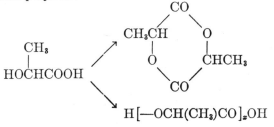

The condensation of amino acids likewise may produce cyclic and/or linear products; the same is true of virtually all polyfunctional condensation reactions. The conversion of cyclic monomers and dimers (or other cyclic low polymers) to chain polymers was discussed in the preceding chapter; the reverse reaction may often occur as well. Thus the alternative ring and chain products which may be produced by condensation of a bifunctional monomer usually are interconvertible, but with varying degrees of facility.

The prime factor governing the course followed by a bifunctional reaction is the size of the ring (or rings) which can be obtained through intramolecular condensation. If the ring size is less than five atoms or more than seven, the product, under ordinary conditions, will consist almost entirely of open chain polymer. If a ring containing five annular atoms can be formed, this will be the product exclusively; if of six or seven atoms, either or both ring and chain polymer are likely to be formed. Larger rings are formed only under special conditions, e.g., by conducting the condensation at high dilution[38] where opportunities for intermolecular reaction are less favorable, or by heating the polymer in vacuum in the presence of catalysts, the cyclic product being continually removed as it is formed through intramolecular cyclization.[39]

The success of the high dilution method for preparing ring compounds from bifunctional monomers which cyclize with difficulty arises from the fact that monomeric ring formation is unimolecular in the reactant, whereas the formation of higher condensates and chain polymers proceeds bimolecularly. Stoll, Rouvé, and Stoll-Comte[40] define a cyclization constant C as the ratio of the uni- and bimolecular rate constants, k_1 and k_2 respectively, for these two processes. The rate of cyclization is given by k_1c, and the rate of chain polymerization by k_2c^2, where c is the concentration of reactant. The ratio R of cyclic monomer to chain polymer is found experimentally to be inversely proportional to the concentration in accordance with the relationship derived in this manner:

$$R = k_1/k_2c = C/c$$

The cyclization constant C, which may be evaluated from the observed ratio of the two products at a given concentration c, affords a measure of the tendency for a given bifunctional compound to cyclize. A plot of log C vs. the ring size n for the lactonization of ω-hydroxy acids

$$HO(CH_2)_{n-2}\text{---}COOH \rightarrow \begin{matrix} O \\ \diagup \| \\ (CH_2)_{n-2} \\ \diagdown \| \\ CO \end{matrix} + H_2O$$

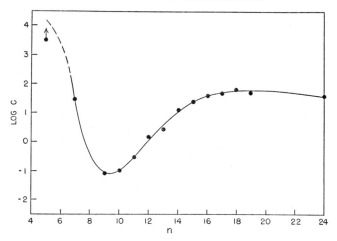

FIG. 6.—Log of the cyclization constant C for the lactonization of ω-hydroxy acids vs. ring size. The point for $n=5$ represents a minimum value. (Data from Stoll and Rouvé.[41])

carried out in benzene in the presence of a fixed concentration of strong acid catalyst is shown in Fig. 6.[41] The cyclization constant falls by a factor of 10^{-5} from a maximum for the five-membered ring to a minimum apparently at $n=9$. C then rises slowly to a broad maximum value occurring around $n=18$, where C is almost 10^3 times larger than at $n=9$. Beyond this maximum a gradual decrease in C (which is obscured on the semilog plot) sets in with further increase in the size of the unit. The ease of formation of cyclic esters by heating various polyesters of the formula

$$[-O(CH_2)_2O-CO(CH_2)_nCO-]_x$$

(including ethylene carbonate) in the presence of a catalyst is depicted qualitatively in Fig. 7.[42] Also shown are the relative stabilities with respect to polymerization of the cyclic anhydrides of the dibasic acids. The location of the' minimum is not the same within different series of cyclic compounds, but the forms of the curves are similar.

The difficulty with which rings of less than five atoms are formed is readily explained by the strain imposed on the valence angles. Five-atom rings are virtually strainless (in a symmetrical five-atom ring the bond angle is 108°); in all larger rings valence angle strain can be relieved entirely through the assumption of nonplanar forms (Sachse-Mohr theory), except for such obstructions as may arise from steric interferences between substituents. Nevertheless, bifunctional condensations involving units of more than seven atoms do not ordinarily

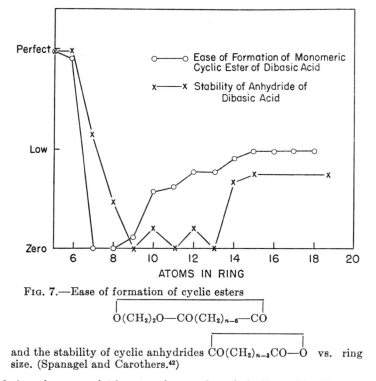

FIG. 7.—Ease of formation of cyclic esters

$$\overline{O(CH_2)_2O—CO(CH_2)_{n-6}—CO}$$

and the stability of cyclic anhydrides $\overline{CO(CH_2)_{n-3}CO—O}$ vs. ring size. (Spanagel and Carothers.[42])

yield rings in appreciable quantity, and, as is indicated in Fig. 6 and 7, rings of about eight to twelve members are formed with exceptional difficulty, even at high dilution or in a vacuum at elevated temperatures in the presence of a catalyst.

The reason for this minimum in the ease of ring formation is made clear by an examination of scale models for rings composed mainly of —CH₂— groups. In the range from eight to about twelve members, it will be observed that many of the hydrogen atoms are forced to occupy positions within the interior of the ring, where they are crowded together.[43,44] Repulsions between these hydrogen atoms discourage arrangement of the bifunctional molecule in a form conducive to ring formation. It will be noted further that the interference between hydrogen atoms leaves little freedom of choice of configuration for the ring; in order to form the ring structure, it is necessary to arrange the bonds in a fairly specific pattern. A structural unit of this size is capable of assuming many configurations, virtually any one of which is acceptable for the formation of a unit in a polymer chain. Consequently, the formation of the required configuration for ring closure is unfavorable statistically as well as energetically.[45]

As the ring size is increased above twelve atoms, the number of permissible ring configurations increases and it no longer remains necessary to crowd hydrogen atoms within the ring.[44]　The ease of ring formation increases in this range.　However, in ordinary bifunctional condensations (no diluent) the primary product from monomers (or dimers) of twelve or more members (having tetrahedral valence angles) is almost exclusively linear polymer.　This results from the statistical improbability that the ends of a long chain of atoms, connected by valence bonds about which there is free rotation, will meet.　Although various stable ring configurations are possible for long chains, the total number of other configurations is disproportionately larger; i.e., the ring configurations represent but a small fraction of the total of all possible configurations.　It can be shown from statistical considerations[46] that the probability that the two ends of a very long chain will occupy positions adjacent to each other varies approximately as the inverse *three-halves* power of the chain length or number of chain atoms (see Chap. VIII).　The probability that the end group of a given molecule is adjacent to the end group of another will vary as their concentration, and hence inversely as the *first* power of the chain length. Consequently, intramolecular reaction gradually becomes increasingly improbable as the length of the bifunctional chain increases, as in fact is indicated by the data shown in Fig. 6 for the formation of rings of more than eighteen members.

　　The principles set forth above account reasonably well for the course of bifunctional condensations under ordinary conditions and for the relative difficulty of ring formation with units of less than five or more than seven members.　They do not explain the formation of cyclic monomers from five-atom units to the total exclusion of linear polymers. Thus[32] γ-hydroxy acids condense exclusively to lactones such as I, γ-amino acids give the lactams II, succinic acid yields the cyclic anhydride III, and ethylene carbonate and ethylene formal occur only in the cyclic forms IV and V.

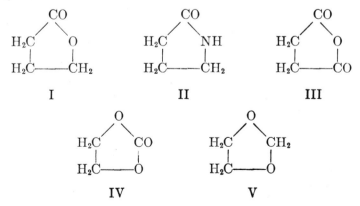

Furthermore, formation of these products occurs with much greater ease than linear polymerization of units of six or more chain members, and they are more stable to hydrolysis or other ring-opening reactions.

The steric and configurational factors discussed above would suggest that five-membered rings should form somewhat more readily than rings of six or seven members, but they offer no explanation for the total exclusion of intermolecular condensation, nor do they explain the much greater rate of intramolecular reaction of five-membered units as compared with the rates of intermolecular reaction of larger units. A possible partial explanation for these peculiarities of five-membered

TABLE IX.—RING FORMATION VS. CHAIN POLYMERIZATION FROM SIX- AND SEVEN-MEMBERED UNITS

Type	Structural unit	Products of bifunctional condensation	
		Six-membered unit	Seven-membered unit
ω-Hydroxy acid self ester	$-O(CH_2)_nCO-$	Both ring and polymer spontaneously interconvertible[47]	Chiefly ring; interconvertible[48]
α-Hydroxy acid self ester	$-OCHCO-OCHCO-$ $\quad\quad CH_3 \quad\quad CH_3$	Linear polymer probably is the primary product;[47] interconversion is easy	
Alkylene carbonate	$-(CH_2)_nOCOO-$	Both ring and polymer; easily interconvertible[47,49]	Linear polymer only[49]
Dibasic acid anhydride	$-(CH_2)_nCO-O-CO-$	Ring only	Linear polymer; convertible to ring[50]
Alkylene formal	$-(CH_2)_nOCH_2O-$	Ring only[51]	Both ring and polymer; interconvertible[51]
Self amide of ω-amino acid	$-NH(CH_2)_nCO-$	Ring only[32]	Both ring and polymer; interconvertible at high temperatures[52]
Self amide of α-amino acid	$-NHCHCO-NHCHCO-$ $\quad\quad R \quad\quad\quad R$	Ring usually predominates, but some linear polymer can be formed; interconversion is difficult	
Alkylene sulfide	$-CH_2CH_2S-CH_2CH_2S-$	Linear polymer; converted with some difficulty to ring[53]	
Alkylene ether-sulfide	$-CH_2CH_2O-CH_2CH_2S-$	Linear polymer[54]	

ring closure may be found from further consideration of hydrogen repulsions. In these rings the hydrogen atoms occur around the periphery; hence their mutual repulsions are minimized. (In the open chain, planar-zigzag form each hydrogen atom, though slightly farther removed from its nearest neighbors, is surrounded by a greater number of near-neighbor hydrogen atoms than in the planar ring form.)

Bifunctional monomers capable of forming six- or seven-membered rings condense variably, depending upon the particular monomer. The products normally obtained in the absence of diluent in various representative bifunctional condensations are listed in Table IX for unit lengths of six and seven members. The term interconvertibility refers to the reversible transformation between the ring and the linear polymer. Several of the six-membered units (Table IX) prefer the ring form exclusively, but most of them yield both products, or at any rate the ring and chain products are readily interconvertible. Seven-membered units either yield linear polymers exclusively, or, if the cyclic monomer is formed under ordinary conditions, it is convertible to the linear polymer.

Substitution of carbonyl, oxygen, or other atoms or groups for methylene should be expected to modify ring-forming tendencies. Valence angles and bond lengths are altered somewhat, but perhaps of greater importance is the decrease in the number of interfering hydrogens as the methylene members are replaced. Replacement of some of the methylene groups by less bulky chain members, such as oxygen for example, should diminish both the overwhelming tendency toward intramolecular reaction of five-membered units and the difficulty of forming rings of eight to twelve members. Experimental observations confirm the latter of these predictions.[55]

The preceding discussion applies only to structural units in which the chain atoms consist principally of carbon or other similarly bonded elements from the first row of the periodic table. The greater bond lengths and modified bond angles for larger atoms lead to rather different circumstances. In the cyclodimethylsiloxane series

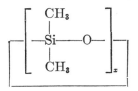

for example, the eight-membered ring ($x = 4$) is obtained in the highest yield, only traces of the cyclic trimer (six-membered ring) being present at equilibrium.[56] The reluctance with which the latter is formed has been attributed to the large value of the Si—O—Si bond angle,[57] with the consequence that valence angles in rings of less than eight members are under strain.* Unlike the cyclic carbon compounds (see Fig. 6) there is no minimum in the ease of cyclization with increase in ring size for $x > 4$, and the optimum ring size ($x = 4$) is not overwhelmingly preferred, in contrast to the five- and six-membered carbon rings. The equilibrium abundance of each species is a uniformly decreasing function of the ring size from $x = 4$ to 8.[56] The greater length of the Si—O bond and the large Si—O—Si angle may alleviate the repulsions between substituents (in this case —CH₃), which in rings consisting of —CH₂— groups have been suggested as being responsible for both the strong preference for five- and six-membered rings and for the severe difficulty of forming rings of eight to twelve members. The gradual decrease in abundance of the dimethylsiloxane rings with increasing ring size from $x = 4$ is qualitatively in accordance with expectation if this trend is solely due to the statistical decrease in probability of ring closure with increasing length of chain, specific steric factors being of minor importance.

7. SUMMARY AND CONCLUSIONS: THE PRINCIPLE OF EQUAL REACTIVITY OF ALL FUNCTIONAL GROUPS

The combined results of kinetic studies on condensation polymerization reactions and on the degradation of various polymers by reactions which bring about chain scission demonstrate quite clearly that the chemical reactivity of a functional group does not ordinarily depend on the size of the molecule to which it is attached. Exceptions occur only when the chain is so short as to allow the specific effect of one end group on the reactivity of the other to be appreciable. Evidence from a third type of polymer reaction, namely, that in which the lateral substituents of the polymer chain undergo reaction without alteration in the degree of polymerization, also support this conclusion. The velocity of saponification of polyvinyl acetate, for example, is very nearly the same as that for ethyl acetate under the same conditions.[58]

On the basis of these results, together with the assurance provided by theoretical considerations as discussed earlier in this chapter, we may conclude that *at all stages of the polymerization the reactivity of every*

---

* If the O—Si—O angle is taken to be 110°, as surely must be very nearly correct, it is necessary to assume an Si—O—Si angle exceeding 130° in order to account for the probable strain in the six-membered cyclic trimer. However, Sauer and Mead's value of 160° for the latter angle almost certainly is too large. (L. K. Frevel, private communication, finds 130° ±10° from X-ray diffraction.) It follows that all higher silicone rings are puckered.

like functional group is the same. The only exceptions to this state-ment may occur among the very lowest species in the polymeric mix-ture as already noted, and these deviations from it are generally minor. This *principle of equal reactivity of all functional groups* is of the utmost importance to the analysis of the constitution of polymers, for it per-mits the application of simple statistical considerations to the problem of the distribution of the bonds, formed during polymerization, among the structural units present in the system. This subject will be dis-cussed in Chapters VIII and IX. The principle of equal reactivity also simplifies immensely the kinetic treatment of polymerization reac-tions, for it permits total disregard of the complex array of *molecules* participating in the polymerization. Instead, the polymerization process may be regarded merely as a reaction between functional groups without differentiation according to the sizes of the molecules to which they are attached. In other words, the velocity coefficient may be considered to be the same regardless of the sizes of the mole-cules involved in a given process. Rate expressions are most conven-iently set down, therefore, in terms of the concentrations of functional groups—not of molecules—and a single velocity coefficient may be used for all reactions in which the same chemical process is involved.

The principle of equal reactivity as enunciated above asserts merely that all like functional groups remaining unreacted at *any given stage* of the process enjoy equal opportunities for reaction. It has *not* been stated that the level of reactivity (i.e., the velocity coefficient) neces-sarily remains constant as the reaction progresses. Indeed, the velocity coefficient for polyesterification has been shown to change as the reac-tion progresses from 0 to 80 percent conversion. Since monoesterifica-tions exhibit a similar perturbance, it has been concluded that the size of the polymer molecules formed does not alter the course of the reaction; hence the principle of equal reactivity is not violated.

A shift in the velocity constant such as is observed in bulk esterifica-tion is the exception rather than the rule. A source of more general concern is the enormous increase in viscosity which accompanies poly-merization. Both theory and experimental results indicate that this factor usually is of no importance except under the extreme conditions previously mentioned. Consequently, the velocity coefficient *usually* remains constant throughout the polymerization (or degradation) process. Barring certain abnormalities which enter when the velocity coefficient is sensitive to the environmental changes accompanying the polymerization process, application of the ordinary methods of chemical kinetics to polymerizations and other processes involving polymer molecules usually is permissible.

REFERENCES

1. N. Menschutkin, *Ann.*, **197**, 220 (1879); *Ber.*, **17**, 846 (1884); *J. prakt. Chem.*, [2] **29**, 422, 437 (1884); *Ber.*, **31**, 1423 (1898).
2. B. V. Bhide and J. J. Sudborough, *J. Indian Inst. Science*, **8A**, 89 (1925).
3. D. P. Evans, J. J. Gordon, and H. B. Watson, *J. Chem. Soc.* **1938**, 1439.
4. P. C. Haywood, *J. Chem. Soc.*, **121**, 1904 (1922).
5. A. Kailan, *Z. physik. Chem.*, **85**, 706 (1913).
6. J. Meyer, *Z. physik. Chem.*, **66**, 81 (1909).
7. J. Meyer, *Z. physik. Chem.*, **67**, 257 (1909).
8. J. G. Kirkwood and F. H. Westheimer, *J. Chem. Phys.*, **6**, 506, 513 (1938).
9. F. H. Westheimer and M. W. Shookhoff, *J. Am. Chem. Soc.*, **61**, 555 (1939).
10. F. H. Westheimer and M. W. Shookhoff, *J. Am. Chem. Soc.*, **62**, 269 (1940).
11. A. Skrabal and E. Singer, *Monatsh.*, **41**, 339 (1920).
12. E. Rabinowitch and W. C. Wood, *Trans. Faraday Soc.*, **32**, 1381 (1936); E. Rabinowitch, *ibid.*, **33**, 1225 (1937).
13. K. J. Laidler, *Chemical Kinetics* (McGraw-Hill Book Co., New York, 1950).
14. H. Goldschmidt, *Ber.*, **29**, 2208 (1896); A. C. Rolfe and C. N. Hinshelwood, *Trans. Faraday Soc.*, **30**, 935 (1934).
15. P. J. Flory, *J. Am. Chem. Soc.*, **61**, 3334 (1939); **62**, 2261 (1940).
16. N. Ivanoff, *Bull. soc. chim. France*, [5], **17**, 347 (1950).
17. R. H. Kienle, P. A. van der Meulen, and F. E. Petke, *J. Am. Chem. Soc.*, **61**, 2258, 2268 (1939); R. H. Kienle and F. E. Petke, *ibid.*, **62**, 1053 (1940); **63**, 481 (1941).
18. P. J. Flory, U.S. Patent 2,244,192 (1941).
19. K. Freudenberg, W. Kuhn, W. Dürr, F. Bolz, and G. Steinbrunn, *Ber.*, **63**, 1510 (1930); K. Freudenberg and G. Blomqvist, *ibid.*, **68**, 2070 (1935); W. Kuhn, *ibid.*, **63**, 1503 (1930).
20. W. Kuhn, *Z. physik. Chem.*, **A159**, 368 (1932).
21. M. L. Wolfrom, J. C. Sowden, and E. N. Lassettre, *J. Am. Chem. Soc.*, **61**, 1072 (1939).
22. E. Husemann and G. V. Schulz, *Z. physik. Chem.*, **B52**, 1 (1942); G. V. Schulz and E. Husemann, *ibid.*, **B52**, 23 (1942); T. Kleinert and V. Mössmer, *Monatsh.*, **79**, 442 (1948).
23. G. V. Schulz, *J. Polymer Sci.*, **3**, 365 (1948).
24. W. Kuhn, C. C. Molster, and K. Freudenberg, *Ber*, **65**, 1179 (1932); K. Freudenberg, G. Piazolo, and C. Knoevenagel, *Ann.*, **537**, 197 (1939).
25. G. Scatchard, J. L. Oncley, J. W. Williams, and A. Brown, *J. Am. Chem. Soc.*, **66**, 1980 (1944).
26. A. Matthes, *J. prakt. Chem.*, **162**, 245 (1943).
27. P. J. Flory, *J. Am. Chem. Soc.*, **62**, 2255 (1940).
28. P. J. Flory, *J. Am. Chem. Soc.*, **64**, 2205 (1942); *J. Chem. Phys.*, **12**, 425 (1944).

29. M. M. Brubaker, D. D. Coffman, and F. C. McGrew, U.S. Patent 2,339,237 (1944). L. F. Beste and R. C. Houtz, *J. Polymer Sci.*, **8**, 395 (1952).

30. M. D. Stern and A. V. Tobolsky, *J. Chem. Phys.*, **14**, 93 (1946).

31. W. Patnode and D. F. Wilcock, *J. Am. Chem. Soc.*, **68**, 358 (1946); D. W. Scott, *ibid.*, **68**, 2294 (1946).

32. W. H. Carothers, *Chem. Revs.*, **8**, 353 (1931).

33. W. H. Carothers and J. W. Hill, *J. Am. Chem. Soc.*, **54**, 1559 (1932).

34. H. Staudinger and F. Berndt, *Makromol. Chem.*, **1**, 22 (1947).

35. J. R. Whinfield, *Nature*, **158**, 930 (1946); Brit. Patent 578,079 (1946).

36. P. J. Flory and F. S. Leutner, U.S. Patents 2,589,687 and 2,589,688 (1952).

37. A. J. Barry, *J. Applied Phys.*, **17**, 1020 (1946); D. W. Scott, *J. Am. Chem. Soc.*, **68**, 1877 (1946).

38. P. Ruggli, *Ann.*, **392**, 92 (1912); K. Ziegler and co-workers, *Ann.*, **504**, 94 (1933); **513**, 43 (1934); G. Salomon, *Helv. Chim. Acta*, **17**, 851 (1934).

39. J. W. Hill and W. H. Carothers, *J. Am. Chem. Soc.*, **55**, 5031 (1933); **57**, 925 (1935); E. W. Spanagel and W. H. Carothers, *ibid.*, **57**, 929 (1935); **58**, 654 (1936).

40. M. Stoll, A. Rouvé, and G. Stoll-Comte, *Helv. Chim. Acta*, **17**, 1289 (1934).

41. M. Stoll and A. Rouvé, *Helv. Chim. Acta*, **18**, 1087 (1935).

42. E. W. Spanagel and W. H. Carothers, *J. Am. Chem. Soc.*, **57**, 929 (1935).

43. M. Stoll and G. Stoll-Comte, *Helv. Chim. Acta*, **13**, 1185 (1930).

44. W. H. Carothers and J. W. Hill, *J. Am. Chem. Soc.*, **55**, 5043 (1933).

45. G. Salomon, *Trans. Faraday Soc.*, **34**, 1311 (1938); G. M. Bennett, *ibid.*, **37**, 794 (1941).

46. H. Jacobson and W. H. Stockmayer, *J. Chem. Phys.*, **18**, 1600 (1950).

47. W. H. Carothers, G. L. Dorough, and F. J. Van Natta, *J. Am. Chem. Soc.*, **54**, 761 (1932).

48. F. J. Van Natta, J. W. Hill, and W. H. Carothers, *J. Am. Chem. Soc.*, **56**, 455 (1934).

49. W. H. Carothers and F. J. Van Natta, *J. Am. Chem. Soc.*, **52**, 314 (1930).

50. J. W. Hill, *J. Am. Chem. Soc.*, **52**, 4110 (1930).

51. J. W. Hill and W. H. Carothers, *J. Am. Chem. Soc.*, **57**, 925 (1935).

52. W. H. Carothers and G. J. Berchet, *J. Am. Chem. Soc.*, **52**, 5289 (1930).

53. W. Mansfeld, *Ber.*, **19**, 696 (1886).

54. J. C. Patrick, *Trans. Faraday Soc.*, **32**, 347 (1936); S. M. Martin, Jr., and J. C. Patrick, *Ind. Eng. Chem.*, **28**, 1144 (1936).

55. K. Ziegler and H. Holl, *Ann.*, **528**, 143 (1937).

56. D. W. Scott, *J. Am. Chem. Soc.*, **68**, 2294 (1946).

57. R. O. Sauer and D. J. Mead, *J. Am. Chem. Soc.*, **68**, 1794 (1946). See also, W. L. Roth, *ibid.*, **69**, 474 (1947); *Acta Krist.*, **1**, 34 (1948); E. H. Aggarwal and S. H. Bauer, *J. Chem. Phys.*, **18**, 42 (1950).

58. S. Lee and I. Sakurada, *Z. physik. Chem.*, **A184**, 268 (1939).

CHAPTER IV

Polymerization of Unsaturated Monomers

by Free Radical Mechanisms

THE processes by which unsaturated monomers are converted to polymers of high molecular weight exhibit the characteristics of typical chain reactions. They are readily susceptible to catalysis, photo-activation, and inhibition. The quantum yield in a photoactivated polymerization in the liquid phase may be of the order of 10^3 or more,[1] expressed as the number of monomer molecules polymerized per quantum absorbed. The efficiency of certain inhibitors is of a similar magnitude, thousands of monomer molecules being prevented from polymerizing by a single molecule of the inhibitor.[2]

Other common features of the so-called vinyl polymerizations suggest that the active center of the kinetic chain is retained by a single polymer molecule throughout the course of its growth. Thus it has been pointed out in Chapter II that the partially polymerized mixture consists of high molecular weight polymer and unchanged monomer with virtually no constituents at intermediate stages of growth. In fact, polymer molecules formed at the very outset, even during the first fraction of a percentage conversion, are comparable in molecular weight to those present in the aggregate of polymer at an advanced stage of the process.[3] It is at once apparent that individual polymer molecules grow to maturity while most of the monomers remain intact. If, on the contrary, the active centers responsible for execution of the conversion of monomer to polymer were transferred indiscriminately at almost every step from one molecule to another, all molecular species would participate in the process of chemical combination with other species at all stages of the process. All polymer molecules would then grow more or less simultaneously; lower species, such as dimers, trimers, and tetramers, would be prevalent at early stages of the polymerization, and these would advance in size as the process continued.

Not only are intermediates of this nature generally undetectable, but in the case of styrene the known dimer and trimer are themselves quite inert toward further polymerization.[4]

These considerations together with the mere fact of the extremely high molecular weight of the polymer molecules produced lead inescapably to the conclusion that a given molecule is formed by consecutive steps of a *single* chain process set off by the creation of an active center of one sort or another. Evidently, the active center is in some way retained by a growing polymer chain from each addition of a monomer molecule to the next. The course of the growth of an individual polymer molecule must occupy only a minute fraction of the time required for the over-all conversion, and the polymer molecule thus produced is not susceptible (with reservations; see Chap. VI) to further growth during subsequent stages of the polymerization process.

The following free radical chain mechanism, first suggested by Taylor and Bates[5] to explain the polymerization of ethylene induced by free radicals in the gas phase and independently proposed by Staudinger[6] for liquid phase polymerizations, offers an explanation for the above general characteristics of vinyl polymerizations.

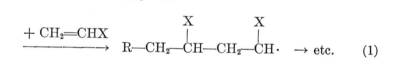

where X is a substituent which may be C_6H_5, Cl, Br, $OCOCH_3$, $COOCH_3$, or H. Disubstituted monomers such as vinylidene chloride or methyl methacrylate are, of course, to be included. The chain propagation step consists essentially of free radical attack at one of the doubly bonded carbon atoms of the monomer. One electron of the double bond pairs with the odd electron of the free radical to form a bond between the free radical and this carbon atom; the remaining electron of the double bond shifts to the other carbon atom, which then becomes a free radical. In this way, the active center shifts uniquely to the newly added monomer, which is thereby rendered capable of adding another monomer, etc. Like all chain reactions, the over-all polymerization involves at least two other processes: *chain initiation,* which, according to the above scheme must depend on a reaction which introduces free radicals into the system, and *chain termination,* whereby the terminal radical on a growing chain is destroyed or otherwise rendered inactive. The nature of these processes depends on the conditions employed, and generally they are more obscure than the growth process.

An impressive number of substances capable of generating free radicals have been shown to be potent accelerators for the polymerization of typical vinyl monomers such as styrene, methyl methacrylate, butadiene, and vinyl acetate. The most commonly employed initiators (often referred to inaccurately as catalysts) are organic peroxides, such as benzoyl peroxide. These are known to decompose slowly at temperatures of 50° to 100°C with release of free radicals as follows:[7,8,9]

$$(RCOO)_2 \;\to\; 2RCOO\cdot \;\to\; 2R\cdot + 2CO_2 \qquad (2)$$

Organic hydroperoxides, such as t-butyl hydroperoxide, $(CH_3)_3C\!-\!O\!-\!OH$, likewise induce polymerization in vinyl monomers through the action of free radicals formed as primary intermediates in their decomposition. The following compounds, or classes of compounds, also are effective polymerization initiators at temperatures where they undergo slow thermal decomposition by mechanisms which are believed to involve the release of free radicals as indicated:

N-nitrosoacylanilides[10]

$$C_6H_5\!-\!\underset{\underset{NO}{|}}{N}\!-\!CO\!-\!R \;\to\; C_6H_5\cdot + N_2 + R\!-\!\overset{\overset{O}{\|}}{C}\!-\!O\cdot \;\to\; R\cdot + CO_2$$

p-Bromobenzenediazo hydroxide[11]

$$BrC_6H_4\!-\!N\!=\!N\!-\!OH \;\to\; BrC_6H_4\cdot + N_2 + HO\cdot$$

Triphenylmethylazobenzene[12]

$$(C_6H_5)_3C\!-\!N\!=\!N\!-\!C_6H_5 \;\to\; (C_6H_5)_3C\cdot + N_2 + C_6H_5\cdot$$

Aliphatic azobisnitriles[13,14] such as azo-bis-isobutyronitrile

$$(CH_3)_2\overset{\overset{CN}{|}}{C}\!-\!N\!=\!N\!-\!\overset{\overset{CN}{|}}{C}(CH_3)_2 \;\to\; (CH_3)_2\overset{\overset{CN}{|}}{C}\!-\!N\!=\!N\cdot + \cdot\overset{\overset{CN}{|}}{C}(CH_3)_2$$

$$\longrightarrow\; (CH_3)_2\overset{\overset{CN}{|}}{C}\cdot + N_2$$

The last-mentioned compound has been chosen in preference to benzoyl peroxide in many of the more recent kinetic investigations on account primarily of the freedom of its decomposition from side reactions such as radical-induced decomposition (see p. 113). It may also serve as a photoinitiator under the influence of near ultraviolet radiation, which is absorbed with dissociation into radicals. Tetraphenylsuccinonitrile in dilute solution at 100°C dissociates *reversibly* into free radicals

$$
\begin{array}{cc}
\text{CN} & \text{CN} \\
| & | \\
(C_6H_5)_2C & \!\!\!\!-\!\!\!-C(C_6H_5)_2
\end{array}
\;\rightleftarrows\; 2(C_6H_5)_2\overset{\displaystyle\text{CN}}{\underset{|}{C}}\cdot
$$

The degree of dissociation is very small but the diphenylcyanomethyl radical is sufficiently reactive to induce polymerization in styrene.[15] Methyl radicals or hydrogen atoms bring about polymerization of vinyl monomers in the gas phase.[16] Hydrogen peroxide in the presence of ferrous ions initiates polymerization[17] in the aqueous phase or in aqueous emulsions through generation of hydroxyl radicals according to the Haber-Weiss mechanism

$$
H_2O_2 + Fe^{++} \;\rightarrow\; HO^- + Fe^{+++} + HO\cdot
$$

Final and conclusive evidence for the free radical mechanism is afforded by detection in the polymer of radical fragments from the initiator. Polystyrene and poly-(methyl methacrylate) prepared with bromobenzoyl peroxide, for example, contain bromine which cannot be removed by repeated reprecipitation of the polymers.[18] Both bromophenyl and bromobenzoate groups are attached to the polymer chains,[19] their proportions depending somewhat on the polymerization conditions.[20] Thus both the benzoate radical, which is the primary product of the peroxide decomposition, and the phenyl radical formed from it by loss of carbon dioxide initiate polymer chains. Fragments from N-nitrosoacylanilides[10] have likewise been identified in the polymer prepared by using them as initiators. Radioactive initiators such as potassium persulfate[21] containing S^{35} or azo-bis-isobutyronitrile[22] containing C^{14} yield polymers in which the concentrations of radical fragments have been measured quantitatively by radiochemical techniques. Poly-(methyl methacrylate)[17] and polystyrene[23] prepared in an aqueous medium using hydrogen peroxide and ferrous ion contain hydroxyl end groups, as should be expected according to the mechanism of free radical generation. Free radical fragments usually are found to occur in the approximate ratio of one or two per molecule.[17,18,21-24] (see Sec. 1c).

Although all sources of reactive free radicals which have been tried initiate the polymerization of unsaturated monomers, the converse of this statement, namely, that all initiators are free-radical-producing substances, is not true. Thus, strong acids (in the Lewis sense) such as $AlCl_3$, BF_3, and $SnCl_4$, which are characterized by a strong affinity for a pair of electrons, bring about rapid polymerization of certain monomers. These polymerizations also proceed by chain mechanisms. The propagating center is, in this case, a positively charged ion—a carbonium ion—which differs from the free radical center by the absence of the odd electron of the latter. Otherwise the chain propagation mechanism is similar. The regenerative aspect of the free radical chain is retained; with each addition of monomer a new carbonium ion center succeeds the one to which the monomer adds, exactly as in reaction (1). Alkali metals, their hydrides, and certain metal alkyls induce an analogous *anionic* polymerization. In this case, the propagating center possesses a pair of unshared electrons conferring a single net negative charge on the terminal carbon. Ionic polymerizations will be discussed in the following chapter, the present one being concerned exclusively with reaction mechanisms and their kinetic interpretation in free radical polymerizations. Many of the concepts to be developed here are applicable to ionic polymerizations also.

1. THE CONVERSION OF MONOMER TO POLYMER

1a. Kinetic Scheme for Polymerization in the Presence of an Initiator.—The process of chain initiation may be considered to involve two steps, the first being the decomposition of the initiator I (e.g., benzoyl peroxide or azo-bisnitrile) to yield a pair of free radicals $R\cdot$

$$I \xrightarrow{k_d} 2R\cdot \tag{3}$$

and the second the addition of a monomer M to a *primary* radical $R\cdot$ to yield a chain radical

$$R\cdot + M \xrightarrow{k_a} M_1\cdot \tag{4}$$

The adduct, $R—CH_2CH\cdot$, formed from a vinyl monomer is here repre-
$$\overset{|}{X}$$
sented by $M_1\cdot$. Since the free radicals have odd numbers of electrons, they will be formed in pairs, either or both of which may initiate polymerization according to reaction (4). Not all of the radicals released in step (3) necessarily yield chain radicals according to (4), however; some of them may be lost through side reactions.

The growth of polymer molecules by successive addition of monomers according to reaction (1) to the radicals $M_1\cdot$ and to their successors may be represented by

$$M_1\cdot + M \xrightarrow{k_p} M_2\cdot$$

$$M_2\cdot + M \xrightarrow{k_v} M_3\cdot$$

or in general

$$M_x \cdot + M \xrightarrow{k_p} M_{x \cdot +1} \tag{5}$$

The same reaction rate constant k_p is written for each propagation step under the assumption that the radical reactivity is independent of the chain length, in accordance with the conclusion reached in the preceding chapter.

Bimolecular reaction between a pair of chain radicals accounts for annihilation of active centers. Two obvious processes by which this may occur are chain *coupling* (or combination)

$$\text{—CH}_2\underset{\underset{X}{|}}{\text{CH}}\cdot + \cdot\underset{\underset{X}{|}}{\text{CH}}\text{CH}_2\text{—} \rightarrow \text{—CH}_2\underset{\underset{X}{|}}{\text{CH}}\text{—}\underset{\underset{X}{|}}{\text{CH}}\text{CH}_2\text{—} \tag{6}$$

and *disproportionation* through transfer of a hydrogen atom

$$\text{—CH}_2\underset{\underset{X}{|}}{\text{CH}}\cdot + \cdot\underset{\underset{X}{|}}{\text{CH}}\text{CH}_2\text{—} \rightarrow \text{—CH}_2\underset{\underset{X}{|}}{\text{CH}}_2 + \underset{\underset{X}{|}}{\text{CH}}\text{=CH—} \tag{6'}$$

The presence of two hydroxyl groups per molecule in poly-(methyl methacrylate) and in polystyrene, each polymerized in aqueous media using the hydrogen peroxide–ferrous ion initiation system, has been established[17,23] by chemical analysis and determination of the average molecular weight. Poly-(methyl methacrylate) polymerized by azo-bis-isobutyronitrile labeled with radioactive C^{14} has been shown to contain radical fragments, $(CH_3)_2\overset{\overset{\displaystyle CN}{|}}{C}$—, in the ratio of two per molecule.[22]* Similar studies on polystyrene polymerized in emulsion using potassium persulfate containing radioactive S^{35} also reveal two initiator residues per molecule.[21] These observations provide direct evidence for the conclusion that coupling is the dominant chain-terminating process and that disproportionation occurs, at most, to a minor extent only. Detailed analysis of kinetic data on the rate and degree of polymerization, although possibly more susceptible to errors of interpretation, leads to the same conclusion.[25,26,27] It is interesting to note

* Recently Bevington, Melville, and Taylor (Meeting of the International Union of Pure and Applied Chemistry, Symposium on Macromolecules, Stockholm, 1953) have reported precise radiochemical determinations of the number of initiator fragments in polymers formed by the action of C^{14} labeled azo-bis-isobutyronitrile. In the case of styrene polymerization at 25°C they found 2 fragments per polymer molecule, indicating termination by coupling; with methyl methacrylate an average of 1.2 was found. The latter result indicates a preponderance of termination by disproportionation in this case.

in this connection[25] that α-alkyl-substituted benzyl radicals undergo coupling with no evidence for disproportionation.[28] Moreover, ethyl radicals combine in the gas phase at moderate temperatures to yield butane, and only minor proportions of ethylene and ethane are formed by disproportionation.[29]

For purposes of kinetic analysis, the termination step may be written

$$M_x \cdot + M_y \cdot \xrightarrow{k_{tc}} M_{x+y} \tag{7}$$

Allowing for possible cases in which disproportionation should not be disregarded, we also write

$$M_x \cdot + M_y \cdot \xrightarrow{k_{td}} M_x + M_y \tag{7'}$$

where M_x, M_y, and M_{x+y} represent inactive polymer molecules having the numbers of units indicated by the subscript in each case.* For most purposes the bimolecular nature of the termination process alone is important, irrespective of whether coupling or disproportionation assumes dominance. We shall therefore represent the termination rate constant by k_t, except where it is desired to emphasize specifically which of the alternative bimolecular processes is involved. Annihilation of the odd electron free radicals must occur by reactions involving pairs for the same reason that they must be formed in pairs; only in this way is it possible to destroy them without forming other odd electron molecules which retain free radical character. Of course, the chain radical may, by reaction with another molecule or conceivably by some internal rearrangement, be converted to, or replaced by, a radical of such low activity as to avoid further addition of monomer. Substances which effect such changes in the chain radical fall in the class of inhibitors to be discussed later. For the present we assume they are absent, hence that annihilation of a free radical active center may only occur bimolecularly between pairs.

Eqs. (3), (4), (5), and (7) describe the mechanism of an initiated free radical polymerization in a form amenable to general kinetic treatment. The rate of initiation of chain radicals according to Eqs. (3) and (4) may be written

$$R_i = (d[M\cdot]/dt)_i = 2fk_d[I] \tag{8}$$

where $[M\cdot]$ represents the total concentration of all chain radicals ir-

* Termination involving reaction between a chain radical and a primary radical may be neglected inasmuch as the concentration of primary radicals ordinarily will be maintained at a very low level owing to their rapid reaction with monomer according to reaction (4).

respective of size, and $(d[M\cdot]/dt)_i$ represents the rate of step (4). The initiator concentration is given by $[I]$, k_d is the rate constant for its decomposition into two radicals $R\cdot$ per molecule, and the factor f represents the fraction of these primary radicals which initiate chains according to step (4). When f is close to unity, the initiation rate R_i should be independent of the monomer concentration $[M]$. If, however, a large fraction of the primary radicals tend to disappear through other reactions, the fraction f may increase with monomer concentration and the rate of initiation should then depend accordingly on $[M]$. The efficiency usually is nearly independent of $[M]$, and the kinetic treatment consequently is simplified (see below).

Organic peroxides and hydroperoxides decompose in part by a self-induced radical chain mechanism whereby radicals released in spontaneous decomposition attack other molecules of the peroxide.[7,8] The attacking radical combines with one part of the peroxide molecule and simultaneously releases another radical. The net result is the wastage of a molecule of peroxide since the number of primary radicals available for initiation is unchanged. The velocity constant k_d we require refers to the *spontaneous decomposition only* and not to the total decomposition rate which includes the contribution of the chain, or induced, decomposition. Induced decomposition usually is indicated by deviation of the decomposition process from first-order kinetics and by a dependence of the rate on the solvent, especially when it consists of a polymerizable monomer. The constant k_d may be separately evaluated through kinetic measurements carried out in the presence of inhibitors[8,9] which destroy the radical chain carriers. The aliphatic azo-bis-nitriles offer a real advantage over benzoyl peroxide in that they are not susceptible to induced decomposition.

The rate of chain termination according to either (7) or (7') may be written

$$R_t = - (d[M\cdot]/dt)_t = 2k_t[M\cdot]^2 \tag{9}$$

where the factor 2 enters as a result of the disappearance of two radicals at each incidence of the termination reaction. Under all ordinary conditions the concentration of chain radicals will assume a value such that the rate of their disappearance through termination will equal their rate of creation. In this "steady state," therefore, $R_t = R_i$, and we have, according to Eqs. (8) and (9)

$$[M\cdot] = (fk_d[I]/k_t)^{1/2} \tag{10}$$

which expresses the steady-state concentration of chain radicals. The rate of the propagation step (5) is

$$R_p = k_p[M][M\cdot] \tag{11}$$

Substituting from Eq. (10)

$$R_p = k_p(fk_d[I]/k_t)^{1/2}[M] \tag{12}$$

Since the number of monomer molecules reacting in step (4) is insignificant compared with those consumed in the propagation step (5), provided the chain length is great, R_p may also be identified with the rate of polymerization; i.e., to an approximation which ordinarily is negligible

$$-d[M]/dt = R_p$$

We shall use R_p to represent the rate of polymerization as well as the rate of propagation, therefore. According to Eq. (12), the rate of polymerization should vary as the square root of the initiator concentration. If f is independent of the monomer concentration, which will almost certainly be true if f is near unity, the conversion of monomer to polymer will be of the first order in the monomer concentration. On the other hand, if f should be substantially less than unity, it may then depend on the concentration of monomer; in the extreme case of a very low efficiency, f might be expected to vary directly as $[M]$, whereupon the chain radical concentration becomes proportional to $[M]^{1/2}$ and the polymerization should be three-halves order in monomer.

The preceding reaction scheme may be adapted to photochemical polymerizations merely by replacing $k_d[I]$ with the intensity of active radiation-absorbed I_{abs}, which may be expressed in moles of light quanta (einsteins) absorbed per liter per second. The factor f then is to be interpreted as the quantum yield for chain initiation, i.e., as the number of pairs of chain radicals generated per quantum absorbed. Let us consider first the case in which the active radiation is absorbed by the monomer directly. If only a small fraction of the light is absorbed,* I_{abs} will be proportional to the product of the incident light intensity I_0 and the monomer concentration, i.e., $I_{abs} = \epsilon I_0[M]$, where ϵ is the molar absorption coefficient for the active radiation. Upon replacing $k_d[I]$ of Eq. (12) with I_{abs}, the rate expression under these conditions becomes

$$R_p = k_p(f\epsilon I_0/k_t)^{1/2}[M]^{3/2} \tag{13}$$

Thus, the rate should be proportional to the three-halves power of the

---

* If most of the light is absorbed in passing through the reaction chamber, I_{abs} will vary through the cell, nonuniform conditions will prevail, and simple analysis of the over-all kinetics will be impossible.

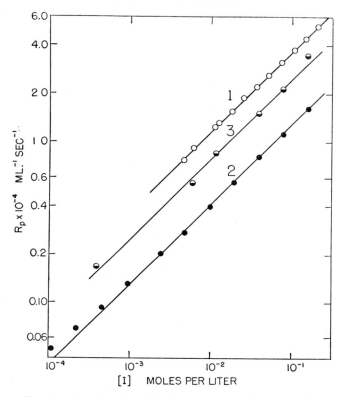

FIG. 8.—Log-log plot of initial polymerization rates R_p in moles/l./sec. against the initiator concentration $[I]$ in moles/l. Line 1, methyl methacrylate using azo-bis-iso-butyronitrile at 50°C.[26] Line 2, styrene with benzoyl peroxide at 60°C.[25] Line 3, methyl methacrylate with benzoyl peroxide at 50°C.[31]

concentration of monomer and to the square root of the incident intensity.

If the radiation is absorbed by a photosensitizer which decomposes into radicals on photoactivation, or which brings about initiation in some other manner, then $I_{abs} = \epsilon c_s I_0$ for sufficiently small concentrations c_s of the sensitizer. The rate expression resembles Eq. (12); R_p is proportional to $c_s^{1/2}$ and to $I_0^{1/2}$.

1b. Dependence of Initial Rates on the Concentrations of Initiator and Monomer.—The predicted proportionality between the rate of polymerization and the square root of the initiator concentration is abundantly confirmed by experimental measurements on numerous monomer-initiator pairs.* Initial polymerization rates for methyl

* For a more detailed bibliography on this subject see Ref. 30, pp. 103 ff.

methacrylate and for styrene are plotted in Fig. 8 against the initiator concentration using logarithmic scales.   The straight lines have been drawn with the theoretical slope of one-half.   Excellent agreement is maintained over a range of two-hundred-fold in the initiator concentration.   The deviations from the square root relationship at the lowest concentrations of benzoyl peroxide used by Mayo, Gregg, and Matheson[25] in the polymerization of styrene at 60°C (curve 2 of Fig. 8) are caused by an appreciable contribution of spontaneous, or thermal, initiation.   Initial rates for the benzoyl-peroxide-initiated polymerizations of vinyl acetate in benzene[32] and of d-sec-butyl α-chloroacrylate in dioxane[33] and for the polymerization of styrene with azo-bis-iso-butyronitrile[27] are accurately proportional to the square roots of the initiator concentrations.   Arnett[26] has shown further that the rates of polymerization of methyl methacrylate using different aliphatic azo-

TABLE X.—INITIAL RATES OF POLYMERIZATION; DEPENDENCE ON
MONOMER CONCENTRATION

$[M]$ moles/liter	$[I] \times 10^3$ moles/liter	Initial rate $R_p \times 10^4$ moles/l./sec.	$R_p/[I]^{1/2}[M]$ $\times 10^4$
Methyl methacrylate in benzene using azo-bis-isobutyronitrile at 77°C (Arnett[26])			
9.04[a]	0.235	1.93	14.0
8.63	.206	1.70	13.7
7.19	.255	1.65	14.4
6.13	.228	1.29	14.0
4.96	.313	1.22	13.9
4.75	.192	0.937	14.3
4.22	.230	.867	13.6
4.17	.581	1.30	13.0
3.26	.245	0.715	14.0
2.07	.211	.415	13.8
Methyl methacrylate in benzene using benzoyl peroxide at 50°C (Schulz and Harborth[37])			
9.44[a]	41.3	1.66	0.87
7.55	41.3	1.31	.86
5.66	41.3	0.972	.85
3.78	41.3	.674	.88
1.89	41.3	.334	.81
0.944	41.3	.153	.80
Styrene in benzene using benzoyl peroxide at 60°C (Mayo et al.[25])			
8.35[a]	4.00	0.255	0.48
5.845	4.00	.1734	.47
3.339	4.00	.0930	.44
0.835	4.00	.01855	.35
.418	4.00	.00838	.32

[a] Concentration of undiluted monomer.

bis-nitriles at the same concentration and temperature are proportional to the square roots of the decomposition rate constants k_d over a hundredfold range in the latter. These observations provide specific confirmation for the bimolecular nature of the termination process, as is apparent from the derivation of Eqs. (10) and (12). Additional support comes from the observed proportionality between the rate and the square root of the intensity of illumination in the direct photopolymerization of vinyl acetate[34,35] and of methyl methacrylate[36] (see Eq. 13).

Data illustrating the relationship of the initial rate to the concentration of monomer at fixed initiator concentration are given in Table X for styrene in benzene and for methyl methacrylate polymerized at various concentrations in the same solvent. If the efficiency f of utilization of primary radicals is independent of the monomer concentration, the quantity $R_p/I^{1/2}[M]$ given in the last column should be constant (see Eq. 12). For the azo initiator with methyl methacrylate this is indeed true within the exceptionally small experimental error exhibited by these data. The results for methyl methacrylate with benzoyl peroxide suggest a possible small decrease in f with dilution over the somewhat wider range covered in this case. A definite decrease in f with dilution is shown by styrene with benzoyl peroxide; a similar trend was indicated in the earlier work of Schulz and Husemann[38] on the same system in toluene at 27°C. The rate of polymerization of d-sec-butyl α-chloroacrylate by benzoyl peroxide in dioxane solution has been observed (polarimetrically) to be accurately proportional to the concentration of monomer.[33,39]

1c. Initiator Efficiencies.—The strict adherence to first-order kinetics (with respect to monomer) observed in several instances indicates an efficiency of utilization of primary radicals which is independent of dilution. It may be inferred from this observation that the efficiency probably is near unity. In other cases where the kinetics point to a decrease in f with dilution this decrease is rather small, and efficiencies near unity for the undiluted monomer are by no means precluded.

Several other methods have been demonstrated for determining the efficiency f. The most direct of these depends on a quantitative assay of the polymer for initiator fragments, which may then be compared with the amount of initiator decomposed. Evans,[23] in the work to which previous reference was made on the polymerization of styrene by hydrogen peroxide and ferrous ions in aqueous solution, showed not only that each polymer molecule thus formed contained two hydroxyl groups, but also that the number of moles of hydroxyl groups found in the polymer was very nearly equivalent to the moles of peroxide decomposed. Since according to the Haber-Weiss mechanism one hydroxyl radical is released for each peroxide molecule decomposed, an efficiency near unity is indicated.   Arnett and Peterson[22] determined efficiencies of initiation by C^{14}-labeled azo-bis-isobutyronitrile by comparing the radioactive content of the polymer with the amount of initiator disappearing in the course of the polymerization.   Efficiencies for initiation in methyl methacrylate, vinyl acetate, styrene, vinyl chloride, and acrylonitrile ranged from about 0.6 to 1.0, increasing in the order mentioned.   Implicit in this method is the assumption that the initiator decomposes exclusively by a spontaneous mechanism, wastage of initiator by induced decomposition being negligible.   This condition appears to be fulfilled by the azo nitriles.   Otherwise it is necessary to determine by some means the portion of the decomposition which is induced under the conditions of polymerization and to deduct this from the total initiator decomposed.

Direct quantitative determination of the number of initiator fragments combined with the polymer is feasible only under rather exceptional circumstances.   A more generally useful method depends on the determination of the molecular weight by an appropriate method.* The number of polymer molecules may then be calculated, and, assuming termination by coupling, the number of combined primary radicals taken to be twice the number of molecules.   Before the results of this calculation are accepted, it is necessary to make certain that no appreciable generation of chain ends occurs by processes other than chain initiation.   As will be shown shortly, chain transfer reactions often are responsible for a significant fraction of the chain ends in the polymer.   Hence it may be necessary to introduce appropriate corrections for the increase in the number of molecules resulting from processes of this nature.   The application of this method for estimating f will appear in Section 2c.   It suffices to mention here that the efficiencies thus obtained[27] for benzoyl peroxide in undiluted styrene and in methyl methacrylate are in the range from 0.6 to 1.0, in agreement with the indications of other observations.   Analysis of molecular weight data for azo-nitrile-initiated methyl methacrylate polymers leads to an efficiency slightly greater than one-half,[26] in agreement with the results obtained by the more direct tracer methods with C^{14} (see Table XI).

* The number average molecular weight is required.   This is obtained directly from measurements of a colligative property, such as the osmotic pressure, of dilute polymer solutions (see Chap. VII).   It is often more convenient to establish an empirical correlation between the osmotic molecular weight and the dilute solution viscosity, i.e., the so-called intrinsic viscosity, and then to estimate molecular weights from measurements of the latter quantity on the products of polymerization.

A third method for determining the efficiency depends on the use of certain inhibitors (see Sec. 4) which have been shown to react stoichiometrically with radical chains. The stable free radical 2,2-diphenyl-1-picrylhydrazyl is such an inhibitor for vinyl acetate[40] and for styrene,[41] one inhibitor molecule being consumed by each chain radical. When a very small quantity of this inhibitor is added to vinyl acetate containing benzoyl peroxide, no polymerization whatever can be detected until practically all of the inhibitor has been consumed by chain radicals generated from the decomposition of the initiator.[40] The number of moles of radicals released by the initiator during the inhibition interval as calculated from its rate of uninduced decomposition, i.e., from $2k_d$, is equal to the moles of inhibitor added, within the experimental error in evaluating the former quantity. During the similar inhibition of the azo-nitrile-initiated polymerization of styrene at 30° and at 60° C, the number of radicals calculated to have been released by the initiator (i.e., $2k_d$) agrees with the number of 2,2-diphenyl-1-picrylhydrazyl radicals within 10 percent.[41] Duroquinone[40] and benzoquinone,[42] though less reactive than the picrylhydrazyl, may be used also for the purpose of "counting" the number of radicals initiated in vinyl acetate, and in styrene under suitable circumstances. One of each of these inhibitor molecules is consumed per chain radical (see Sec. 4a), and the results obtained likewise lead to efficiencies close to unity for vinyl acetate. Benzoquinone is not a quantitatively effective inhibitor for methyl methacrylate[43] nor for the *thermal* polymerization of styrene[44] (where the radical concentration is very low).

The inhibition method has found wide usage as a means for determining the rate at which chain radicals are introduced into the system either by an initiator or by illumination. It is, however, open to criticism on the ground that some of the inhibitor may be consumed by primary radicals and, hence, that actual chain radicals will not be differentiated from primary radicals some of which would not initiate chains in the absence of the inhibitor. This possibility is rendered unlikely by the very low concentration of inhibitor (10^{-4} to 10^{-3} molar). The concentration of monomer is at least 10^4 times that of the inhibitor, yet the reaction rate constant for addition of the primary radical to monomer may be less than that for combination with inhibitor by only a factor of 10^{-2} to 10^{-3}. Hence most of the primary radicals may be expected to react with monomer even in the presence of inhibitor, the action of the latter being confined principally to the termination of chain radicals of very short length.[42]

Let us consider the mechanism of initiation in greater detail, taking benzoyl peroxide as an example. It may be assumed to dissociate initially into benzoate radicals, as previously indicated by reaction (2).

Owing to restrictions on the motion of molecules in the liquid state, a pair of radicals formed in solution may execute many oscillations in their respective "cages" consisting of surrounding molecules before they diffuse apart.[45] Thus, recombination with restoration of the original molecule during the interval of their existence as immediate neighbors in the liquid may well occur.   The rate of decomposition would then depend not only on the rate of fission of the initiator into radicals but also on the rate at which they subsequently diffuse apart. It is, of course, also possible that the pair of benzoate radicals may combine to yield decomposition products, e.g., phenyl benzoate and carbon dioxide, instead of the reactant.

The decomposition according to these ideas may be represented by the scheme

$$(C_6H_5COO)_2 \underset{(a')}{\overset{(a)}{\rightleftharpoons}} [2C_6H_5COO\cdot] \overset{(b)}{\longrightarrow} [C_6H_5COOC_6H_5 + CO_2]$$
$$\text{or other products}$$

$$(c) \quad \updownarrow \quad (c')$$

$$C_6H_5COO\cdot + C_6H_5COO\cdot$$

$$(d) \quad \downarrow$$

$$C_6H_5\cdot + CO_2$$

where substances remaining within the "cages" of the initial benzoate radicals are shown in brackets.   The decomposition of other initiators may be represented by a corresponding scheme.   (Radical-induced decomposition has been omitted.)   If radical recombination occurred at every collision, the rate of recombination, step (a'), would be about 10^{13} sec.$^{-1}$.   This is extremely unlikely.   Judging from the termination rate constant (see Sec. 3b) for polymer chain radicals, a more probable value for the rate of combination of neighboring radicals would be of the order of 10^9 sec.$^{-1}$; or the mean life of a pair of neighboring radicals might reasonably be of the order of 10^{-9} sec.   Diffusion out of the cage is estimated[45] to occur at a small fraction of the collision rate—smaller by a factor of 10^{-2} or 10^{-3} perhaps.   The mean duration of existence of a pair of radicals as first neighbors should therefore be 10^{-11} to 10^{-10} sec.; hence diffusion may well be a more rapid process than recombination, although this may not invariably be true.   After a single "jump" out of the initial cage, there remains a considerable probability that the radical pair will return to adjacent positions (step c') and thus be exposed again to the possibility of recombination.   After two or more jumps apart a re-encounter becomes improbable, and separate future existences are practically assured for the pair of radicals.   Before they lose completely their identity as a pair, i.e., before they become separated by a distance equaling the mean distance to another radical in the solution, some 10^6 jumps will occur, however (for a radical concentration such as usually prevails during a polymerization or initiator decomposition experiment, i.e., about 10^{-8} molar, for which the mean distance between nearest radicals is about 10^4Å). During the interval (which may be of the order of a second) between the separation of a radical from its initial partner and its combination with another radical at random, one or the other or both of them probably will have lost carbon dioxide. Consequently, recombination of radicals generated from different initiator molecules usually spells production of a decomposition product rather than restoration of a peroxide molecule. Similar considerations should apply to other initiators.

In the presence of monomer the radicals released in step (c), or those converted by (d), are intercepted by monomer in the course of their diffusion through the solution and long before they have the opportunity to combine with other radicals. The primary radicals usually are more reactive than the chain radicals formed from them by the addition of monomer;[8] hence, under all conditions yielding polymer molecules averaging more than a few units in length, virtually complete utilization of those primary radicals which are separated by diffusion from their original partners should be assured. The addition of primary radicals to monomer is much slower ($< 10^6$ sec.$^{-1}$ per radical in pure monomer) than diffusion; hence events during the life of the initial cage should not be much influenced by monomer (except as monomer may affect the physical factors governing diffusion). The prevalence of initiation rates which are approximately equal to twice the rate of spontaneous decomposition of the initiator is to be expected on this basis. A process such as (b) would, of course, lower the efficiency, but the extent to which it occurs should not, in first approximation, be altered by dilution. The fact that efficiencies are usually near unity argues against the normal occurrence of this process. It may, however, be invoked to explain the rather surprising efficiency of only 0.5 to 0.6 for azo-bis-isobutyronitrile in methyl methacrylate.[22,25] More difficult to explain are the indications of a *decrease* in the efficiency of benzoyl peroxide with dilution of styrene, according to the deviation of the initial polymerization rate in solution from first-order kinetics (see Table X). No fully satisfactory explanation appears to have been offered.*

* Matheson's[46] widely accepted explanation in terms of the "cage effect" appears to be based on an unrealistic interpretation of the rates of the processes involved.

1d. Evaluation of Parameters.—The quantity $fk_dk_p^2/k_t$ may be calculated from the rate of polymerization and the concentrations of monomer and initiator, for according to Eq. (12)

$$fk_dk_p^2/k_t = R_p^2/[I][M]^2 \tag{14}$$

If the rate constant k_d for the spontaneous decomposition of the initiator is known, and its efficiency has been established, the important ratio k_p^2/k_t may then be evaluated. With greater generality, Eq. (8) may be combined with (12) to give

$$k_p^2/k_t = 2R_p^2/R_i[M]^2 \tag{15}$$

R_i may be evaluated from the rate of decomposition of the initiator supplemented by knowledge of the efficiency f, or from the rate of consumption of an inhibitor when added to the system under conditions otherwise identical with those giving the rate R_p in the absence

TABLE XI.—COMPARISON OF VALUES OF k_p^2/k_t

Monomer	Initiator[a]	Method for assigning initiation rate[b]	Temp. °C	$(k_p^2/k_t) \times 10^3$ l./mole/sec.	Ref.
Styrene	Bz₂O₂	MW	27	0.105	41[c]
Styrene	Azo	I	30	.125	41
Styrene	Azo	Inhib.	30	.115	41
Styrene	Bz₂O₂	MW	50	.39	41[c]
Styrene	Bz₂O₂	MW	60	1.19[d]	25, 41
Styrene	Azo	MW	60	1.18	27
Styrene	Bz₂O₂	I	60	0.95 (1.58)	25
Styrene	Azo	I	60	.76 (0.95)	41
Styrene	Azo	Inhib.	60	.74	41
Vinyl acetate	Azo	I	25	30.4	47
Vinyl acetate	Azo	Inhib.	25	34.8	47
Vinyl acetate	Photo.	Inhib.	25	31.3	48
Vinyl acetate	Azo	I	50	125	47
Vinyl acetate	Azo	Inhib.	50	119	47

[a] Initiators: Azo =azo-bis-isobutyronitrile, Bz₂O₂ =benzoyl peroxide, Photo =direct photochemical initiation.

[b] Determinations based on the rate of decomposition of the initiator assuming $f=1$ are indicated by I. (Values given in parentheses have been calculated using $f=0.60$ and 0.80 for styrene-Bz₂O₂ and for styrene-azo, respectively, as indicated by other work.[8,22]) Those from the inhibition method are indicated by Inhib., and those from analysis of molecular weights (Sec. 2e) by MW.

[c] Calculated by Matheson et al.[41] from the data of Schulz and Husemann.[38]

[d] The value given is derived in Section 2c from the results of Mayo, Gregg, and Matheson.[25] It differs slightly from that given by Matheson et al. (1.10×10^{-3}).[41]

of the inhibitor. Then k_p^2/k_t can be calculated from R_p and the monomer concentration. Results obtained by these two methods as applied to styrene and to vinyl acetate polymerized at several temperatures and with various initiators (including ultraviolet radiation in one instance) are compared in Table XI. The ratio k_p^2/k_t may also be derived from an analysis of the dependence of the degree of polymerization on the rate according to a method set forth in Section 2c, and several values obtained in this manner for styrene are included in Table XI. The efficiency has been taken equal to unity in applying the first method. The agreement obtained for vinyl acetate constitutes evidence supporting this assumption. The results for styrene suggest that the efficiency may be somewhat less than unity.

Expressing the rate constant according to the Arrhenius equation, e.g.

$$k_p = A_p e^{-E_p/RT} \tag{16}$$

where E_p and A_p are the energy of activation and the frequency factor, respectively, for chain propagation, and combining Eq. (16) with the corresponding expression for k_t, we have

$$k_p/k_t^{1/2} = (A_p/A_t^{1/2})e^{-(E_p - E_t/2)/RT} \tag{17}$$

By plotting $\log(k_p/k_t^{1/2})$ against $1/T$ in the customary fashion, $E_p - E_t/2$ and $A_p/A_t^{1/2}$ may be determined. The results for styrene treated in this manner yield[41]

$$E_p - E_t/2 = 6.5 \text{ kcal./mole}$$

and for vinyl acetate[47]

$$E_p - E_t/2 = 4.7 \text{ kcal./mole}$$

Independent estimates of these quantities can be obtained from the temperature coefficients of photopolymerization. If the rate of photoinitiation is assumed to be independent of the temperature, the increase in rate must be due entirely to the change in $k_p/k_t^{1/2}$ (see Eq. 13), hence the slope of the plot of $\log R_p$ against $1/T$ for the photochemical polymerization should yield $E_p - E_t/2$. Burnett[49] reported the value 5.5 kcal./mole for styrene, and Burnett and Melville[34] found 4.4 kcal./mole for vinyl acetate, in satisfactory agreement with the values given above.

The temperature coefficient of the rate of a polymerization induced by a thermally decomposing initiator must depend according to Eq. (12) both on the temperature coefficient of $k_p/k_t^{1/2}$ and on that of k_d. Upon substituting Arrhenius expressions for each of the rate constants occurring in this equation, we obtain

$$\ln(R_p/[I]^{1/2}M) = \ln[A_p(fA_d)^{1/2}/A_t^{1/2}] - [E_p + (E_d - E_t)/2]/RT \quad (18)$$

where f is assumed independent of temperature. Hence the apparent activation energy E_a obtained from the slope of a plot of the logarithm of the rate against $1/T$ will be related to the individual activation energies as follows

$$E_a = E_d/2 + (E_p - E_t/2) \quad (19)$$

The activation energy for the spontaneous decomposition of benzoyl peroxide is 30 (± 1) kcal. per mole,[8,9] and the same value applies also within experimental error to the azo nitrile.[9 13] The apparent activation energy for the polymerization of styrene initiated by either is about 22 kcal. per mole, therefore.

1e. The Course of the Conversion of Monomer to Polymer; Autoacceleration.—The concentration $[I]$ of the initiator usually does not change very much during the course of the polymerization. Hence, if the efficiency f is independent of the monomer concentration, first-order kinetics may be expected for the conversion of monomer to polymer. Experimentally, the polymerization of styrene in toluene solution in the presence of benzoyl peroxide proceeds approximately as a first-order reaction according to the results of Schulz and Husemann[38] shown in Fig. 9. The benzoyl-peroxide-initiated polymerization of d-sec-butyl α-chloroacrylate in dioxane and of vinyl-l-β-phenylbutyrate in dioxane[39] are quite accurately of the first order up to high conversions.

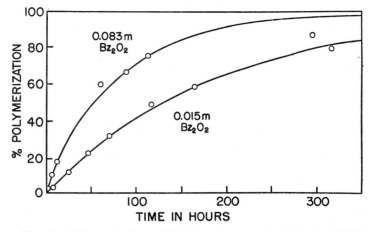

FIG. 9.—Polymerization of 40 percent styrene in toluene at 50°C in the presence of the amounts of benzoyl peroxide indicated. (Schulz and Husemann.[38])

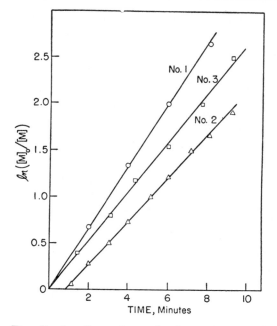

Fig. 10.—Benzoyl per-
oxide–initiated polymerization of vinyl-l-β-phenyl-
butyrate in dioxane at
60°C plotted as a first-
order reaction. $[M]_0$ and
$[M]$ represent concentrations of monomer initially
and at time t, respectively.
In experiments 1, 2, and
3, respectively, $[M]_0 = 2.4$,
7.28, and 5.97 g. of mon-
omer per 100 cc. of dioxane.
(Results of Marvel, Dec,
and Cooke[39] obtained polarimetrically.)

Results for the polymerization of the latter of these monomers are
shown in Fig. 10.

Polymerization of an undiluted monomer or of a monomer in concentrated solution yields results of a quite different character.[50] Shown
in Fig. 11 are the polymerization curves obtained by Schulz and Harborth[37] for methyl methacrylate, undiluted and at various dilutions in
benzene, with benzoyl peroxide as initiator. Volume changes measured in a dilatometer were used to follow the course of the process.
This widely used method is capable of high accuracy owing to the
marked decrease in volume which accompanies the polymerization of
vinyl monomers. For all monomer concentrations up to 40 percent,
the curves showing the extent of reaction plotted against the time are
roughly independent of the initial concentration, as is required for a
first-order process. (An order slightly greater than the first is indicated both by the shapes of the curves and by their displacement with
concentration.) At higher initial monomer concentrations a marked
acceleration in rate is observed to occur at an advanced stage of the
polymerization. The conversion at which this pronounced deviation
from the first-order course sets in corresponds in each case to a polymer
concentration of about 25 percent. Simultaneously, the average
molecular weight of the polymer being formed was observed to increase
sharply, although this increase was not proportionate to the accelera-

tion of the rate (owing to the limiting effect of chain transfer on the degree of polymerization; see Sec. 2c). The autoacceleration may lead to a large rise in temperature[31] if, under the conditions used, dissipation of the heat of polymerization is inadequate. The anomaly is not eliminated, however, when isothermal conditions are maintained.[37,50] The temperature rise was negligible in the experiments of Schulz and Harborth[37] shown in Fig. 11, for example.

Autoacceleration in the rate of polymerization occurs also with other monomers.[51] It is far more marked with methyl acrylate or acrylic

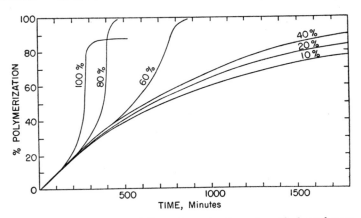

FIG. 11.—The course of the polymerization of methyl methacrylate at 50°C in the presence of benzoyl peroxide at various concentrations of monomer in benzene. (Schulz and Harborth.[37])

acid. It sets in before a conversion of 1 percent has been reached in the polymerization of the former without diluent.[51] It may easily end in explosion with either. Autoacceleration occurs also with styrene and vinyl acetate, but to a lesser degree than with methyl acrylate or methyl methacrylate. It is probably responsible, however, for the abnormally low apparent kinetic orders observed for the conversion of the undiluted monomers, styrene and vinyl acetate, to polymers.[52]

In quest of an explanation for this phenomenon, one is led to conclude either that the combination of constants occurring in the rate equation (12) must undergo a large increase when autoacceleration occurs or that a totally different mechanism of polymerization must take over. We should obviously prefer the former alternative if it will lead to a satisfactory explanation of the facts. An increase in $k_d f$ seems unlikely; autoacceleration is not a function of the initiator. This leaves us with the ratio $k_p/k_t^{1/2}$, which will be required to increase by as much as a hundredfold. An increase in the propagation rate constant k_p or a decrease in the termination constant (or both) would account for the phenomenon.

Norrish and Smith,[50] Trommsdorff,[53] and, later, Schulz and Harborth[37] concluded that the autoacceleration in the rate of polymerization of undiluted methyl methacrylate and the simultaneous increase in average molecular weight of the polymer being formed can only be explained by the assumption of a decrease in the termination rate. They attributed this decrease in k_t to the high viscosity of the medium when the concentration of previously formed polymer molecules is large. Chain termination involving combination of two free radicals is a very fast reaction which occurs at an extraordinarily large fraction of the collisions (about one in 10^4; see below) between radicals. When the growing polymer molecules bearing free radical end groups are embedded in the highly viscous mixture consisting of a large proportion of polymer molecules, the rate of termination may be controlled by the rate of diffusion, as discussed in the previous chapter. Although the intrinsic reactivity of the radical presumably has not suffered, its opportunities for reaction with another radical are diminished, and this is manifested by a reduction in k_t. The concentration of active radicals (see Eq. 10) will therefore rise, and the rate of consumption of monomer must increase proportionately (Eq. 11). The mobilities of the polymer molecules, or rather of the chain ends bearing free radicals, are ample to maintain the normal rate of termination only in the presence of a sufficiency of a diluent, or of monomer. The propagation process, on the other hand, should be insensitive to the viscosity of the medium for two reasons. In the first place, its rate constant is much lower, i.e., the fraction of collisions between radical and monomer which are fruitful is only of the order of 10^{-9}; hence maintenance of an equilibrium population of monomers near each radical places a lesser demand on diffusion processes. Secondly, one of the reactants is a monomer molecule the movements of which are not seriously impeded by the long polymer chains in the mixture.

It will be observed from Fig. 11, however, that the autoacceleration phase ends appreciably short of complete conversion of monomer to polymer and that the degree of conversion at the upper asymptote of the curve is higher the greater the dilution. Pure poly-(methyl methacrylate) is a glass below about 90°C (see Table II, p. 52). At any lower temperature, the mixture of polymer and monomer (or of polymer, monomer, and diluent) transforms to a glass when, in the course of the polymerization, the percentage of polymer reaches a certain rather well-defined (high) value. Even the monomer molecules are frozen in their positions in the glassy state. The virtual cessation of polymerization short of completion for monomers such as styrene or methyl methacrylate at temperatures below the glass transition temperatures T_g of their polymers (about 100°C for polystyrene) occurs, therefore as a result of transformation of the reaction mixture to an immobile glass.

The above explanation of autoacceleration phenomena is supported by the manifold increase in the initial polymerization rate for methyl methacrylate which may be brought about by the addition of poly-(methyl methacrylate) or other polymers to the monomer.[53,54] It finds further support in the suppression, or virtual elimination, of autoacceleration which has been observed when the molecular weight of the polymer is reduced by incorporating a chain transfer agent (see Sec. 2f), such as butyl mercaptan, with the monomer.[51] Not only are the much shorter radical chains intrinsically more mobile, but the lower molecular weight of the polymer formed results in a viscosity at a given conversion which is lower by as much as several orders of magnitude. Both factors facilitate diffusion of the active centers and, hence, tend to eliminate the autoacceleration. Final and conclusive proof of the correctness of this explanation comes from measurements of the absolute values of individual rate constants (see p. 160),[51,55] which show that the termination constant does indeed decrease a hundredfold or more in the autoacceleration phase of the polymerization, whereas k_p remains constant within experimental error.

The susceptibility of the polymerization of a given monomer to autoacceleration seems to depend primarily on the size of the polymer molecules produced.[51] The high propagation and low termination constants for methyl acrylate as compared to those for other common monomers lead to an unusually large average degree of polymerization ($> 10^4$), and this fact alone seems to account for the incidence of the decrease in k_t at very low conversions in this case.

Ramifications of the phenomenon of autoacceleration in polymerization are of much more than passing importance. A polymer which possesses only a very limited degree of mobility, as for example a space network polymer formed through cross-linking reactions, may under certain conditions acquire the capacity to induce the polymerization of monomer brought into contact with it, and this capacity may be retained for a considerable period of time. Proliferous polymerization of this sort is not uncommon among the dienes such as chloroprene,[56] particularly when the polymerization is induced with ultraviolet radiation. Melville[57] has found that methyl methacrylate photopolymerized with radiation of about 2600Å will cause additional quantities of monomer brought into contact with it to polymerize without the assistance of additional radiation or initiators. The potential activity is retained for as long as 2 days. Evidently some of the active centers (i.e., radicals) produced during proliferous polymerization enjoy extremely long life expectancies owing to the hindrance to their bimolecular self-destruction offered by the structures of which they are a part. This is an extreme case of suppressed mobility.

1f. Kinetics of Thermal Polymerization.—Attempts to measure rates of polymerization in the absence of added initiators often have yielded conflicting results on account of the sensitivity of the reaction to traces of materials which may accelerate or retard it. In the case of styrene, however, a purely thermal polymerization has been demonstrated.[58] Reproducible rates devoid of spurious induction periods are obtained provided the monomer is purified with extreme care and the polymerization is conducted with rigorous exclusion of oxygen, which has been shown to be an inhibitor while the peroxides it produces may act as initiators.[59] Rates determined in different laboratories stand in reasonably good agreement.[58,60,61] Walling and Briggs[62] succeeded in obtaining reproducible thermal polymerization of methyl methacrylate in the temperature range from 100 to 150°C after adopting an elaborate purification procedure supplemented by the incorporation of a small amount of hydroquinone, which acts as an antioxidant without retardation of polymerization. The rate observed was about one-fiftieth of that for styrene at the same temperature. Bamford and Dewar[63] attributed the low thermal rate they observed for methyl methacrylate to an inhibitor spontaneously produced in methyl methacrylate, but Mackay and Melville[43,64] failed to confirm this. The reality of a strictly thermal polymerization with these two monomers seems well established, although the value of the thermal rate for the latter of them may be in doubt. Carefully purified vinyl acetate[65] and vinyl chloride[66] do not polymerize detectably at 100°C.

Initial rates for the thermal polymerization of styrene in various

TABLE XII.—THERMAL POLYMERIZATION OF STYRENE IN TOLUENE AT 100°C

(Schulz, Dinglinger, and Husemann[60])

$[M]$ mole/l.	Initial rate $R_p \times 10^7$ mole/l./sec.	$(R_p/[M]^2) \times 10^7$ l./mole/sec.
5.82	224	6.6
3.88	93.5	6.2
1.94	21.5	5.7
0.97	4.9	5.2
.605	1.9	5.2

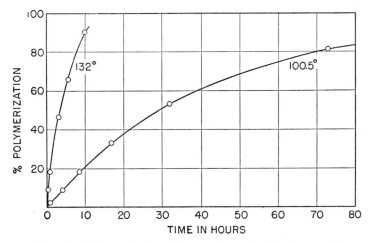

FIG. 12.—Polymerization of pure styrene in the absence of
oxygen at 100.5° and 132°C. (Schulz and Husemann.[67])

solvents[60,61,67] are approximately proportional to the square of the
monomer concentration. A typical set of results is shown in Table
XII. Moreover, the rate is not much influenced by the nature of the
solvent (provided the solution remains homogeneous). The polymerization of undiluted styrene, on the other hand, follows a course
which seems to be between zero and first order[58,67] according to data
such as are shown in Fig. 12. This discrepancy between bulk and
solution polymerization doubtless is a manifestation of incipient autoacceleration. The depression of the termination rate caused by the
diminished mobility at higher conversions in bulk polymerization may
be presumed to affect the apparent kinetic order. Autoacceleration
is less pronounced in thermal polymerization, probably because of the
increased mobility at the higher temperatures required.

In the search for a plausible mechanism for initiation in thermal
polymerization, it is necessary to reject unimolecular processes such
as the opening of the double bond to form a monomeric diradical

$$CH_2{=}CHX \ \rightarrow \ \cdot CH_2{-}\overset{\displaystyle X}{\overset{\displaystyle |}{C}}H \cdot \quad\quad (20)$$

in consideration of the large energy required—about 50 kcal. Dissociation to a pair of radicals, or to a radical and a hydrogen atom, would
require even greater energy. The bimolecular analog of reaction (20)

$$2CH_2{=}CHX \ \rightarrow \ \cdot \overset{\displaystyle X}{\overset{\displaystyle |}{C}}H{-}CH_2{-}CH_2{-}\overset{\displaystyle X}{\overset{\displaystyle |}{C}}H \cdot \quad\quad (21)$$

is a much more attractive possibility from this point of view, for it should be endothermic to the extent of only 20 to 30 kcal.[68] Furthermore, it leads to satisfactory over-all kinetics. Thus the rate of initiation should be of the second order.

$$R_i = 2k_i[M]^2$$

where k_i is the velocity constant for reaction (21). If the rate of termination is assumed to be bimolecular, as was previously found for monoradical-initiated polymerizations, then it follows from Eq. (9) and the steady-state condition $R_i = R_t$ that

$$[M\cdot] = (k_i/k_t)^{1/2}[M] \tag{22}$$

The rate of polymerization consequently is given by

$$R_p = k_p[M\cdot][M] = k_p(k_i/k_t)^{1/2}[M]^2 \tag{23}$$

The proposed mechanism leads to a second-order rate for thermal conversion of monomer in agreement with experiment.

The above bimolecular diradical mechanism unfortunately is open to objections which weigh heavily against it. In the first place, a diradical must almost inevitably cyclize at some early stage in its growth. It should be most vulnerable to cyclization at the trimer stage. Even if it survived to grow larger than a trimer, statistical calculations[69] taking into consideration the relative rates of chain growth and termination show that the probability of growth of a diradical to a long chain polymer is fantastically small. Instead, rings consisting of comparatively small numbers of units should be formed by intramolecular coupling.[69] (Termination, incidentally, would then enter as a first-order process, and this would *raise* the over-all order for the polymerization.) It seems necessary, therefore, to reject a diradical initiation mechanism and to search for a process by which a pair of monoradicals could be formed, while at the same time preserving the second-order over-all kinetics. None has yet been found. Hence, in applying the rate equation (23) to thermal polymerizations, we are obliged to admit that the nature of the process to which k_i applies is unknown.*

From measured rates of thermal polymerization, and the previously evaluated ratios k_p^2/k_t, we may assign values to k_i. Thus from the data given in Table XI extrapolated to 100°C, $k_p^2/k_t \cong 8.5 \times 10^{-3}$ for styrene

---

* F. R. Mayo (private communication) has found evidence that thermal polymerization of styrene may actually be of a higher order than second, i.e., about five-halves order. This would suggest a termolecular initiation step. Generation of a pair of monoradicals in this manner, i.e., from three monomer molecules, would be acceptable from the standpoint of energy considerations.

at this temperature. Taking $R_p/[M]^2 = k_p(k_i/k_t)^{1/2} = 6 \times 10^{-7}$ according to the data shown in Table XII, we obtain $k_i = 4.2 \times 10^{-11}$ liter/mole/sec. The thermal polymerization of methyl methacrylate is much slower; the initial rate in pure monomer at 100°C is only 6.2×10^{-7} |mole/liter/sec. (compare with Table XII).[62] From the separately evaluated k_p^2/k_t for this monomer, the rate constant for thermal initiation at 100°C is found to be 0.35×10^{-15},[62] which is much smaller than that for styrene.* This rate of thermal generation of radicals in methyl methacrylate is extremely small.

The apparent activation energy obtained from the slope of a plot of the logarithm of the rate R_p of polymerization against $1/T$ will be related to the individual activation energies, in analogy with Eq. (19), as follows:

$$E_a = E_p + (E_i - E_t)/2 \qquad (24)$$

Results for styrene[61,67] yield $E_a = 21$ kcal. Since $E_p - E_t/2$ was found previously to be 6.5 kcal., we conclude that the activation energy E_i for thermal initiation in styrene is 29 kcal., which would be quite acceptable for the process (21), already rejected on other grounds. For methyl methacrylate,[62] $E_a = 16$ kcal. and $E_p - E_t/2 = 5$ kcal. Hence $E_i = 22$ kcal. These initiation reactions are very much slower than is normal for other reactions with similar activation energies. The extraordinarily low frequency factors A_i apparently are responsible. For methyl methacrylate, A_i is less than unity.[62] Interpreted as a bimolecular process, this would imply initiation at only one collision in about 10^{13} of those occurring with the requisite energy!

2. CHAIN LENGTHS

2a. Kinetic Chain Length and Degree of Polymerization.—The kinetic chain length ν, representing the average number of monomers reacting with a given active center from its initiation to its termination, will be given by the ratio of the rate of propagation R_p to the rate of initiation R_i. Or, since R_i must equal the rate of termination R_t under steady-state conditions

$$\nu = R_p/R_i = R_p/R_t \qquad (25)$$

* A great deal of confusion exists in the definitions chosen by different authors (and by the same authors on different occasions) for the rate constants for initiation and termination. The factor 2 expressing the fact that two radicals are created or destroyed in the respective processes is sometimes incorporated in the rate constants. Here we have consistently taken k_d, k_i, and k_t to represent the rate constants for the *reactions* as ordinarily written, hence the 2 is *not* included in the rate constant. Results expressed otherwise have been converted to this basis.

Then from Eqs. (9) and (11)

$$\nu = (k_p/2k_t)[M]/[M\cdot] \tag{26}$$

Eliminating the radical concentration using Eq. (11)

$$\nu = (k_p^2/2k_t)[M]^2/R_p \tag{27}$$

Eqs. (26) and (27) apply irrespective of the nature of the initiation process; it is required merely that the propagation and termination processes be of the second order. They emphasize the very general inverse dependence of the kinetic chain length on the radical concentration and therefore on the rate of polymerization. The kinetic chain length may be calculated from the ratios k_p^2/k_t as given in Table XI and the rate of polymerization. Thus, for pure styrene at 60°C

$$\nu \cong 0.030/R_p$$

and for pure vinyl acetate at 50°C.

$$\nu \cong 5.5/R_p$$

At the same rate of polymerization, the kinetic chain length for vinyl acetate is over a hundred times that for styrene on account of the greater speed of propagation relative to termination for vinyl acetate. At a convenient rate of 10^{-4} moles/liter/sec., for example, each radical chain generated consumes on the average about 5×10^4 vinyl acetate units under the conditions stated.

Although we shall prefer the general equation (27) in later discussion, the expressions obtained specifically for initiated and for thermal polymerizations are of interest also. In the former case, the radical concentration is given by Eq. (10); hence from Eq. (26)

$$\nu = [k_p/2(fk_dk_t)^{1/2}][M]/[I]^{1/2} \tag{28}$$

For a polymerization with second-order initiation as has been indicated for thermal initiation, it follows from Eqs. (22) and (26) that

$$\nu = k_p/2(k_ik_t)^{1/2} \tag{29}$$

i.e., the kinetic chain length should be independent of the concentration in this case.

If reactions other than those considered so far did not take place to an appreciable extent, the average degree of polymerization would be directly related to the kinetic chain length ν. Assuming termination by combination of radicals, as is indicated by experiments previously cited, each polymer molecule formed in a monoradical-initiated polymerization would consist of two kinetic chains grown from two other-

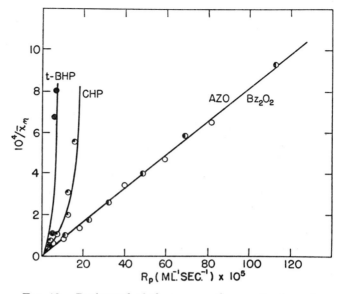

Fig. 13.—Reciprocal of the average degree of polymerization $\bar{x}_n$ plotted against the rate of polymerization R_p at 60°C for undiluted methyl methacrylate using azo-bis-isobutyronitrile (Azo), benzoyl peroxide (Bz_2O_2), cumene hydroperoxide (CHP), and t-butyl hydroperoxide (t-BPH). (Baysal and Tobolsky.[27])

wise unrelated primary radicals, and the number average degree of polymerization $\bar{x}_n$ should equal 2ν.* At intermediate rates of polymerization of undiluted styrene with benzoyl peroxide, $\bar{x}_n$ is found to be approximately equal to 2ν, calculated according to Eqs. (27) or (28), but at very low or very high initiator concentrations large deviations appear.[25] Degrees of polymerization of methyl methacrylate (undiluted monomer) are very nearly equal to 2ν for certain initiators but not for others.[27] The degrees of polymerization obtained for vinyl acetate are much less than twice the large kinetic chain lengths for this monomer.[47]

If the degree of polymerization were directly proportional to the kinetic chain length, then according to Eq. (27) it should increase with $[M]^2$ and decrease inversely with the rate R_p. For the purpose of

* If termination were to occur in part by disproportionation, $\bar{x}_n$ would still be proportional to ν but the constant of proportionality would lie between one and two, its exact value depending on the fraction of termination which occurs by combination. The assumption that the reactions considered above suffice for the interpretation not only of the polymerization rate but also of the degree of polymerization requires in any case that $\bar{x}_n$ be proportional to ν.

illustrating the observed dependence of $\bar{x}_n$ for methyl methacrylate[27] on the rate at the constant monomer concentration of the undiluted monomer, reciprocal degrees of polymerization (derived from intrinsic viscosity measurements calibrated against osmotic molecular weights; see Chap. VII) are plotted in Fig. 13 against the rate of polymerization at 60°C. Individual points represent results for different concentrations of the various initiators. The polymerizations were carried to low conversions so that the rate and concentrations remained substantially at their initial values. For two of the initiators used, benzoyl peroxide and the azo nitrile, a direct proportionality is observed, which indicates that $\bar{x}_n$ is indeed proportional to ν of Eq. (27). At the same polymerization rates with the hydroperoxides as initiators, $1/\bar{x}_n$ is much greater, and $\bar{x}_n$ is not proportional to ν. The quantity $1/\bar{x}_n$ is a measure of the number of polymer molecules; more exactly, it is the number of polymer molecules per monomer unit polymerized. It is apparent, then, that in the presence of the hydroperoxides additional processes occur which increase the number of molecules formed.

Reciprocal degrees of polymerization for the polystyrenes obtained

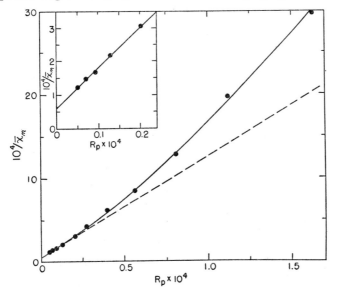

FIG. 14.—Reciprocal initial degree of polymerization plotted against the initial rate for undiluted styrene at 60°C using benzoyl peroxide. The dashed straight line corresponds to the line drawn through the points for low concentrations of initiator shown in the inset using an enlarged scale. (From the results of Mayo, Gregg, and Matheson.[25])

by Mayo, Gregg, and Matheson[25] in their thorough study of the benzoyl-peroxide-initiated polymerization over an exceptionally wide range in initiator concentration are shown in Fig. 14. (The rates for the same polymerizations are those shown in Fig. 8, line 2, as a function of the initiator concentration.) Here two deviations from Eq. (27) are displayed: in the first place, $1/\bar{x}_n$ does not extrapolate to zero at zero rate, which indicates a molecule-forming process independent of the rate; and, secondly, $1/\bar{x}_n$ increases at higher rates more rapidly than in direct proportion to R_p.

More glaring discrepancies appear when the monomer concentration is varied by the addition of a diluent. The degree of polymerization in second-order thermal polymerization of styrene, instead of being unaffected by dilution in accordance with Eq. (29), invariably decreases with dilution.[60,61] The magnitude of the decrease depends very much on the particular solvent (see Sec. 2d). The average molecular weights of poly-(vinyl acetate)s prepared in benzene solutions using benzoyl peroxide were found by Kamenskaya and Medvedev[32] to decrease more rapidly with dilution than Eq. (28) would predict. The use of certain other solvents gave polymers of lower molecular weight.

2b. Kinetics of Chain Transfer.—The results which have been cited reveal a lack of correlation between $\bar{x}_n$ and 2ν which is the rule rather than the exception. The former frequently is less than the latter, and the discrepancy may be large; particularly is this true for polymerization in a solvent. In other words, the number of moles of polymer generated in the polymerization of a mole of monomer tends to exceed $1/2\nu$—sometimes by a wide margin. Processes evidently occur which increase the number of polymer molecules produced beyond the one to be expected per pair of initiating radicals. Whatever the nature of these processes may be, they evidently do not affect the rate appreciably, because the discordance between $\bar{x}_n$ and 2ν is not in any way reflected in deviations of the rate from simple chain kinetics. A reaction in which the free radical center is somehow transferred from a growing chain to another molecule (e.g., solvent or monomer) would explain these observations.[68] The growth of the chain previously bearing the free radical would thereby be terminated, and the molecule acquiring the radical should be capable of starting a new chain, which would grow at the same rate. A prominent mechanism for *chain transfer* reactions of this nature consists in removal by the chain radical of a hydrogen atom from the molecule which intervenes, i.e., the *transfer agent*. The chain molecule loses its capacity for further growth, and the transfer agent acquires radical character in consequence of the transfer process. Thus the solvent, represented by SH, may react with the growing radical as follows:

$$M_x\cdot + SH \xrightarrow{k_{tr,S}} M_xH + S\cdot \qquad (30)$$

and the solvent radical $S\cdot$, presumed to be sufficiently reactive to add a monomer,* proceeds to grow a new chain.   A halogen atom if present, or possibly a labile group, may be similarly transferred instead of a hydrogen atom, with equivalent results.   The initiator may also function as a transfer agent.   This may take place in the case of benzoyl peroxide according to the following mechanism in which the peroxide undergoes a radical-induced decomposition:

$$M_x\cdot + (C_6H_5COO)_2 \xrightarrow{k_{tr,I}} M_x{-}OCOC_6H_5 + C_6H_5COO\cdot \qquad (31)$$

The monomer must also be considered a potential transfer agent.   It may perform the function of the solvent in reaction (30), an atom being transferred from the monomer to saturate the radical.   Possibly a proton may be transferred from the β carbon atom of the radical chain to the unsaturated monomer as follows:

$$\begin{matrix} X & & X & & & X \\ | & & | & & & | \\ {-}CH_2{-}CH\cdot + CH_2{=}CH & \rightarrow & {-}CH{=}CHX + CH_3CH\cdot \end{matrix} \qquad (32)$$

Chain transfer may also occur with a previously formed polymer molecule.   The latter type of transfer produces no net increase in the number of molecules; hence it need not concern us here.   Its effect on the polymer structure will be considered in Chapter VI.

The total number of (linear) polymer molecules may be equated to the number of pairs of chain ends, and the average degree of polymerization is the ratio of the total number of monomers polymerized to this number of pairs.   The average degree of polymerization of the polymer formed in any small time interval will be given by the ratio of the rate of polymerization R_p to the rate of production of pairs of chain ends.   Each incidence of one of the preceding chain transfer steps produces a pair of chain ends and, hence, increases by one the number of polymer molecules.   On the assumption that termination occurs exclusively by coupling, the only other source of chain ends is at the initiation step.   If, then, all of the *kinetic* chains start from monoradicals released by an initiator, the reciprocal of the degree of polymerization should be given by

---

* If it is not so reactive, SH is an inhibitor or retarder (see Sec. 4) rather than a transfer agent.

$$1/\bar{x}_n = \{fk_d[I] + k_{tr,M}[M][M\cdot] + k_{tr,S}[S][M\cdot] \\ + k_{tr,I}[I][M\cdot]\}/R_p \quad (33)$$

where $k_{tr,M}$, $k_{tr,S}$, and $k_{tr,I}$, are the velocity constants of chain transfer with monomer, solvent, and initiator, respectively. To the extent that chain termination may occur by disproportionation, the additional term $k_{td}[M\cdot]^2$ should be included in the numerator. Thermal diradical initiation would not contribute to the number of pairs of chain ends, but a monoradical initiation process not dependent on the initiator would require an additional term in Eq. (33). The radical concentrations occuring in Eq. (33) may be eliminated by use of Eq. (11), and the initiator concentration may be eliminated in favor of the rate of polymerization by Eq. (12). Then, with further substitutions as indicated below,

$$1/\bar{x}_n = C_M + C_S[S]/[M] + (k_t/k_p^2)R_p/[M]^2 \\ + C_I(k_t/k_p^2 fk_d)R_p^2/[M]^3 \quad (34)$$

where the *transfer constants* are defined by

$$C_M = k_{tr,M}/k_p; \quad C_S = k_{tr,S}/k_p; \quad C_I = k_{tr,I}/k_p \quad (35)$$

2c. Chain Transfer in Undiluted Monomer; Evaluation of Parameters.—In the absence of solvent the second term vanishes, leaving

$$1/\bar{x}_n = C_M + (k_t/k_p^2)R_p/[M]^2 + C_I(k_t/k_p^2 fk_d)R_p^2/[M]^3 \quad (36)$$

This quadratic in R_p is of the form required by the data for styrene-benzoyl peroxide shown in Fig. 14. The first term, corresponding to the intercept, represents the creation of chain ends through transfer with monomer. It occurs to an extent which is independent of the polymerization rate. The second term corresponds to $1/2\nu$ according to Eq. (27); it represents the pairs of ends created at the initiation step.* Its coefficient is given by the initial slope of the line in Fig. 14. The third term, which accounts for the curvature at higher rates, represents the contribution of chain transfer with benzoyl peroxide. This becomes more prominent at higher rates because of the larger amounts of the initiator which are present. The marked rise in the curves for

---

* Thermal initiation makes an appreciable contribution to the polymerization rate for styrene at very low initiator concentrations, as we have pointed out earlier. Since the rate R_p includes contributions from thermal as well as from "catalytic" initiation, the second term in Eq. (36) remains valid *provided the thermal initiation involves monoradicals.* Diradical initiation, if it occurred, would introduce a deviation, since it produces no chain ends.

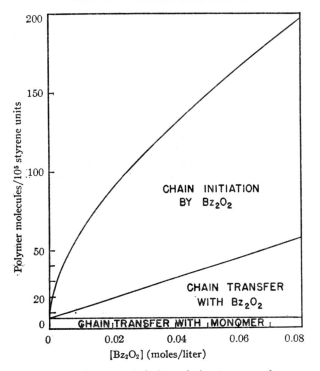

Fig. 15.—Sources of chain ends in styrene polymerization with benzoyl peroxide at 60°C.   (Mayo, Gregg, and Matheson.[25])

the hydroperoxides in Fig. 13 is likewise due to transfer of the methyl methacrylate chains with these initiators.[27]   Chain transfer with the azo nitrile and with benzoyl peroxide is negligible for this monomer.

Over the range in which the rate of polymerization is proportional to the square root of the initiator concentration, R_p may be replaced in Eq. (36) with $[I]^{1/2}$, the coefficients of the terms being appropriately altered.   The contributions of the various sources of chain ends in the polymerization of styrene with benzoyl peroxide at 60°C are shown in Fig. 15 as functions of the initiator concentration.[25]   The uppermost curve represents the total number of polymer molecules per unit, and the differences between successive curves represent the contributions of the separate processes indicated.

The initial linear portion of the curve shown in Fig. 14 (see inset) is given by the equation

$$1/\bar{x}_n = 0.60 \times 10^{-4} + 12.0\ R_p$$

Hence, by comparison with Eq. (36), evidently $C_M = 0.60 \times 10^{-4}$, i.e., chain radicals at 60°C prefer to add monomer rather than undergo transfer with monomer by the factor $(1/0.60) \times 10^4$. Knowing the value of C_M, one may replot the quantity $(1/\bar{x}_n - C_M)/R_p$ against R_p. The slope of the resulting linear plot is equal to the coefficient of the last term in Eq. (36). In this way, the complete expression

$$1/\bar{x}_n = 0.60 \times 10^{-4} + 12.0\, R_p + 4.2 \times 10^4\, R_p^2$$

is obtained for the polymerization of pure styrene with benzoyl peroxide at 60°C, the rate being expressed in moles/liter/sec. If the monomer concentration (8.35 molar in pure styrene) is inserted in the above expression, the more general relationship

$$1/\bar{x}_n = 0.60 \times 10^{-4} + 8.4 \times 10^2\, R_p/[M]^2 + 2.4 \times 10^7\, R_p^2/[M]^3 \qquad (37)$$

is obtained for styrene-benzoyl peroxide at 60°C. Then by comparison with Eq. (36)

$$k_t/k_p^2 = 840 \qquad (38)$$

and

$$C_I(k_t/k_p^2 f k_d) = 2.4 \times 10^7 \qquad (39)$$

in units of moles, liters, and seconds. Thus, measurements of initial degrees of polymerization over a wide range in rates at constant temperature permit an evaluation of k_p^2/k_t which is independent of any assumptions regarding either the efficiency of initiation or the efficiency of termination by an inhibitor. The result expressed in Eq. (38) has been included in Table XI (p. 122) as an example of the "molecular weight" method for determining this ratio.

The transfer constant C_I may be isolated from Eq. (39) by utilizing the rate of polymerization and Eq. (12). According to the result of Mayo, Gregg, and Matheson[25] on the polymerization of styrene with benzoyl peroxide at 60°C (see Fig. 8)

$$R_p = 4.0 \times 10^{-4}\, [I]^{1/2}$$

Hence from Eq. (12) with $[M] = 8.35$ moles/liter

$$f k_d k_p^2/k_t = R_p^2/[I][M]^2 = 2.29 \times 10^{-9} \qquad (40)$$

Multiplying Eq. (39) by (40)

$$C_I = 0.055$$

i.e., styrene chain radicals add styrene monomers about eighteen times as rapidly as they would transfer with benzoyl peroxide if its concentration were equal to that of the monomer. From Eqs. (38) and (40)

$$fk_d = 1.92 \times 10^{-6} \qquad (41)$$

If the rate constant k_d for spontaneous decomposition of the initiator is known, the efficiency f of initiation may be determined. (This is a refinement of the molecular weight method set forth in Sec. 1c) The spontaneous decomposition rate constant for benzoyl peroxide in styrene, according to Swain, Stockmayer, and Clarke[8] is 3.2×10^{-6} sec.$^{-1}$ at 60°C.* Hence the efficiency of initiation of the polymerization of styrene by benzoyl peroxide at 60°C is indicated to be about 0.60.

2d. Chain Transfer with Solvents.—In the presence of a solvent or any added substance with which the radical chains may transfer, the second term of the general equation (34) usually makes a major contribution to the total number of molecules present. The need for simplification in order to extricate this term from the other three is apparent. By keeping the initiator concentration low or, preferably, through the choice of an initiator such as an aliphatic azo nitrile which is not itself susceptible to chain transfer, the last term in Eq. (34) is rendered negligible. The third term on the right-hand side may be kept constant by so adjusting the initiator concentration as to keep $R_p/[M]^2$ constant while the concentration $[S]$ of the solvent is varied. In the case of a thermal polymerization following second-order kinetics, $R_p/[M]^2$ is constant with dilution and no adjustment is needed. Subject to the conditions described, whereby the fourth term in Eq. (34) can be neglected and the third term is held constant, we may write, therefore

$$1/\bar{x}_n = (1/\bar{x}_n)_0 + C_S[S]/[M] \qquad (42)$$

where $(1/\bar{x}_n)_0$, the reciprocal degree of polymerization in the absence of a solvent (or a regulator added in small quantity), represents the first and third terms of Eq. (34).

Reciprocal degrees of polymerization of polystyrenes prepared by thermal polymerization[70] at 100°C in hydrocarbon solvents are plotted against $[S]/[M]$ in Fig. 16. Conversions were sufficiently low to permit the assumption of constancy in this ratio, which is taken equal to its initial value. The linearity of plots such as these, including those for numerous other monomer-solvent pairs which have been investigated,[70,73] affords the best confirmation for the widespread occurrence of chain transfer and for the bimolecular mechanisms assumed. It is

---

* Bawn and Mellish[9] found the similar value, 2.8×10^{-6} sec.$^{-1}$, for k_d for the decomposition of benzoyl peroxide in toluene at 60°C.

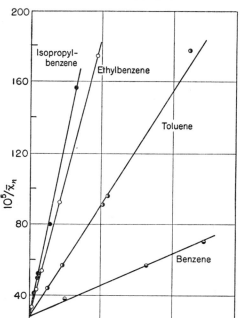

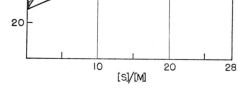

FIG. 16.—Effects of several hydrocarbon solvents on the degree of polymerization of styrene at 100°C. (Gregg and Mayo.[70])

significant that the solvent has little specific influence on the rate,[70,74,75] in contrast to its effect on the degree of polymerization; that is, the rates at equal concentrations in different solvents are about the same, but this is by no means true of the degrees of polymerization, as is clear from Fig. 16.

Transfer constants for polystyrene chain radicals at 60° and 100°C, obtained from the slopes of these plots and others like them, are given in the second and third columns of Table XIII. Almost any solvent is susceptible to attack by the propagating free radical. Even cyclohexane and benzene enter into chain transfer, although to a comparatively small extent only. The specific reaction rate at 100°C for transfer with either of these solvents is less than two ten-thousandths of the rate for the addition of the chain radical to styrene monomer. A fifteenfold dilution with benzene was required to halve the molecular weight, i.e., to double $1/\bar{x}_n$ from its value $(1/\bar{x}_n)_0$ for pure styrene (see Fig. 16).* Other hydrocarbons are more effective in lowering the degree of polymerization through chain transfer.

In all of the examples included in Table XIII, with the probable exceptions of benzene and chlorobenzene,† chain transfer evidently takes

* According to the experiments shown $(1/\bar{x}_n)_0 \cong 2.5 \times 10^{-4}$ at 100°C, which is somewhat larger than the value of C_M for pure styrene indicated by other experiments at this temperature. The presence in the monomer of impurities which cause chain transfer would account for this discrepancy.

† Mayo[71] has suggested that the attacking radical may add to the benzene nucleus as follows:

TABLE XIII.—TRANSFER CONSTANTS FOR VARIOUS SUBSTANCES WITH STYRENE[a]

Transfer agent	Transfer constants $C_S \times 10^4$		$E_{tr,s} - E_p$ in kcal.[b]	$\text{Log}(A_{tr,s}/A_p)$[b]	Reference
	At 60°C	At 100°C			
Cyclohexane	0.024	0.16	13	3	70
Benzene	.018	.184	15	4	70
Toluene	.125	.65	10	2	70
Ethylbenzene	.67	1.62	6	−0.5	70
iso-Propylbenzene	.82	2.0	6	−0.5	70
tert-Butylbenzene	.06	0.55	14	4	70
Triphenylmethane	3.5	8.0	5	0	70
Chlorobenzene		ca. 0.5			71
n-Butyl chloride	0.04	.37	14	4	75
n-Butyl bromide	.06	.35	11	2	75
n-Butyl iodide	1.85	5.5	7	1	75
Methylene chloride	0.15				75
Chloroform	.5				75
Ethylene dichloride	.32		15	5	76
Ethylene dibromide		6.6	10	2.5	76
Tetrachloroethane		18			76
Carbon tetrachloride	90	180	5	1	74
Carbon tetrabromide	13,600	23,500	3	2	75
Pentaphenylethane	20,000				70

[a] For further data on chain transfer between styrene and a wider variety of substances, see Gregg and Mayo.[75]

[b] The activation energy differences and frequency factor ratios have been calculated from measurements of C_S at several temperatures and in most cases not merely from the 60° and 100°C results shown here.

place through removal of an atom (H or Cl) from the solvent after the manner indicated by reaction (30). Among aromatic hydrocarbons, those having benzylic hydrogens are most reactive. The reactivity increases with increasing substitution at the α-carbon (compare toluene, ethylbenzene, and isopropylbenzene), the effect of a phenyl substituent being much greater than that of a methyl. This is in accordance with

$$\text{M}\cdot + \text{C}_6\text{H}_6 \rightarrow \begin{matrix} \text{M} \\ \text{H} \end{matrix} \text{(including resonance structures)}$$

when more reactive benzylic hydrogens are not available. This intermediate may then complete the chain transfer process by donating the hydrogen atom to a monomer molecule. The latter step resembles the process (32) for direct transfer from a chain radical to a monomer. The former step is involved in the action of certain aromatic inhibitors (see Sec. 4a).

the established influence of substituents in diminishing the strength of the C—H bond.[70] It is to be noted that t-butylbenzene, having no benzylic hydrogens, is comparatively nonreactive in spite of the nine available hydrogens in its methyl groups.

The alkyl chlorides and bromides are not much more reactive in chain transfer than cyclohexane, suggesting that the chlorine atom may actually be no more susceptible to radical attack in these compounds than the hydrogens. The somewhat greater transfer constants for methylene chloride, ethylene dichloride, and chloroform may be due to increased reactivity of the remaining hydrogens caused by halogen substitution.[75] Carbon tetrachloride and carbon tetrabromide are much more reactive, however, for reasons not altogether clear.[75] Analysis of polystyrenes prepared in the presence of carbon tetrachloride reveals the occurrence of four chlorine atoms per molecule of polymer.[74,77] This observation supports the view that fragments from one solvent molecule occur as end groups in the polymer, which therefore may be represented by the formula

$$CCl_3 \left[\begin{array}{c} C_6H_5 \\ | \\ -CHCH_2- \end{array} \right]_x Cl$$

in accordance with the assumed transfer mechanism

$$CCl_3-M_x \cdot + CCl_4 \;\rightarrow\; CCl_3-M_x-Cl + CCl_3 \cdot$$

$$CCl_3 \cdot \;\xrightarrow{+M}\; CCl_3-M \cdot \;\xrightarrow{+M}\; \text{etc.} \;\rightarrow\; CCl_3M_x \cdot$$

and with a similar mechanism which has been established for additions to double bonds involving free radicals.[70]

The results of chain transfer studies with different polymer radicals are compared in Table XIV. Chain transfer constants with hydrocarbon solvents are consistently a little greater for methyl methacrylate radicals than for styrene radicals. The methyl methacrylate chain radical is far less effective in the removal of chlorine from chlorinated solvents, however. Vinyl acetate chains are much more susceptible to chain transfer than are either of the other two polymer radicals. As will appear later, the propagation constants k_p for styrene, methyl methacrylate, and vinyl acetate are in the approximate ratio 1:2:20. It follows from the transfer constants with toluene, that the rate constants $k_{tr,s}$ for the removal of benzylic hydrogen by the respective chain radicals are in the ratio 1:3.5:6000. Chain transfer studies offer a convenient means for comparing radical reactivities, provided the absolute propagation constants also are known.

TABLE XIV.—COMPARISON OF CHAIN TRANSFER CONSTANTS
FOR DIFFERENT RADICALS AT 80°C

Solvent	$C_S \times 10^4$		
	Styrene[a]	Methyl-methacrylate[b]	Vinyl acetate
Benzene	0.059	0.075	
Cyclohexane	.066	.10	
Toluene	.31	.52	92[c]
Ethylbenzene	1.08	1.35	
iso-Propylbenzene	1.30	1.90	
Ethyl acetate		0.24	
Carbon tetrachloride	130	2.39	
Tetrachloroethane	18[d]	0.20	10⁴[e]

[a] Calculated for 80°C according to the Arrhenius equation using constants of Table XIII.

[b] From Basu, Sen, and Palit.[72]

[c] From Nozaki.[73]

[d] At 100°C.

[e] From Stockmayer, Clarke, and Howard.[73] Converted to 80°C from measurements at 60°C.

2e. Dependence of the Degree of Polymerization on Temperature.—
An increase in the temperature of polymerization nearly always lowers the molecular weight of the polymer produced regardless of whether the molecular chain length is controlled by chain transfer or by chain termination. The increase in the chain transfer constant with temperature (Table XIII) is a direct measure of this negative temperature coefficient if other processes contribute negligibly to the generation of chain ends. From a plot of $\log C_S$ against $1/T$ (or by plotting $-\log \bar{x}_n$ against $1/T$ if chain transfer is the dominant molecular weight controlling process), the activation energy difference $E_{tr,S} - E_p$ and the frequency factor ratio $A_{tr,S}/A_p$ may be evaluated from the slope and intercept, respectively. These are given in the fourth and fifth columns of Table XIII. Low transfer constants, and therefore low rate constants $k_{tr,S}$, go hand in hand with high activation energies $E_{tr,S}$, as would be expected. The frequency factor for transfer tends to exceed that for propagation, hence it is the higher activation energy for chain transfer as compared with monomer addition which is responsible for the generally slower rates of the former.

If the degree of polymerization is controlled principally by chain termination so that $\bar{x}_n$ is proportional to the kinetic chain length, the temperature coefficient of the average molecular weight will depend

on the initiating process.   If an initiator is used, then it follows from
Eq. (28) that

$$(d \ln \bar{x}_n/dT) = [(E_p - E_t/2) - E_d/2]/RT^2$$

The activation energy E_d for the decomposition of the initiators commonly used is of the order of 30 kcal., while $(E_p - E_t/2)$ assumes values
from 4 to 7 kcal.   Hence, insofar as it is termination-controlled, the
temperature coefficient of the degree of polymerization is negative.
Similarly for thermal polymerizations, the corresponding quantity

$$(E_p - E_t/2) - E_i/2$$

(see Eq. 29) is negative and the same conclusion should apply.   From a
slightly different point of view, an increase in temperature increases the
radical population out of proportion to the increase in k_p/k_t of Eq. (26).
The incidence of termination is more frequent at higher temperatures
and, in spite of the (small) increase in k_p with temperature, shorter
chains are grown.

Photochemically induced polymerizations in which the molecular
weight is termination-controlled are exceptions to the rule that the
average degree of polymerization diminishes with temperature.   The
rate of initiation at fixed intensity of illumination is virtually independent of temperature, hence $d \ln \bar{x}_n/dT = (E_p - E_t/2)/RT^2$, which is
a positive quantity.

2f. Regulators.—Carbon tetrachloride, and especially carbon tetrabromide and pentaphenylethane, are many times more reactive with
styrene chain radicals than are the other substances listed in Table
XIII.   Small proportions of these substances bring about large depressions in the molecular weight of the polymer.   The latter two actually exceed styrene in reactivity toward the chain radical; i.e., C_S
exceeds unity.   Compounds whose transfer constants are of the order
of unity or greater are especially useful in controlling the molecular
weight of the polymer.   Small quantities only are required to depress
the molecular weight to almost any desired level.   When used for this
purpose they are referred to as regulators or modifiers.   The aliphatic
mercaptans are widely used in polymerizations involving butadiene or
other diolefins, as for example in the preparation of synthetic rubbers,
in order to reduce the polymer chain length to the range required for
subsequent processing.   Approximately one mole of mercaptan is consumed for each mole of polymer formed, and sulfur occurs in the polymer in the approximate ratio of one atom per molecule.[79]   In complete analogy with the transfer processes discussed above, the following reactions are indicated:

$$M_x\cdot + RSH \;\rightarrow\; M_x\!-\!H + RS\cdot \;\xrightarrow{\;+M\;}\; RS\!-\!M\cdot \;\xrightarrow{\text{etc.}}$$

Since the rate of polymerization is not depressed by the mercaptan, the transfer radical $RS\cdot$ evidently reacts readily with monomer to start a new chain.

Transfer constants for regulators may be determined by application of the previously discussed "molecular weight" method, which depends on the use of Eq. (42). The concentration ratio $[S]/[M]$ is assumed to be constant in this method, hence the conversion must be kept very low, unless C_S is near unity. For transfer constants greater than about five, impractically low conversions are required. A second method[80,81] depends on determination of the amount of regulator reacting. The rate of reaction of regulator is given by

$$- d[S]/dt = k_{tr,S}[S][M\cdot] \tag{43}$$

Division by Eq. (11) for the polymerization rate $-d[M]/dt$ yields

$$d[S]/d[M] = (k_{tr,S}/k_p)[S]/[M]$$

Or

$$d\log[S]/d\log[M] = C_S \tag{44}$$

Hence the slope of the plot of $\log[S]$ against $\log[M]$ must equal C_S. The concentration $[S]$ of unreacted mercaptan regulator may be determined by amperometric titration with silver nitrate,[80,82] or the combined sulfur in the polymer may be determined radiochemically[83] using mercaptan containing S^{35}, and the unreacted mercaptan concentration $[S]$ calculated by difference. Transfer constants for carbon

TABLE XV.—CHAIN TRANSFER CONSTANTS FOR MERCAPTANS AT 60°C

Monomer	Transfer agent	Method[a]	C_{RSH}
Styrene	n-Butyl mercaptan[83]	2	22
Styrene	n-Dodecyl mercaptan[80]	1	15
Styrene	n-Dodecyl mercaptan[83]	2	19
Styrene	t-Butyl mercaptan[80]	1	3.7
Styrene	t-Butyl mercaptan[80]	2	3.6
Styrene	Ethyl thioglycolate[80]	2	58
Methyl methacrylate	n-Butyl mercaptan[83]	2	0.67
Methyl acrylate	n-Butyl mercaptan[83]	2	1.7
Vinyl acetate	n-Butyl mercaptan[83]	2	48

[a] Method 1 refers to determination from the average degree of polymerization using Eq. (42), and 2 refers to determination from the consumption of mercaptan using Eq. (44).

tetrachloride may be determined similarly by analysis of the polymer for chlorine.[74]

Transfer constants for mercaptans with several monomers are given in Table XV. Results for the two methods described above are in satisfactory agreement. The rate of reaction with mercaptan relative to the rate of monomer addition (i.e., the transfer constant) varies considerably for different chain radicals. Temperature coefficients of the transfer constants for mercaptans are very small,[80] which fact indicates that the activation energy for removal of a hydrogen atom from the sulfhydryl group of a mercaptan is nearly equal to that for monomer addition.

3. ABSOLUTE VALUES OF THE RATE CONSTANTS FOR INDIVIDUAL STEPS

Four distinct reactions occur in the composite polymerization process: initiation, propagation, termination, and chain transfer; the last mentioned may include transfer with each of the several components present in the polymerizing system. Analysis of experimental results on the steady state rate of polymerization and on the degree of polymerization affords the necessary information for evaluating the rate constant characterizing only one of these processes, namely, chain initiation, for which values of fk_d (or k_i for thermal initiation) have been presented. Other rate constants so far have appeared only in certain ratios such as k_p^2/k_t and the various chain transfer constants $C = k_{tr}/k_p$. While we have succeeded in assigning numerical values to these ratios, the velocity constants have so far evaded evaluation. It is, in fact, quite impossible to arrive at a complete resolution of the velocity constants for the individual steps by measuring only rates of polymerization under steady state conditions and degrees of polymerization. An additional measurement of a suitable independent parameter is required. One that fulfills the need is the average "lifetime" τ_s of an active center, i.e., the average time from the creation, or initiation, of a radical chain to its ultimate annihilation, disregarding possible intervening transfer processes. This lifetime must equal the "population" $[M\cdot]$ of chain radicals divided by their rate of destruction. That is,

$$\tau_s = [M\cdot]/2k_t[M\cdot]^2 = 1/2k_t[M\cdot] \tag{45}$$

Or, from Eq. (11)

$$\tau_s = (k_p/2k_t)[M]/R_p \tag{46}$$

If τ_s were somehow determined at the polymerization rate R_p, the ratio

k_p/k_t could be calculated from Eq. (46). Having previously obtained k_p^2/k_t, it becomes possible at once to solve these ratios for the individual velocity constants.

3a. Methods for Determining the Mean Lifetime of an Active Center.[34,55,84]—We have previously taken it for granted that experimental measurements are confined to steady-state conditions, where the rate of annihilation of radicals is almost exactly equal to the rate of their formation. Thus it was that we were able to equate R_i to R_t to obtain Eq. (10). At the very outset of the polymerization, however, the radical concentration is zero, and a finite time must be required before it reaches (or closely approaches) the steady state level. Since the rate of polymerization is proportional to the radical concentration, the interval preceding arrival at the steady state must be characterized by an acceleration in the polymerization.

In order to formulate an answer to the obviously important question of the length of this interval of acceleration and to ascertain under what conditions it may be long enough to observe experimentally, we shall examine the non-steady-state interval from the point of view of reaction kinetics. Let us suppose, however, that the polymerization is photoinitiated, with or without the aid of a sensitizer. It is then possible to commence the generation of radicals abruptly by exposure of the polymerization cell to the active radiation (usually in the near ultraviolet), and the considerable period required for temperature equilibration in an otherwise initiated polymerization can be avoided. Then the rate of generation of radicals (see p. 114) will be $2fI_{abs}$, and the rate of their destruction $2k_t[M\cdot]^2$. Hence

$$d[M\cdot]/dt = 2fI_{abs} - 2k_t[M\cdot]^2 \qquad (47)$$

At the steady state this net rate is zero and $[M\cdot]_s = (fI_{abs}/k_t)^{1/2}$, where the subscript s refers specifically to the steady state. Hence the above equation can be written

$$d[M\cdot]/dt = 2k_t([M\cdot]_s^2 - [M\cdot]^2)$$

which gives on integration

$$\ln\{(1 + [M\cdot]/[M\cdot]_s)/(1 - [M\cdot]/[M\cdot]_s)\} = 4k_t[M\cdot]_s(t - t_0) \quad (48)$$

In this and subsequent equations t_0 enters as a constant of integration such that $[M\cdot]=0$ at $t=t_0$. According to Eq. (45), $2k_t[M\cdot]_s$ on the right-hand side of Eq. (48) may be replaced by the reciprocal of the steady state lifetime τ_s. Making this substitution and rearranging, we obtain the compact relation

$$\tanh^{-1}([M\cdot]/[M\cdot]_s) = (t - t_0)/\tau_s \qquad (49)$$

or, since the rate is proportional to the radical concentration, we may write

$$R_p/(R_p)_s = [M \cdot]/[M \cdot]_s = \tanh[(t - t_0)/\tau_s] \qquad (50)$$

Throughout these formulas t may be identified with the duration of the illumination. If $[M \cdot] = 0$ at $t = 0$, then $t_0 = 0$; but if $[M \cdot] = [M \cdot]_2 > 0$ at the start of the illumination, then

$$t_0/\tau_s = - \tanh^{-1}([M \cdot]_2/[M \cdot]_s)$$

and

$$\tanh^{-1}([M \cdot]/[M \cdot]_s) - \tanh^{-1}([M \cdot]_2/[M]_s) = t/\tau_s \qquad (51)$$

The curve OAE in Fig. 17, calculated according to Eq. (50), shows the course of the rise in the radical concentration following the commencement of illumination when the initial radical concentration is zero, i.e., $t_0 = 0$. Observation of the rate of polymerization as a function of the time during the interval preceding the steady state (i.e., for $t < {\sim}2\tau_s$) would provide information suitable for the evaluation of τ_s.

When the light is turned off, the radical concentration decays according to the expression

$$d[M \cdot]/dt' = - 2k_t[M \cdot]^2$$

where t' measures the duration of the dark interval. Upon integration,

$$1/[M \cdot] - 1/[M \cdot]_1 = 2k_t t' \qquad (52)$$

where $[M \cdot]_1$ is the radical concentration at the beginning of the dark interval. Multiplying by $[M \cdot]_s$ and again substituting $1/\tau_s$ for $2k_t[M \cdot]_s$

$$[M \cdot]_s/[M \cdot] - [M \cdot]_s/[M \cdot]_1 = t'/\tau_s \qquad (53)$$

The radical decay according to this equation is depicted by curve ABF in Fig. 17. Observation of the decay in the polymerization rate immediately following cessation of illumination offers an alternative method for determining τ_s, for it follows from Eq. (53) that the slope of the ratio $(R_p)_s/R_p$ plotted against t' should equal $1/\tau_s$.

Bamford and Dewar[55] have adapted the latter method to the deduction of values for τ_s, and hence to the determination of the individual rate constants k_p and k_t. They chose to observe the rate of polymerization by measuring the increase in viscosity with time, using for this purpose a specially designed reaction cell equipped with a viscometer. Having established by separate experiments the relation-

ship between the intrinsic viscosity and the molecular weight, and also
the dependence of the molecular weight on the rate, they succeeded in
converting observed increases in the viscosity to extents of polymerization.   By observing the rate for steady illumination and the subsequent
rate for the photochemical aftereffect, they were thus able to arrive
at values for the radical concentration ratio $[M\cdot]/[M\cdot]_s$ as a function
of time, and hence to obtain τ_s.   It is necessary, however, for satisfactory treatment of the data to convert the observed increase in viscosity
to the effect which the increment of polymer would impart in the limit
of infinite dilution in monomer.   The required extrapolation to in-

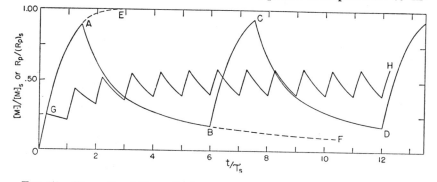

FIG. 17.—The ratio of the radical concentration to its steady-state concentration as a function of the time during alternate light and dark periods of durations
t and rt, respectively. Curve OABCD represents $t/\tau_s = 1.5$ and $r = 3$; curve OGH,
with $r = 3$ and $t/\tau_s = 1/4$, approaches a condition of oscillation about an average
radical concentration $[\overline{M}\cdot]/[M\cdot]_s = 1/2$.

finite dilution can be carried out accurately only for moderate increases in viscosity.   Consequently, the rate measurements must be
confined to very low conversions—usually less than 1 percent.   Moreover, τ_s should be at least 10 minutes in order for the photochemical
aftereffect to persist over a sufficient interval of time for completion of
a series of viscosity determinations.   The rate of polymerization must
therefore be made quite small (see Eq. 46) by using low intensities of
illumination.   The rate of the competing thermal polymerization of
some monomers sets a lower limit on the photochemical rate it is
feasible to use.   The viscosity method on the whole is rather in-
flexible, and the reliability of the results obtained appears to suffer
from the complexities of analysis of the data.

Usually, although not always, the approach to the steady state following commencement of the illumination or the decay during a subsequent dark period is too rapid for satisfactory measurements.   In such
cases it is nevertheless possible to observe average rates of polymeriza-

tion over a succession of many cycles of regularly alternating dark and light periods. If the time for a cycle is very long compared with τ_s, the amount of polymerization will simply be proportional to the total time of illumination. In this extreme case (not shown in Fig. 17) the rate vs. time curve will consist of a series of humps of nearly rectangular proportions; if the ratio r of dark to light periods is, for example, equal to three, so that the sample is illuminated only one-fourth of the time, the rate will be very nearly one-fourth of the steady illumination rate. If, however, the length of the cycle is reduced so that decay is incomplete during the dark phase and the steady-state concentration is not reached during the illumination phase, as in curve OABCD in Fig. 17, the radical concentration and the rate averaged over an integral number of cycles will considerably exceed one-fourth of their steady-state values. This is true because decay during the dark phase is not so rapid as the accumulation of radicals during illumination. If the flash time is shortened, the segmented curve OGH is obtained where the average radical concentration (and rate) is nearly one-half of the steady-state value. An extremely short flashing cycle would maintain the radical concentration at an approximately constant level. It would be equivalent to steady illumination at an intensity $1/(1+r)$ times that actually used. Since the steady-state radical concentration is proportional to $I_{abs}^{1/2}$, then with $r+1=4$ the rate would be one-half that for steady illumination. Hence the rate should rise from one-fourth to one-half of $(R_p)_s$ as the flashing frequency $1/(t+rt)$ is increased from a very low to a very high value compared with $1/\tau_s$.

The problem posed by intermittent illumination in photochemically induced chain reactions involving bimolecular termination was first treated by Briers, Chapman, and Walters[86,87] with particular reference to gaseous systems. Their method is applicable here as well. After a sufficient number of flashing cycles has been completed, the radical concentration will oscillate uniformly in successive cycles. Letting $[M\cdot]_1$ be the radical concentration at the end of a flash of duration t and at the beginning of a dark period of duration $t'=rt$, and letting $[M\cdot]_2$ be the radical concentration at the beginning of a flash and at the end of a dark period, we have from Eqs. (51) and (53)[34,55]

$$\tanh^{-1}([M\cdot]_1/[M\cdot]_s) - \tanh^{-1}([M\cdot]_2/[M\cdot]_s) = t/\tau_s \qquad (51')$$

and

$$[M\cdot]_s/[M\cdot]_2 - [M\cdot]_s/[M\cdot]_1 = rt/\tau_s \qquad (53')$$

For given values of the ratio $r=t'/t$ and of t/τ_s, the maximum and minimum radical concentration ratios $[M\cdot]_1/[M\cdot]_s$ and $[M\cdot]_2/[M\cdot]_s$, respectively, can be calculated from these equations. The average radical concentration $\overline{[M\cdot]}$ throughout a cycle will be

$$\overline{[M\cdot]} = \left\{ \int_0^t [M\cdot]dt + \int_0^{rt} [M\cdot]dt' \right\} \Big/ (t + rt)$$

where $[M\cdot]$ in the first integral covering the illumination phase is given by Eq. (51), and in the second integral for the dark phase of the cycle $[M\cdot]$ is given by Eq. (53). The result obtained upon evaluating these integrals is[34,55,87]

$$\overline{[M\cdot]}/[M\cdot]_s$$

$$= (r+1)^{-1}\left\{1+(\tau_s/t)\ln\left[\frac{[M\cdot]_1/[M\cdot]_2+[M\cdot]_1/[M\cdot]_s}{1+[M\cdot]_1/[M\cdot]_s}\right]\right\} \qquad (54)$$

Thus, having obtained from Eqs. (51′) and (53′) the ratios occurring in the argument of the logarithm of Eq. (54), the ratio of the average radical level to that for the steady state may be calculated. It is thereby possible to relate $\overline{[M\cdot]}/[M\cdot]_s$, which is equal to $\overline{R_p}/(R_p)_s$, to the quantity t/τ_s for a fixed value of the dark-to-light period ratio r. The curve obtained by plotting the average radical concentration ratio against $\log(t/\tau_s)$ for $r=3$ is shown in Fig. 18. The radical concentration ratio falls from one-half for rapid flashing rates (small t/τ_s) to one-

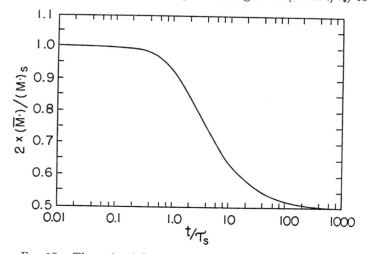

FIG. 18.—The ratio of the average radical concentration for intermittent illumination to its value for steady illumination plotted against $\log(t/\tau_s)$. The ratio r of dark to light intervals is three. (Matheson, Auer, Bevilacqua, and Hart.[55])

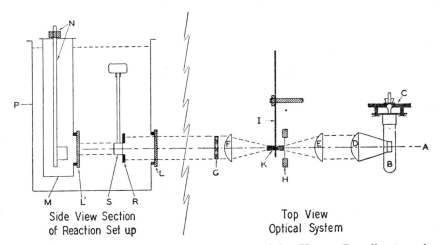

Side View Section
of Reaction Set up

Top View
Optical System

FIG. 19.—Schematic diagram of apparatus used by Kwart, Broadbent, and
Bartlett[48] for polymerization with intermittent illumination.

fourth for slow flashing, as we have seen previously must be true.   By
matching experimentally measured polymerization rates for different
flashing frequencies (i.e., for different values of t) to the theoretical
curve, it is possible to establish the value of τ_s.*

The required intermittency of illumination is achieved by inter-
posing in the path of the light beam a rotating sector consisting
of a disc from which one or more sectors of equal size have been
cut out.   The disc is driven by a synchronous motor with reduction
gears to afford the desired constant flashing frequencies.   A diagram
of the apparatus developed by Kwart, Broadbent, and Bartlett[48] is
shown in Fig. 19.   Radiation emitted by the 85-watt mercury arc B
is focused at K by quartz lenses D and E.   The iris opening H serves
to eliminate stray light.   As a result of the formation of an image of
the source at K, which is very small compared with the opening in the
sector I, the transition from dark to light or from light to dark in the
cell S occurs within a very small fraction of the duration t of the flash.
The quartz lense F collimates the beam passing through the cell.   G
is a filter for isolation of the desired portion of the lamp spectrum.   P
is a water thermostat bath; N is a phototube for monitoring the intensity of the radiation; L and L' are quartz windows.   The poly-

---

* It is assumed in the analysis given above that the rate of polymerization in
darkness is entirely negligible.   If the dark rate is appreciable, τ_s may nevertheless be evaluated from rates of polymerization at different flashing frequencies.[55,87]   The equations required are considerably more complicated,
however.

merization is followed dilatometrically by observing the change in level of the monomer in a capillary (not shown in detail) attached to the quartz cell S. From a knowledge of the (large) difference in densities of polymer and monomer, changes in level in the capillary of the cell are readily converted to amounts of polymerization. Carefully purified monomer and the photosensitizer (if one is used) are sealed in the cell, from which oxygen has been excluded. The rate of polymerization is then measured alternately with constant and with sectored radiation, the rate of rotation of the sector being changed as desired in each successive period. In this way a series of rate ratios $\overline{R_p}/(R_p)_s$ is obtained at fixed sector ratio r, the sole variable being the sector speed which defines the flash time t. The rate ratios are then plotted against log t. The theoretical curve for the same value of r, such as the curve in Fig. 18 for $r=3$, is superposed on the experimental points by shifting it along the abscissa scale until the best fit is obtained. Since the abscissa scale for the theoretical curve represents log $t-$log τ_s, the displacement of one scale relative to the other gives log τ_s.

TABLE XVI.—EVALUATION OF ABSOLUTE RATE CONSTANTS IN THE POLYMERIZATION OF VINYL ACETATE AT 25°C BY KWART, BROADBENT, AND BARTLETT[48]

	Run 2	Run 3
$R_i \times 10^9$	1.11	7.29
$(R_p)_s \times 10^4$	0.450	1.19
$(k_p^2/k_t) \times 10^2$	3.17	3.37
τ_s	4.00	1.50
$(k_p/k_t) \times 10^5$	3.35	3.32
$k_p \times 10^{-3}$	0.94	1.01
$k_t \times 10^{-7}$	2.83	3.06
$[M \cdot] \times 10^8$	0.44	0.54

All units in moles, liters, seconds.

3b. Experimental Results.—The data of Kwart, Broadbent, and Bartlett[48] shown in Table XVI and Fig. 20 for the photoinitiated polymerization of vinyl acetate at 25°C illustrate the method. The initiation rate $R_i = 2fI_{abs}$ for steady illumination was first established as a function of the *relative* incident light intensity $I_{0,rel}$ and of the concentration c_s of the photosensitizer, di-t-butyl peroxide. The inhibition method was used, duroquinone being the inhibitor; it had previously been shown to act efficiently down to very low concentrations. Thus, the proportionality constant between R_i and the product $c_s I_{0,rel}$ was determined. Two sets of polymerization rate measure-

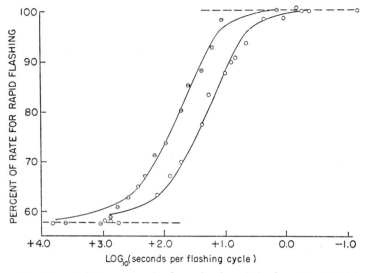

Fig. 20.—Relative rates of polymerization of vinyl acetate at 25°C as a function of the flashing frequency ($r=2$). The two sets of experimental points are for different intensities (runs 2 and 3 of Table XVI). The theoretical curves have been matched to the experimental points by horizontal displacement. (Kwart, Broadbent, and Bartlett.[48])

ments, indicated by run 2 and run 3 in Table XVI, were then carried out at different sensitizer concentrations and incident light intensities. The initiation rates as calculated from the proportionality previously established are given in the first line of Table XVI. Rates of polymerization $(R_p)_s$ were then measured for steady illumination, and k_p^2/k_t values as given in the third line of the table were calculated using Eq. (15). Finally, the rates were measured for intermittent illumination using a rotating disc from which a 120° sector was cut out giving $r=2$. Flashing frequencies were varied over a range of 10^5 by adjusting the speed of rotation of the sector. These results plotted as the percentage of the rate for rapid flashing are shown in Fig. 20 by the two sets of points for the respective initiation rates. The abscissa scale is reversed from that used in Fig. 18, and the calculated rate for slow flashing is $1/(r+1)^{1/2}=1/\sqrt{3}$, or 57.7 percent of the rapid flashing rate. The curves in Fig. 20 have been calculated from Eqs. (51′), (53′), and (54) for $r=2$, and each has been displaced laterally to secure optimum fit with the two sets of experimental points. Values of τ_s were deduced from these shifts in the manner indicated above. Using Eq. (46), the ratio k_p/k_t has been calculated from τ_s and R_p. Individual values of k_p and k_t given near the foot of Table XVI were computed from the two ratios k_p^2/k_t and k_p/k_t. The steady state radical concentrations given in the last line of the table were calculated from τ_s and k_t using Eq. (45).

The rate of growth of an individual vinyl acetate radical, $k_p[M]$ in undiluted monomer at 25°C is about 10^4 sec.$^{-1}$, according to these results. The kinetic chain length ν, equal to the product of this rate and the mean lifetime (compare Eqs. 26 and 45), is thus about 4×10^4 at steady illumination under the conditions of run 2. The average degree of polymerization[48] (not included in Table XVI) is much less than 2ν; at 25°C it is about 3.5×10^3. Thus chain transfer must occur on an average of about ten times during the life of a kinetic chain, and $C_M \cong 1/\bar{x}_n \cong 2.5 \times 10^{-4}$ (see p. 138). The absolute value of the rate constant $k_{tr,M}$ for chain transfer with monomer may be calculated from the C_M and the absolute value of k_p; i.e., for transfer of the vinyl acetate radical with its monomer at 25°C, $k_{tr,M} = C_M k_p = 0.25$, an unusually large value for transfer of a radical with its monomer. In the case of styrene at 60°C, for example, $k_{tr,M} \cong 0.01$ as obtained from $C_M = 0.6 \times 10^{-4}$ (see Sec. 2c) and k_p as given in Table XVII.

Absolute propagation and termination constants for the polymerization of various monomers, together with activation energies and frequency factors where these are known, are assembled in Table XVII. The data shown, with the exception of those for butadiene and isoprene, have been derived with the aid of lifetimes determined by the rotating sector method. The propagation constants given for butadiene and isoprene were obtained from emulsion polymerization experiments in which the rate of growth per particle (see Chap. V) was measured. Most of the other results quoted are due to Matheson and co-workers. Constants obtained for vinyl acetate by Kwart, Broadbent, and Bartlett[48] are almost identical with those of Matheson.[47] Melville and his collaborators[64,91] also applied a rotating sector method to the study of the polymerization of the first three monomers listed, but with results which in general cannot be reconciled quantitatively with those of Matheson and Bartlett and their co-workers. Bamford and co-workers,[63,85] have arrived at values for k_p and k_t for the same three monomers by following the photochemical aftereffect viscometrically. The results thus obtained do not correlate well with those from intermittent illumination experiments in which the rate of polymerization was followed dilatometrically over a wider range of conversion.

The propagation rate constants for the seven monomers represented in Table XVII vary over a range of nearly one-hundred-fold. Since both reactants, radical and monomer, differ in each instance, no con-

TABLE XVII.—ABSOLUTE RATE CONSTANTS FOR CHAIN PROPAGATION AND
TERMINATION[a]

Monomer	k_p		E_p kcal.	A_p $\times 10^{-7}$	$k_t \times 10^{-7}$		E_t kcal.	A_t $\times 10^{-9}$
	30°	60°			30°	60°		
Vinyl acetate[47,48]	1,240	3,700	7.3	24	3.1	7.4	5.2	210
Styrene[41]	55	176	7.8	2.2	2.5	3.6	2.4	1.3
Methyl methacrylate[55]	143	367	6.3	0.51	0.61	0.93	2.8	0.7
Methyl acrylate[51,b]	720	2,090	~7.1	~10	0.22	0.47	~5	~15
Butyl acrylate[88]	14				.0009[c]			
Methacrylonitrile[89]	~21				~2.7			
Butadiene[90,d]		100	9.3	12				
Isoprene[90,d]		50	9.8	12				

[a] All rate constants are in units of l. mole⁻¹ sec.⁻¹. Rate constants quoted for vinyl acetate at 30°C have been interpolated from reported values at 25°C. Measurements at temperatures other than 30° and 60°C have been included for the calculation of activation energies and frequency factors.

[b] Owing to the necessity for extrapolating measurements on methyl acrylate to a conversion of less than 1 percent in order to avoid the pronounced autoacceleration occurring with this monomer, the data are of lower accuracy than for most of the other monomers investigated.

[c] Results of Melville and Bickel[88] have been recalculated assuming termination by coupling and using the present notation.

[d] Propagation constants for butadiene and isoprene were determined from rate of polymerization per particle in emulsion polymerization.

clusions with reference to the separate reactivities of either the radicals or the monomers alone can be drawn from these results. Copolymerization studies, to be discussed in the following chapter, provide suitable additional data for separate appraisal of the characteristic reactivities of monomers and of chain radicals. An order of reactivity of chain radicals also may be deduced from chain transfer constants for different radicals with the same transfer agent, provided the rate constants k_p also are known (see Sec. 2e). The conclusion to be drawn from the more detailed discussion of monomer and radical reactivities in Chapter V is that the more reactive monomers (e.g., styrene and butadiene) give the least reactive radicals, and vice versa. The reactivity of a monomer is enhanced by a substituent offering resonance stabilization, but such a substituent effects an even greater *decrease* in the reactivity of the chain radical. Consequently, the propagation constants for the reactive monomer styrene with a styrene radical turns out to be less (see Table XVII) than that for the much less reactive vinyl acetate monomer with the very much more reactive vinyl acetate radical; the latter lacks appreciable resonance stabilization. For the present we shall be concerned only with the reaction rate constants k_p for a monomer with its own radical, without attempting to resolve effects originating in the separate reactants.

Except for the dienes, the activation energies E_p given in Table XVII are within 1 kcal. of 7 kcal. per mole. The frequency factors A_p vary over a fiftyfold range, which suggests that steric effects may be somewhat more important than the activation energy. In support of this statement, methyl methacrylate with two substituents on the same ethylenic carbon exhibits the lowest frequency factor while the comparatively unhindered monomers (and their radicals) such as vinyl acetate and the dienes exhibit significantly larger values. Although data are lacking from which to estimate A_p for butyl acrylate, steric hindrance by its comparatively large substituent group seems to offer the only plausible explanation for the low value of k_p. The effect on the termination constant is even more marked, as might have been expected from the presence of a large substituent on each of the pair of free radical carbons which combine in the termination reaction.

It may be significant also that the A_p values are lower than would be calculated from the collision rate by factors less than 10^{-3}. This has been attributed by Harman and Eyring[92] to a singlet-triplet transition in the unpairing of electrons of the double bond of the monomer, such a transition being of low probability. Baxendale and Evans[93] prefer to consider that the low values for A_p are caused by the loss in degrees of freedom of the monomer in the activated state as compared with its previous situation as an independent molecule in the liquid. It may be shown that a similar loss of degrees of freedom should occur in other combination reactions, and in particular in the termination process for which the A_t are much larger. Hence the explanation offered by Baxendale and Evans for the low values of A_p must be considered unsatisfactory.

Evans, Gergely, and Seaman[94] call attention to the correlation between activation energy and heat of reaction. The greater the decrease in energy from reactants to products for a reaction of a given type, the lower should be the activation energy. Variations in the activation energy should be less than the differences in heats of reaction for different reactants, however. Evans and co-workers estimate (from consideration of the intersection of potential energy curves for the initial and final states) that differences in the former should be somewhat less than half the differences in the heats of reaction. In the addition of a monomer to a radical of its own kind, the product radical and the reactant radical are the same; hence the heat of reaction represents simply the energy difference between a structural unit in the polymer chain and a monomer molecule. As will be discussed more fully in Chapter VI, two important factors affect the heat evolved in polymerization of a monomer: resonance stabilization of the ethylenic group of the monomer and steric repulsions between substituents. Both of these factors tend to lower the heat evolved. Steric repulsions are particularly large in disubstituted units such as methyl methacrylate. At the distance of separation of the reactants in the activated complex, however, the steric repulsions should not be expected to affect the energy very much. If these repulsions are neglected, the heat of polymerization ΔH_p should be about -20 kcal. per mole for all unconjugated monomers (see Chap. VI). For styrene and butadiene, $-\Delta H_p$ should be lower by about 2 or 3 kcal. owing to resonance stabilization of the monomer which is not present in the polymer unit. The higher values of E_p for styrene, butadiene, and isoprene can be explained in this way, although the observed increment for styrene probably is not experimentally significant. Methyl methacrylate, for reasons which are not apparent, exhibits the lowest value for E_p.

Activation energies for chain termination are smaller than for chain propagation, but they are significantly greater than zero. This might not have been anticipated inasmuch as methyl radicals seem to combine in the gas phase without measurable activation energy.[95]

Matheson and co-workers found no significant changes in the rate constants k_p and k_t for styrene and methyl methacrylate when the rate of polymerization was changed sixfold by changing the initiation rate. Since chain transfer is of minor importance in these monomers, the degrees of polymerization vary approximately inversely with the polymerization rate (see Eq. 36); hence the constancy of k_p and k_t with the rate of polymerization at a given temperature leads to the conclusion that the rates of these processes are independent of the chain length. With the occurrence of autoacceleration at higher conversions, however, τ_s increases simultaneously with the rise in rate. Determination of k_p^2/k_t and of k_p/k_t (from τ_s and R_p) in the region of accelerated rate shows that whereas k_p remains at its initial value within experimental error, k_t has decreased by a factor of 100 or more in some cases.[41,51,55] These results confirm the explanation previously given for autoacceleration.

A superficially related dependence of k_t on the medium has been observed by Norrish and Smith[50] working with methyl methacrylate, and by Burnett and Melville[35] with vinyl acetate. Rates in poor solvents are high, and determination of τ_s by the rotating sector method reveals what appears to be a decrease in k_t in the poor solvents. This apparent decrease in k_t accounts for the increased rate of polymerization. Actually, precipitation of the polymer seems to be responsible for the effect. The growing radicals become imbedded in precipitated droplets, presumably of very small size. The termination reaction is suppressed owing to isolation of the chain radical in one droplet from that in another. This "gel effect" is fairly common in systems yielding polymer which is not soluble in the reaction medium.[96] It is more closely related to the phenomenon of emulsion polymerization to be discussed in the following chapter than to autoacceleration in homogeneous polymerizing systems, however. The one feature in common is the decrease in (apparent) k_t caused by the reaction environment. However, the environmental factors obviously differ.

Absolute values of the rate constants for other reactions occurring in polymerizations and copolymerizations are readily obtained once k_p and k_t have been determined. Rate constants for chain transfer between growing radical and various species may be assigned absolute values simply by multiplying the appropriate transfer constant C (see Tables XIII, XIV and XV) by k_p, as previously indicated. By comparison of k_{tr} for different radicals with the same transfer agent, a direct measure of radical reactivity may be obtained. The chain transfer studies conducted up to the present time have been concerned with a variety of transfer agents and relatively few radicals; hence they are of limited use for this purpose. As will appear in Chapter V, k_p and k_t are essential for the assignment of absolute values to the rate constants for the cross propagation and termination processes which assume importance in copolymerizations.

4. INHIBITION AND RETARDATION OF POLYMERIZATION

If there is added to the monomer a substance which reacts with chain radicals to yield either nonradical products or radicals of such low reactivity as to be incapable of adding monomer, the normal growth of polymer chains will be suppressed. The substance is designated an *inhibitor* if it is so effective as to reduce the polymerization rate substantially to zero. If, however, it acts less efficiently so that the rate and the degree of polymerization are reduced without total suppression of the polymerization, the substance is referred to as a *retarder*. The difference is, of course, one of degree. Results of Schulz[97] illustrating inhibition and retardation in the thermal polymerization of styrene are shown in Fig. 21. In the presence of 0.10 percent of benzoquinone, curve II, no measurable polymerization occurs throughout the *induction period*, during which the quinone is consumed by the thermally generated radicals. Thereafter the polymerization proceeds according to its normal course in the absence of inhibitor (curve I). A greater quantity of nitrobenzene (curve III) depresses the rate but without introducing an induction period. Nitrosobenzene (curve IV) causes inhibition, but following the induction period the rate remains lower than that for the pure monomer. The product resulting from the action of chain radicals on nitrosobenzene evidently is a retarder.

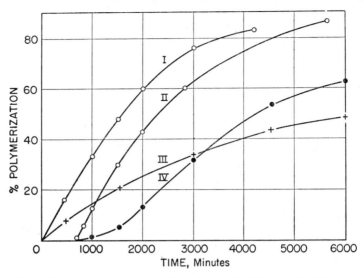

Fig. 21.—A comparison of the effects of 0.1 percent of benzo-quinone (curve II), 0.5 percent of nitrobenzene (curve III), and 0.2 percent of nitrosobenzene (curve IV) on the thermal polymerization of styrene at 100°C. Curve I represents the polymerization of pure styrene. (Results of Schulz.[97])

4a. Reaction Mechanisms.—If the inhibitor is itself a free radical, the product of reaction with the chain radical will have no unpaired electron; hence it almost certainly will be a stable molecule incapable of adding monomer. However, the free radical chosen to inhibit must be of low reactivity, for otherwise it may initiate chains as well as terminate them. The stable free radical 2,2-diphenyl-1-picryl-hydrazyl

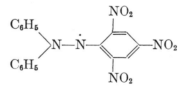

is an extremely effective inhibitor of polymerization which gives no evidence of initiation.[40] Even at concentrations less than 10^{-4} molar, it totally inhibits the polymerization of either vinyl acetate[40] or sty-rene.[41] Each hydrazyl radical consumed stops one chain radical.[40,41] Triphenylmethyl also is an inhibitor, but when added to pure styrene the induction period observed at 100°C is much shorter than it should be if each triphenylmethyl radical disappeared by combining with a thermally generated styrene chain.[44] As many as 77 of these radicals may be consumed for each chain radical which would have appeared in the absence of inhibitor. Apparently triphenylmethyl is also an initiator for this comparatively reactive monomer at elevated temperatures. It produces an induction period because it terminates chains with greater dispatch than it initiates them.[44]

The inhibitors more commonly used are molecules which in one way or another react with active chain radicals to yield product radicals of low reactivity. The classic example is benzoquinone.* As little as 0.01 percent causes virtual total suppression of polymerization of styrene or other monomers. This is true of both thermal and initiated polymerizations. Results of Foord[99] for the inhibition of thermal polymerization of styrene by benzoquinone are shown in Fig. 22. The

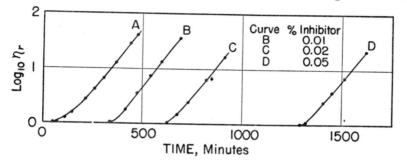

FIG. 22.—Inhibition of the thermal polymerization of styrene at 90°C by benzoquinone. The log of the viscosity relative to that of pure monomer is here used as a measure of polymerization. The small induction period in the absence of quinone presumably was caused by spurious inhibitors present in the monomer. (Results of Foord.[99])

length of the induction period, during which quinone is converted to products having little or no inhibitory action,[97,98] is proportional to the amount of quinone initially present.[34,40,41,43,98,99,100,101,102] The same applies to inhibition by 2,2-diphenyl-1-picrylhydrazyl and other quinone inhibitors.[103] The rate of consumption of the inhibitor is therefore *independent of the inhibitor concentration;* it depends only on the rate of generation of radicals, irrespective of whether they are generated by the action of an initiator or by a thermal process, which proceeds independently of the inhibitor in either case. Thus, the consumption of inhibitor is of the zero order. (The length of the induc-

---

* Hydroquinone often is used as a monomer "stabilizer." In the absence of oxygen it is not an inhibitor, however, and its action in the presence of oxygen is the result of its oxidation to quinone. See Ref. 98.

tion period is not, however, proportional to the initial concentration of an inhibitor like triphenylmethyl because of its simultaneous action as an initiator.)

Cohen[20] has presented evidence that quinone inhibits mainly by capturing styrene chain radicals rather than by intercepting radicals released by the initiator. The primary radicals (e.g., $C_6H_5COO\cdot$ or $C_6H_5\cdot$) usually are considerably more reactive than chain radicals; hence they may be expected to add one of the more numerous monomer molecules prior to a fatal encounter with an inhibitor present in very small concentration. The chain radicals, being less reactive, add monomer at a comparatively moderate rate only; hence there will be many collisions with inhibitor molecules during the interval between successive monomer addition steps, even when the concentration of the inhibitor is very low. If the rate constant k_z for reaction of the chain radical with inhibitor is enough greater than k_p, interception of the chain radical by inhibitor at a very early stage will be assured, its small concentration notwithstanding. A minute amount of inhibitor may then effectively halt the polymerization of the monomer.

Several mechanisms whereby a quinone may react with chain radicals are set forth below:

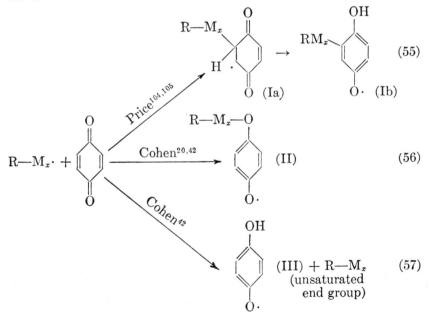

A shift of the hydrogen atom in radical Ia converts it to the more stable hydroquinone radical Ib. Owing to resonance (the numerous resonance structures having been omitted above), the similar inhibitor radicals Ib, II, and III should be relatively stable and therefore not readily able to regenerate active chain radicals by adding monomer. They may be expected to disappear instead by reaction between pairs of these inactive radicals, in partial analogy with normal bimolecular chain termination in the uninhibited polymerization.* If annihilation reactions of this nature assume dominance in regard to the fate of inhibitor radicals, termination of a chain radical may be scored at the occurrence of one of the reactions (55), (56), or (57) rather than at subsequent annihilation of the inhibitor radical.

Let us examine in greater detail the annihilation† reactions between pairs of inhibitor radicals. Disproportionation may be expected to play a more important role than coupling. Thus a hydrogen atom may be transferred between two radicals such as Ib or III to yield quinone and hydroquinone, either or both of which will bear nuclear substituents if Ib radicals are involved. Of course, the radical Ib may transfer its hydrogen atom to a molecule of quinone, giving a hydroquinone radical III and a substituted quinone. Hydroquinone is one of the products of quinone inhibition.[98] Hydroquinone monoethers could be formed by hydrogen transfer to a radical II from hydroquinone or from the radicals Ib or III, provided, of course, the various reactions between chain radicals and quinone occur simultaneously. Characterization of the inhibition products obtained with styrene[20] and with allyl acetate[102] by spectroscopic and chemical methods shows the presence of both nuclear- and oxygen-substituted products; hence the reaction (56) shown above must occur as well as (55) and perhaps also (57). The proportions of nuclear and oxygen addition certainly will be altered by nuclear substitution of the quinone used as inhibitor. Thus duroquinone (i.e., tetramethylbenzoquinone) appears to react predominantly according to reaction (56).[40]

Disappearance of pairs of inhibitor radicals by disporportionation leads to the regeneration of a molecule of benzoquinone or a substituted quinone. This molecule may terminate a subsequent chain. Hence, if the inhibitor radicals disappear exclusively by disproportionation,

* It is possible, of course, that an inhibitor radical may terminate another chain radical. In view of the low concentration of inhibitor radicals compared to the concentration of inhibitor molecules, reaction of a chain radical with inhibitor should occur with far greater frequency than with an inhibitor radical, hence the latter reaction normally makes only a minor contribution to the destruction of radicals in the system (see Sec. 4b).

† We avoid reference to reactions between pairs of inhibitor radicals as terminations in order to avert confusion with the reaction between chain radical and inhibitor, which is in fact the essential terminating step.

the ratio of chains terminated to quinone molecules consumed should equal two. Supporting evidence has been obtained by Cohen,[42] who found that the number of initiator molecules decomposing during the induction period in the benzoquinone-inhibited, benzoyl-peroxide-initiated polymerization of styrene was very nearly equal to the number of molecules of quinone; i.e., two radicals are stopped for each molecule of benzoquinone consumed. In the case of duroquinone in vinyl acetate, however, one molecule of inhibitor is consumed for each chain radical, which suggests that the radicals of type II believed to be formed in this case disappear by combination.

In sharp contrast to the results of Cohen cited above on the benzoquinone inhibition of the polymerization of styrene containing benzoyl peroxide, Mayo and Gregg[44] found that 17 or more quinone molecules disappeared for each chain radical generated in the *thermal* polymerization of styrene at 100°C. Schulz and Kämmerer[101] showed further that small traces of benzoquinone lower the average molecular weight of the polystyrene formed much more than they decrease the polymerization rate.   These facts indicate beyond question that benzoquinone is a transfer agent as well as an inhibitor in the polymerization of styrene. A radical formed in one of the reactions (55), (56), or (57), although quite unreactive, may nevertheless occasionally add a molecule of styrene, thereby regenerating the kinetic chain.* The fact that the molecular weight is reduced indicates that radical III, or one formed by hydrogen exchange with Ia or Ib, is involved in chain regeneration. The addition of monomer to radical Ib or to II would result in copolymerization of the quinone with no greater reduction in molecular weight than the decrease in rate, contrary to the results of Schulz and Kämmerer at very low concentrations of quinone. The superficial discrepancy between Cohen's results and those of Mayo and Gregg is due to two differences in the conditions prevailing in their experiments.   In the first place, those of Mayo and Gregg were conducted thermally at a somewhat higher temperature (100° compared to ca. 70°C) where reaction with monomer should be more rapid.   Secondly, the use of an initiator by Cohen greatly increased the steady-state concentration of inhibitor radicals, thus favoring their mutual reaction with one another.

In view of the potential complications of chain transfer by benzoquinone, it cannot be considered a suitable inhibitor for establishing

---

* Chlorine-substituted quinones, and especially chloranil (i.e., tetrachloroquinone), are predominantly transfer agents rather than inhibitors. Nuclear substitution as in the first step of reaction (55) may be involved with subsequent transfer of a chlorine atom to a monomer molecule.[105]

the rate of initiation in styrene according to the method described in Section 2c, although in less reactive monomers like vinyl acetate it seems to be satisfactory for this purpose. In general, the mere occurrence of a well-defined induction period should not be accepted as sufficient evidence for the existence of a fixed stoichiometric relationship between inhibitor consumed and chain radicals destroyed.

Aromatic nitrocompounds (see Fig. 21) such as nitrobenzene and the dinitrobenzenes diminish the rate of polymerization of styrene without suppressing it altogether and without introducing an induction period;[97,106] i.e., they are typical retarders.[107] Larger quantities are required to produce significant reductions in the rate, and the retardation persists throughout the polymerization. Price[104,105,106] suggests that the chain radical adds to the aromatic nucleus

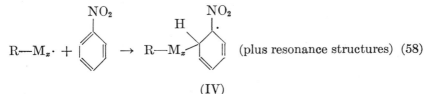

$$R\!-\!M_x\cdot \; + \; \langle NO_2 \rangle \;\rightarrow\; R\!-\!M_x \quad \text{(plus resonance structures)} \quad (58)$$

(IV)

Bartlett and co-workers concluded that addition to the nitro group also occurs.[40,102] Price and Read[105] found that several m-dinitrobenzene molecules were combined with the polymer for each fragment from the p-bromobenzoyl peroxide used as initiator in the retarded polymerization of styrene. They inferred that the radical corresponding to IV transfers its hydrogen atom to a molecule of styrene as follows:

$$(IV) + C_6H_5CH\!=\!CH_2 \;\rightarrow\; R\!-\!M_x\!-\!\langle NO_2 \rangle \; + \; CH_3\!-\!CH\cdot \;\underset{C_6H_5}{|} \quad (59)$$

The sequence of reactions (58) and (59) corresponds to the mechanism proposed[71] for chain transfer with benzene (see p. 142, footnote). The experimental result of Price and Read does not, however, rule out the possibility that the inhibitor radical (e.g., IV, or a possible successor formed by intramolecular rearrangement) may occasionally add monomer, thus giving rise to limited copolymerization of the retarder.

The dinitrobenzenes display the characteristics of inhibitors for the more reactive vinyl acetate chain radicals.[40] Two radicals are terminated during the induction period by each molecule of dinitrobenzene, indicating disappearance of inhibitor radicals by a disproportionation reaction.

Oxygen, a well-known inhibitor of polymerization,[30,59] yields a peroxide radical of rather low reactivity[104]

$$M_x \cdot + O_2 \rightarrow M_x—O—O \cdot \tag{60}$$

but capable nevertheless of adding a monomer (styrene, methyl methacrylate, or vinyl acetate[59]) to regenerate a normal chain radical with the ultimate production of polymer containing oxygen[108,109]

$$M_x—O—O \cdot + M \rightarrow M_x—O—O—M \cdot \tag{61}$$

Polymeric products having compositions approaching that for the 1:1 addition of oxygen to monomer, i.e.

$$[—M—O—O—]_x$$

are obtained.[59,109] Barnes and co-workers[59] have confirmed this structure by hydrogenation and identification of the resulting glycol. The degrees of polymerization are low—about 10 to 40. Bovey and Kolthoff[109] found that the consumption of oxygen in the well-agitated emulsion polymerization of styrene was independent of the oxygen pressure. The rate depended directly on the initiator ($K_2S_2O_8$) concentration, however, and was about one-thousandth of the polymerization rate in the absence of oxygen. These facts, together with the occurrence of genuine induction periods in thermal or photochemical polymerizations conducted in the presence of limited quantities of oxygen, demonstrate that whereas reaction (60) supersedes monomer addition in speed, reaction (61) must be very slow indeed.

At the conclusion of the induction period due to oxygen, polymerization sets in at a rate exceeding that for pure monomer under the same conditions. The polymeric peroxides apparently furnish a source of free radicals. Oxygen therefore combines the roles of inhibitor, co-monomer, and (indirectly) of initiator.

The hazards of a rigid classification of substances which may modify the course of a free radical polymerization are well illustrated by the examples of inhibitors and retarders which have been cited. The distinction between an inhibitor or retarder, on the one hand, and a co-monomer or a transfer agent, on the other, is not sharply defined. Moreover, if the substance is a free radical, it is potentially either an initiator or an inhibitor, and it may perform both functions as in the case of triphenylmethyl. If the substance with which the chain radicals react is a molecule rather than a radical, three possibilities may arise: (i) The adduct radicals may be completely unreactive toward monomer. They must then disappear ultimately through mutual interaction, and we have a clear-cut case of either *inhibition* or *retarda-*

tion, depending on the efficiency of the substance in combining with the chain radicals. (ii) The adduct may add monomer, reluctantly perhaps, with *copolymerization* as the result (e.g., oxygen in styrene). (iii) Transfer of an atom or group may occur at some step in the sequence from interception to radical regeneration, and thus bring about *chain transfer* (e.g., benzoquinone in styrene). Although alternatives (ii) and (iii) lead to regeneration of the active chain radical, retardation may nevertheless be observed if the reaction of monomer with the inhibitor radical is slow.

4b. Kinetics of Inhibition and Retardation.—The various reactions which may be brought about by an inhibiting or retarding substance, represented by Z, may be reduced to the following simple scheme:

$$M\cdot + Z \xrightarrow{k_z} Z\cdot \tag{62}$$

$$Z\cdot + M \xrightarrow{k_{zp}} M\cdot \tag{63}$$

$$2Z\cdot \xrightarrow{k_{zt}} \text{Nonradical products} \tag{64}$$

which will suffice for treating the rates of consumption of monomer and inhibitor. Chain radicals and inhibitor radicals are represented here simply by $M\cdot$ and $Z\cdot$, respectively, without further description. If the regeneration reaction (63) assumes importance, the question arises as to whether a radical transfer occurs in the sequence consisting of steps (62) and (63); if so, we have a case of chain transfer rather than copolymerization. This question may be settled independently by determining the degree of polymerization; it is of no concern to the analysis of rates, however. If the radical annihilation reaction (64) occurs by the coupling of inhibitor radicals, each inhibitor molecule, apart from those wasted through occurrence of the radical regeneration reaction (63), will be capable of stopping one chain radical. If the mechanism is one of disproportionation, each inhibitor may stop two chains. Finally, the possibility of the termination reaction

$$M\cdot + Z\cdot \quad \rightarrow \quad \text{(nonradical products)} \tag{65}$$

must not be overlooked entirely, although its occurrence is normally improbable because of the concentration sequence

$$[M\cdot] \ll [Z\cdot] \ll [Z]$$

in the presence of an inhibitor or retarder of consequence. The normal terminating reaction involving two chain radicals is assumed to be totally unimportant for the same reason.

Neglecting reaction (65) and considering only (62), (63), and (64), we have under steady-state conditions at an initiation rate R_i

$$d[M\cdot]/dt = R_i - k_z[M\cdot][Z] + k_{zp}[Z\cdot][M] = 0 \qquad (66)$$

$$d[Z\cdot]/dt = k_z[M\cdot][Z] - k_{zp}[Z\cdot][M] - 2k_{zt}[Z\cdot]^2 = 0 \qquad (67)$$

Solution of these equations for the two radical concentrations yields

$$[M\cdot] = \{R_i + k_{zp}(R_i/2k_{zt})^{1/2}[M]\}/k_z[Z] \qquad (68)$$

$$[Z\cdot] = (R_i/2k_{zt})^{1/2} \qquad (69)$$

The rate of consumption of inhibitor is

$$- d[Z]/dt = k_z[M\cdot][Z] - yk_{zt}[Z\cdot]^2 \qquad (70)$$

where y may be set equal to zero if the annihilation process (64) occurs by combination; if this reaction occurs by disproportionation with release of a molecule of inhibitor, y will equal unity. Then from Eq. (68)

$$- (d[Z]/dt) = (1-y/2)R_i + k_{zp}(R_i/2k_{zt})^{1/2}[M] \qquad (71)$$

The rate of consumption of initiator should therefore be constant (i.e., of zero order) under given conditions. This deduction applies regardless of whether or not inhibitor radicals may undergo regeneration (reaction 63). It emphasizes again that the observation of a well-defined induction period of duration proportional to the amount of inhibitor initially present offers no assurance of a simple stoichiometric ratio between radicals stopped and inhibitor consumed. It will be observed that the rate of consumption of inhibitor, and therefore the length of the induction period for a given amount of inhibitor, depends exclusively on R_i if $k_{zp}=0$; if $k_{zp}\neq0$, it depends also on $k_{zp}/k_{zt}^{1/2}$.

If cross termination reactions represented by (65) were to assume importance, the rate of consumption of inhibitor would not be independent of its concentration. The almost universal observation of proportionality between the length of the inhibition period and the amount of inhibitor available confirms our expectation that such processes are unimportant.

When the concentration of inhibitor is reduced sufficiently, appreciable growth of chain radicals will proceed and the rate of polymerization will no longer be negligible. Polymerization is then competitive with consumption of the inhibitor, and the rate of polymerization increases toward its normal value as the last traces of inhibitor are con-

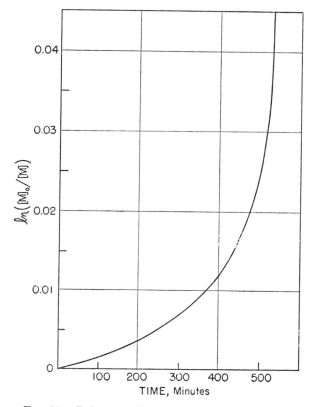

FIG. 23.—Polymerization of vinyl acetate at 45°C in the presence of 9.3×10^{-4} molar duroquinone and 0.2 molar benzoyl peroxide. (Bartlett, Hammond, and Kwart.[102])

sumed. This transition region is evident in Foord's results in Fig. 22. It is shown more clearly in Fig. 23, representing the polymerization of vinyl acetate in the presence of the mild inhibitor, duroquinone.[102] It is possible to deduce the rate constant ratio k_z/k_p from the relative rates of reaction of inhibitor and monomer in this region. Division of Eq. (70) by the rate of consumption of monomer yields

$$d[Z]/d[M] = (k_z/k_p)[Z]/[M]$$

where we assume $y = 0$. Upon integrating

$$\log([Z]/[Z]_0) = (k_z/k_p) \log([M]/[M]_0) \tag{72}$$

If k_z/k_p is very large, as will be true for a good inhibitor, the inhibitor concentration will decrease by many orders of magnitude according to

Eq. (72) before the monomer concentration changes appreciably; in other words, if $k_z/k_p \gg 1$ the inhibitor will be almost entirely exhausted before appreciable polymerization sets in.

The value of k_z/k_p can be obtained, assuming y is zero, from the slope of log $[Z]$ plotted against log $[M]$. Application of this method obviously requires data taken within the range where measurable polymerization is occurring, yet short of complete exhaustion of inhibitor, i.e., in the region of transition between complete inhibition and assumption of the normal rate (see Fig. 23). In the case of a powerful inhibitor, its concentration in this transition region may be too small to measure accurately. Direct measurements of inhibitor concentration may be obviated, however, by making use of the linear relation between inhibitor concentration and the time[40] (under conditions where R_i is constant)

$$[Z] = [Z]_0 - \text{Const.} \times t \qquad (73)$$

which must hold as long as the polymerization rate does not exceed a small fraction of the normal rate (i.e., until normal termination claims an appreciable fraction of the chain radicals). Values of k_z/k_p which have been obtained through the use of Eq. (72), with or without the aid of Eq. (73), are given in Table XVIII.

TABLE XVIII.—INHIBITION RATE RATIOS[a]

Inhibitor or retarder	k_z/k_p		
	Vinyl acetate, 45°	Styrene, 90°	Allyl acetate 80°
Benzoquinone		560	50
p-Xyloquinone		43	
Duroquinone	90	0.67	4.1
Nitrobenzene	20		
o-Dinitrobenzene	95		
m-Dinitrobenzene	105		
p-Dinitrobenzene	265		

[a] Data from Bartlett and co-workers.[40,102] Ratios for styrene calculated by them from the results of Breitenbach and Breitenbach.[103]

4c. Allylic Polymerization; Autoinhibition.—Discussion of the interesting features of the polymerization of allylic compounds has been withheld from earlier portions of this chapter because of the peculiar interplay of both chain transfer and inhibition in the over-all process. The rate of polymerization of allyl acetate is abnormally low; so is its degree of polymerization; and in exception to all monomers discussed so far, its rate is approximately proportional to the *first power* of the initiator concentration.[110]   Chain propagation is postulated to occur in the normal manner as follows:

$$M_x\cdot + CH_2{=}CH{-}CH_2{-}OCOCH_3$$

$$\rightarrow M_x{-}CH_2{-}\overset{\cdot}{C}H{-}CH_2{-}OCOCH_3 \qquad (74)$$

Lacking resonance stabilization, the chain radicals doubtless are very reactive, but owing to the corresponding lack of resonance structures in the transition state allyl acetate is a relatively unreactive monomer. These factors are conducive to the occurrence of the competitive reaction

$$M_x\cdot + CH_2{=}CH{-}CH_2OCOCH_3$$

$$\rightarrow M_xH + \overset{\cdot}{C}H_2{\cdots}CH{\cdots}\overset{\cdot}{C}H{-}OCOCH_3 \qquad (75)$$

in which an allylic radical is formed by removal of an *alpha* hydrogen from the monomer.   These radicals, being too well stabilized by resonance to add monomer, disappear through bimolecular combination. Bartlett and Altschul[110] refer to reaction (75) as a "degradative chain transfer"; the process itself is like ordinary chain transfer, but the transfer radical is too unreactive to regenerate a new chain.   In effect, then, reaction (75) is equivalent to termination by an inhibitor, which in this case is the monomer itself.   The same reactants are involved in the propagation and in the inhibition reactions; hence they are kinetically equivalent.   The length of the chain (about 14 at 80°C) must equal the ratio of the rate constants for the propagation (74) and termination (75) processes; it is independent of other factors such as the concentration of initiator.   Since the chain length is independent of the initiation rate, the rate of polymerization is directly proportional to the concentration of initiator and not to its square root as in other vinyl polymerizations.

Isopropenyl acetate[111] and allyl chloride[110] behave similarly.   In the polymerization of the latter monomer degradative chain transfer occurs more readily by removal of the chlorine atom to yield the unsubstituted allyl radical $\overset{\cdot}{C}H_2{\cdots}CH{\cdots}\overset{\cdot}{C}H_2$, which manages to add monomer occasionally.   This is indicated by the formation of about three polymer molecules, having an average degree of polymerization of six units, for each molecule of benzoyl peroxide decomposing.

Olefins having α-methylenic hydrogens, e.g., octene-1, propylene, and isobutylene, polymerize with difficulty in the presence of free radicals and yield products of low molecular weight. Nozaki[78] suggests that they are likewise subject to autoinhibition by reactions analogous to (75). The fact that other monomers such as methyl methacrylate and methacrylonitrile which also have α-hydrogens polymerize normally he attributes to the stabilizing effects of their substituents. The substituent group reduces the reactivity of the radical thereby decreasing the tendency for chain transfer, while at the same time it enhances the reactivity of the monomer with respect to addition.

REFERENCES

1. H. S. Taylor and A. A. Vernon, *J. Am. Chem. Soc.*, **53**, 2527 (1931); H. W. Melville, *Proc. Roy. Soc.* (London), **A167**, 99 (1938).
2. K. Jeu and H. N. Alyea, *J. Am. Chem. Soc.*, **55**, 575 (1933).
3. H. Staudinger and H. W. Kohlschütter, *Ber.*, **64**, 2091 (1931); G. V. Schulz and E. Husemann, *Z. physik. Chem.*, **B36**, 184 (1937).
4. H. Staudinger and W. Frost, *Ber.*, **68**, 2351 (1935).
5. H. S. Taylor and J. R. Bates, *J. Am. Chem. Soc.*, **49**, 2438 (1927); H. S. Taylor and W. H. Jones, *ibid.*, **52**, 1111 (1930).
6. H. Staudinger, *Die hochmolekularen organischen Verbindungen* (Julius Springer, Berlin, 1932), p. 151.
7. K. Nozaki and P. D. Bartlett, *J. Am. Chem. Soc.*, **68**, 1686 (1946); *ibid.*, **69**, 2299 (1947). See also B. Barnett and W. E. Vaughan, *J. Phys. Chem.*, **51**, 926, 942 (1947).
8. C. G. Swain, W. H. Stockmayer, and J. T. Clarke, *J. Am. Chem. Soc.* **72**, 5426 (1950).
9. C. E. H. Bawn and S. F. Mellish, *Trans. Faraday Soc.*, **47**, 1216 (1951).
10. A. T. Blomquist, J. R. Johnson, and H. J. Sykes, *J. Am. Chem. Soc.*, **65**, 2446 (1943).
11. C. C. Price and D. A. Durham, *J. Am. Chem. Soc.*, **64**, 2508 (1942).
12. G. V. Schulz, *Naturwissenschaften*, **27**, 659 (1939).
13. F. M. Lewis and M. S. Matheson, *J. Am. Chem. Soc.*, **71**, 747 (1949).
14. C. G. Overberger, M. T. O'Shaughnessy, and H. Shalit, *J. Am. Chem. Soc.*, **71**, 2661 (1949).
15. G. V. Schulz and G. Wittig, *Naturwissenschaften*, **27**, 387, 456 (1939); G. V. Schulz, *Z. Electrochem.*, **47**, 265 (1941).
16. T. T. Jones and H. W. Melville, *Proc. Roy. Soc.* (London), **A187**, 37 (1946); H. W. Melville and R. F. Tuckett, *J. Chem. Soc.*, **1947**, 1211; F. A. Raal and C. J. Danby, *ibid.*, **1949**, 2219.
17. J. H. Baxendale, M. G. Evans, and G. S. Park, *Trans. Faraday Soc.*, **42**, 155 (1946); J. H. Baxendale, S. Bywater, and M. G. Evans, *J. Polymer Sci.*, **1**, 237 (1946).
18. C. C. Price, R. W. Kell, and E. Krebs, *J. Am. Chem. Soc.*, **64**, 1103 (1942); C. C. Price and B. E. Tate, *ibid.*, **65**, 517 (1943); J. W. Breitenbach and H. Schneider, *Ber.*, **76B**, 1088 (1943).
19. P. D. Bartlett and S. G. Cohen, *J. Am. Chem. Soc.*, **65**, 543 (1943).
20. S. G. Cohen, *J. Polymer Sci.*, **2**, 511 (1947).

21. W. V. Smith, *J. Am. Chem. Soc.*, **71**, 4077 (1949).
22. L. M. Arnett and J. H. Peterson, *J. Am. Chem. Soc.*, **74**, 2031 (1952).
23. M. G. Evans, *J. Chem. Soc.*, **1947**, 266.
24. H. F. Pfann, D. J. Salley, and H. Mark, *J. Am. Chem. Soc.*, **66**, 983 (1944).
25. F. R. Mayo, R. A. Gregg, and M. S. Matheson, *J. Am. Chem. Soc.*, **73**, 1691 (1951).
26. L. M. Arnett, *J. Am. Chem. Soc.*, **74**, 2027 (1952).
27. D. H. Johnson and A. V. Tobolsky, *J. Am. Chem. Soc.*, **74**, 938 (1952); B. Baysal and A. V. Tobolsky, *J. Polymer Sci.*, **8**, 529 (1952).
28. M. S. Kharasch, H. C. McBay, and W. H. Urry, *J. Org. Chem.*, **10**, 401 (1945).
29. E. W. R. Steacie, *Atomic and Free Radical Reactions* (Reinhold Publishing Corp., New York, 1946), pp. 329–336; D. J. LeRoy and A. Kahn, *J. Chem. Phys.*, **15**, 816 (1947).
30. L. Küchler, *Polymerisationskinetik* (Springer Verlag, Berlin-Göttingen-Heidelberg, 1951).
31. G. V. Schulz and F. Blaschke, *Z. physik. Chem.*, **B51**, 75 (1942).
32. S. Kamenskaya and S. Medvedev, *Acta Physicochim.* (U.R.S.S.), **13**, 565 (1940).
33. C. C. Price and R. W. Kell, *J. Am. Chem. Soc.*, **63**, 2798 (1941).
34. G. M. Burnett and H. W. Melville, *Proc. Roy. Soc.* (London), **A189**, 456 (1947).
35. G. M. Burnett and H. W. Melville, *Proc. Roy. Soc.* (London), **A189**, 494 (1947).
36. K. S. Bagdasaryan, *J. Phys. Chem.* (U.S.S.R.), **21**, 25 (1947).
37. G. V. Schulz and G. Harborth, *Makromol. Chem.*, **1**, 106 (1947).
38. G. V. Schulz and E. Husemann, *Z. physik. Chem.*, **B39**, 246 (1938).
39. C. S. Marvel, J. Dec, and H. G. Cooke, *J. Am. Chem. Soc.*, **62**, 3499 (1940).
40. P. D. Bartlett and H. Kwart, *J. Am. Chem. Soc.*, **72**, 1051 (1950).
41. M. S. Matheson, E. E. Auer, E. B. Bevilacqua, and E. J. Hart, *J. Am. Chem. Soc.*, **73**, 1700 (1951).
42. S. G. Cohen, *J. Am. Chem. Soc.*, **67**, 17 (1945); **69**, 1057 (1947).
43. M. H. Mackay and H. W. Melville, *Trans. Faraday Soc.*, **46**, 63 (1950).
44. F. R. Mayo and R. A. Gregg, *J. Am. Chem. Soc.*, **70**, 1284 (1948).
45. J. Franck and E. Rabinowitch, *Trans. Faraday Soc.*, **30**, 120 (1934).
46. M. S. Matheson, *J. Chem. Phys.*, **13**, 584 (1945).
47. M. S. Matheson, E. E. Auer, E. B. Bevilacqua, and E. J. Hart, *J. Am. Chem. Soc.*, **71**, 2610 (1949).
48. H. Kwart, H. S. Broadbent, and P. D. Bartlett, *J. Am. Chem. Soc.*, **72**, 1060 (1950).
49. G. M. Burnett, *Quarterly Reviews*, **4**, 292 (1950).
50. R. G. W. Norrish and R. R. Smith, *Nature*, **150**, 336 (1942).
51. M. S. Matheson, E. E. Auer, E. B. Bevilacqua, and E. J. Hart, *J. Am. Chem. Soc.*, **73**, 5395 (1951).
52. A. C. Cuthbertson, G. Gee, and E. K. Rideal, *Proc. Roy. Soc.* (London), **A170**, 300 (1939).

53. E. Trommsdorff, *Colloquium on High Polymers* (Freiburg, 1944); E. Trommsdorff, H. Köhle, and P. Lagally, *Makromol. Chem.*, 1, 169 (1948).

54. Z. A. Rogovin and L. A. Tsaplina, *J. Applied Chem.* (U.S.S.R.), 20, 875 (1947).

55. M. S. Matheson, E. E. Auer, E. B. Bevilacqua, and E. J. Hart, *J. Am. Chem. Soc.*, 71, 497 (1949).

56. W. H. Carothers, I. Williams, A. M. Collins, and J. E. Kirby, *J. Am. Chem. Soc.*, 53, 4203 (1931).

57. H. W. Melville, *Proc. Roy. Soc.* (London), A163, 511 (1937).

58. C. Walling, E. R. Briggs, and F. R. Mayo, *J. Am. Chem. Soc.*, 68, 1145 (1946).

59. C. E. Barnes, *J. Am. Chem. Soc.*, 67, 217 (1945); C. E. Barnes, R. M. Elofson, and G. D. Jones, *ibid.*, 72, 210 (1950).

60. G. V. Schulz, A. Dinglinger, and E. Husemann, *Z. physik. Chem.*, B43, 385 (1939).

61. H. Suess, K. Pilch, and H. Rudorfer, *Z. physik. Chem.*, A179, 361 (1937); H. Suess and A. Springer, *ibid.*, A181, 81 (1937).

62. C. Walling and E. R. Briggs, *J. Am. Chem. Soc.*, 68, 1141 (1946).

63. C. H. Bamford and M. J. S. Dewar, *Proc. Roy. Soc.* (London), A197, 356 (1949).

64. M. H. Mackay and H. W. Melville, *Trans. Faraday Soc.*, 45, 323 (1949).

65. J. W. Breitenbach and R. Raff, *Ber.*, 69, 1107 (1936); C. Cuthbertson, G. Gee, and E. K. Rideal, *Nature*, 140, 889 (1937).

66. J. W. Breitenbach and W. Thury, *Experientia*, 3, 281 (1947).

67. G. V. Schulz and E. Husemann, *Z. physik. Chem.*, B36, 184 (1937).

68. P. J. Flory, *J. Am. Chem. Soc.*, 59, 241 (1937).

69. B. H. Zimm and J. K. Bragg, *J. Polymer Sci.*, 9, 476 (1952).   See also R. N. Haward, *Trans. Faraday Soc.*, 46, 204 (1950).

70. R. A. Gregg and F. R. Mayo, *Faraday Soc. Discussions*, 2, 328 (1947).

71. F. R. Mayo, *J. Am. Chem. Soc.*, 65, 2324 (1943).

72. S. Basu, J. N. Sen, and S. R. Palit, *Proc. Roy. Soc.* (London), A202, 485 (1950).

73. W. H. Stockmayer, J. T. Clarke, and R. O. Howard, to be published.

74. R. A. Gregg and F. R. Mayo, *J. Am. Chem. Soc.*, 70, 2373 (1948).

75. R. A. Gregg and F. R. Mayo, *J. Am. Chem. Soc.*, 75, 3530 (1953).

76. C. H. Bamford and M. J. S. Dewar, *Faraday Soc. Discussions*, 2, 314 (1947).

77. J. W. Breitenbach and A. Maschin, *Z. physik. Chem.*, A187, 175 (1940).

78. K. Nozaki, *Faraday Soc. Discussions*, 2, 337 (1947).

79. R. H. Snyder, J. M. Stewart, R. E. Allen, and R. J. Dearborn, *J. Am. Chem. Soc.*, 68, 1422 (1946); F. T. Wall, F. W. Banes, and G. D. Sands, *ibid.*, 68, 1429 (1946).

80. R. A. Gregg, D. M. Alderman, and F. R. Mayo, *J. Am. Chem. Soc.*, 70, 3740 (1948).

81. W. V. Smith, *J. Am. Chem. Soc.*, **68**, 2059 (1946).
82. I. M. Kolthoff and W. E. Harris, *Ind. Eng. Chem.*, Anal. Ed., **18**, 161 (1946).
83. C. Walling, *J. Am. Chem. Soc.*, **70**, 2561 (1948).
84. C. G. Swain and P. D. Bartlett, *J. Am. Chem. Soc.*, **68**, 2381 (1946).
85. C. H. Bamford and M. J. S. Dewar, *Proc. Roy. Soc.* (London), **A192**, 309 (1948); *Faraday Soc. Discussions*, **2**, 310 (1947); G. Dixon-Lewis, *Proc. Roy. Soc.* (London), **A198**, 510 (1949).
86. F. Briers, D. L. Chapman, and E. Walters, *J. Chem. Soc.*, 562 (1926).
87. W. A. Noyes, Jr., and P. A. Leighton, *The Photochemistry of Gases* (Reinhold Publishing Corp., New York, 1941), p. 202.
88. H. W. Melville and A. F. Bickel, *Trans. Faraday Soc.*, **45**, 1049 (1949).
89. A. Copperman and M. T. O'Shaughnessy, Meeting, *American Chemical Society*, Sept. 1952.
90. M. Morton, P. P. Salatiello, and H. Landfield, *J. Polymer Sci.*, **8**, 215, 279 (1952).
91. H. W. Melville and L. Valentine, *Trans. Faraday Soc.*, **46**, 210 (1950).
92. R. A. Harman and H. Eyring, *J. Chem. Phys.*, **10**, 557 (1942).
93. J. H. Baxendale and M. G. Evans, *Trans. Faraday Soc.*, **43**, 210 (1947).
94. M. G. Evans, J. Gergely, and E. C. Seaman, *J. Polymer Sci.*, **3**, 866 (1948).
95. R. Gomer and G. B. Kistiakowsky, *J. Chem. Phys.*, **19**, 85 (1951).
96. W. I. Bengough and R. G. W. Norrish, *Proc. Roy. Soc.* (London), **A200**, 301 (1950).
97. G. V. Schulz, *Chem. Ber.*, **80**, 232 (1947).
98. J. W. Breitenbach, A. Springer, and K. Horeischy, *Ber.*, **71**, 1438 (1938); **74**, 1386 (1941).
99. S. G. Foord, *J. Chem. Soc.*, 48 (1940).
100. G. Goldfinger, I. Skeist, and H. Mark, *J. Phys. Chem.*, **47**, 578 (1943).
101. G. V. Schulz and H. Kämmerer, *Chem. Ber.*, **80**, 327 (1947).
102. P. D. Bartlett, G. S. Hammond, and H. Kwart, *Faraday Soc. Discussions*, **2**, 342 (1947).
103. J. H. Breitenbach and H. L. Breitenbach, *Ber.*, **75**, 505 (1942); *Z. physik. Chem.*, **A190**, 361 (1942).
104. C. C. Price, *Mechanisms of Reactions at Carbon-Carbon Double Bonds* (Interscience Publishers, New York, 1946); *Annals, N.Y. Acad. Sci.*, **44**, 351 (1943).
105. C. C. Price and D. H. Read, *J. Polymer Sci.*, **1**, 44 (1946).
106. C. C. Price and D. A. Durham, *J. Am. Chem. Soc.*, **65**, 757 (1943); C. C. Price, *ibid.*, **65**, 2380 (1943).
107. I. M. Kolthoff and F. A. Bovey, *J. Am. Chem. Soc.*, **70**, 791 (1948); *Chem. Rev.*, **42**, 491 (1948).
108. H. Staudinger and L. Lautenschläger, *Ann.*, **488**, 1 (1931).
109. F. A. Bovey and I. M. Kolthoff, *J. Am. Chem. Soc.*, **69**, 2143 (1947).
110. P. D. Bartlett and R. Altschul, *J. Am. Chem. Soc.*, **67**, 812, 816 (1945).
111. R. Hart and G. Smets, *J. Polymer Sci.*, **5**, 55 (1950).

CHAPTER V

Copolymerization, Emulsion Polymerization, and Ionic Polymerization

1. THE COMPOSITION OF ADDITION COPOLYMERS[1,2,3]

THE array of reactions required to represent the copolymerization of two or more monomers increases geometrically with the number of monomers participating. The variety of chain radicals to be considered is equal to the number of monomers present; the reaction characteristics of a chain radical are determined almost entirely by the terminal monomer unit, the nature of those preceding it in the chain being of little importance. In the copolymerization of two monomers therefore, two chain radicals must be differentiated. Addition of the two monomers to each of these introduces four simultaneously occurring propagation reactions. Three different chain-terminating reactions between pairs of radicals must be considered, and a variety of chain transfer steps are possible. Chains may be initiated by "activating" either monomer, but the alternative possibilities for initiation of chain radicals are of little consequence in the presence of an initiator which acts efficiently on either monomer.

If the chains are long, the composition of the copolymer and the arrangement of units along the chain are determined almost entirely by the relative rates of the various chain propagation reactions. On the other hand, the rate of polymerization depends not only on the rates of these propagation steps but also on the rates of the termination reactions. Copolymer composition has received far more attention than has the rate of copolymerization. The present section will be confined to consideration of the composition of copolymers formed by a free radical mechanism.

1a. Kinetics of Chain Propagation in Copolymerization.—The chain-propagating reactions occurring when two monomers M_1 and M_2 are present may be written

$$
\left.\begin{array}{c}
M_1 \cdot + M_1 \overset{k_{11}}{\rightarrow} M_1 \cdot \\[4pt]
M_1 \cdot + M_2 \overset{k_{12}}{\rightarrow} M_2 \cdot \\[4pt]
M_2 \cdot + M_2 \overset{k_{22}}{\rightarrow} M_2 \cdot \\[4pt]
M_2 \cdot + M_1 \overset{k_{21}}{\rightarrow} M_1 \cdot
\end{array}\right\} \tag{1}
$$

where $M_1\cdot$ and $M_2\cdot$ represent chain radicals having monomers M_1 and M_2, respectively, as their terminal, free-radical-bearing units. Departing from previous notation, we use the subscript here to designate the type of unit rather than the number x of units in the chain. The first subscript attached to the rate constant refers to the reacting radical, and the second to the monomer. The subscript p previously used to designate propagation is omitted.

Radicals of type $M_1\cdot$ are formed by primary initiation and by reaction (2,1) above. They are destroyed by the reaction (1,2) and in termination reactions. At the steady state, the rates of generation and of disappearance of these radicals are practically equal. If the chains are long, initiation and termination are of exceedingly rare occurrence compared with the reactions (1), and it suffices therefore to consider the latter only for the present where we are concerned merely with the relative concentrations of the two types of chain radicals. The steady-state condition reduces in this approximation to

$$
k_{21}[M_2\cdot][M_1] = k_{12}[M_1\cdot][M_2] \tag{2}
$$

The same equation may, of course, be derived by similar application of the steady-state condition to radicals of type $M_2\cdot$. The rates of consumption of monomers M_1 and M_2 are

$$
-d[M_1]/dt = k_{11}[M_1\cdot][M_1] + k_{21}[M_2\cdot][M_1] \tag{3}
$$

$$
-d[M_2]/dt = k_{12}[M_1\cdot][M_2] + k_{22}[M_2\cdot][M_2] \tag{4}
$$

Using Eq. (2) to eliminate one of the radical concentrations and dividing Eq. (3) by (4), we obtain

$$
\frac{d[M_1]}{d[M_2]} = \left(\frac{[M_1]}{[M_2]}\right)\left(\frac{r_1[M_1]/[M_2] + 1}{[M_1]/[M_2] + r_2}\right) \tag{5}
$$

where r_1 and r_2 are *monomer reactivity ratios* defined by

$$
\left.\begin{array}{c}
r_1 = k_{11}/k_{12} \\[4pt]
r_2 = k_{22}/k_{21}
\end{array}\right\} \tag{6}
$$

Thus r_1 represents the ratio of the rate constants for the reaction of a radical of type 1 with monomer M_1 and with monomer M_2, respectively. The monomer reactivity ratio r_2 similarly expresses the relative reactivity of an $M_2\cdot$ radical toward an M_2 compared with an M_1 monomer. The quantity $d[M_1]/d[M_2]$ given by Eq. (5) represents the ratio of the two monomers in the increment of polymer formed when the ratio of unreacted monomers is $[M_1]/[M_2]$. The former ratio obviously differs in general from the latter; hence the unreacted monomer ratio will change as the polymerization proceeds, and this will give rise to a continually changing composition of the polymer being formed at each instant.

The compositions of the monomer feed and of the polymer formed may be expressed as mole fractions instead of the mole ratios used above. To this end we let F_1 represent the fraction of monomer M_1 in the increment of copolymer formed at a given stage in the polymerization. Then

$$F_1 = d[M_1]/d([M_1] + [M_2]) = 1 - F_2 \tag{7}$$

Letting f_1 and f_2 represent the mole fractions of unreacted monomers M_1 and M_2 in the feed, i.e.

$$f_1 = [M_1]/([M_1] + [M_2]) = 1 - f_2$$

we obtain from Eq. (5)

$$F_1 = (r_1 f_1^2 + f_1 f_2)/(r_1 f_1^2 + 2f_1 f_2 + r_2 f_2^2) \tag{8}$$

The composition of the increment of polymer formed at a monomer composition specified by $f_1 (=1-f_2)$ is readily calculated from Eq. (8) if the monomer reactivity ratios r_1 and r_2 are known. Again it is apparent that the mole fraction F_1 in general will not equal f_1; hence both f_1 and F_1 will change as the polymerization progresses. The polymer obtained over a finite range of conversion will consist of the summation of increments of polymer differing progressively in their mole fractions F_1.

Curves calculated from Eq. (8) for various values of the parameters r_1 and r_2 are shown in Figs. 24 and 25. The ordinate (F_1) represents the composition of the increment of copolymer formed from the monomer mixture having the composition (f_1) given along the abscissa axis. Fig. 24 treats the case in which the two radicals display the same preference for one of the monomers over the other. That is

$$k_{11}/k_{12} = k_{21}/k_{22}$$

or

$$r_1r_2 = 1 \tag{9}$$

Eqs. (5) and (8) reduce in this case to

$$d[M_1]/d[M_2] = r_1[M_1]/[M_2]$$

or

$$d \ln[M_1]/d \ln[M_2] = r_1 \tag{10}$$

and

$$F_1 = r_1f_1/(r_1f_1 + f_2) \tag{11}$$

The values of $r_1 = 1/r_2$ are indicated with each curve in Fig. 24. The straight line for $r_1 = 1$ represents the trivial case in which $k_{11} = k_{12}$ and $k_{22} = k_{21}$; i.e., that in which the two monomers are equally reactive with each radical. (The reactivities of the two radicals might differ however; i.e., k_{11} need not equal k_{22}.) The polymer composition is equal to the monomer composition throughout the range in this case.

Wall[4] first called attention to the close analogy between the copolymer-monomer mixture composition relationships and vapor-liquid

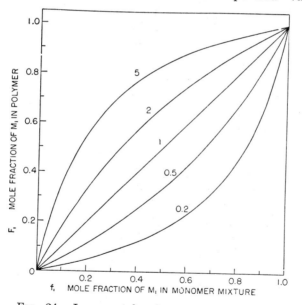

FIG. 24.—Incremental polymer composition (mole fraction F_1) plotted against the monomer composition (mole fraction f_1) for ideal copolymerizations ($r_1 = 1/r_2$). Values of r_1 are indicated.

equilibria in binary systems. He introduced the term *ideal copoly-merization* for the case $r_1r_2 = 1$, in recognition of the analogy to vapor-liquid equilibria for ideal liquid mixtures. The monomer reactivity ratio r_1 in Eq. (11) corresponds to the ratio of the vapor pressures (P_1°/P_2°) of the pure components of the ideal mixture, and F_1 and f_1 to the mole fractions of component 1 in the vapor and liquid, respectively, at equilibrium. When $r_1 > 1$, the polymer ("vapor") is richer in M_1 than is the monomer feed; hence the residual mole fraction f_1 must diminish as the polymerization ("distillation") proceeds. For $r_1 < 1$, the opposite holds.

It is also apparent that the sequence of monomer units in an ideal copolymer must necessarily be random. That is to say, the likelihood of occurrence of an M_1 unit immediately following an M_2 unit is the same as for an M_1 to follow an M_1 unit. The probability of either unit at any place in the chain is always equal to its mole fraction in an ideal copolymer.*

If the two radicals display different selectivities in their choice of monomers, then $r_2 \neq 1/r_1$, or

$$r_1r_2 = (k_{11}k_{22}/k_{12}k_{21}) \neq 1$$

The critical quantity r_1r_2 represents the ratio of the product of the rate constants for reaction of a radical with a monomer of the same kind to the product of the rate constants for the cross reactions. If $r_1r_2 > 1$, the tendency for radicals of a given kind to regenerate themselves by the addition of a like unit exceeds their tendency for alternation. The two curves with lowest initial slopes (at $f_1 = 0$) in Fig. 25 are examples of this as yet hypothetical case. Such a copolymer would contain sequences of like units in greater abundance than in a random copolymer of the same composition, and this tendency favoring sequences should be greater the larger the product r_1r_2. In the extreme case that both k_{12} and k_{21} are zero, the two monomers might polymerize simultaneously, but a mixture of two polymers would be obtained rather than a copolymer containing both units. No example of this sort is known. In fact, there is no established instance of a free-radical-propagated copolymerization for which $r_1r_2 > 1$. Although ideal copolymerization ($r_1r_2 = 1$) may be closely approached in a few instances, the product r_1r_2 is almost always less than unity (*see below*). In practice, therefore, the cross monomer addition reactions dominate additions of a like monomer. Four examples of this case ($r_1r_2 < 1$)

---

* This applies only to the increment of copolymer formed over a narrow range of conversion and not to the total product, which consists in general of increments formed at progressively changing monomer ratios.

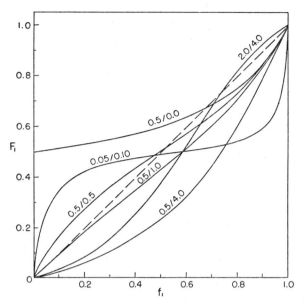

FIG. 25.—Incremental polymer compositions F_1 as functions of the monomer composition f_1 for the values of the reactivity ratios indicated (r_1/r_2). The broken straight line represents $F_1 = f_1$ (i.e., $r_1 = r_2 = 1$).

are shown in Fig. 25. If both r_1 and r_2 are very small, corresponding to relatively very small rate constants for the 1,1 and the 2,2 reactions, a copolymer is obtained in which the monomers alternate with near perfect regularity along the chain. Copolymerization of maleic anhydride with either stilbene[5] or allyl acetate[6] closely approaches this case. The olefin-sulfur dioxide copolymers[7] provide even more striking examples.* The mole fraction F_1 is then very nearly one-half for all monomer compositions. These examples are extremes. The reciprocal of the product $r_1 r_2$ furnishes an index of the alternation tendency in any case.

It will be observed that several of the curves shown in Fig. 25 cross the broken line representing $F_1 = f_1$. At the point of intersection the composition of polymer being formed coincides with that of the mono-

---

* It has been shown by Barb[8] and by Dainton and Ivin[8] that a 1:1 complex formed from the unsaturated monomer (n-butene or styrene) and sulfur dioxide, and not the latter alone, figures as the comonomer reactant in vinyl monomer-sulfur dioxide polymerizations. Thus the copolymer composition may be interpreted by assuming that this complex copolymerizes with the olefin, or unsaturated monomer. The copolymerization of ethylene and carbon monoxide may similarly involve a 1:1 complex (Barb,[8] 1953).

mer mixture; hence polymerization proceeds without change in composition. In further analogy with two-component distillation, Wall[4] designates these critical mixtures as *copolymerization azeotropes*. Setting $F_1 = f_1$ in Eq. (8) and solving, we find for the critical concentration

$$(f_1)_c = (1 - r_2)/(2 - r_1 - r_2) \qquad (12)$$

The value of $(f_1)_c$ lies within the permissible range $0 < f_1 < 1$ only if both r_1 and r_2 are greater than unity or if both are less than unity. If one of the reactivity ratios exceeds unity while the other is less than unity, no critical composition exists (see Fig. 25). In the as yet unknown case in which both ratios are greater than unity, which is illustrated by the $r_1 = 2$, $r_2 = 4$ curve in Fig. 25, the copolymer ("vapor") shows a greater departure from the azeotropic composition than does the monomer mixture. This behavior corresponds to that of a binary liquid mixture exhibiting a maximum boiling point (i.e., a strong negative deviation from ideality).

Instances of the opposite case in which both r_1 and r_2 are less than unity are numerous. Three of the curves of Fig. 25 are of this type. The copolymer composition always is more nearly that of the azeotrope than is the monomer composition. The azeotrope thus corresponds to a minimum boiling binary liquid mixture. As the polymerization progresses, the compositions, f_1 and F_1, depart increasingly from that of the azeotrope. The final increment of polymer formed at complete conversion would consist of pure polymer 1 or 2, depending on the location of the initial composition relative to that of the azeotrope. If the monomer reactivity ratios are much less than unity, as in the case $r_1 = 0.05$ and $r_2 = 0.10$ shown in Fig. 25, the mixture is strongly azeotropic; the copolymer composition approximates that of the azeotrope over a wide range in f_1, and the two units tend to alternate regularly along the chain.

If one of the monomers is very much more reactive than the other, the two monomers tend to polymerize consecutively; the first polymer formed will consist predominantly of the more reactive monomer, and the other monomer polymerizes only after nearly all of the former has been consumed. Styrene (1)-vinyl acetate (2) is such a system, r_1 being greater than unity while r_2 is very small (Table XIX); $r_1 r_2$ is less than unity. No curve corresponding to this case is included in Fig. 25.

Copolymerization equations for systems of more than two monomers have been derived, and several experimental studies of copolymerizations involving three monomers have been reported.[1] Six reactivity ratios are required for treatment of the composition in a three-component system. We shall be content to confine attention to binary systems.

(The average composition of the copolymer produced over a finite range of conversion is most readily calculated by the method of Skeist.[9] Let $[M]=[M_1]+[M_2]$. The number of moles of M_1 polymerized out of a total of $-d[M]$ moles of monomers converted to polymer is $-F_1 d[M]$. Meanwhile f_1 changes by df_1 and the number of moles of unreacted M_1 changes from $f_1[M]$ to (f_1+df_1) $([M]+d[M])$. Since the decrease in the number of moles of M_1 must obviously equal the number appearing in the newly formed polymer

$$f_1[M] - (f_1 + df_1)([M] + d[M]) = - F_1 d[M]$$

giving

$$d[M]/[M] = df_1/(F_1 - f_1) \qquad (13)$$

This result may be converted to the integral from the initial feed composition $(f_1)_0$ to some value f_1; i.e.

$$\ln([M]/[M]_0) = \int_{(f_1)_0}^{f_1} df_1/(F_1 - f_1) \qquad (14)$$

For given values of r_1 and r_2, F_1 may be calculated as a function of f_1 through the use of Eq. (8). The indicated integration may then be performed graphically to give the degree of conversion (i.e., $1-[M]/[M]_0$) required for a change in feed composition from $(f_1)_0$ to f_1. Through repetition of this process for suitably chosen values of f_1, it is possible to construct the relationship between f_1 and the degree of conversion. The integral composition of the aggregate of polymer formed up to a given conversion is easily calculated from the change in f_1 over the interval. Also, the incremental composition of the copolymer formed may be expressed graphically as a function of the conversion with the aid of Eq. 5 or 8.)

1b. Evaluation of Monomer Reactivity Ratios.[1,2,3]—Several procedures are available for determination of the essential parameters r_1 and r_2. All depend, of course, on careful analysis of the copolymer formed from monomer mixtures at a series of compositions. Since the composition changes with conversion, it is necessary either to limit the copolymerization to conversion of a very small fraction of the monomer mixture or to treat the integral composition, using, for example, the method outlined above. The calculations required by the latter method are excessively cumbersome, and it is usually preferred therefore to conduct experiments which meet the requirement of the former. The mole fraction composition of each of a series of copolymers prepared by polymerization to low conversion may be plotted against the composition of the monomer mixture from which it was prepared. The theoretical curve giving the best fit to the experimental points may then be found by trial selections of r_1 and r_2. Although in principle operable, such troublesome curve fitting is clearly unattractive, and in practice it is superseded by other methods. The theoretical curves in Fig. 25 show, however, that the fit obtained will be most sensitive to changes in r_1 and r_2 at extremes in the composition; the set of curves tends to converge in the middle composition region. Experiments made at low and at high concentrations of one of the monomers will therefore be most useful in fixing the values of the reactivity ratios. It is also apparent that the copolymer composition must be determined with great accuracy if the reactivity ratios are to be defined within narrow limits. These considerations apply irrespective of the particular procedure used to derive r_1 and r_2 from the composition data.

A widely used method for deriving r_1 and r_2 from suitable copolymer composition data consists in substituting the copolymer and monomer compositions for a single copolymerization in Eq. (8) (or in the equivalent equation 5) and plotting r_2 as a function of r_1. This is done for each of several copolymerizations. If there were no experimental error, all of the lines would intersect at the same point, which would represent the proper values of r_1 and r_2. An average point of intersection is taken to represent the best experimental pair of r_1 and r_2 values. Theoretical composition curves calculated from r_1 and r_2 determined in this way are found[2] to agree with the experimentally observed copolymer composition throughout the range in monomer composition. It seems justified to conclude, therefore, that the theoretical treatment is correct and, in particular, that the same rate constant ratios apply at all compositions.

The method of Fineman and Ross[10] represents a considerable improvement in the direction of straightforward analysis of copolymerization data. In their method Eq. (8) is rearranged to the linear form

$$f_1(1 - 2F_1)/(1 - f_1)F_1 = r_2 + [f_1^2(F_1 - 1)/(1 - f_1)^2 F_1]r_1 \qquad (15)$$

where f_2 has been replaced by $1 - f_1$. The quantity occurring on the left-hand side of the equation and the coefficient of r_1 occurring on the right may be calculated from the compositions F_1 and f_1 for each copolymerization. The former is then plotted against the latter and a straight line is drawn through the points, using the method of least squares. The intercept and slope of the line are equal to r_2 and r_1,

Table XIX.—Results of Copolymer Composition Studies on Typical
Binary Monomer Mixtures

Monomer 1	Monomer 2	°C	r_1	r_2	$r_1 r_2$	Ref.
Styrene	Butadiene	60	0.78	1.39	1.08	11
Styrene	p-Methoxystyrene	60	1.16	0.82	0.95	12
Styrene	Vinyl acetate	60	55	.01		13
Vinyl acetate	Vinyl chloride	60	0.23	1.68	.39	13
Vinyl acetate	Diethyl maleate	60	.17	0.043	.007	5
Maleic anhydride	Isopropenyl acetate	75	.002	.032	6×10^{-5}	14
Methyl acrylate	Vinyl chloride	50	9.0	.083	0.75	15

respectively. The equation may, of course, be rearranged to give r_1 as an intercept and r_2 as a slope.

Results for a few of the numerous copolymerization systems which have been investigated are given in Table XIX.* The first two monomer pairs listed behave approximately ideally, i.e., $r_1 \cong 1/r_2$ within experimental error. They are exceptional in this respect. At the other extreme, maleic anhydride-isopropenyl acetate yields an $r_1 r_2$ product of the order of 10^{-4}. Both reactivity ratios being very much less than unity, the pair is strongly azeotropic. A copolymer very close to 1:1 in composition, the two units alternating with nearly perfect regularity, is obtained over a wide range in monomer composition. Vinyl acetate–diethyl maleate is somewhat less strongly azeotropic; the vinyl acetate radical exhibits an appreciable tendency to add vinyl acetate monomer in the presence of the maleate ester.

For the remaining three systems, styrene–vinyl acetate, vinyl acetate–vinyl chloride, and methyl acrylate–vinyl chloride, one reactivity ratio is greater than unity and the other is less than unity. They are therefore nonazeotropic. Furthermore, since both r_1 and $1/r_2$ are either greater than or less than unity, both radicals prefer the same monomer. In other words, the same monomer—styrene, vinyl chloride, and methyl acrylate in the three systems, respectively—is more reactive than the other with respect to either radical. This preference is extreme in the styrene–vinyl acetate system where styrene is about fifty times as reactive as vinyl acetate toward the styrene radical; the vinyl acetate radical prefers to add the styrene monomer by a factor of about one hundred as compared with addition of vinyl acetate. Hence polymerization of a mixture of similar amounts of styrene and vinyl acetate yields an initial product which is almost pure polystyrene. Only after most of the styrene has polymerized is a copolymer formed

* The reader is referred to Refs. 1, 2, and 3 for comprehensive tabulations of monomer reactivity ratios.

TABLE XX.—RELATIVE REACTIVITIES ($1/r$) OF MONOMERS WITH VARIOUS POLYMER RADICALS (ca. 60°C)

Monomer	Butadiene[a]	Styrene	Isopropenyl acetate	Vinyl acetate	Allyl acetate	Allyl chloride	Vinyl chloride	Chloroprene	2,5-Dichlorostyrene	Methyl methacrylate	Vinylidene chloride	Methyl acrylate	β-Chloroethyl acrylate	Methacrylonitrile	Acrylonitrile	Diethyl fumarate	Maleic anhydride
Chloroprene	16	19		>50	>25[b]			(1.0)		12		12	12	25	22	40	>100
1,1-Diphenylethylene	2.2	3.5[b]						0.32							36		
2,5-Dichlorostyrene		1.3							(1.0)	2.3					5		
Butadiene	(1.0)								2.2	4.0							
α-Methylstyrene									0.3[b]	2.0		10			20		
Styrene	0.7	(1.0)		70	>50	>30	30	0.3	0.5[b]	2.2	>20	7		17	17	14	50
Phenylacetylene		1.9										20			20		
Methyl methacrylate	1.3	3.5	60	>50	>50	20	10[b]	0.12	0.44	(1.0)	12	5.5		6	4		50
Methyl vinyl ketone		4		18				0.16	0.5[b]	1.5	10	1.6		1.5	5.5		
Methacrylonitrile	2.8	2.4		>30						0.74	4	5.5		(1.0)	1.6		
Acrylonitrile	3	1.7		>50	>50	4	12	0.19		0.74	1.8[b]	1.6	1.1		(1.0)	22	
β-Chloroethyl acrylate		1.3									2.7	1.1	(1.0)		1.1		
Methyl acrylate	1.3	0.54						0.09	0.3	0.4	1.0	(1.0)			0.30		
Vinylidene chloride	0.5	0.4					3.5			0.07	(1.0)	1.2	1.2		0.33		
Methyl vinyl sulfone		0.2					0.5					3					
Methyl vinyl sulfide		0.05															
Methallyl chloride				3.5			0.5			0.14	0.9	0.18			0.2		
Isobutylene			4	1.4							0.7						
Methallyl acetate		0.014		1.7						0.01							
Vinyl chloride		0.05		1.0	2.2		(1.0)			0.07	0.5	0.11	0.25	0.08		2.1	120
Allyl chloride		0.03		0.3		(1.0)	0.5			0.02	0.26	0.11					
Vinyl acetate		0.02	1.0	(1.0)	2.2	1.5	0.9			0.05	0.25	0.2				2.3	300
Allyl acetate		0.01		5.3	(1.0)		0.45			0.04	0.17	0.3	0.18				500
Isopropenyl acetate			(1.0)														
Vinyl ethyl ether		0.01					0.5[c]			0.03	0.3						
3,3,3-Trichloropropene		0.14															
Maleic anhydride		>20		18	>130						0.1				0.2	0.90	(1.0)
Diethyl fumarate		3.3		90			3.5			0.15	0.08	0.36			0.082	(1.0)	
Diethyl maleate		0.17		6			8	0.15			0.025				0.16		
Crotonic acid		0.05		3			1.3			0.05	0.03	0.03			0.12		
Trichloroethylene		0.06		1.5						0.01					0.08		
trans-Dichloroethylene		0.03		1.0											0.05		
cis-Dichloroethylene		0.005		0.17											0.015		
Tetrachloroethylene		0.005		0.16								0.001			0.002		

[a] Includes isoprene results which are indistinguishable from those for butadiene:

[b] Newer values[2] not included by Mayo and Walling.[1]

[c] Vinyl isobutyl ether.

which contains a comparable proportion of vinyl acetate. When the proportion of styrene monomer is very small, it acts as a retarder for the polymerization of vinyl acetate. Vinyl acetate radicals are frequently converted to styrene radicals owing to the large value of $1/r_2$ and in spite of the low concentration of styrene. The less reactive styrene radical in turn adds vinyl acetate reluctantly; hence growth of the chain is delayed.[16,17]

The effect of temperature on the monomer reactivity ratio is fairly small. In those few cases examined with sufficient accuracy,[18] the ratio nearly always changes toward unity as the temperature increases —a clear indication that a difference in activation energy is responsible, in part at least, for the difference in rate of the competing reactions. In fact, the difference in energy of activation seems to be the dominant factor in these reactions; differences in entropy of activation usually are small, which suggests that steric effects ordinarily are of minor importance only.

1c. **Monomer Reactivity in Relation to Structure.**—The reciprocal of a reactivity ratio r expresses the relative reactivity of the "unlike" and "like" monomers with a given radical. By comparing the reciprocal reactivity ratios $1/r$ for a series of monomers with the same radical, it is possible to arrange these monomers in the order of their reactivities with the given radical. If the relative reactivities of monomers were the same toward different radicals, the same order of arrangement of the monomers would be obtained on selecting different radicals. The mere fact that r_2 rarely is equal to $1/r_1$—and often the difference is very great indeed—shows that the same order of monomer reactivity cannot be expected to obtain for different radicals. The relative reactivities of a series of monomers depends not only on the monomers but also on the reacting radical. It is readily apparent, nevertheless, from examination of the extensive array of copolymerization data available[1,2] that some monomers (e.g., styrene) are consistently more reactive than others (e.g., vinyl acetate). Mayo and Walling[1] have arranged the 34 monomers listed in Table XX in an approximate order of decreasing *average* reactivity by comparing $1/r$ values for the series of monomers with each of the radicals given at the head of a column in this table. The relative reactivities $(1/r)$ in each column decrease from top to bottom, but the order of decrease is irregular owing to the specificity of addition of monomers by radicals.

Mayo and Walling[1] draw the following conclusions from their tabulation (Table XX):

The effects of 1-substituents in increasing the reactivity of monomers towards attacking radicals are in the order, $-C_6H_5 > -CH=CH_2 >$

—COCH$_3$ >—CN >—COOR >—Cl >—CH$_2$Y >—OCOCH$_3$ >—OR. The effect of a second 1-substituent is roughly additive. 2-Chlorobutadiene and 2,3-dichlorobutadiene [not included in Table XX] are the most reactive monomers examined. A methyl group usually increases reactivity (methyl methacrylate >methyl acrylate, methacrylonitrile >acrylonitrile, methallyl >allyl derivatives) and two chlorine atoms are nearly as effective as a carbalkoxy group.

The above order of reactivity can be correlated with the stabilization

$$\overset{\text{X}}{\underset{|}{}}$$

of the product radical M$_x$—CH$_2$CH· due to added opportunities for resonance offered by the substituent.[1] Three quinoid resonance structures may be written for the substituted benzyl radical obtained in the polymerization of styrene, for example:

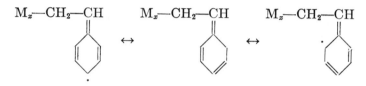

As a result, the radical is stabilized to the extent of about 20 kcal.[1] Two structures of nearly equal energies can be written for the allyl radical resulting from the addition of butadiene:

$$M_x—CH_2—CH=CH—CH_2· \quad \leftrightarrow \quad M_x—CH_2—CH—CH=CH_2$$

The amount of the resonance stabilization is similar to that for the benzyl radical. In radicals formed from monomers having C=O or C≡N groups conjugated with the carbon-carbon double bonds, the corresponding resonance structures

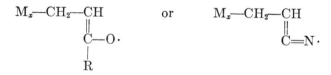

describe states of higher energy than that for the normal radical struc-

ture M$_x$—CH$_2$—$\overset{\text{X}}{\underset{|}{\text{CH}}}$· hence the resonance energy is smaller. When the monomer contains no unsaturated group conjugated with the ethylenic double bond, as when X = Cl or OR, the only resonance structures are ones of considerably higher energy representing either polar or non-bonded forms, e.g.[1]

$$\begin{array}{c} H \\ | \\ M_x\text{—}CH_2\text{—}\overset{\displaystyle}{C}\!:\!^- \\ | \\ \overset{\displaystyle}{Cl}\cdot^+ \end{array} \qquad \text{and} \qquad \begin{array}{c} H \\ | \\ M_x\text{—}CH_2\text{—}C \\ \| \\ O \quad\cdot R \end{array}$$

Stabilization may then be only 1 or 2 kcal. per mole.*

Substituents tend to stabilize the monomer also. The additional resonance structures which are introduced by the presence of the substituent contain fewer bonds than are present in the structure normally written for the monomer; hence they represent much higher energy states. Resonance stabilization by the substituent is therefore much less in the monomer than in the corresponding radical. In styrene or butadiene, for example, the resonance stabilization due to conjugation amounts to about 3 kcal. per mole. The energy change for the addition of a monomer to a radical may be expressed as the algebraic sum of a normal energy change $(\Delta H_p)_0$ which would be observed in the absence of all conjugation (or other) effects of the substituents plus corrections due to the latter. Thus

$$-\Delta H_p = -(\Delta H_p)_0 + U_P - U_M - U_A \qquad (16)$$

where U_A, U_M, and U_P are the resonance stabilization energies for the attacking radical, the monomer, and the product radical, respectively.† Positive values of the U's correspond to stabilization (which is contrary to the usual thermodynamic choice of sign); the heat of reaction ΔH_p is reckoned as negative for the exothermic polymerization process. In the comparison of a series of monomers with a given radical, U_A is fixed and the quantity $U_P - U_M$ is variable. As has been indicated above, when U_P is increased by incorporation of a stabilizing substituent (e.g., $-C_6H_5$), U_M also increases but by a much smaller amount. Thus, the stabilizing effect on the product radical is compensated only to a limited extent by stabilization of the monomer.

The rate of the reaction depends not on the heat of reaction, but on the energy of the transition state, or activated complex, relative to

* That it is not entirely negligible, however, is indicated by the fact (demonstrated by structure studies; see Chap. VI) that radical addition occurs preferentially—presumably at the unsubstituted carbon. Addition at the substituted carbon would yield a radical on the unsubstituted carbon; hence the substituent would be without effect.

† The frequently large steric effects due to repulsion between substituents (see Chap. VI) are neglected here, inasmuch as we are interested in the energy change accompanying monomer addition only insofar as it is related to the transition state. The distance of separation in the transition state presumably is too great for large steric repulsions.

the energy of the reactants. This energy of activation can be correlated, however, with the heat of reaction within a series of similar
reactions. The basis for this correlation may be seen from the following considerations. The monomer addition reaction consists of the
formation of a bond between the reactants and the simultaneous opening of the double bond:

$$C \cdot + C\!\!=\!\!C \quad \rightarrow \quad [C\text{----}C\!\!=\!\!C] \quad \rightarrow \quad C\!\!-\!\!C\!\!-\!\!C \cdot$$
$$\text{transition state}$$

In the nonbonding initial electronic state the radical is repelled as it
approaches the carbon atom of the monomer, and the potential energy
therefore increases as the distance of separation decreases. This
state of affairs is represented schematically by curve I of Fig. 26.
Curve II represents the potential energy of the final state as a function
of the length of the newly formed bond. We may picture the transition
from the initial to the final bond arrangement as taking place at the
point of intersection A.[19] The transition from one state to the other is
not so sharply defined, however, for resonance between the two states
will cause the two curves to round off from one into the other. The
associated resonance energy will lower the energy of the transition

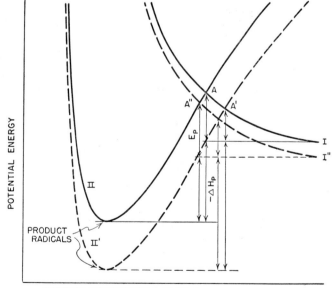

FIG. 26.—Illustrative representation of the potential energy
of the radical-monomer pair as a function of their distance of
separation. See text for further explanation.

state. We may, however, disregard the energy of resonance between the initial and final electronic configurations (but not other sources of resonance stabilization), since a similar contribution is involved in each case that will be considered.* For present purposes it is sufficient[19] to consider the displacement of the point A above the initial asymptote of curve I (at large distance of separation) to represent the activation energy E_p. The heat $-\Delta H_p$ evolved in the reaction is the difference between the asymptote of curve I and the minimum in curve II.

If the newly formed radical is resonance stabilized, the lower curve II' replaces curve II. On the assumption, for purposes of illustration, that the same curve I applies to the reactants, the energy of the transition state, represented by the intersection A', is lower than in the previous case. The activation energy is correspondingly reduced, *but not as much as* $-\Delta H_p$ *is increased*. If, now, the reactant radical enjoys stabilization which is not present in the product radical (e.g., if an unconjugated monomer adds to a relatively stable radical), the reaction may be described by curve I'' intersecting curve II. The intersection A'' occurs at a lower absolute level, but the difference between A'' and A is substantially less than the difference between the initial (asymptotic) energies for cases I'' and I. The activation energy is therefore *greater* than for the I and II reaction. The heat $-\Delta H_p$ evolved is much smaller.

In the light of these arbitrarily drawn examples, one may generalize as follows: If changes of structure have the effect merely of shifting the potential energy curves in the vertical direction without major alteration in their shapes, then the change in activation energy E_p must parallel the change in ΔH_p. The greater the heat evolved $(-\Delta H_p)$, the lower the activation energy E_p, but the differences in E_p are smaller than those in ΔH_p. Corresponding to Eq. (16), we write therefore

$$E_p = (E_p)_0 - \alpha[(U_P - U_M) - U_A] \tag{17}$$

where α is a factor which Evans and co-workers[19] estimate to be about 0.4.

Addition of a monomer of the same kind reproduces a product radical equivalent to the attacking radical, hence $U_P = U_A$ and the activation energy depends only on the term αU_M. In the polymerization of

---

* In expressing the potential energy as a function of one coordinate in Fig. 26, we neglect changes in the length of the C=C bond of the monomer as it is converted to a single bond. For a more accurate representation, another coordinate should be added to represent this length. Potential energy surfaces[2] would then replace the curves of Fig. 26.

styrene, for example, resonance stabilization due to conjugation of the phenyl group with the double bond is about 3 kcal.; hence the activation energy may be *increased* by 1 or 2 kcal. on this account. We have called attention to this predicted effect of a conjugated substituent on the propagation step in the preceding chapter (p. 158).

When the additions of each of a series of monomers with a given radical are compared as in Table XX, U_A is fixed and the quantity $U_P - U_M$ given in parentheses in Eq. (17) varies. As previously noted, the effect of a substituent which yields a large U_P is compensated to some extent by stabilization, U_M, of the monomer, although the latter is smaller than the former. For styrene or butadiene the value of $U_P - U_M$ is about 16 to 18 kcal.; hence the activation energy for the addition of styrene should be in the neighborhood of 6 kcal. less than for the addition of an unconjugated monomer to the same radical. In line with this interpretation is the fact that the average reactivity of styrene monomer exceeds that of vinyl acetate toward the same radical by a factor of about 50 in the neighborhood of 60°C. As previously noted, the $-OCOCH_3$ substituent of the latter monomer, though much less effective than the phenyl group, nevertheless contributes stabilization to a significant extent.

The effect of a conjugating substituent in the monomer may be summarized by observing that its influence is much greater in the product radical than in the monomer. In the activated complex, which is intermediate in character between reactants and product, resonance stabilization is substantially greater than in the monomer reactant, though less than in the product radical. The substituent therefore lowers the activation energy for the process, and enhances thereby the reactivity of the monomer.

1d. Radical Reactivity in Relation to Structure.—The monomer reactivity ratio, being a ratio of rate constants for different monomers with the *same* radical, offers no basis for comparing reactivities of *different* radicals with the same monomer. For the latter purpose it is necessary to utilize absolute progagation rate constants. Division of the absolute propagation constant k_{11}, determined in the manner described in the preceding chapter, by the reactivity ratio r_1 gives the rate constant k_{12} for reaction of radical 1 with the monomer 2. In Table XXI absolute rate constants are assembled for the reactions of four radicals with each of the corresponding monomers. The number of radicals which may be thus compared is unfortunately limited to those few for which absolute propagation constants (see Chap. IV) are known.

The order of reactivity of the radicals is the reverse of that for the monomers; styrene radical is the least and vinyl acetate radical the

TABLE XXI.—ABSOLUTE PROPAGATION RATE CONSTANTS AT 60°C (l. mole^{-1}, sec.$^{-1}$)
(From Mayo and Walling[1])

Monomer	Radical			
	Styrene	Methyl methacrylate	Methyl acrylate	Vinyl acetate
Styrene	176	789	11,500	~370,000
Methyl methacrylate	338	367		~250,000
Methyl acrylate	235		2,100	~ 37,000
Vinyl acetate	3.2	18.3	233	3,700

most reactive. Resonance stabilization depresses the activity of the radical in accordance with expectation. Moreover, the effect of the substituent in depressing the reactivity of the radical exceeds its enhancement of the reactivity of the monomer; the styrene radical is about 1000 times less reactive (disregarding specific alternation effects) than the vinyl acetate radical with the same monomer, whereas styrene monomer is only about 50 times as reactive as vinyl acetate toward the same radical (Table XX). These observations are in complete accord with Eq. (17). For a chosen monomer reacting with different radicals, $U_P - U_M$ is fixed and U_A varies over a wider range than does the quantity $U_P - U_M$ in the opposite case in which rates for different monomers with the same radical are compared. Hence the smaller range observed in the latter case is due to the compensating effect of U_M on U_P, as previously noted.

1e. Alternating Effects.—The previous discussion of monomer reactivity as a characteristic inherent in the monomer and independent of the radical with which it reacts was an admitted oversimplification. If this premise were literally true, the reactivities within a series of monomers like those listed in Table XX would follow the same order with different radicals. Striking exceptions are apparent. As explained earlier, the order chosen is only an approximate average order for different radicals. If the relative reactivities of different radicals were independent of the monomer, rate constants in each column in Table XXI would be in the same ratio to one another. This obviously is not quantitatively correct, although an unmistakable correspondence is evident.

The deviation of r_1r_2 from unity has already been cited as a measure both of alternating tendency and of specificity in the radical-monomer reactions. This product of the reactivity ratios approaches unity only in those cases in which the monomer substituents are similar to one another in their electron-attracting or releasing capacities. Devia-

TABLE XXII.—Monomer Reactivity Ratio Products (50 to 80°)[a]

(From Mayo and Walling[1])

Monomer (e)	α-MeS	Bd	St	IPAc	VAc	VCl	2-ClBd	2,5-DClSt	MMA	VdCl	MA	MVK	β-ClEA	MAN	AN	DEF
α-Methylstyrene (−0.6)																
Butadiene (−0.8)																
Styrene (−0.8)	1.0															
Isopropenyl acetate																
Vinyl acetate (−0.3)			1.0													
Vinyl chloride (0.2)			0.34	0.55	0.63											
2-Chlorobutadiene			0.2		0.4	1.0										
2,5-Dichlorostyrene			.2		.16		0.7									
Methyl methacrylate (0.4)			0.07		.19	.24		.5	1.0							
Vinylidene chloride (0.6)			<.1		.16				0.61							
Methyl acrylate (0.6)			.04		.14				0.9	0.5	0.8					
Methyl vinyl ketone (0.7)			.10			<.12										
β-Chloroethyl acrylate (0.9)			.06									0.9				
Methacrylonitrile (1.0)			.006		.06	<.24			.43		.24		1.1			
Acrylonitrile (1.1)			.006		.02	<.25			.34		.24	.015		0.9		
Diethyl fumarate (1.5)					.004				.56		.17			.34	.56	
Maleic anhydride			.02		.004	.0002	.00006		.13		.06	.004		.24	.13	.6

[a] Numbers in parentheses refer to Price's[20] values for e in the Q,e scheme.

tions from copolymerization ideality are greatest when this dissimilarity is extreme. The relative reactivity $(1/r)$ of a given monomer is considerably greater than would be indicated by its position in Table XX with a radical bearing a dissimilar substituent in the above sense. For example, methyl vinyl ketone displays an abnormally high reactivity with the styrene radical (i.e., it is 3.5 times as reactive as styrene monomer which occupies a higher position in the table), but with the acrylonitrile radical its reactivity is quite normal. The polar properties of the substituent provide what appears to be a satisfactory criterion of similarity.

In Table XXII, from Mayo and Walling,[1] the monomers are arranged in a diagonal order such that the $r_1 r_2$ product tends to be smaller the farther apart the monomer pair in the sequence. The $r_1 r_2$ products are given below one monomer and opposite the other. Monomers with electropositive (i.e., electron-releasing) substituents occur at the top and those with electronegative (i.e., electron-attracting) substituents at the foot of the table. Thus it appears that specificity in the radical-monomer reaction is favored by dissimilarity in the polarization properties of the substituent. The correlation[2] is sufficiently regular to suggest the existence of a semiquantitative relationship.

Price[21] was the first to suggest that the factor of specificity in monomer addition is owing to electrostatic interaction of net charges on the monomer double bond and on the radical arising from polarization by the substituent. Alfrey and Price[22] proposed that the rate constant be written, in analogy with Hammett's equation for the effects of nuclear substituents on the reactivity of aromatic compounds, as follows:

$$k_{12} = P_1 Q_2 \exp(-e_1 e_2) \tag{18}$$

where P_1 and Q_2 relate to the reactivities of the radical $M_1\cdot$ and the monomer M_2, respectively, and e_1 and e_2 are considered to be proportional to the residual electrostatic charges in the respective reacting groups. (Specifically, if ϵ_1 and ϵ_2 represent the charges, D the dielectric constant, and k the Boltzmann constant, $e_1 e_2 = \epsilon_1 \epsilon_2 / r^* D k T$, where r^* is the distance separating the charges in the activated complex.) A further assumption, which can be justified only by the success of the final scheme in interpreting relative reactivities, is that the same e_j applies both to the monomer M_j and to the radical $M_j\cdot$. Then

$$k_{11} = P_1 Q_1 \exp(-e_1^2) \tag{19}$$

$$r_1 = (Q_1/Q_2) \exp\left[-e_1(e_1 - e_2)\right] \tag{20}$$

$$r_2 = (Q_2/Q_1) \exp\left[-e_2(e_2 - e_1)\right] \tag{21}$$

$$r_1 r_2 = \exp\left[-(e_1 - e_2)^2\right] \tag{22}$$

Hence by assigning two parameters, a Q and an e, to each of a set of monomers, it should be possible according to this scheme to compute reactivity ratios r_1 and r_2 for any pair.   In consideration of the number of monomer pairs which may be selected from n monomers—about $n^2/2$—the advantages of such a scheme over copolymerization experiments on each pair are obvious.   Price[20] has assigned approximate values to Q and e for 31 monomers, based on copolymerization of 64 pairs.   The latitude of uncertainty is unfortunately large; assignment of more accurate values is hampered by lack of better experimental data.   Approximate agreement between observed and predicted reactivity ratios is indicated, however.

It will be observed that the product $r_1 r_2$ depends only on the magnitude of the difference in e for the two reactants.   The e values assigned by Price[20] are included in parentheses after the monomers in Table XXII.

The Q,e scheme is vulnerable to criticism on theoretical grounds.   In the first place, there is no a priori justification for the assumption that the electrostatic charge should be the same on both the monomer and on the radical derived from it.   Furthermore, if the alternating effect is electrostatic in origin, it should depend on the reaction medium, i.e., on its dielectric constant.   This has not been found to be the case.[1] Even in solvents of widely differing dielectric constant the reactivity ratios are the same within experimental error.[23]   In the light of these criticisms, the electrostatic basis for justifying the above equations from a theoretical point of view cannot be accepted without serious reservations.   The Q,e scheme is preferably to be regarded as semiempirical.   Although it should not, perhaps, be expected to hold accurately, it has proved useful as a basis for correlating an extensive array of otherwise incoherent data in a remarkably satisfactory fashion.

Mayo and Walling,[1] who have given a penetrating critique of the Q,e scheme, point out that it represents in essence merely a transcription to equation form of the reactivity series of Table XX and the "polarity series" of Table XXII.   Regardless of the manner of interpretation adopted, it is apparent that monomer reactivity in copolymerization depends on two factors.   One of these relates to the intrinsic characteristics of the monomer (and of the activated complex produced from it as well) as they tend to favor its addition to a radical.   As we have seen, the capacity for resonance stabilization in the transition state is of foremost importance in determining the general level of monomer reactivity.   The second factor has to do with the specificity in the reaction between monomer and radical. The greater the dissimilarity in the polarization of the two reacting centers, the more favorable the reaction. This second factor modifies the first to an extent which appears to depend semiquantitatively on the degree of dissimilarity in this respect.

2. THE RATE OF ADDITION COPOLYMERIZATION

The rate of copolymerization in a binary system depends not only on the rates of the four propagation steps but also on the rates of initiation and termination reactions. To simplify matters, the rate of initiation may be made independent of the monomer composition by choosing an initiator which releases primary radicals that combine efficiently with either monomer. The spontaneous decomposition rate of the initiator should be substantially independent of the reaction medium, as otherwise the rate of initiation may vary with the monomer composition. 2-Azo-bis-isobutyronitrile meets these requirements satisfactorily.[17] The rate R_i of initiation of chain radicals of both types $M_1\cdot$ and $M_2\cdot$ is then fixed and equal to $2fk_d[I]$, or twice the rate of decomposition of the initiator I if the efficiency f is equal to unity (see Chap. IV). The relative proportion of the two types of chain radicals created at the initiation step is of no real importance, for they will be converted one into the other by the two cross-propagation reactions of the set (1). Melville, Noble, and Watson[24] presented the first complete theory of copolymerization suitable for handling the problem of the rate. The theory was reduced to a more concise form by Walling,[17] whose procedure is followed here.

Two steady state conditions apply: one to the total radical concentration and the other to the concentrations of the separate radicals $M_1\cdot$ and $M_2\cdot$. The latter has already appeared in Eq. (2), which states that the rates of the two interconversion processes must be equal (very nearly). It follows from Eq. (2) that the ratio of the radical population, $[M_1\cdot]/[M_2\cdot]$, is proportional to the ratio k_{21}/k_{12} of the cross propagation reaction rate constants. The steady-state condition as applied to the total radical concentration requires that the combined rate of termination shall be equal to the combined rate of initiation, i.e., that

$$R_i = 2k_{t11}[M_1\cdot]^2 + 2k_{t12}[M_1\cdot][M_2\cdot] + 2k_{t22}[M_2\cdot]^2 \qquad (23)$$

where k_{t11} and k_{t22} are the bimolecular termination constants for reactions between like radicals, and k_{t12} applies to the cross reaction. In conformity with the notation used in the preceding chapter, the k_t's refer to the termination *processes,* and the coefficient 2 applies because two radicals are destroyed at each incidence of the process.* If $[M_2\cdot]$ is eliminated from Eq. (23) through the use of Eq. (2), then on solving for $[M_1\cdot]$ we obtain

$$[M_1\cdot] = (R_i/2)^{1/2}\{k_{t11} + (k_{t12}k_{12}/k_{21})([M_2]/[M_1]) \\ + (k_{t22}k_{12}^2/k_{21}^2)[M_2]^2/[M_1]^2\}^{-1/2} \qquad (24)$$

The total rate of polymerization, obtained by adding Eqs. (3) and (4) and then eliminating $[M_2\cdot]$ through the use of Eq. (2), is

$$- d([M_1] + [M_2])/dt = [M_1\cdot]\{k_{11}[M_1] + 2k_{12}[M_2] \\ + (k_{22}k_{12}/k_{21})[M_2]^2/[M_1]\} \qquad (25)$$

Substituting Eq. (24) in (25)

$$- \frac{d([M_1] + [M_2])}{dt} \\ = \frac{(r_1[M_1]^2 + 2[M_1][M_2] + r_2[M_2]^2)(R_i^{1/2}/\delta_1)}{\{r_1^2[M_1]^2 + 2(\phi r_1 r_2 \delta_2/\delta_1)[M_1][M_2] + (r_2\delta_2/\delta_1)^2[M_2]^2\}^{1/2}} \qquad (26)$$

where

$$\delta_1 = (2k_{t11}/k_{11}^2)^{1/2}; \qquad \delta_2 = (2k_{t22}/k_{22}^2)^{1/2}$$

$$\phi = k_{t12}/2k_{t11}^{1/2}k_{t22}^{1/2}$$

The δ's will be recognized as reciprocal square roots of the familiar $k_p^2/2k_t$ ratio of prominence in Chapter IV. The symbol ϕ represents the ratio of half the cross termination rate constant to the geometric mean of the constants for terminations involving like radicals. The factor $\frac{1}{2}$ enters into this ratio because the reaction between unlike radicals is statistically favored by a factor of two compared with reaction between like species. Thus at equal concentrations of the two radicals there will be twice as many 1,2 collisions as either 1,1 or 2,2 collisions. Hence if the probability of reaction at a collision of each kind were the same, we would have $k_{t12}=2k_{t11}=2k_{t22}$. The quantity ϕ therefore compares the cross termination rate constant to the geometric mean of the rate constants for the like reactions with appropriate correction for the a priori advantage of the former. A value of ϕ less than unity would signify a comparatively unfavorable reaction

---

* Our notation differs from that of Walling[1,17] whose k_{t1} and k_{t2} are equivalent to our $2k_{t11}$ and $2k_{t22}$ respectively. The k_{t12} used here is equivalent to the same symbol in Walling's notation.

between unlike radicals; a value exceeding unity corresponds to preference for cross termination.*

If one monomer only is present, Eq. (26) reduces to

$$- d \ln [M_1]/dt = R_i^{1/2}/\delta_1 \tag{27}$$

which corresponds to Eq. (IV–15). The ratio of the homopolymerization rates, $-d \ln [M_1]/dt$ and $-d \ln [M_2]/dt$, under equivalent initiation conditions equals the ratio δ_2/δ_1. Thus rate measurements on the pure monomers yield two of the quantities, $R_i^{1/2}/\delta_1$ and δ_2/δ_1, required in Eq. (26). The monomer reactivity ratios r_1 and r_2 can be derived from copolymer composition studies, which leaves undetermined only the ratio ϕ. A single measurement of the rate of copolymerization is therefore sufficient in principle to establish the value of ϕ, provided the other quantities mentioned have been determined. Preferably, the rate should be measured for a series of monomer compositions and the average taken of the values of ϕ computed at each composition. Typical results of Walling[17] on the styrene–methyl methacrylate system are compared in Fig. 27 with theoretical curves calculated from Eq.

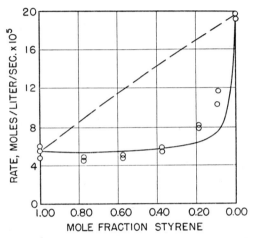

FIG. 27.—The rate of copolymerization of styrene and methyl methacrylate at 60°C in the presence of azo-bis-isobutyronitrile (1 g./l.) plotted against the mole fraction of styrene. Broken line has been calculated from Eq. (26) assuming $\phi = 1$. Solid line represents calculated curve for $\phi = 13$. (Walling.[17])

(26) with $\phi = 1$ and 13. The values chosen for r_1 and r_2 are those given in Table XX, and the rates for the pure monomers define the quantities $R_i^{1/2}/\delta_1$ and δ_2/δ_1. The curve for $\phi = 1$ happens to be nearly linear in this particular case. This is not typical. The occurrence of rates below the $\phi = 1$ line is typical, however, and it means that ϕ is greater than unity, i.e., that the cross termination rate exceeds the geometric mean for mutual termination of like radicals.

Results derived from copolymerization rate studies are summarized

---

* De Butts[25] has succeeded in integrating the copolymerization rate equation to obtain relationships between the monomer concentrations and the time.

TABLE XXIII.—CROSS TERMINATION RATE RATIOS

System	Investigators	ϕ	r_1r_2
Styrene-methyl methacrylate	Walling[17]	13	0.24
Styrene-methyl methacrylate	Melville and Valentine[26]	14	
Styrene-methyl acrylate	Walling[17,1]	50	.14
Styrene-butyl acrylate	Arlman and Melville[27]	~100	.07
Styrene-p-methoxystyrene	Melville et al.[28]	1	.95
Methyl methacrylate-p-methoxystyrene	Melville et al.[28]	~20	.09

in Table XXIII.   The cross termination ratio ϕ exceeds unity for every system investigated except the similar pair, styrene–p-methoxystyrene. Moreover ϕ is greater the smaller the value of r_1r_2, i.e., preference for cross termination seems to go hand in hand with preference for the cross propagation reactions.   Differences in polarity may be responsible for both.[26,28] It will be observed, however, that ϕ varies over a wider range than r_1r_2, indicating that the termination reaction is more sensitive to polar effects than is the propagation process.*

Equation (26) offers no simple generalizations concerning the rate of a copolymerization compared to the rates of polymerization of the pure monomers at the same initiation rates.   A strong alternating tendency (low r_1r_2) signifies a more rapid chain propagation owing to higher rate constants for the cross propagation steps.   From the examples cited, however, this accelerated chain growth appears to be more than offset by the large concomitant cross termination rate. Hence, at a fixed rate of initiation, the copolymerization of monomers well separated from one another in the polarity series (Table XXII) is likely to be slower than the mean of the rates for the separate polymerizations.

In thermal polymerization where the rate of initiation may also vary with composition, an abnormal cross *initiation* rate may introduce a further contribution to nonadditive behavior.   The only system investigated quantitatively is styrene–methyl methacrylate, rates of thermal copolymerization of which were measured by Walling.[17]   The rate ratios appearing in Eq. (26) are known for this system from studies on the individual monomers, from copolymer composition studies, and from the copolymerization rate at fixed initiation rate.   Hence a single measurement of the thermal copolymerization rate yields a value for R_i.   Knowing k_{i11} and k_{i22} from the thermal initiation rates for either monomer alone (Chap. IV), the bimolecular cross initiation rate constant k_{i12} may be calculated.   At 60°C it was found to be 2.8 times that

* Actually, $\sqrt{r_1r_2}$ should be compared with ϕ.   On this basis the greater sensitivity of the cross termination reaction is even more marked.

for pure styrene and about 3000 times that for pure methyl methacrylate at the same temperature; the thermal cross initiation rate constant is thus nearly 100 times the geometric mean of the constants for the pure monomers. This preference for the cross initiation process has been attributed also to polar interactions in the transition state.[1,17] The thermal initiation reaction appears to be even more sensitive than the termination reaction to polar interactions between dissimilar reactants. As a consequence of this fact, the rate of a *thermal* copolymerization, in contrast to copolymerization at fixed initiation rate, may readily exceed that of either monomer alone. The extremely rapid (sometimes explosive) copolymerization of maleic anhydride with monomers such as butadiene, styrene, and vinyl acetate appears to represent an extreme case of favorable cross initiation. Thermal polymerization is unknown in maleic anhydride alone, and its occurrence has not been established in either pure butadiene or pure vinyl acetate. Maleic anhydride is a strong electron acceptor monomer, whereas those with which it polymerizes spontaneously are electron donors; hence polar interaction should be very large indeed, and this may account for the rapidity of the thermal copolymerizations mentioned.[1]

3. EMULSION POLYMERIZATION

Vinyl monomers may be polymerized at favorable rates in an aqueous medium containing an emulsifier and a water-soluble initiator. A typical simple "recipe" would consist of the following ingredients with their proportions indicated in parts by weight: 100 of monomer, 180 of water, 2 to 5 of a fatty acid soap, and 0.1 to 0.5 of potassium persulfate. Cationic soaps (e.g., dodecylamine hydrochloride) may be used instead of the fatty acid soap, and various other initiators may replace the persulfate (e.g., hydrogen peroxide and ferrous ion, or a water-soluble organic hydroperoxide).

Not only is emulsion polymerization characteristically more rapid than bulk or solution polymerization at the same temperature, but surprisingly enough the average molecular weight may be much greater than could be obtained at the same rate in bulk polymerization. According to Eq.(IV–27), twice the kinetic chain length ν for polymerization in a *homogeneous* system is given by

$$2\nu = (k_p^2/k_t)[M]^2/R_p$$

and this represents the maximum average degree of polymerization $\bar{x}_n$ obtainable at the rate R_p. Monomers for which k_p^2/k_t is low, e.g., isoprene, methacrylonitrile, and to a lesser degree butadiene and styrene (see Table XVII, p. 158), require that the rate of polymerization be intolerably low if chain lengths exceeding 10^4 are to be obtained by homogeneous polymerization. It may be shown that this equation does not apply to emulsion polymerization, where a high rate is not incompatible with a high degree of polymerization. The basis for this difference lies in the fact that in a typical emulsion polymerization system the initiator radicals are generated in the aqueous phase, whereas growth and termination occur in the dispersed oil phase. A further advantage of emulsion polymerization consists in the ease with which the emulsion may be handled, in contrast to the extremely viscous solution of polymer in unreacted monomer or solvent obtained by bulk or solution polymerization. This advantage is especially important in commercial scale operation.

3a. Qualitative Theory.—When soap is added to water in excess of a generally low critical concentration, micellar aggregates are formed.[29,30] Depending on the particular soap and on other factors, these may consist of from 50 to 100 or more soap molecules.[31] The structure of the smaller (50 to 100 Å) micelles occurring at low ionic strengths (i.e., at low salt concentrations) remains unsettled at present. McBain[29] argued for lamellar micelles consisting of two layers of parallel soap molecules having their polar "heads" in the outer surfaces and their hydrocarbon tails meeting at the central plane of the micelle. Hartley,[30] on the other hand, proposed a spherical micelle with polar groups on the surface and hydrocarbon chains arranged somewhat irregularly in the interior. Energetic considerations[31] favor this latter structure, which is more closely related to the much larger rodlike micelles prevalent at higher salt concentrations.[32] According to light-scattering[32] and streaming birefringence measurements,[33] these rodlike micelles range from 1000 to 3000 Å in length, depending on the salt concentration; their diameters approximate twice the length of the detergent molecule. Presumably the polar heads of the soap molecules are distributed more or less evenly over the surface. The addition of inorganic salts lowers the electrostatic energy of the polar layer and thus allows formation of a larger (i.e., longer) micelle.[31]

The solubility in water of a monomer, or other sparingly soluble organic compound, is increased considerably by the presence of soap or detergent micelles.[34,35] X-ray and light-scattering measurements[32] show that the micelles increase in size when monomer is added—a clear indication that the "solubilized" monomer occurs in the micelles, presumably among the hydrocarbon chains occupying the interior. The proportion of monomer initially present in an emulsion polymerization mixture is much greater than can be accommodated by the soap micelles; hence most of the monomer occurs as droplets much larger in size (a micron or more in diameter) which may be stabilized by a portion of the soap. The picture thus emerges that while most of the monomer initially is present in macroscopic droplets, most of the soap occurs in micelles. More important, the micelles present a much greater total surface area than the droplets, although the total volume of micelles is considerably less than that of the droplets.

The immediate question which must be resolved in advance of inquiry into the detailed mechanism of emulsion polymerization concerns the locus of polymerization in these evidently complex systems. Conceivably it might be in the monomer droplets, in the aqueous phase, within the micelles, or at an interface; or different steps of the process might occur in different loci. The monomer droplets are ruled out as a polymerization locus by the fact that upon interruption of the agitation at an intermediate stage of the polymerization, the remaining droplets coalesce to yield a supernatant layer containing no measurable quantity of polymer. (The oil droplets become unstable after polymerization to the extent of several percent owing to the removal of soap from the droplets by the polymer particles which are formed; see below.) Moreover, Harkins[34] reported direct observations of individual monomer droplets which diminished in size as the polymerization progressed, without leaving a residue of polymer exceeding 0.1 percent of the initial volume of the droplet. Such observations led early investigators to conclude that polymerization occurs in the "aqueous phase," although no clear distinction was drawn between the water solution and the micelles. Polymerization can be effected without soap of any kind, or even in the absence of droplets of liquid monomer. Butadiene or styrene, for example, may be supplied via the vapor phase to the water solution containing initiator. Hence initiation of polymer chains in a homogeneous aqueous medium containing no emulsifier is possible, and, moreover, the chains grow somehow in spite of the low concentration of dissolved monomer. It is necessary to remark, however, that the polymerization is very slow in the absence of soap and that in the presence of soap it increases with soap concentration. Typical results are shown in Fig. 28.

After only a small percentage of the monomer has been converted to polymer (in the presence of emulsifier), the initially low surface tension of the aqueous emulsion rises rather abruptly, indicating a decrease in the soap concentration in the aqueous phase of the emulsion. The soap concentration is then too low to maintain micelles, which may therefore be abandoned as a locus for further polymerization beyond this point. As additional evidence of the depletion of soap in the aqueous phase, monomer droplets are no longer stable, and upon discontinuing agitation a supernatant monomer layer is readily formed.

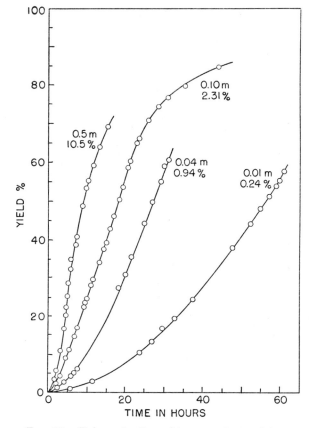

Fig. 28.—Polymerization of isoprene in emulsion at 50°C using 0.3 g of $K_2S_2O_8$ per 100 g. of monomer, and with the amounts of soap (potassium laurate) indicated in weight percent and in molality m. (Harkins[34].)

As was mentioned above, no appreciable quantity of polymer is to be found in this layer. The polymer which has been formed occurs in numerous polymer particles, which at this stage may be no more than 200 to 400 Å in diameter. These particles appear to have acquired nearly all of the soap, which is largely adsorbed on the polymer particle surfaces even at this early stage of the polymerization. The polymer particles may be observed by electron microscopy or by light-scattering measurements on the emulsion from which unreacted monomer has been removed by vaporization ("stripping"). As the polymerization progresses further, the polymer particles grow in size, but their number remains sensibly constant. Over this portion of the polymerization, i.e., from several percent to 60 to 80 percent conversion, the rate is nearly constant, as is shown by the curves in Fig. 28 with the exception of that for the lowest soap concentration. At 60 to 80 percent conversion all monomer has been absorbed by the polymer particles, and none remains to form a supernatant layer. The rate then decreases rapidly with further conversion. The final emulsion consists of discrete polymer particles having an average diameter in the range 500 to 1,500 Å,* i.e., larger than the initial micelles (40 to 80 Å) but smaller than the initial monomer droplets (10,000 Å or more).

Harkins[34] concluded from a critical appraisal of the characteristics of emulsion polymerization as outlined briefly above that it is necessary to postulate two principal loci of polymerization, one operative at the outset and the other later on when the character of the emulsion has been materially altered. The micelles, in which monomer is present in relative abundance compared with the remainder of the initial solution, provide a favorable environment for the initiation of chains by radicals generated in the surrounding aqueous medium. As polymerization proceeds in a micelle thus activated, more monomer is supplied to it by the surrounding solution. Consequently the polymer chain continues to grow beyond the size of the original micelle before termination is brought about by the action of a second radical. Polymer particles may also be generated in the aqueous phase to an extent depending on the solubility of the monomer and on the amount of soap. In the case of styrene at a soap concentration of 1 percent or more, the micelles represent the major locus of polymer particle formation, however.

As more and more micelles are converted to polymer particles which quickly grow to sizes surpassing the initial micelles, the soap is rapidly acquired by the polymer particles; the soap concentration in the aqueous phase falls to a low level and micelles disappear. Further polymerization takes place almost exclusively in existing polymer particles. It is not surprising that the polymer particles rather than monomer droplets represent the predominant locus of polymerization when it is considered that the former present a very much greater surface than the large (and unstable) monomer droplets. It follows that radicals generated in the aqueous phase should be efficiently absorbed by the polymer particles in preference to the monomer droplets. Radicals entering the polymer particles initate polymerization of the monomer dissolved in the polymer particle. Monomer diffuses through the aqueous phase from the monomer droplets to the polymer particles. The monomer droplets thus serve as reservoirs supplying the polymer

---

* A spherical particle of diameter 500 Å and unit density would have a "molecular weight" of about 4×10^7, i.e., large enough to accommodate many polymer molecules.

particles with monomer to replace that consumed by polymerization. The concentration of monomer in the polymer particles appears to be restricted by the surface tension at the particle-water interface, which opposes swelling of the (small) polymer particles. Otherwise, they might be expected to swell indefinitely until all of the supernatant monomer would be absorbed. Polymer particles in the emulsion polymerization of styrene or butadiene normally contain from 45 to 65 percent of monomer, depending to some extent of course on the available emulsifier. At an advanced stage of the polymerization when the supply of excess monomer has been exhausted, the monomer concentration in the droplets necessarily decreases, and this is reflected in a diminishing rate.

In recapitulation, the Harkins[34] theory differentiates the polymer particle formation stage from the subsequent polymerization. The latter occurs in the polymer particles into which both radicals and monomer diffuse from the surrounding medium. The former stage usually is virtually complete when no more than 2 or 3 percent of the monomer has polymerized. Hence nearly all of the polymer present at a high conversion will have been generated in the polymer particles. In confirmation of his theory, Harkins cited the fact that the greater the quantity of soap the more numerous the particles per unit volume, and therefore the smaller their average size at a given conversion. Also, the increase in the rate with soap concentration can be attributed to the greater number of polymerization loci. Conclusive proof of the Harkins theory is found in quantitative considerations evoked by his hypotheses.

3b. Quantitative Theory.[36,37]—The foregoing qualitative account raises a number of questions concerning the comparative rates of the various processes involved. For example, considering the low solubility of monomer in the aqueous serum, is diffusion of monomer to the polymer particle rapid enough to keep pace with the polymerization? Is it reasonable to suppose that initiator radicals are efficiently captured by the polymer particles (or micelles when present) before other fates befall them? What will be the average radical concentration in a polymer particle? Smith and Ewart[36] developed a theory which answers these questions and offers a basis for quantitative treatment of emulsion polymerization. The salient features of their theory are set forth below.

Under conditions normally employed, the rate ρ of generation of radicals from the initiator is of the order of 10^{13} per cc. per second, and the number of polymer particles is about 10^{14} per cc. (10^{13} to 10^{15} per cc. would include nearly all cases). Hence if all of the initiator radicals enter polymer particles, a particle will acquire a radical once in about 10 seconds, on an average. Now it is readily shown that the diffusion current I into a spherical particle of radius r is given by $I = 4\pi r D \Delta c$, where D is the diffusion constant and Δc is the difference in concentration of the diffusing substance far from the particle and at its surface. Presuming the concentration at the surface to be zero, Δc may be replaced by the equilibrium radical concentration $[R\cdot]_{aq}$ in the aqueous medium. Hence

$$[R\cdot]_{aq} = I/4\pi r D \tag{28}$$

The diffusion constant of a primary radical must be of the order of 10^{-5} cm.^{2} sec.$^{-1}$, the radius r is about 5×10^{-6} cm., and as we have seen $I \sim 10^{-1}$ per second. Hence $[R\cdot]_{aq} \sim 10^8$ radicals per cc. But the radicals are being generated at a rate of 10^{13} cc.$^{-1}$ sec.$^{-1}$; hence the average lifetime of a radical from generation to capture by a polymer particle will be only 10^{-5} sec.[37] The rate of termination by reaction between two radicals in the aqueous phase at the calculated equilibrium concentration, 10^8 radicals per cc., will be given by

$$(R_t)_{aq} = 2k_t[R\cdot]_{aq}^2$$

The termination constants k_t found previously (see Table XVII, p. 158) are of the order of 3×10^7 l. mole^{-1} sec.$^{-1}$ Conversion to the specific reaction rate constant expressed in units of cc. molecule^{-1} sec.$^{-1}$ yields $k_t \cong 5 \times 10^{-14}$. At the radical concentration calculated above, 10^8 per cc., the rate of termination should therefore be only $\sim 10^3$ radicals cc.$^{-1}$ sec.$^{-1}$, which is many orders of magnitude less than the rate of generation of radicals. Hence termination in the aqueous phase is utterly negligible, and it may be assumed with confidence that virtually every primary radical enters a polymer particle (or a micelle). Moreover the average lifetime of a chain radical in the aqueous phase (i.e., 10^{-5} sec.) is too short for an appreciable expectation of addition of a dissolved monomer molecule by the primary radical prior to its entrance into a polymer particle.

We turn next to consideration of the concentration of radicals in a polymer particle. The rate of termination in a particle of volume $\mathcal{U}$ may be written

$$(R_t)_{\text{particle}} = 2k_t n(n-1)/\mathcal{U}$$

where n is the number of radicals in the particle and k_t is again the specific rate constant. A particle of the diameter considered above will possess a volume of about 5×10^{-16} cc.; hence $2k_t/\mathcal{U} \cong 2 \times 10^2$ per second per radical. If the particle contains two radicals, mutual termination is to be expected within an interval which is the reciprocal of twice this quantity, or 2.5×10^{-3} sec. This is much smaller than the interval between radical captures. It is so small in fact that a radical which enters a particle already containing a chain radical may scarcely have an opportunity to add a monomer before it is terminated. (Thus $k_p \sim 100$ l. mole^{-1} sec.$^{-1}$ for styrene or butadiene at 50°C. Hence, at a monomer concentration of 5 molar the rate of addition of monomers to a radical is ~ 500 sec.$^{-1}$, or a mean life between successive addition steps of 2×10^{-3} sec.) To a good approximation, therefore, we may consider that at any given time *a particle will have zero or one radical;* the brief intervals in which it has two radicals are negligible. Following termination by a second radical, a given particle will be dormant until another radical enters, i.e., for a period of about 10 seconds. After another interval of equal average length, a second radical will be captured, whereupon mutual termination of the pair occurs and the cycle is repeated. Any given polymer particle will contain one radical half of the time, and the other half of the time it will contain none. At any given instant, therefore, half of the particles will be activated, and each of these will contain a single growing radical.[36,37]

It remains to show that diffusion of monomer from the aqueous phase, in which its concentration is quite low, is adequate to keep the polymer particle amply supplied. As in the preceding discussion of radical diffusion, we may consider the (styrene) monomer concentration $[M]_{aq}$ required to maintain a current I of some 500/2 monomer molecules per second required by a particle containing a single active radical half of the time. Again taking $D \sim 10^{-5}$ cm.^{2} sec.$^{-1}$, $[M]_{aq}$ according to the equation corresponding to Eq. (28) is about 4×10^{11} molecules per cc., or only about 10^{-9} molar. Since the saturation concentration of monomer in water invariably exceeds this figure by a wide margin, monomer is easily supplied to the polymer particles at the required rate.

The highly important deduction that one-half of the polymer particles will contain a single growing radical and the other half will contain none at any given time was first set forth by Smith and Ewart.[36] They immediately pointed out that the rate of polymerization *per cubic centimeter of water* should then be given by

$$R_p = k_p(N/2)[M] \tag{29}$$

where N is the number of polymer particles in a cubic centimeter of the aqueous phase and $[M]$ represents the monomer concentration in the polymer particles. In other words, the rate should depend principally on the *number of particles*, the monomer concentration being subject only to minor variations as long as excess monomer droplets remain. It should be *independent of the rate of generation of radicals* by the initiator, and it should not depend on the size of the particles (for a fixed number of particles), provided they are not too large.*

Smith[37] carried out a series of crucial experiments specifically designed to test the above predictions. He first subjected a polystyrene latex containing particles whose diameters had been measured with the electron microscope to further emulsion polymerization with added quantities of styrene monomer. No additional emulsifier was provided. Particle diameters measured for various degrees of additional polymerization showed an average growth in volume per particle which was approximately equal to the ratio of final to initial polymer. This relationship held even for a seventyfold increase. It was concluded that the number of particles remains essentially constant during polymerization in the absence of micellar soap. The approximate constancy of the rate of polymerization over a wide range of conversion (see Fig. 28) follows according to Eq. (29) from this constancy in N and from the approximately unchanging concentration of monomer in the polymer particles while the supply of monomer droplets has not been exhausted.

Smith then showed that the rate of polymerization *per particle* in a seed latex containing no micellar soap was independent of the following factors: (1) the particle size over a twenty-five-fold range in particle volume; (2) the number of particles from about 2×10^{12} to 2×10^{14} per cc. (a decrease was observed at higher particle concentrations for reasons not satisfactorily explained); (3) the persulfate initiator concentration over a sixteenfold range. Thus the simple rate expression (29) has been fully confirmed. The most striking feature is the lack of dependence of the rate of emulsion polymerization on the initiator concentration (after cessation of the generation of new polymer particles; see below). An increase in the amount of initiator merely increases the frequency of the alternation between activity and inactivity in a particle, but since it is activated just half of the time regardless of the frequency of radical capture, there is no effect on the rate.

Equation (29) embodies the suggestion of a method for determining the absolute value of the rate constant k_p for the propagation reaction. It may be calculated directly from the rate of the emulsion polymeri-

---

* It will be apparent that if the particle is larger in volume by several orders of magnitude than the size considered above, it may then accommodate two or more radicals without immediate termination. The typical characteristics of emulsion polymerization vanish, and we have what is often referred to as bead polymerization, which resembles ordinary bulk polymerization in its kinetics.

zation in a latex containing a known number of particles. The latter quantity is difficult to determine accurately. The method nevertheless was applied successfully by Smith[37] and subsequently by Morton[38] and co-workers, whose results for butadiene and isoprene are quoted in Table XVII (p. 158).

Each primary radical which enters an inactive particle is presumed to start the growth of a new polymer chain, and this chain is terminated almost immediately following capture of another radical. If it is assumed that chain transfer may be disregarded, the average degree of polymerization under these conditions should equal the ratio of the rate of growth of a chain to the frequency ρ/N of capture of primary radicals; i.e.[36,37]

$$\bar{x}_n = k_p N [M]/\rho \tag{30}$$

Both the rate of polymerization (Eq. 29) and the degree of polymerization should vary directly with the number N of particles, but the degree of polymerization, unlike the rate, depends inversely on the rate ρ of radical generation. In support of this mechanism of chain termination, Smith[39] showed, using tracer methods with radioactive persulfate in the emulsion polymerization of styrene, that the polymer contained two initiator radicals for each polymer molecule produced. He showed also that the average degree of polymerization at fixed persulfate concentration varied directly as the rate of polymerization in latices containing different concentrations of particles. In other words, the number of polymer molecules formed per second depends only on ρ (hence, on the initiator concentration which was held constant in Smith's experiments) and not on the number of particles. This observation confirms the mechanism postulated for polymer molecule formation. The average molecular weight at very high particle concentrations where $\bar{x}_n$ exceeded 5×10^4 was, however, somewhat lower than would have been predicted on the above basis. The deviation was attributed to chain transfer with styrene monomer, which places an upper limit on the degree of polymerization attainable.

3c. The Number of Particles.—We have thus far considered the number of polymer particles in the emulsion as an independent variable. Ordinarily, the number of polymer particles will depend on conditions prevailing at the outset when they are being created. Specifically, their ultimate number will depend on the emulsifier and its initial concentration, and also on the rate at which primary radicals are generated. Since the number N of particles plays such an important part in determining both the rate and the degree of polymerization, it is desirable to consider briefly Smith and Ewart's[36] theoretical estimation of it.

For the purpose of simplifying the problem, they first chose to assume that substantially all of the primary radicals enter soap micelles until the polymer particles become so numerous as to require all of the soap to cover their surfaces, leaving none to maintain micelles. The rate of generation of particles as long as micelles remain may be equated to ρ under these conditions; after micelles disappear, this rate is of course negligible. The above tactical assumption notwithstanding, the existing polymer particles will acquire a sufficient fraction of the primary radicals to cause each of them to polymerize at the average rate $(k_p/2)[M]$. If the solubility of monomer in the particles is such as to maintain a volume fraction v_M of monomer, the rate μ of increase in volume of a particle will be given by

$$\mu = (k_p/2)[M]V_u/(1 - v_M) \tag{31}$$

where V_u is the volume of a monomer unit. At a time t the volume of a particle formed at time τ will be $\mu(t-\tau)$, and its area, assuming the particle to be spherical, will be

$$a_{\tau,t} = [(4\pi)^{1/2}3\mu(t - \tau)]^{2/3}$$

The combined area $\mathcal{A}_t$ of all particles present at time t is given by the sums of the areas of all particles formed from time $t=0$ to τ. Or, since $\rho d\tau$ particles are generated in the interval τ to $\tau+d\tau$

$$\mathcal{A}_t = [(4\pi)^{1/2}3\mu]^{2/3} \int_0^t (t - \tau)^{2/3}\rho d\tau$$

Upon integration with ρ constant

$$\mathcal{A}_t = (3/5)[(4\pi)^{1/2}3\mu]^{2/3}\rho t^{5/3} \tag{32}$$

Soap is assumed to form a continuous monolayer over the polymer particles at the point of exhaustion of micellar soap. If we let a_s represent the area thus occupied by a gram of soap, the total area of the particles in one cc. at the time t_1 when this occurs will be $c_s a_s$, where c_s is the soap concentration in grams per cc. Equating $\mathcal{A}_t$ at $t=t_1$ to this quantity, we have from Eq. (32)

$$t_1 = [5^{3/5}/3(4\pi)^{1/5}](c_s a_s/\rho)^{3/5}\mu^{-2/5}$$
$$= 0.53(c_s a_s/\rho)^{3/5}\mu^{-2/5} \tag{33}$$

Hence the number of particles should be

$$N = \rho t_1 = 0.53(c_s a_s)^{3/5}(\rho/\mu)^{2/5} \tag{34}$$

The number thus calculated is too large on account of the assumption that all primary radicals enter micelles so long as micelles remain. Making the alternate assumption that the micelles and polymer particles compete for the radicals in proportion to their respective total areas, Smith and Ewart[36] obtained an otherwise equivalent expression except for the lower numerical factor 0.37. Since this assumption introduces an error in the opposite direction,* the number of particles, to an accuracy well within the limitations of other assumptions, may be taken as

$$N = 0.4(c_s a_s)^{3/5}(\rho/\mu)^{2/5} \tag{35}$$

The increase in N, and therefore in the rate as well, with initial soap concentration is thus explained. Quantitative results agree approximately with the predicted three-fifths power dependence.[37] The prediction of an increase in polymerization rate with $\rho^{2/5}$ also has been confirmed by experiments[37] at variable initiator concentrations.† Most important of all, the actual number of particles N calculated from Eq. (35) agrees within a factor of two with that observed. It is thus apparent that the theory of emulsion polymerization developed by Harkins[34] and by Smith and Ewart[36,37] has enjoyed spectacular success in accounting for the unique features of the emulsion polymerization process.

   3d. Concluding Remarks on Heterophase Polymerization.—In order to illustrate the contrasting characteristics of emulsion and bulk phase polymerization, let us compare the rates of polymerization R_p and the degrees of polymerization $\bar{x}_n$ for styrene in the two systems at corresponding rates (R_i and ρ) of primary radical generation. At 60°C the rate constants (see Table XVII, p. 158) are

$$k_p = 176 \text{ l. mole}^{-1} \text{ sec.}^{-1} = 3 \times 10^{-19} \text{ cc. molecule}^{-1} \text{ sec.}^{-1}$$

$$k_t = 3.6 \times 10^7 \text{ l. mole}^{-1} \text{ sec.}^{-1} = 0.6 \times 10^{-13} \text{ cc. molecule}^{-1} \text{ sec.}^{-1}$$

Let the monomer concentration $[M]$ be 5 molar both in the bulk phase and in the emulsion particles. The comparison is set forth below:

---

   * The rate at which a given (small) particle acquires radicals will be proportional to its radius and not to its surface area (compare Eq. 28). Hence the small micelles will compete more favorably with the larger polymer particles than this calculation would lead one to expect.

   † This increase with ρ, and therefore with initiator concentration, is not at variance with the previous assertion that the rate is independent of ρ *for a given number of particles*. We are concerned here with the effect of initiator concentration on the number of particles generated from the original emulsion of monomer.

Bulk Polymerization of Styrene at 60°C

If:

$$R_i = 8 \times 10^{-9} \text{ mole l.}^{-1} \text{ sec.}^{-1} = 5 \times 10^{12} \text{ radicals cc.}^{-1} \text{ sec.}^{-1}$$

then:

$$[M \cdot] = (R_i/2k_t)^{1/2} = 1.05 \times 10^{-8} \text{ mole l.}^{-1} = 0.63 \times 10^{13} \text{ radicals cc.}^{-1}$$

$$R_p = k_p[M \cdot][M] = 0.93 \times 10^{-5} \text{ mole l.}^{-1} \text{ sec.}^{-1}$$

$$= 5.5 \times 10^{15} \text{ molecules cc.}^{-1} \text{ sec.}^{-1}$$

$$\bar{x}_n = 2\nu = k_p(2/k_t R_i)^{1/2}[M] = 2.3 \times 10^3$$

(The average degree of polymerization is taken as twice the kinetic chain length ν, i.e., chain transfer is neglected. Under conditions such that the kinetic chain length exceeds about 10^4, this would not be permissible since $C_M \sim 5 \times 10^{-5}$; see Eq. (IV-37). The initiation rate chosen corresponds roughly to that obtained with 5×10^{-4} molar benzoyl peroxide.)

Emulsion Polymerization of Styrene at 60°C

If:

$$\rho = 5 \times 10^{12} \text{ radicals sec.}^{-1} \text{ (cc. of water)}^{-1}$$

and

$$N = \begin{cases} 10^{13} \text{ particles per cc. of water or} \\ 10^{15} \text{ particles per cc. of water} \end{cases}$$

then:

$$R_p = k_p(N/2)[M] = 440 N = \begin{cases} 4.4 \times 10^{15} \text{ molecules cc.}^{-1} \text{ sec.}^{-1} \\ 4.4 \times 10^{17} \text{ molecules cc.}^{-1} \text{ sec.}^{-1} \end{cases}$$

$$\bar{x}_n = k_p N[M]/\rho = 1.76 \times 10^{-10} N = \begin{cases} 1.76 \times 10^3 \\ 1.76 \times 10^5 \end{cases}$$

(The rates for emulsion polymerization are expressed per cc. of water phase, and the amount of styrene usually is less than the quantity of water. Hence the rates per unit weight of monomer will be somewhat greater than the figures given above.)

Since the same propagation rate constant applies to both bulk and emulsion polymerization, comparable rates of polymerization R_p must obtain when the number of emulsion particles is twice the number of radicals at the steady state in the bulk polymerization. An increase in the bulk rate at the given temperature can only be realized by an increase in the rate of initiation and, thus, an increase in the

radical concentration.   But this would necessarily result in a decrease
in the average degree of polymerization.*   In emulsion polymerization,
on the other hand, the concentration of radicals may be increased merely by increasing the number N of particles per unit volume.   If the
rate ρ of radical generation is kept constant, the degree of polymerization *increases* as well.   Thus the segregation of radicals individually
in emulsion particles makes possible the maintenance of a much higher
radical population without increasing the termination rate.[40]   The termination rate *per particle* actually decreases when N is increased at
fixed ρ.   It is this segregation of radicals in discrete particles which
prevents their termination of one another, and this in turn is responsible for the advantageous features of emulsion polymerization.   The
conditions required for optimum rate and degree of polymerization
are apparent.   A large number of particles per unit volume (and therefore small particle size) is essential for a high rate.   If a high degree of
polymerization is desired, the ratio of N to ρ must be large; i.e., for a
given particle concentration, a low initiation rate is required.   This
would seem to argue for a low initiator concentration if high molecular
weight is an important objective.   However, it must be remembered
that a low value of ρ (i.e., low initiator concentration) during the par-
ticle-generating phase of the polymerization favors a smaller number of
particles.   To this extent the requirement of large N and small ρ during the main portion of the polymerization stand in opposition to one
another.

Polymerizations conducted in nonaqueous media in which the
polymer is insoluble also display the characteristics of emulsion polymerization.   When either vinyl acetate[41] or methyl methacrylate[42]
is polymerized in a poor solvent for the polymer, for example, the rate
accelerates as the polymerization progresses. This acceleration, which
has been called the gel effect,[41] probably is associated with the precipitation of minute droplets of polymer[43] highly swollen with monomer.
These droplets may provide polymerization loci in which a single chain
radical may be isolated from all others.   A similar heterophase polymerization is observed even in the polymerization of the pure monomer
in those cases in which the polymer is insoluble in its own monomer.
Vinyl chloride,[44] vinylidene chloride,[45] acrylonitrile, and methacrylonitrile polymerize with precipitation of the polymer in a finely divided
dispersion as rapidly as it is formed.   The reaction rate increases as
these polymer particles are generated.   In the case of vinyl chloride[44]

* The decrease in degree of polymerization as the rate is increased is less
serious for a monomer such as vinyl acetate or methyl methacrylate for which
k_p^2/k_t is larger.   See Chap. IV.

at least, the rate of polymerization of monomer by benzoyl peroxide is increased by the addition of previously formed polymer. The polymer alone in the absence of initiator is without effect. If the monomer is diluted with a solvent capable of dissolving the polymer, normal polymerization occurs (in the presence of an initiator) without evidence of acceleration.

The above facts support the view that these heterophase polymerizations resemble emulsion polymerization. While the initiator may be soluble in the polymer particles, it is to be remembered that initiator molecules decomposing within the particles will release radicals in pairs; and, if the particles are small, the termination of two radicals will closely follow the generation of a pair. The number of radicals (one or zero) in the particle will therefore remain unchanged. Only the capture of single radicals from the surrounding solution will be significant insofar as the rate is concerned. Hence the isolation of single radicals in discrete particles may bring about a higher level of radical concentration than would exist in a homogeneous system.[40]

4. IONIC POLYMERIZATION

4a. Cationic Polymerization.—Certain vinyl monomers are susceptible to ready polymerization by very small amounts of catalysts of the type used in Freidel-Crafts reactions.[46,47] Effective polymerization catalysts in this class include $AlCl_3$, BF_3, $AlBr_3$, $TiCl_4$, $SnCl_4$, activated clays (attapulgite, montmorillonite, and silicic acid), and in some cases H_2SO_4. All of them are strong acids in the terminology of G. N. Lewis; i.e., they are strong electron acceptors. Isobutylene, styrene, α-methylstyrene, butadiene, and vinyl alkyl ethers are representative of the monomer types readily converted to polymers of high molecular weight by catalysts among those mentioned. Propylene and other olefins may be polymerized in the presence of Friedel-Crafts catalysts, as is well known, but the products tend to be very low in molecular weight. The substituents of the aforementioned monomers are of the electron releasing type (compare Table XXII, p. 196); hence, their doubly bonded carbon atoms should be favorably inclined toward sharing a pair of electrons with an electrophilic reagent. In other words, these monomers are comparatively basic. By analogy with the polar mechanisms of other reactions catalyzed by strongly electrophilic reagents, a mechanism involving carbonium ions is clearly indicated.

The polymerizations under consideration are characterized by high rates at low temperatures. Frequently the polymerization proceeds so rapidly when the catalyst and monomer are brought together that uniform reaction conditions can be neither established nor maintained. Isobutylene,[48] for example, may be polymerized under the influence of BF_3 or $AlCl_3$ at $-100°C$ within a fraction of a second to polymer molecules ranging up to 10^5 in degree of polymerization. In order to prevent excessive rises in temperature, a liquid (e.g., ethylene, propane, or butane) at its boiling point may be used as an "internal refrigerant"; the heat of polymerization is thereby dissipated in vaporizing a portion of the diluent. Temperature coefficients are low, and they may in some cases be negative, the rate actually decreasing with rise in temperature.[47,49] As in polymerizations propagated by free radicals, the average degree of polymerization decreases with rise in temperature. Molecular weights obtained at room temperature and above are much lower than those for radical polymerization, however. Molecular weights of polyisobutylenes as a function of the polymerization temperature[48] are shown in Fig. 29.

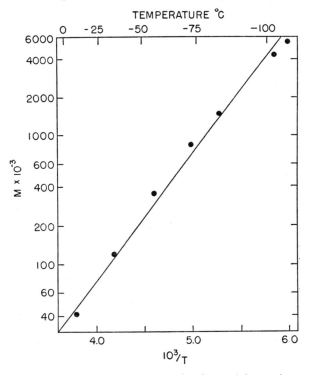

FIG. 29.—Semilog plot of molecular weight against the reciprocal of the polymerization temperature for isobutylene polymerized in the presence of BF_3. Results have been recalculated from the data of Thomas *et al.*[48] The slope of the line corresponds to an activation energy difference of 4.6 kcal./mole.

Two pieces of direct evidence support the manifestly plausible view that these polymerizations are propagated through the action of carbonium ion centers. Eley and Richards[50] have shown that triphenylmethyl chloride is a catalyst for the polymerization of vinyl ethers in *m*-cresol, in which the catalyst ionizes to yield the triphenylcarbonium ion $(C_6H_5)_3C^+$. Secondly, A. G. Evans and Hamann[51] showed that 1,1'-diphenylethylene develops an absorption band at 4340 Å in the presence of boron trifluoride (and adventitious moisture) or of stannic chloride and hydrogen chloride. This band is characteristic of both the triphenylcarbonium ion and the diphenylmethylcarbonium ion. While similar observations on polymerizable monomers* are precluded by intervention of polymerization before a sufficient concentration may be reached, similar ions should certainly be expected to form under the same conditions in styrene, and in certain other monomers also. In analogy with free radical polymerizations, the essential chain-propagating step may therefore be assumed to consist in the addition of monomer to a carbonium ion

$$\text{\small\textasciitilde\textasciitilde\textasciitilde CH}_2\!-\!\underset{\underset{Y}{|}}{\overset{\overset{X}{|}}{C}}{}^+ + \text{CH}_2\!=\!\underset{\underset{Y}{|}}{\overset{\overset{X}{|}}{C}} \quad \rightarrow \quad \text{\textasciitilde\textasciitilde\textasciitilde CH}_2\!-\!\underset{\underset{Y}{|}}{\overset{\overset{X}{|}}{C}}\!-\!\text{CH}_2\!-\!\underset{\underset{Y}{|}}{\overset{\overset{X}{|}}{C}}{}^+ \qquad (36)$$

Chain initiation might conceivably be brought about by polarization of the monomer by an electrophilic catalyst as follows:

$$BF_3 + CH_2\!=\!\underset{\underset{Y}{|}}{\overset{\overset{X}{|}}{C}} \quad \rightarrow \quad F\!-\!\underset{\underset{F}{|}}{\overset{\overset{F}{|}}{B}}{}^-\!-\!CH_2\!-\!\underset{\underset{Y}{|}}{\overset{\overset{X}{|}}{C}}{}^+ \qquad (37)$$

Such a mechanism is open to serious objections both on theoretical and experimental grounds. Cationic polymerizations usually are conducted in media of low dielectric constant in which the indicated separation of charge, and its subsequent increase as monomer adds to the chain, would require a considerable energy. Moreover, termination of chains growing in this manner would be a second-order process involving two independent centers such as occurs in free radical polymerizations. Experimental evidence indicates a termination process of lower order (see below). Finally, it appears doubtful that a halide catalyst is effective without a *co-catalyst* such as water, alcohol, or acetic acid. This is quite definitely true for isobutylene,[52,53] and it may hold also for other monomers as well.

*1,1'-Diphenylethylene is unable to polymerize, presumably owing to the excessive steric requirements of its two large substituents. See Chap. VI.

It is much more likely that initiation involves transfer of a proton, or possibly some other cation, to the monomer. Thus, the mechanism proposed by Evans and Polanyi[54] and others[52,55] to account for the polymerization of isobutylene in the presence of boron trifluoride monohydrate is represented as follows:

$$BF_3 \cdot OH_2 + CH_2{=}C(CH_3)_2 \rightarrow CH_3{-}\overset{\overset{\displaystyle CH_3}{|}}{\underset{\underset{\displaystyle CH_3}{|}}{C}}{}^+ + BF_3OH^- \qquad (38)$$

$$CH_3{-}\overset{\overset{\displaystyle CH_3}{|}}{\underset{\underset{\displaystyle CH_3}{|}}{C}}{}^+ + BF_3OH^- \xrightarrow{+\ M} CH_3{-}\overset{\overset{\displaystyle CH_3}{|}}{\underset{\underset{\displaystyle CH_3}{|}}{C}}{-}CH_2{-}\overset{\overset{\displaystyle CH_3}{|}}{\underset{\underset{\displaystyle CH_3}{|}}{C}}{}^+ + BF_3OH^- \quad (39)$$

$$(CH_3)_3C \left[{-}CH_2{-}\overset{\overset{\displaystyle CH_3}{|}}{\underset{\underset{\displaystyle CH_3}{|}}{C}} \right]_{x-1}^{+} + BF_3OH^-$$

$$\xrightarrow{+\ M} (CH_3)_3C \left[{-}CH_2{-}\overset{\overset{\displaystyle CH_3}{|}}{\underset{\underset{\displaystyle CH_3}{|}}{C}} \right]_{x}^{+} + BF_3OH^- \qquad (40)$$

$$\text{www}CH_2{-}\overset{\overset{\displaystyle CH_3}{|}}{\underset{\underset{\displaystyle CH_3}{|}}{C}}{}^+ + BF_3OH^- \rightarrow \text{www}CH_2{-}\overset{\overset{\displaystyle CH_2}{\|}}{\underset{\underset{\displaystyle CH_3}{|}}{C}} + BF_3 \cdot OH_2 \qquad (41)$$

Initiation occurs through transfer of a cation, in this case a proton, from the catalyst complex to the monomer. Chain propagation proceeds by the previously indicated process (36), except that the anion is here indicated to be in the vicinity of the cationic center. Owing to the large energy required for complete separation of the ion pair in media of low dielectric constant, the gegen ion must remain in close proximity. The nature of this ion may be expected, therefore, to influence the speed of the growth reaction and the incidence of other processes such as chain termination and transfer. In marked contrast with free radical polymerizations where the initiating agent has no effect whatever on the destiny of the propagating center (except when chain transfer with the initiator is important), the cationic terminus remains under the influence of the gegen anion complex throughout its life. Different catalyst complexes may produce chains growing at different rates, and where there is opportunity for structural isomerism of the unit (see Chap. VI) the chain structure of the polymer may conceivably vary somewhat with the particular catalyst used. Definitive evidence along these lines is lacking, however. Finally, in the termination reaction (41) the gegen ion removes a proton leaving an unsaturated terminal unit.[55] Chain transfer to monomer as follows:

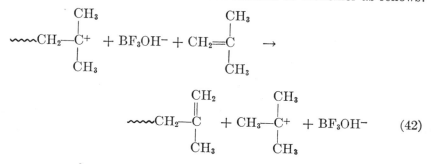

$$\tag{42}$$

may occur also.

Firm evidence for the above mechanism has been obtained from infrared spectra by Dainton and Sutherland.[46,55] They were able to detect both $(CH_3)_3C-$ and $-CH_2-\underset{\underset{CH_3}{|}}{C}=CH_2$ end groups present in considerable abundance in polymers of low degrees of polymerization (ca. 10). The end group $-CH=C(CH_3)_2$, which would be formed by proton removal from the β-methylene group, and the group $-CH_2-\underset{\underset{CH_3}{|}}{\overset{\overset{CH_3}{|}}{C}}-OH$, which would result from transfer of an hydroxyl ion from the associated anion, may also have been present but in much smaller amounts.* When the complex $BF_3 \cdot OD_2$ was used,[55] polymers containing deuterium were formed and the catalyst complex was converted to $BF_3 \cdot OH_2$. This is also in accord with the above mechanism. The catalyst does not combine with the polymer, and many polymer molecules are produced for each catalyst molecule present.[53,56] Studies of this nature are

---

* M. St. C. Flett and P. H. Plesch, J. Chem. Soc., *1952*, 3355, found evidence for the presence of the trisubstituted ethylene end group and also for trichloroacetate end groups in low molecular weight polyisobutylenes prepared at 0°C using TiCl₄ and Cl₃COOH as catalyst and co-catalyst, respectively. For results on the similar polymerization of styrene with TiCl₄, see P. H. Plesch, J. Chem. Soc., *1953*, 1653, 1659, 1662.

necessarily restricted to polymerizations yielding products of rather low molecular weight in order that the number of end groups shall be sufficient to measure.   It is possible of course that other mechanisms might prevail when the chains are several thousand units in length, but there is no evidence to suggest that this is true.

The complexes formed with boron trifluoride are insoluble in the reaction media generally used, hence the over-all process is heterogeneous in character.[57]   Soluble catalysts like $SnCl_4$, $TiCl_4$, or $AlBr_3$ are therefore preferable for kinetic studies.   For the purpose of generalizing the mechanism presented above, let M represent the monomer, A the catalyst and A·SH the complex between the catalyst and the co-catalyst SH.   Then the individual reaction steps may be written

$$A \cdot SH + M \quad \rightarrow \quad HM^+ + AS^- \tag{38'}$$

$$M_x^+ + AS^- + M \overset{k_p}{\rightarrow} M_{x+1}^+ + AS^- \tag{40'}$$

$$M_x^+ + AS^- \overset{k_t}{\rightarrow} M_x + A \cdot SH \tag{41'}$$

$$M_x^+ + AS^- + M \overset{k_{tr}}{\rightarrow} M_x + HM^+ + AS^- \tag{42'}$$

Transfer of a proton has again been assumed in both the initiating and the terminating processes.   Other cations may of course be transferred in some cases.

In a homogeneous system, the rate of initiation should be proportional to the concentration of the catalyst complex (which ordinarily will not exceed about 1 mole percent of the monomer concentration) and also to the concentration of monomer.   Very often the chemical nature of the complex is not known, much less its concentration; hence one is obliged to express the rate of initiation in terms of the total concentration of either the primary catalyst or the co-catalyst, whichever is present in stoichiometric deficiency.   Calling this concentration $[C]$, the rate of initiation should be either $k_i[C]$ or $k_i[C][M]$, depending on whether the initiating complex A·SH is or is not largely converted in step (38').   The termination step should be of the first order, since the terminating agent AS^- is retained in the vicinity of the growing center through electrostatic attraction, and primarily for this reason the kinetics of cationic polymerizations differ from those normally observed for free radical polymerization.   The concentration of active centers in the steady state then becomes

$$[M^+] = (k_i/k_t)[C][M]^{a-1} \tag{43}$$

where a is one or two depending on the initiation kinetics as detailed above. The rate of polymerization is

$$R_p = k_p[M^+][M] = (k_i k_p / k_t)[C][M]^a \tag{44}$$

The average degree of polymerization, assuming dominance of termination (41') over chain transfer (42'), will be

$$\bar{x}_n = R_p / R_t = (k_p / k_t)[M] \tag{45}$$

If, however, chain transfer is the dominant process, then

$$\bar{x}_n = k_p / k_{tr} \tag{46}$$

i.e., the average degree of polymerization should be independent not only of the concentration of the catalyst but also of the monomer concentration.

Results of available kinetic studies seem on the whole to support the mechanism set forth above. The polymerization of styrene with SnCl$_4$ at 25°C in ethylene dichloride according to Pepper[58] proceeds at a rate proportional to the catalyst concentration and to the square of the monomer concentration. The average molecular weight is independent of the catalyst concentration and proportional to the monomer concentration in accordance with Eq. (45), indicating dominance of termination over transfer to monomer. Similar results were obtained by Eley and Richards[50] for the polymerization of vinyl alkyl ethers by SnCl$_4$, AgClO$_4$, and (C$_6$H$_5$)$_3$C—Cl (in m-cresol).* Moisture may be involved in forming the essential complex with SnCl$_4$ in the polymerization of both styrene and vinyl ethers,[51] but this has not been demonstrated definitely to be the case. It is possible that the catalyst combines with the vinyl ether monomer to form a complex capable of initiating polymerization through transfer of a cation to another monomer molecule. Pepper[59] showed also that an increase in the dielectric constant of the reaction medium increased both the rate of polymerization and the degree of polymerization. This is in harmony with expectation; an increase in dielectric constant should increase the rate of initiation (k_i) where separation of charge occurs and decrease the rate of termination (k_t) involving a recombination of ions.

Rates of polymerization of isobutylene in n-hexane by TiCl$_4$ with either water or trichloracetic acid as co-catalyst at −90 to 0°C have been estimated by Plesch[49] from the adiabatic temperature rise. His

---

* The rate of polymerization of vinyl 2-ethylhexyl ether catalyzed by iodine[50] is proportional to $[I_2]^2[M]$ in accordance with an initiation-rate-controlling process

$$2I_2 \rightarrow I^+ + I_3^-$$

followed by

$$I^+ + M \rightarrow I - M^+$$

results are in accord with the above mechanism, the rate being proportional to the co-catalyst concentration (in the presence of an excess of catalyst) and to a power of the monomer concentration between one and two. The degree of polymerization was, however, independent of monomer concentration, thus indicating chain transfer (42′) to monomer. Rather different results were obtained by Norrish and Russell[52] for the polymerization of isobutylene in ethyl chloride at low temperatures using $SnCl_4$ and water. Their isothermally measured rates were proportional to the water concentration (in the presence of an excess of $SnCl_4$), but the degree of polymerization varied inversely with the water concentration. The latter observation is difficult to explain. It suggests that a catalyst hydrate may somehow participate in chain transfer.

4b. Anionic Polymerization.—Monomers with electronegative substituents are readily susceptible to a third type of polymerization which takes place in the presence of reagents capable of generating carbanions. Beaman[60] has shown that sodium in liquid ammonia is particularly effective. When methacrylonitrile is added to such a solution at $-75°C$, immediate polymerization occurs with formation of polymer of high molecular weight. The rapidity of the transformation is reminiscent of the cationic polymerization of isobutylene at low temperatures. Beaman showed that Grignard reagents and triphenylmethylsodium also effect polymerization of methacrylonitrile, but the molecular weights are lower. Acrylonitrile[61] and methyl methacrylate[60] also were polymerized by sodium in liquid ammonia; isobutylene and butadiene refused to polymerize, but styrene yielded some low polymer.[60] As is well known,[62] butadiene and isoprene polymerize readily in the presence of metallic sodium at ordinary temperatures,[63,64,65] but the reaction proceeds at a moderate rate only. Alkali metal alkyls have been widely used in the polymerization of diolefins.[62,63,65] Styrene is converted to low polymers by the amide ion NH_2^-,[66,67] but no polymerization occurs with either butadiene or α-methylstyrene.[66]

Mechanisms depending on carbanionic propagating centers for these polymerizations are indicated by various pieces of evidence: (1) the nature of the catalysts which are effective, (2) the intense colors that often develop during polymerization, (3) the prompt cessation of sodium-catalyzed polymerization upon the introduction of carbon dioxide[65] and the failure of t-butylcatechol to cause inhibition,[66] (4) the conversion of triphenylmethane to triphenylmethylsodium in the zone of polymerization of isoprene under the influence of metallic sodium,[65] (5) the structures of the diene polymers obtained (see Chap. VI), which differ both from the radical and the cationic polymers, and (6) the copolymerization ratios, which likewise differ both from those for radical and those for cationic polymerization (see Sec. 4c).   In analogy with the latter polymerizations, the addition of monomer to the carbanionic centers may occur as follows:

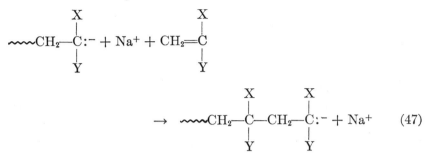

$$\rightarrow \quad \text{\small\textasciitilde\textasciitilde\textasciitilde CH}_2\text{—}\underset{\underset{Y}{|}}{\overset{\overset{X}{|}}{C}}\text{—CH}_2\text{—}\underset{\underset{Y}{|}}{\overset{\overset{X}{|}}{C}}\text{:}^- + \text{Na}^+ \qquad (47)$$

Initiation presumably involves metal alkyls as the primary source of carbanions.   These are immediately available from the Grignard reagents, organosodium compounds, or sodium amide used as catalysts; when the alkali metal itself or its solution in liquid ammonia is used, addition to the monomer may precede actual initiation.[65]

Higginson and Wooding[67] investigated the kinetics of polymerization of styrene by potassium amide in liquid ammonia.   The rate of polymerization was observed to be proportional to the amide ion concentration and to the square of the monomer concentration.   The degree of polymerization (ranging from 5 to 35) increased approximately as the styrene concentration but was independent of the amide ion.   These observations, together with the presence of approximately one nitrogen atom per molecule and the absence of unsaturation, offer strong support for a mechanism consisting of the following steps, which correspond to those indicated above for cationic polymerization:

$$\text{KNH}_2 \quad \rightleftarrows \quad \text{K}^+ + \text{NH}_2^- \qquad (48)$$

$$\text{NH}_2^- + \text{M} \quad \rightarrow \quad \text{NH}_2\text{—M}^- \qquad (49)$$

$$\text{NH}_2\text{—M}_x^- + \text{M} \quad \rightarrow \quad \text{NH}_2\text{—M}_{x+1}^- \qquad (50)$$

$$\text{NH}_2\text{—M}_x^- + \text{NH}_3 \quad \rightarrow \quad \text{NH}_2\text{—M}_x\text{—H} + \text{NH}_2^- \qquad (51)$$

(The gegen ion, K^+, has been dismissed from the scheme in consideration of the comparatively high dielectric constant of the liquid ammonia medium.)   The associated kinetic equations are equivalent to Eqs. (44) and (45).   It will be observed that the termination step (51) is essentially a chain transfer with solvent.   A similar process has been indicated in the sodium-catalyzed polymerization of butadiene and of isoprene in the presence of toluene.[64,65] The mechanism appears also to involve the transfer of a proton from the solvent, i.e.

$$\text{wwwCH}_2\text{—CH}\text{=}\text{CH—CH}_2\text{:}^- + \text{Na}^+ + \text{CH}_3\text{C}_6\text{H}_5$$

$$\rightarrow \text{wwwCH}_2\text{—CH}\text{=}\text{CH—CH}_3 + \text{C}_6\text{H}_5\text{CH}_2\text{:}^- + \text{Na}^+ \quad (52)$$

4c. Ionic Copolymerization.—Ionic copolymerizations tend to be more discriminating than those propagated by free radicals. When, for example, an equimolar mixture of styrene and methyl methacrylate is polymerized by stannic chloride or by boron trifluoride etherate, the product[68] obtained at low conversions is almost pure polystyrene; metallic sodium yields a polymer consisting of over 99 percent methyl methacrylate units; by free radical polymerization a copolymer is obtained whose composition approximates that of the feed.[69] In a mixture of acrylonitrile with methyl methacrylate the former monomer is polymerized by sodium in liquid ammonia almost to the total exclusion of the latter;[61] neither monomer is subject to polymerization by a Friedel-Crafts catalyst,[69] but copolymers are readily obtained using free radicals. Results such as these, even in the absence of other evidence, would demand postulation of different propagation mechanisms for the three polymerization types.

Because of the reluctance of many pairs of monomers to copolymer-

TABLE XXIV.—SUMMARY OF IONIC COPOLYMERIZATION RESULTS[a]

Monomer (M_1)	Monomer (M_2)	r_1	r_2	$r_1 r_2$	$r_1 r_2$ (radical)[b]	Ref.
Cationic Copolymerization						
Styrene	p-Chlorostyrene	2.7 (± .3)	0.35 (± .05)	0.7–1.2	0.8	70
Styrene	2,5-Dichlorostyrene	15 (±2)	.25 (± .15)	1.3–7	.16	71
α-Methylstyrene	p-Chlorostyrene	28 (±2)	.12 (± .03)	2–5		72
o-Chlorostyrene	Anethole	0.03 (± .005)	18 (±3)	0.4–0.7	<.03	73
Styrene	Methyl acrylate	2.2 (± .2)	0.4 (± .2)	.4–1.5	. .14	74
Styrene	Methyl methacrylate	10.5 (± .2)	.1 (± .05)	.5–1.5	.24	75
Styrene	Vinyl acetate	8.2 (± .1)	0–0.03	0–0.25		75
Isobutylene	Butadiene	115 (± 15)	0–0.02			68
Isobutylene	Isoprene	2.5 (± .5)	0.4 (± .1)	0.6–1.5		68
Anionic Copolymerization						
Methyl methacrylate	Methacrylonitrile	.67 (± .2)	5.2 (±1)	2–5	0.43	76
Methyl methacrylate	Vinyl acetate[c]	3.2 (±1)	0.4 (± .2)	0.4–2.5	<.3	74
Methyl methacrylate	Methyl acrylate[c]	0–0.2	4.5 (± .5)	0–1		74
Methyl methacrylate	Acrylonitrile		>25		.24	61
Methyl methacrylate	Styrene	6.4 (± .1)	0.12 (± .05)	0.4–1.1	.24	75
Vinyl butyl sulfone	Acrylonitrile	0.2 (± .1)	1.1 (± .2)			61

[a] Most of the data shown are taken from a compilation by Landler.[74]

[b] See Table XX and Ref. 2.

[c] These results may be subject to considerable error resulting from excessive conversions reached in rapid anionic copolymerization.

ize by an ionic mechanism, determination of reactivity ratios is necessarily restricted to relatively few examples. Results obtained according to the methods discussed in Section 1b are summarized in Table XXIV. We may note first that the $r_1 r_2$ products tend to be closer to unity than for free radical copolymerization (compare fifth and sixth columns), and it is possible that they may exceed unity in a few instances. An $r_1 r_2$ product near unity indicates approach to the case of ideal copolymerization described by Eqs. (9), (10), and (11) and illustrated in Fig. 24 (p. 181). It also signifies that the relative reactivities of the two monomers toward different terminal ionic units are about the same. It is consequently permissible to interpret monomer reactivities without reference to the particular unit which bears the carbonium ion or carbanion. According to the data given in Table XXIV monomer reactivities conform reasonably well to the polarity series given previously in Table XXII. The most electropositive (electron-releasing) substituents, which confer a negative charge on the unsaturated carbons, favor cationic polymerization; those located near the foot of Table XXII, having electronegative substituents which tend to withdraw electrons from the ethylenic group, favor anionic polymerization. Mayo and Walling[68] arrived at the following order of reactivity in cationic polymerization: Vinyl ethers > Isobutylene > α-Methylstyrene > Isoprene > Styrene > Butadiene; and for anionic polymerization: Acrylonitrile > Methacrylonitrile > Methyl methacrylate > Styrene > Butadiene. Styrene and the diolefins are unique in being susceptible to both types of ionic polymerization (and of course to radical polymerization also). They occupy low positions in each series, however. These orders, being based on reactivity ratios, refer to the propagation reaction only; they may not correspond exactly to the sequence which would be obtained for the over-all rate of polymerization for a given catalyst, although the revisions required to convert to the latter basis appear to be minor.

Polarity of the monomer assumes much greater importance in ionic polymerization, as should be expected in view of the comparatively high charge on the propagating center. A small difference in polarity of the monomer when reacting with an ionic propagating center is considerably more significant than in reaction with a free radical. Moreover, the charge on the propagating center is approximately the same for different terminal units; hence the relative reactivities for different monomers should be expected to be nearly independent of the terminal unit. No doubt it is for this reason, pointed out by Landler,[74] that ideal copolymerization is more closely approached by the ionic types than by free-radical-propagated polymerization in which the polarity of the radical is correlated with the polarity of the monomer from which it was formed.

REFERENCES

COPOLYMERIZATION

1. F. R. Mayo and C. Walling, *Chem. Revs.*, **46**, 191 (1950).
2. T. Alfrey, Jr., J. J. Bohrer, and H. Mark, *Copolymerization* (Interscience Publishers, New York, 1952).
3. L. Küchler, *Polymerizationskinetik* (Springer-Verlag, Berlin, 1951), pp. 160–204.
4. F. T. Wall, *J. Am. Chem. Soc.*, **66**, 2050 (1944).
5. F. M. Lewis and F. R. Mayo, *J. Am. Chem. Soc.*, **70**, 1533 (1948).
6. P. D. Bartlett and K. Nozaki, *J. Am. Chem. Soc.*, **68**, 1495 (1946).
7. C. S. Marvel and co-workers, *J. Am. Chem. Soc.*, **57**, 1691 (1935); **57**, 2311 (1935); **59**, 707 (1937).
8. W. G. Barb, *Proc. Roy. Soc.* (London), **A212**, 66 (1952); F. S. Dainton and K. J. Ivin, *ibid.*, **A212**, 96 (1952); W. G. Barb, *J. Am. Chem. Soc.*, **75**, 224 (1953).
9. I. Skeist, *J. Am. Chem. Soc.*, **68**, 1781 (1946).
10. M. Fineman and S. D. Ross, *J. Polymer Sci.*, **5**, 259 (1950).
11. F. M. Lewis, C. Walling, W. Cummings, E. R. Briggs, and W. J. Wenisch, *J. Am. Chem. Soc.*, **70**, 1527 (1948).
12. C. Walling, E. R. Briggs, K. B. Wolfstirn, and F. R. Mayo, *J. Am. Chem. Soc.*, **70**, 1537 (1948).
13. F. R. Mayo, C. Walling, F. M. Lewis, and W. F. Hulse, *J. Am. Chem. Soc.*, **70**, 1523 (1948).
14. M. C. de Wilde and G. Smets, *J. Polymer Sci.*, **5**, 253 (1950).
15. E. C. Chapin, G. E. Ham, and R. G. Fordyce, *J. Am. Chem. Soc.*, **70**, 538 (1948).
16. F. R. Mayo, F. M. Lewis, and C. Walling, *Faraday Soc. Discussions*, **2**, 285 (1947).
17. C. Walling, *J. Am. Chem. Soc.*, **71**, 1930 (1949).
18. F. M. Lewis, C. Walling, W. Cummings, E. R. Briggs, and F. M. Mayo, *J. Am. Chem. Soc.*, **70**, 1519 (1948).
19. M. G. Evans, J. Gergely, and E. C. Seaman, *J. Polymer Sci.*, **3**, 866 (1948); M. G. Evans, *Faraday Soc. Discussions*, **2**, 271 (1947).
20. C. C. Price, *J. Polymer Sci.*, **3**, 772 (1948); *Faraday Soc. Discussions*, **2**, 304 (1947). See also R. G. Fordyce, E. C. Chapin, and G. E. Ham, *J. Am. Chem. Soc.*, **70**, 2489 (1948).
21. C. C. Price, *J. Polymer Sci.*, **1**, 83 (1946).
22. T. Alfrey, Jr., and C. C. Price, *J. Polymer Sci.*, **2**, 101 (1947).
23. C. Walling and F. R. Mayo, *J. Polymer Sci.*, **3**, 895 (1948). See also Ref. 18.
24. H. W. Melville, B. Noble, and W. F. Watson, *J. Polymer Sci.*, **2**, 229 (1947).

25. E. H. de Butts, *J. Am. Chem. Soc.*, **72**, 411 (1950).
26. H. W. Melville and L. Valentine, *Proc. Roy. Soc.*, (London), **A200**, 337, 358 (1950).
27. E. J. Arlman and H. W. Melville, *Proc. Roy. Soc.* (London), **A203**, 301 (1950).
28. E. P. Bonsall, L. Valentine, and H. W. Melville, *J. Polymer Sci.*, **7**, 39 (1951); *Trans. Faraday Soc.*, **48**, 763 (1952).

EMULSION POLYMERIZATION

29. J. W. McBain, in *Advances in Colloid Science*. Vol. I (Interscience Publishers, New York, 1942), p. 124.
30. G. S. Hartley, *Aqueous Solutions of Paraffin-Chain Salts* (Hermann et Cie, Paris, 1936).
31. P. Debye, *J. Phys. Colloid Chem.*, **53**, 1 (1949).
32. P. Debye and E. W. Anacker, *J. Phys. Coll. Chem.*, **55**, 644 (1951).
33. H. A. Scheraga and J. K. Backus, *J. Am. Chem. Soc.*, **73**, 5108 (1951); *J. Colloid Sci.*, **6**, 508 (1951).
34. W. D. Harkins, *J. Am. Chem. Soc.*, **69**, 1428 (1947).
35. H. B. Klevens, *Chem. Revs.*, **47**, 1 (1950).
36. W. V. Smith and R. H. Ewart, *J. Chem. Phys.*, **16**, 592 (1948).
37. W. V. Smith, *J. Am. Chem. Soc.*, **70**, 3695 (1948).
38. M. Morton, P. P. Salatiello, and H. Landfield, *J. Polymer Sci.*, **8**, 111, 215, 279 (1952).
39. W. V. Smith, *J. Am. Chem. Soc.*, **71**, 4077 (1949).
40. R. N. Haward, *J. Polymer Sci.*, **4**, 273 (1949).
41. G. M. Burnett and H. W. Melville, *Proc. Roy. Soc.* (London), **A189**, 494 (1947).
42. R. G. W. Norrish and R. R. Smith, *Nature*, **150**, 336 (1942).
43. G. V. Schulz and G. Harborth, *Angew, Chem.*, **59A**, 90 (1947).
44. W. I. Bengough and R. G. W. Norrish, *Proc. Roy. Soc.* (London), **A200**, 301 (1950).
45. J. D. Burnett and H. W. Melville, *Trans. Faraday Soc.*, **46**, 976 (1950).

IONIC POLYMERIZATION

46. D. C. Pepper, *Sci. Proc. Roy. Dublin Soc.*, **25**, 131 (1950) [a discussion ed. by D. C. Pepper]. See also P. H. Plesch, ed., *Cationic Polymerization and Related Complexes* (W. Heffer and Son, Cambridge, 1953).
47. P. H. Plesch, *Research*, **2**, 267 (1949).
48. R. M. Thomas and co-workers, *J. Am. Chem. Soc.*, **62**, 276 (1940).
49. P. H. Plesch, *J. Chem. Soc.*, **1950**, 543.
50. D. D. Eley and A. W. Richards, *Trans. Faraday Soc.*, **45**, 425, 436 (1949); D. D. Eley and D. C. Pepper, *Trans. Faraday Soc.*, **43**, 112 (1947).
51. A. G. Evans and S. D. Hamann, *Sci. Proc. Roy. Dublin Soc.*, **25**, 139 (1950).
52. R. G. W. Norrish and K. E. Russell, *Trans. Faraday Soc.*, **48**, 91 (1952).

53. A. G. Evans and G. W. Meadows, *Trans. Faraday Soc.*, **46**, 327 (1950).
54. A. G. Evans and M. Polanyi, *J. Chem. Soc.*, **1947**, 252.
55. F. S. Dainton and G. B. B. M. Sutherland, *J. Polymer Sci.*, **4**, 37 (1949).
56. R. O. Colclough, *J. Polymer Sci.*, **8**, 467 (1952).
57. A. G. Evans, G. W. Meadows, and M. Polanyi, *Nature*, **160**, 869 (1947).
58. D. C. Pepper, *Trans. Faraday Soc.*, **45**, 404 (1949).
59. D. C. Pepper, *Trans. Faraday Soc.*, **45**, 397 (1949).
60. R. G. Beaman, *J. Am. Chem. Soc.*, **70**, 3115 (1948).
61. F. C. Foster, *J. Am. Chem. Soc.*, **74**, 2299 (1952).
62. L. Küchler, Ref. 3, pp. 246–250.
63. K. Ziegler and co-workers, *Ann.*, **511**, 13, 45, 64 (1934).
64. J. L. Bolland, *Proc. Roy. Soc.* (London), **A178**, 24 (1941).
65. R. E. Robertson and L. Marion, *Can. J. Research*, **26B**, 657 (1948).
66. J. J. Sanderson and C. R. Hauser, *J. Am. Chem. Soc.*, **71**, 1595 (1949).
67. W. C. E. Higginson and N. S. Wooding, *J. Chem. Soc.*, **1952**, 760.
68. F. R. Mayo and C. Walling, Ref. 1, pp. 277–281.
69. C. Walling, E. R. Briggs, W. Cummings, and F. R. Mayo, *J. Am. Chem. Soc.*, **72**, 48 (1950).
70. T. Alfrey, Jr., and H. Wechsler, *J. Am. Chem. Soc.*, **70**, 4266 (1948); C. G. Overberger, L. H. Arond, and J. J. Taylor, *ibid.*, **73**, 5541 (1951).
71. R. E. Florin, *J. Am. Chem. Soc.*, **71**, 1867 (1949).
72. G. Smets and L. de Haas, *Bull. soc. chim. Belges*, **59**, 13 (1950).
73. T. Alfrey, Jr., L. H. Arond, and C. G. Overberger, *J. Polymer Sci.*, **4**, 539 (1949).
74. Y. Landler, *J. Polymer Sci.*, **8**, 63 (1952).
75. Y. Landler, *Compt. rend.*, **230**, 539 (1950).
76. F. C. Foster, *J. Am. Chem. Soc.*, **72**, 1370 (1950).

CHAPTER VI

The Structure of Vinyl Polymers

1. ARRANGEMENT OF UNITS IN POLYMERS DERIVED FROM MONOVINYL MONOMERS

ADDITION of a free radical to a vinyl monomer may occur in either of two ways:

$$M_x + CH_2{=}CHX \nearrow \quad M_x{-}CH_2{-}\overset{\displaystyle X}{\underset{\displaystyle |}{C}}H\cdot \quad \text{(I)} \tag{1}$$

$$\searrow \quad M_x{-}\overset{\displaystyle X}{\underset{\displaystyle |}{C}}H{-}CH_2\cdot \quad \text{(II)} \tag{2}$$

The relative rates of these alternative processes should depend on the relative stabilities of the product radicals (I) and (II).[1] In the product (I), the substituent occurs on the carbon atom bearing the unpaired electron, and in this position it is able to provide resonance structures in which the unpaired electron appears on the substituent. The substituent consequently has the effect of stabilizing the radical, the extent of such stabilization depending, of course, on the capacity of the substituent for resonance. In product radical (II), the substituent is situated on the beta carbon atom, where it is unavailable for participation in resonating structures involving the odd electron. Consequently, the product radical (I) ordinarily will be more stable than (II), and its formation therefore is more probable. If the substituent X possesses unsaturation conjugated with the free radical carbon, as for example when X is phenyl, resonance stabilization may be fairly large. The addition product (I) in this case is a substituted benzyl radical. Comparison of the C—I bond strengths in methyl iodide and in benzyl iodide,[2] and a similar comparison of the C—H bond strengths in methane and toluene,[3] indicate that a benzyl radical of type (I) is favored by resonance stabilization in the amount of 20 to 25 kcal.

231

per mole.   The product radical (II) is a β-phenylethyl radical analog
which should not differ greatly in stability from an ethyl, or methyl,
radical.   Hence the formation of radical (I) should be favored over (II)
by an energy difference of about this magnitude.   Resonance stabilization in the transition state for the process of monomer addition will be
substantially less than the resonance energy of the product.   Nevertheless, the activation energy for reaction (1) should be lower than that
for (2) by about 8 to 10 kcal., which is enough to cause (1) to occur to
the virtual exclusion of (2) when $X = -C_6H_5$.   Resonance stabilization energies for other substituents ($-CH_3$, $-OCOCH_3$, $-COCH_3$,
$-COOCH_3$), though generally smaller than for phenyl, are by no
means negligible.

A further factor which may favor process (1) rather than (2) is the
steric hindrance to the approaching radical offered by the substituent
X.   Although the 1,2-substituted structure for the final polymer chain
usually would be less hindered than the regular 1,3 sequence (see Sec.
3a), the reverse is likely to be true in the transition state for the monomer addition reaction.   Moreover, owing to the fact that the bonds of
a free radical tend to lie in a plane, steric repulsions between the atoms
and groups directly attached to the carbon bearing the unpaired electron and the approaching molecule will be greater in the transition state
than would be the case for tetrahedral symmetry at the same distance
of separation of the approaching group.   The importance of steric factors will depend, of course, on the size of the substituent.

Successive addition of monomer molecules one after another in accordance with the preferred process (1) will produce a polymer chain
bearing the substituents on alternate carbon atoms[4]

$$-CH_2-CH-CH_2-CH-CH_2-CH-, \text{ etc.} \qquad (III)$$
$$\qquad\quad \underset{X}{|}\qquad\quad \underset{X}{|}\qquad\quad \underset{X}{|}$$

This structure in which successive units are oriented in the same direction is referred to as the head-to-tail or the 1,3 structure.   At the other
extreme there is to be considered the head-to-head, tail-to-tail, or
1,2-1,4 structure

$$-CH_2-CH-CH-CH_2-CH_2-CH-CH-CH_2-, \text{ etc.} \qquad (IV)$$
$$\qquad\quad \underset{X}{|}\ \underset{X}{|}\qquad\qquad\quad \underset{X}{|}\ \underset{X}{|}$$

in which the orientation of the units alternates regularly along the
chain.   This arrangement of units is scarcely to be expected from any
polymerization process proceeding by addition of monomer molecules
one at a time, unless, perhaps, the substituents X were to exert excessive

attractive forces on one another. If the substituent X were of such a nature as to direct addition of the free radical to the substituted carbon atom (reaction 2), the identical structure (III) would result. Thus the head-to-tail product should be formed whenever the structure of the monomer is such as to cause one process to occur to the virtual exclusion of the other. If the directive influence of the substituent X is small or negligible, a more or less random polymer consisting of both 1,3 and 1,2-1,4 arrangements could be expected.

Prevalence of the head-to-tail arrangement in vinyl polymers is abundantly confirmed by determinations of polymer structures.* Staudinger and Steinhofer[5] found that destructive distillation of polystyrene at about 300° yielded 1,3-diphenylpropane, 1,3,5-triphenylpentane, and 1,3,5-triphenylbenzene

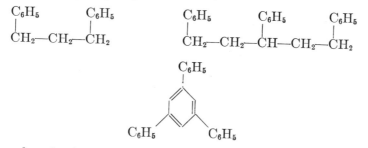

No product having phenyl groups on adjacent carbons was isolated.

Marvel, Sample, and Roy[6] concluded that cyclopropane rings were formed when a dilute solution of poly-(vinyl chloride) in dioxane was treated with zinc, which removes halogen atoms from alternate carbon atoms. Only 84 to 86 percent of the chlorine could be removed, however, a result which was attributed to the occasional isolation of a lone substituent between reacted neighbors. The structure of the product was presumed to be

$$CH_2-\underset{\diagup}{C}H-\underset{\diagdown}{C}H-CH_2-\underset{\diagup}{C}H-\underset{\diagdown}{C}H-CH_2-\underset{|}{C}H-CH_2-\underset{\diagup}{C}H-\underset{\diagdown}{C}H-\text{, etc.}$$

with CH$_2$, CH$_2$, Cl, CH$_2$ groups
(at random)

Statistical calculations[7] lead to the prediction that a fraction $1/e^2$, or 13.5 percent, of the halogen should fail to react owing to isolation between reacted pairs of units as indicated above. The observed extent of halogen removal is in good agreement with this figure. If, on the

* This subject has been reviewed by C. S. Marvel, in *The Chemistry of Large Molecules*, ed. by R. E. Burk and O. Grummit (Interscience Publishers, New York, 1943), Chap. VII.

other hand, the units were oriented at random, then assuming that only 1,2 and 1,3 (but not 1,4) pairs of halogens are removable, a fraction $1/e$, or 18.4 percent, should have remained in the polymer.[7]

If the head-to-tail structure is assumed for the polymer of methyl vinyl ketone

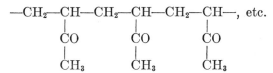

it should be susceptible to internal aldol condensation to a product consisting of sequences of condensed cyclohexene rings interrupted at random intervals by statistically isolated groups.

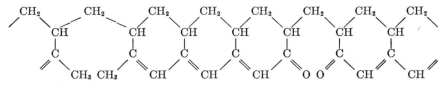

Thus Marvel and Levesque[8] found that from 79 to 85 percent of the oxygen was removed in this process, to be compared with the theoretically calculated[9] figure of 81.6 percent (fraction unreacted equal to $1/e$) for intramolecular reaction of this type in a head-to-tail polymer. A head-to-head, tail-to-tail arrangement, consisting of 1,4-diketone structures, should be expected to yield furan rings

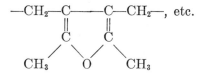

with loss of only 50 percent of the total oxygen. A random polymer should lose an intermediate percentage of oxygen through the occurrence of both types of condensation.[9] Further evidence[10] for the head-to-tail structure is afforded by the absence of furan derivatives among products of cracking poly-(methyl vinyl ketone). Small amounts of 3-methyl-2-cyclohexenone-1

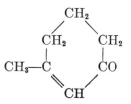

were found, however. The head-to-tail structure is indicated also for poly-(methyl isopropenyl ketone).[10]

The arrangement of units in poly-(vinyl alcohol), $(-CH_2-\overset{\overset{\displaystyle OH}{\displaystyle |}}{CH}-)_x$ may be investigated in a more quantitative fashion through the use of reagents such as periodic acid or lead tetracetate which attack only 1,2 glycol structures. Marvel and Denoon[11] found none of the former reagent to be consumed within experimental error by poly-(vinyl alcohol) obtained by hydrolyzing poly-(vinyl acetate). As little as 2 percent of the 1,2 structure should have been detectable according to their procedure. It is evident, therefore, that nearly all of the structural units are oriented in head-to-tail sequence even in the polymerization of vinyl acetate in which the substituent $(X = -OCOCH_3)$ offers only limited resonance stabilization to the free radical.

A far more sensitive criterion for the presence of head-to-head arrangements in poly-(vinyl alcohol) is the decrease in molecular weight caused by addition of periodic acid or periodate ion to an aqueous solution of the polymer.[12] Wherever a 1,2-glycol structure occurs, the chain is split as follows:

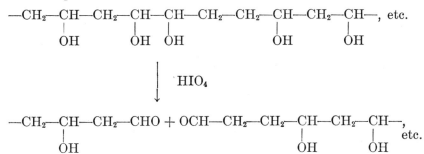

The reaction is complete within a few minutes at room temperature, and the unmistakable decrease in viscosity may be observed even without resort to quantitative measurement. It is apparent therefore that the polymer contains a minor proportion of head-to-head linkages. The number of such interunit linkages should equal the increase in the number of molecules brought about by the reagent. The increase in the number of molecules per gram mole (44 g.) of structural units is given by

$$\delta = 44(1/\overline{M}_n - 1/\overline{M}_n^0) \tag{3}$$

where $\overline{M}_n^{0'}$ and $\overline{M}_n$ are the average (number average) molecular weights before and after degradation by periodic acid. Hence, δ represents the fraction of the interunit linkages joining units in the head-to-head arrangement in the initial polymer.   The value of δ will be determined by the ratio of the velocity constants for the competitive reactions (2) and (1), and if this ratio is small it may be equated to δ, i.e.

$$\delta = k_p' / k_p \qquad (4)$$

where k_p is the velocity constant for chain propagation via the preferred addition of monomer (presumed to be reaction 1) and k_p' the velocity constant for the alternative addition process.

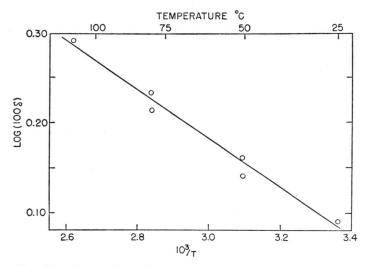

FIG. 30.—Log of the mole percent ($\delta \times 100$) of head-to-head linkages in poly-(vinyl alcohol) plotted against the reciprocal of the absolute temperature of polymerization of the poly-(vinyl acetate) from which the former was prepared by hydrolysis. (Flory and Leutner.[12])

When poly-(vinyl alcohol)s prepared by hydrolysis of poly-(vinyl acetate)s polymerized at various temperatures from 25 to 110°C are treated with periodic acid, the initial molecular weight of 50,000 or more falls to a value in the range from 4000 to 6000.[12]   Through the use of Eq. (3) it is found that the percentage of head-to-head addition varies from 1 to 2 percent depending on the temperature T of polymerization.   An Arrhenius plot of log δ against $1/T$, δ being considered to represent a ratio of velocity constants in accordance with Eq. (4), is shown in Fig. 30.   The straight line drawn through the points is represented by

$$\delta = 0.10 \exp(-1300/RT)$$

from which it is inferred that the head-to-head addition (2) requires an activation energy 1300 cal. greater than does head-to-tail addition (1). Since the exponential factor is of the order of 0.1 within the temperature range employed, it is seen that the preferred addition is favored about equally by steric and energy factors. As already pointed out, the energy factor should be greater in the case of a substituent which more effectively stabilizes the free radical.

The head-to-tail arrangement in poly-(vinyl alcohol) is further confirmed by its X-ray diffraction pattern in the crystalline state. Likewise, analysis of the X-ray diffraction of crystalline poly-(vinylidene chloride), $(-CH_2-CCl_2-)_x$, and of crystalline (stretched) polyisobutylene, $[-CH_2-C(CH_3)_2-]_x$, shows the units to be arranged in these cases also in the expected head-to-tail forms.

Analogous principles should apply to ionically propagated polymerizations. The terminus of the growing chain, whether cation or anion, can be expected to exhibit preferential addition to one or the other carbon of the vinyl group. Polyisobutylene, normally prepared by cationic polymerization, possesses the head-to-tail structure, as already mentioned. Polystyrenes prepared by cationic or anionic polymerization are not noticeably different from free-radical-polymerized products of the same molecular weights, which fact indicates a similar chain structure irrespective of the method of synthesis. In the polymerization of 1,3-dienes, however, the structure and arrangement of the units depends markedly on the chain-propagating mechanism (see Sec. 2b).

Vinyl polymers in which the structural unit bears a single substituent

$$\overset{\displaystyle X}{\underset{\displaystyle |}{}}$$

$(-CH_2-CH-)$ are subject to another type of dissymmetry. The substituted carbon atom of the unit is asymmetric and, as pointed out in Chapter II, both d and l forms ordinarily can be expected to occur in more or less random sequence along the chains. In the fully extended polymer molecule having the chain atoms in the planar zigzag arrangement, substituents of one type of unit will occur above the plane and those of the other below it. This structural irregularity seems to be responsible for the infrequency of occurrence of crystallinity among vinyl polymers possessing asymmetric atoms. Prevalence of d,l heterogeneity is indicated by the fact that crystallinity is almost unknown among polymers in which the (single) substituent X is large. The sole possible exceptions appear to be the polyvinyl ethers $(X = OR)$ prepared under carefully regulated conditions by cationic polymerization at very low temperatures.[13] In contrast to polymers prepared from the same monomers under other conditions, these substances undergo crystallization when cooled sufficiently (see Table II, p. 52). The plausible suggestion has been made[13] that successive structural units tend to alternate regularly between the d and l configurations in the crystallizable polymers.*

2. STRUCTURE AND ARRANGEMENT OF UNITS IN POLYMERS FROM DIENES

2a. Diene Polymers Formed by Free Radical Mechanisms.—Various structural and configurational arrangements of the units in polymers obtained from 1,3-dienes are possible. To take a comparatively simple example, consider the addition of butadiene to the chain free radical represented by $M_n\cdot$. Addition to a terminal carbon atom of the diene may be taken for granted in view of the resonance stabilization of the product radical.

$$M_x\cdot + CH_2 = CH - CH = CH_2$$

$$\rightarrow \begin{Bmatrix} M_x - CH_2 - CH = CH - CH_2\cdot \\ M_x - CH_2 - CH - CH = CH_2 \end{Bmatrix} \qquad (5)$$

Now this resonance hybrid may react by the addition of another monomer at either carbon 4 or 2. The former alternative leads to the 1,4 structure

$$-CH_2 - CH = CH - CH_2 - \qquad (V)$$

for the unit in question, and the latter to the 1,2 structure

$$\begin{array}{c} -CH_2 - CH - \\ | \\ CH \\ \| \\ CH_2 \end{array} \qquad (VI)$$

Thus, whether the remaining ethylenic bond of a given unit is incorporated *within* the chain (V) or in a pendant vinyl group (VI) *external* to the chain is governed by the manner of addition of the succeeding unit.[4] The 1,4 unit may occur as either the *cis* or the *trans* isomer, i.e.

---

* The opposite might be true; i.e., sequences of units of like configuration might be preferred. The polymers would, of course, fail to exhibit optical rotation inasmuch as sequences of the d and of the l units must occur with equal frequency.

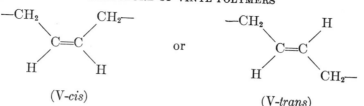

(V-*cis*)   (V-*trans*)

and the 1,2 unit possesses an asymmetric carbon atom which may be either d or l.

A chemical method developed by Kolthoff and Lee[14] for quantitatively determining the proportion of 1,4 and 1,2 units in a polybutadiene involves measurement of the rate of oxidation of the residual double bond by perbenzoic acid.

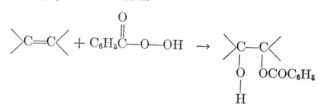

Quantitative differentiation between the two types of unit depends on the roughly 25-fold greater rate of reaction of the symmetrically substituted ethylenic group of the 1,4 unit.

Chemical methods for structure determination in diene polymers have in large measure been superseded by infrared absorption techniques. By comparing the infrared absorption spectra of polybutadiene and of the olefins chosen as models whose ethylenic structures correspond to the respective structural units, it has been possible to show [15,16] that the bands occurring at 910.5, 966.5, and 724 cm.$^{-1}$ are characteristic of the 1,2, the *trans*-1,4, and the *cis*-1,4 units, respectively. Moreover, the proportion of each unit may be determined within 1 or 2 percent from measurements of the absorption intensity in each band.* The extinction coefficients characteristic of each structure must, of course, be known; these may be assigned from intensity measurements on model compounds. Since the proportions of the various units depend on the rates of competitive reactions, their percentages may be expected to vary with the polymerization temperature. The 1,2 unit occurs to the extent of 18 to 22 percent of the total, almost independent of the temperature,[15,16] in free-radical-polymerized (emulsion or mass) polybutadiene. The ratio of *trans*-1,4 to *cis*-1,4, however,

---

* Richardson and Sacher[16] preferred to determine only the 1,2 and the *trans*-1,4 units from the infrared intensities and to obtain the *cis*-1,4 by difference.

decreases as the temperature of polymerization is raised; the percentage of the total which occur as *trans*-1,4 units decreases from 78 percent at $-20°C$ to about 40 percent at $100°C$.[15,16] An Arrhenius plot of the ratio of *trans* to *cis* is shown in Fig. 31, from the slope of which we arrive at a figure of 3.2 kcal. per mole for the difference between the activation energies of *cis*- and *trans*-1,4 addition.[16] The effect of a decrease in polymerization temperature in bringing about an increase in the tendency for the polybutadiene to crystallize[17] is a consequence of preferential formation of the more symmetrical *trans* unit.

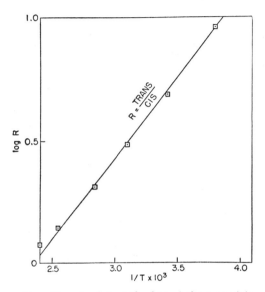

Copolymerization of butadiene with another monomer such as acrylonitrile, methacrylonitrile, styrene, or methyl vinyl ketone increases perceptibly the proportion of 1,4 units at the expense of 1,2.[18] The ratio of *cis* to *trans* is not materially altered, however. One is obliged to conclude that the co-monomer shows a greater preference for addition to the terminal carbon (i.e., to carbon 4) of the resonating chain radical than does the butadiene monomer. Foster and Binder[18] correlated this selectivity with differences in polarity of the co-monomers.

Fig. 31.—A plot of the log of the *trans/cis* ratio for polybutadiene against the reciprocal of the absolute temperature of polymerization by a free radical mechanism. (Results of Richardson[16] obtained by infrared analysis.)

In the case of an unsymmetrical diene such as isoprene, different orientations of the structural units are possible depending on which end of the diene unites with the chain radical.[4] Of the two competing reactions (6) and (7) shown on page 241, the former would appear to be the more probable one on account of the influence of the methyl substituent in stabilizing to some extent one of the resonance hybrid structures which are shown.

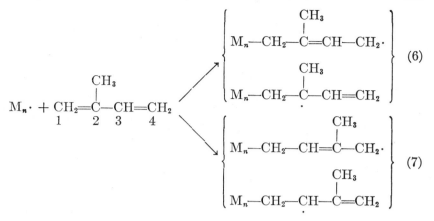

$$(6)$$

$$(7)$$

Dominance of either process would yield a polymer in which successive 1,4 units

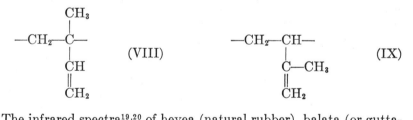

(VII-*cis*)   (VII-*trans*)

are joined in head-to-tail sequence. Reaction (6) may lead to the formation of a certain proportion of 1,2 units (VIII); reaction (7) permits formation of 3,4 units (IX).

The infrared spectra[19,20] of hevea (natural rubber), balata (or gutta-percha), the latter both in the crystalline (α) and the amorphous forms, and of synthetic polyisoprene are compared in Fig. 32.[20] The hevea and balata (amorphous) spectra offer calibrations for *cis*-1,4 and *trans*-1,4 structures, respectively, in the synthetic polymer. Owing to the presence of the methyl substituent, however, the spectral difference between the *cis* and *trans* forms is slight; both absorb at about 840

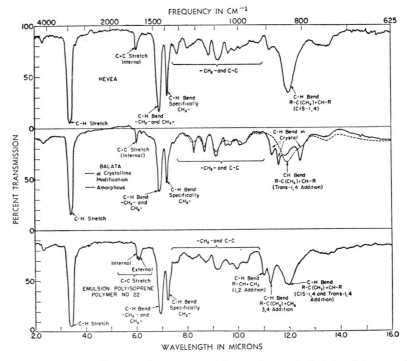

Fig. 32.—Infrared absorption spectra of hevea (natural *cis*-1,4-polyisoprene), of balata (natural *trans*-1,4=polyisoprene), and of synthetic polyisoprene (emulsion). (Richardson and Sacher.[20])

cm.[-1], and differentiation must depend on the difference in extinction coefficients in this band, with a consequent limitation on accuracy. The bands occurring in synthetic polyisoprene at 909 and at 888 cm.[-1] identify the pendant vinyl group of the 1,2 unit and the pendant isopropenyl group of the 3,4 unit, respectively. From intensities measured at higher resolution, using calibrations based on model olefins, the percentages of these units may be determined with high accuracy.[20] Polyisoprenes prepared by free radical polymerization at temperatures from −20° to 120°C contain from 5 to 6 percent of each of the 1,2 and the 3,4 units.[20] These results are confirmed by perbenzoic acid oxidation, which indicates about 13 percent of pendant unsaturated units (1,2 and 3,4).[14] The infrared absorption intensities at 840 cm.[-1] indicate an increase in the proportion of *cis*-1,4 with the polymerization temperature; in the vicinity of 0°C or at lower polymerization temperatures it is no more than 10 percent of the total, but at 100°C it is about 30 percent.[20]

This discussion of the structures of diene polymers would be incomplete without reference to the important contributions which have accrued from applications of the ozone degradation method. An important feature of the structure which lies beyond the province of spectral measurements, namely, the orientation of successive units in the chain, is amenable to elucidation by identification of the products of ozone cleavage. The early experiments of Harries[21] on the determination of the structures of natural rubber, gutta-percha, and synthetic diene polymers through the use of this method are classics in polymer structure determination. On hydrolysis of the ozonide of natural rubber, perferably in the presence of hydrogen peroxide, carbon atoms which were doubly bonded prior to formation of the ozonide are converted to $\diagdown$C=O, —CHO, or —COOH groups. The principal products obtained are levulinic aldehyde and acid, which are those to be expected from a pair of adjoining 1,4 units connected head to tail.

$$\text{etc., } -CH_2-\underset{\underset{\displaystyle CH_3}{|}}{C}=CH-CH_2-CH_2-\underset{\underset{\displaystyle CH_3}{|}}{C}=CH-CH_2-$$

$$\downarrow \quad O_3, \; H_2O_2$$

$$O=CH-CH_2-CH_2COCH_3$$

Employing optimum conditions, Pummerer[22] obtained a combined yield of levulinic aldehyde and acid accounting for 90 percent of the rubber. If *either* member of a pair of consecutive units happened to be other than a 1,4 unit, or if the pair were not connected in the head-to-tail arrangement, the product of ozone degradation would *not* be levulinic aldehyde (or acid). The above result demonstrates, therefore, that at least 95 percent of the units are 1,4, arranged head to tail. Ozone degradations are never quantitative, and it is generally conceded that natural rubber consists of 1,4 structural units in head-to-tail sequence to the virtual exclusion of all other structural arrangements. The absence of 1,2 and 3,4 units is confirmed by the infrared spectrum[19,20] and by perbenzoic acid titration,[14] and the *cis*-1,4 structure in head-to-tail arrangement is confirmed by X-ray diffraction of highly crystalline stretched rubber. Ozonolysis of the natural *trans* isomer, gutta-percha, yields similar results.[21]

Ozonolysis of polyisoprene[21,23] formed by free radical polymerization yields much levulinic aldehyde and acid and only very small amounts of acetonylacetone and succinic acid; the latter products may be formed from head-to-head and tail-to-tail pairs of 1,4 units.

Successive 1,4 units in the synthetic polyisoprene chain evidently are preponderantly arranged in head-to-tail sequence, although an appreciable proportion of head-to-head and tail-to-tail junctions appears to be present as well. Apparently the growing radical adds preferentially to one of the two ends of the monomer. Which of the reactions (6) or (7) is the preferred process cannot be decided from these results alone, however. Positive identification of both 1,2 and 3,4 units in the infrared spectrum shows that both addition reactions take place during the polymerization of isoprene. The relative contributions of the alternative addition processes cannot be ascertained from the proportions of these two units, however, inasmuch as the product radicals formed in reactions (6) and (7), may differ markedly in their preference for addition in one or the other of the two resonance forms available to each. We may conclude merely that structural evidence indicates a preference for oriented (i.e., head-to-tail) additions but that the 1,4 units of synthetic polyisoprene are by no means as consistently arranged in head-to-tail sequence as in the naturally occurring polyisoprenes.

Unlike polybutadiene, polyisoprene prepared at low temperatures shows little or no inclination to crystallize either on stretching or cooling. This may seem surprising in view of the even greater preponderance of *trans*-1,4 units in polyisoprene than in polybutadiene. The explanation for the contrasting behavior in this respect between low temperature synthetic polyisoprene, on the one hand, and gutta-percha and low temperature polybutadiene, on the other, probably is to be found in the appreciable occurrence of head-to-head and tail-to-tail sequences of 1,4 units of the former.

At least 90 percent of free-radical-polymerized 2,3-dimethylbutadiene consists of 1,4 units according to ozone degradation experiments.[24] Successive substitution of the methyl groups on carbons 2 and 3 of butadiene is seen to increase the proportion of 1,4 units formed. In polychloroprene no less than 97 percent of the structure consists of 1,4

$$\overset{\displaystyle Cl}{\underset{\displaystyle |}{}}$$

units,[25] $-CH_2-C=CH-CH_2-$, doubtless in head-to-tail arrangement.[4] When the temperature of polymerization is lowered, the resulting polymer tends to crystallize more readily, and its ultimate extent of crystallization is enhanced.[26] Since the crystalline regions of polychloroprene are known to consist of *trans* units, an increase in the proportion of this geometric isomer with decrease in the polymerization temperature again is indicated.

2b. Diene Polymers Formed by Ionic Mechanisms.—Early results on the ozone degradation of various synthetic diene polymers revealed marked differences between polymers obtained by sodium-induced polymerization rather than by "thermal" (i.e., free radical) polymerization. Sodium-polymerized butadiene,[27] isoprene,[22] and dimethylbutadiene[24] yield in all cases large amounts of unidentifiable materials, little or none of the fragment characterizing pairs of consecutive 1,4 units being isolated. Perbenzoic acid titration of polybutadiene prepared by polymerization with sodium at 50°C indicates that 59 percent of the units possess external double bonds;[14] that is, 59 percent of the structure consists of 1,2 units VI, compared with 18 to 20 percent for free-radical-polymerized butadiene. Infrared absorption measurements confirm these observations. Similarly, polyisoprene prepared by sodium polymerization at 50°C contains about 50 percent of 1,4 units,[14,20] and according to the infrared spectrum these are almost exclusively in the *trans* configuration. Most of the remaining units are 3,4, but a minor proportion (ca. 5 percent) are 1,2.[20] An increase in the temperature of sodium polymerization of either butadiene or isoprene *increases* the proportion of 1,4 units at the expense of the side vinyl group content (1,2 or 1,2 and 3,4 units).[14,20]

As noted in the preceding chapter, so-called sodium polymerizations proceed by an anionic mechanism. Chain propagation reaction (5) for butadiene and reactions (6) and (7) for isoprene are replaced by analogous processes which involve carbanions (e.g., $M_x:^-$) instead of free radicals. Corresponding resonance structures are involved, the unpaired electron of the radical being replaced in each case by an unshared pair representing a single net negative charge. In view of this electrostatic charge, polar factors should assume much greater importance than in the case of free radical polymerization. It is not surprising that the structures produced by anionic polymerization of dienes differ so markedly from those produced by free radical polymerization. The greater abundance of 1,2 (or 3,4) units formed in anionic polymerization may, for example, be due to preference for the resonance structure (in the case of butadiene)

$$M_x\!-\!CH_2\!-\!\overset{\_}{\underset{\cdot\cdot}{C}}H\!-\!CH\!=\!CH_2$$

over the 1,4 structure

$$M_x\!-\!CH_2\!-\!CH\!=\!CH\!-\!\overset{\_}{C}H_2\!:$$

because of the more favorable location of the charge when it is imbedded within the unit, as in the former case, rather than at the end of the chain, as in the latter.

Cationic polymerization of dienes using boron trifluoride or aluminum chloride as catalysts seems also to favor the *trans*-1,4 structure, although 1,2 and 3,4 units also are present.[28] These catalysts also cause cyclization of the structural units, with a consequent decrease in the unsaturation in the polymer.

3. STERIC HINDRANCE IN POLYMER CHAINS

3a. Interactions between Substituents of Successive Units.—Formulas for polymer chains as ordinarily expressed within the limitations of printing in two dimensions suggest nothing unique about the spatial relations between atoms and groups of neighboring units. Scale models of polymer chains bearing bulky substituents on alternate chain atoms, such as occur in many vinyl polymers of the head-to-tail type III, reveal, however, that the neighboring substituents tend to interfere with one another to a rather marked degree. If X = H in formula III, an enormous variety of chain configurations can be realized through rotations about the single bonds joining successive chain atoms, but, if X is a large group such as phenyl or acetate, the number of possible configurations is greatly reduced.

If the chain is fully extended so that all of the chain atoms lie in a plane (planar zigzag form) as in formula X

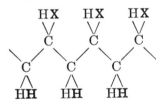

(X)

alternate carbon atoms are spaced only 2.52 Å apart. Hydrogen atoms and substituents shown in boldface type lie above the plane of the carbon atoms, the others below. The symmetries of the asymmetric carbon atoms are indicated arbitrarily to be the same for all of the units shown. Even a substituent as small as methyl with an effective van der Waals radius of about 2.0 Å,[29] corresponding to a distance between centers of 4.0 Å when a pair is in contact, could not be accommodated in this configuration. The situation is more favorable, however, between substituents on asymmetric carbons of opposite symmetry (*d,l*) such that the substituents lie on opposite sides of the plane of the zigzag, rather than as shown in formula X. In polystyrene the phenyl substituent, with a thickness of 3.7 Å and a width of some 6 Å,[29] is too large to permit the fully extended form (X) for successive units which happen to have the same symmetry. Suitable rotations about carbon-carbon bonds of the chain backbone relieve these difficulties, but rotation of the phenyl substituent is hampered for nearly all configurations of the chain. Excessive coiling of the chain, on the other hand, tends to bring about interference between a given substituent and its second neighboring substituents in either direction. Quite generally, the variety of chain configurations which are possible for a head-to-tail vinyl polymer chain having a large substituent on each unit is limited to a degree which is not made evident by customary two-dimensional formulas.

The steric difficulties in chains of mono-substituted units are minor compared with those which occur when alternate atoms of the chain bear two substituents.

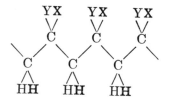

(XI)

Not only are steric interferences intolerably severe in the fully extended configuration XI; it is usually impossible to eliminate them altogether through bond rotation, as may be seen from models. Even when the substituents are no larger than methyl, as in polyisobutylene (X = Y = CH$_3$), steric interferences are so great as to preclude construction of a scale model for any configuration of the chain[30] if the normal C—C and C—H distances and the usually accepted van der Waals radii are used.[29] If Stuart models are used, in which the van der Waals radius for hydrogen is taken as 0.9 Å instead of the more widely accepted value of about 1.2 Å,[29,]* construction of the polyisobutylene chain model becomes possible[31] provided that the configuration chosen is a spiral obtained by so rotating successive units as to allow a methyl group of one unit to fit between the two methyl groups of the succeeding unit. The resulting narrowly restricted structure is shown in Fig. 33. Despite the reduction in the methyl (and methylene) group domain, the tolerable range of rotation about the chain bonds is very small. The permitted spiral configuration is comparatively in-

* The van der Waals radius for carbon is taken as 1.35 Å in the Stuart models, which is also about 0.3 Å too small, but this is of minor importance compared with the hydrogen radius in determining the spatial character of the methyl group.

FIG. 33.—Stuart model of a section of the polyisobutylene chain. (Evans and Tyrrall.[32])

flexible owing to the compact interlocking of the methyl groups.[31] The actual chain of course will not be as stiff as this model would indicate, for the domain of each atom or group is ill defined and considerable compression is possible. On the other hand, repulsive forces between nonbonded atoms and groups extend beyond the domains assigned to the atoms and groups. Hence in other less crowded structures some steric repulsion may be expected between hydrogens, for example, which are close together though not in actual contact in the given model.

In poly-(vinylidene chloride) (X = Y = Cl) steric repulsions should also be expected to be severe, though not so large as in polyisobutylene owing to the smaller van der Waals radius, 1.8 Å,[29] of the covalent chlorine atom. It is of interest to note that neither of these polymers chooses to crystallize in the fully extended configuration. Highly stretched polyisobutylene yields well-ordered crystalline regions in which each bond has been rotated by 22.5° from the planar zigzag configuration, giving a spiral having a period of sixteen chain atoms (eight units) per revolution.[33] Crystalline poly-(vinylidene chloride) gives a repeat distance along the chain axis of 4.67 Å for two structural units,[34] which is appreciably smaller than the 5.04 Å required for the fully extended chain. The difference seems to be due to rotations about chain bonds to an undetermined extent,[35] which would in any case lessen the steric repulsions between chlorine atoms.*

The alternate head-to-head, tail-to-tail structures XII for polymers of disubstituted monomers

$$\left[\begin{array}{c} CH_3 \\ | \\ -Si-O- \\ | \\ CH_3 \end{array}\right]_x$$

* The polydimethylsiloxane chain $\left[\begin{array}{c} CH_3 \\ | \\ -Si-O- \\ | \\ CH_3 \end{array}\right]_x$, though of the type XI, appears to be remarkably free of steric interferences between substituents. This is doubtless related to the greater length (1.65 Å) of the Si—O bond, and perhaps also to a larger value of the Si—O—Si angle. See Chap. X.

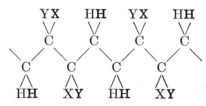

$$(XII)$$

should be comparatively free from steric repulsions between substituents.[31] As the formulas ordinarily are written, greater steric interference might have been expected in this head-to-head (XII) than in the head-to-tail (XI) structure, but construction of models shows indisputably that the reverse is the case. The head-to-tail structure is formed from monomers such as isobutylene and vinylidene chloride in spite of the excessive steric repulsion, and consequently higher energy (*see below*) of this arrangement. The reason for this violation of the generalization that the reaction yielding the most stable product usually is the favored one of two or more competitive processes would seem to lie in the fact that the steric repulsions in the transition state for monomer addition are very small compared with those eventually sustained by the substituent buried within the chain.

No mention has been made of diene polymers in this discussion. Among 1,4 units the substituents are well separated from one another, and the steric repulsions between them should therefore be negligible. A succession of 1,2 units of isoprene (VIII), on the other hand, would form a chain structure of the sterically hindered type XI.

3b. Steric Interaction and the Heat of Polymerization.—The heats of formation at 25°C of the gaseous normal paraffin hydrocarbons $H—(CH_2)_n—H$, with the exception of methane, are adequately represented by the formula[36]

$$\Delta H_f = -10.41 - 4.92_6 n \tag{8}$$

in kcal. per mole. On the reasonable assumption that this formula may be extrapolated to values of n in the polymer range, the first term may be neglected and we have for the heat of formation of the simplest example of a vinyl polymer, linear polyethylene* $(—CH_2—CH_2—)_x$

$$\Delta H_f = -9.85x \tag{9}$$

where x represents the degree of polymerization. The heats of formation of branched-chain saturated hydrocarbons are found to be lower

---

* As it is ordinarily prepared, polyethylene contains an appreciable number of nonlinear units. Its total structure therefore is not as simple as here represented; see below.

than the straight chain isomers. Thus, the heat of an isomerization of the chain skeleton of the type

$$C—C—C—C—C—C \quad \rightarrow \quad C—C—\overset{\overset{\displaystyle C}{|}}{C}—C—C$$

is about -0.8 (± 0.4) kcal., and for a process such as

$$C—C—C—C—C—C—C \quad \rightarrow \quad C—C—\overset{\overset{\displaystyle C}{|}}{\underset{\underset{\displaystyle C}{|}}{C}}—C—C$$

it is about -2.0 (± 1) kcal.[37]   These two heats of isomerization will be referred to as Δ_1 and Δ_2, respectively. If the heat of formation of a molecule could be expressed as the sum of the energies of its individual bonds, steric interactions such as were discussed in the preceding section being neglected, the heats of isomerization of polyethylene to polypropylene

$$(—CH_2—CH_2—)_x \quad \rightarrow \quad \left(—CH_2—\overset{\overset{\displaystyle CH_3}{|}}{CH}—\right)_{2x/3}$$

and to polyisobutylene

$$(—CH_2—CH_2—)_x \quad \rightarrow \quad \left(—CH_2—\overset{\overset{\displaystyle CH_3}{|}}{\underset{\underset{\displaystyle CH_3}{|}}{C}}—\right)_{x/2}$$

would be negative; for the former process ΔH per structural unit should equal Δ_1 and for the latter it should equal Δ_2.   In other words, the above assumption leads to the conclusion that the heats of formation of these polymeric hydrocarbons should be less (algebraically) than that of linear polyethylene.   The heat of formation of polypropylene is not accurately known, but the heat of combustion of polyisobutylene[38] and other evidence (see below) shows that its heat of formation actually is several kcal. per unit higher than that of the isomeric straight chain hydrocarbon as calculated according to Eq. (8) or (9).   This discrepancy has been attributed to steric repulsions between substituents. Rossini[39] has pointed out that consideration must be given to interactions between nonbonded atoms, or groups, in order to account for the heats of formation of various isomeric branched chain hydrocarbons,[37] such as for example 2,2,4-trimethylpentane and 2,2,4,4-tetramethylpentane having the carbon skeletons

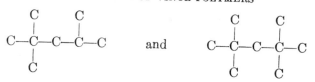

respectively, which are suggestive of the polyisobutylene chain. These steric interactions, though appreciable, are much smaller than in polyisobutylene, where each pair of substituents is situated between neighboring pairs in either direction along the chain.*

An estimate of the energy of steric repulsion may be obtained by comparing the observed heat of polymerization with the value calculated on the assumption that the interaction energy between substituents is zero. For the latter purpose, we consider the following steps:[40]

$$CH_2{=}CXY \xrightarrow[\Delta H_h]{+H_2} CH_3{-}CHXY \xrightarrow[\Delta H_e]{-H_2} \left[-CH_2{-}\overset{\displaystyle X}{\underset{\displaystyle Y}{\overset{|}{\underset{|}{C}}}}- \right]$$

(XIII)   (XIV)

where formula XIV represents a unit situated in the polymer chain. The heat of polymerization is given by

$$\Delta H_p = \Delta H_h + \Delta H_e \qquad (10)$$

The heats of hydrogenation ΔH_h for a number of unsaturated compounds are available from the work of Kistiakowsky and collaborators,[41,42] and in a few cases from accurate heats of formation of the monomer and the reference compound XIII.[43,44] ΔH_e can be estimated from the generalizations set forth above. If $X = Y = H$, ΔH_e according to Eq. (8) is 10.41 kcal. The heat of hydrogenation of ethylene[41] is -32.73 kcal. at 25°C; hence the heat of polymerization of ethylene gas to gaseous polymer should be -22.3 kcal. per mole. The small correction required to convert this result to the liquid state may be

---

* The heats of isomerization Δ_1 and Δ_2 for the simpler isomerizations of the chain skeleton previously indicated are variable even in the absence of multiple branches. Thus values of Δ_1 and Δ_2 will be found[37] to depend both on the length of the carbon chain and on the position of the methyl substituents along the chain. If the side chain is an ethyl instead of a methyl group, the Δ's are nearer zero than the values chosen (for methyl substituents). These variations very probably arise also from steric interactions. The values assigned above to Δ_1 and Δ_2 have been selected arbitrarily for methyl side groups with the object of securing energy corrections applicable to the skeletal structures indicated when steric interactions are at a minimum.

TABLE XXV.—CALCULATED HEATS OF POLYMERIZATION
(in kcal. per mole of monomer at 25°C)[a]

Monomer	Structural unit	ΔH_e calc'd	$-\Delta H_h$ observed[b]	$-\Delta H_p$ calc'd
Ethylene	—CH$_2$—CH$_2$—	10.4	32.7	22.3
Propylene	—CH$_2$—CH— (CH$_3$)	9.6	29.8	20.2
1-Butene	—CH$_2$—CH— (C$_2$H$_5$)	9.6	30.1	20.5
1-Pentene	—CH$_2$—CH— (C$_3$H$_7$)	9.6	30.0	20.4
Isobutylene	—CH$_2$—C— (CH$_3$)(CH$_3$)	9.2	28.1	18.9
Butadiene	cis —CH$_2$—CH=CH—CH$_2$—	10.4	27.8	17.4
	trans —CH$_2$—CH=CH—CH$_2$—	10.4	28.6	18.2
	—CH$_2$—CH— (CH=CH$_2$)	9.6	26.1	16.5
Isoprene	—CH$_2$—C=CH—CH$_2$— (CH$_3$) (cis or trans)	10.4	28.3	17.9
Styrene	—CH$_2$—CH— (C$_6$H$_5$)	9.6	28.3	18.7
Vinyl acetate	—CH$_2$—CH— (OCOCH$_3$)	9.6	30.9	21.3
Vinyl ethyl ether	—CH$_2$—CH— (OC$_2$H$_5$)	9.6	26.5	16.9
Methyl methacrylate	—CH$_2$—C— (COOCH$_3$)(CH$_3$)	9.2	28.4	19.2

[a] The calculations actually refer to the gaseous state. We may, however, disregard the small correction to the liquid state inasmuch as heats of vaporization of the polymer unit and of the monomer will be approximately the same.

[b] The heats of hydrogenation of the olefins are derived from measured heats of hydrogenation by Kistiakowsky and co-workers[41] (see also Ref. 43). Those of the dienes have been obtained from heats of formation of butadiene[44] and of isoprene[46] and the heats of formation of the corresponding mono-olefins.[43,44] Heats of hydrogenation of the last four monomers listed in the table are from direct measurements by Kistiakowsky and co-workers.[42]

neglected, and the values of ΔH_p calculated in this manner can be considered to apply to the polymerization of *liquid* monomer to *amorphous* (i.e., noncrystalline) polymer.

If X is an alkyl group and Y = H, the reference compound XIII is a straight chain hydrocarbon, and the second process above consists in its transformation to a branched unit (XIV) having a single substituent. Hence

$$\Delta H_e = 10.4 + \Delta_1 \cong 9.6 \text{ kcal.}$$

If both X and Y are alkyl groups, the reference compound is a singly branched hydrocarbon, and it is converted to a doubly branched structural unit in the hypothetical dehydrogenation process; hence

$$\Delta H_e = 10.4 + \Delta_2 - \Delta_1 \cong 9.2 \text{ kcal.}$$

The heats of polymerization of various monomers calculated from measured heats of hydrogenation and values for ΔH_e estimated in this manner are given in Table XXV.*

In the case of butadiene, the reference compound (XIII) to be formed by hydrogenation in the above scheme is CH_3—CH=CH—CH_3 (*cis* or *trans*) for 1,4 polymerization and CH_3—CH_2—CH=CH_2 for 1,2 polymerization. The heat of hydrogenation of butadiene to one of these is given by the difference between the heats of formation of the reference compound[43] and of butadiene.[44] In order to evaluate ΔH_e, we observe that the heats of hydrogenation of the reference compound and of the structural unit should be very nearly the same according to experimental data concerning the dependence of the heat of hydrogenation on olefin structure.[41] Hence the energy difference between the two should be the same as that between the corresponding saturated reference compound and the structural unit, or 10.4 kcal., for the *cis*- and *trans*-1,4 units, and 9.6 kcal. for the 1,2 unit which possesses the side chain vinyl group. The heat of 1,4 polymerization of isoprene has been calculated similarly from the heats of formation of isoprene[46] and of 2-methyl-2-butene[44] and the heat of hydrogenation of the latter.[41,44] The difference between the heat contents of the *cis* and *trans* forms of the 1,4-isoprene structural unit should be quite negligible.

No rules are available for estimating the heats of formation of the structural units of the last four examples given in Table XXV. However, it is a reasonable assumption that variations in the nature of the substituent affect the reference compound XIII and the structural

* More extensive calculation of heats of polymerization are given by Roberts.[46]

TABLE XXVI.—COMPARISON OF OBSERVED HEATS OF POLYMERIZATION WITH
ESTIMATES NEGLECTING STERIC INTERACTION

Monomer	Structural unit	Experimental method[a]	$-\Delta H_p$ obs. in kcal./mole	$-\Delta H_p$ calc'd	Difference between obs. and calc'd ΔH_p
Styrene[b]	C_6H_5 —CH$_2$—CH—	Isothermal calorimetric[48]	16.1 (±0.2)	18.7	2.6
Styrene[b]		Heats of combustion[49]	16.7 (±0.2)	18.7	2.0
Vinyl acetate	OCOCH$_3$ —CH$_2$—CH—	Isothermal calorimetric[50]	21.3 (±0.2)	21.3	0
Acrylonitrile	CN —CH$_2$—CH—	Isothermal calorimetric[50]	17.3 (±0.5)	(18)[c]	(0)
Methyl acrylate	COOCH$_3$ —CH$_2$—CH—	Isothermal calorimetric[50]	18.7 (±0.2)	(20)	(1)
Acrylic acid	COOH —CH$_2$—CH—	Adiabatic calorimetric[32]	18.5 (±0.3)	(20)	(1.5)
Isobutylene	CH$_3$ —CH$_2$—C— CH$_3$	Adiabatic calorimetric[31]	12.8	18.9	6.1
Isobutylene		Heats of combustion[38]	12.3 (±0.2)	18.9	6.6
Methyl methacrylate[d]	COOCH$_3$ —CH$_2$—C— CH$_3$	Isothermal calorimetric[47]	13.0 (±0.2)	19.2	6.2
Phenyl methacrylate	COOC$_6$H$_5$ —CH$_2$—C— CH$_3$	Isothermal calorimetric[51]	12.3 (±0.2)	(19)	(7)
Methacrylic acid	COOH —CH$_2$—C— CH$_3$	Adiabatic calorimetric[32]	15.8 (±0.2)	(19)	(3)
α-Methylstyrene	C$_6$H$_5$ —CH$_2$—C— CH$_3$	Heats of combustion[52]	9.0 (±0.2)	(18)	(9)
Vinylidene chloride[e]	Cl —CH$_2$—C— Cl	Isotherma calorimetric[50]	14.4 (±0.5)	(19)	(5)
Isoprene	CH$_3$ —CH$_2$—C=CH—CH$_2$— (cis)	Heat of combustion of natural rubber[45]	17.9 (±1.5)	17.9	(0)
Polybutadiene	ca. 20% cis, 60% trans, 20% 1,2	Heat of combustion[53]	17.3 (±0.2)	17.7	0.4

[a] The various methods have been applied at different temperatures: isothermal calorimetric method usually at 77°, adiabatic method near room temperature, and heats of combustion at 25°C. Corrections to a common temperature would in most cases be smaller than the experimental error.
[b] The heats of polymerization of o-chloro-, p-chloro-, 2,5-dichloro-, and p-ethylstyrene fall in the range from 16.0 to 16.5 kcal. per mole (Tong and Kenyon[48]).
[c] The estimated figure for acrylonitrile given in column 5 is based on the assumption of conjugation between double and triple bonds in the monomer.
[d] Heats of polymerization of other esters of methacrylic acid[47] are as follows: n-butyl, 13.5; cyclohexyl, 12.2; benzyl, 13.4 kcal. per mole.
[e] The observed heat of polymerization of vinylidene chloride includes whatever heat is evolved on crystallization of the polymer. The magnitude of this heat is not known, but it may have increased the observed $-\Delta H_p$ as much as 1 or 2 kcal. per mole
Emulsion polybutadiene prepared at 50°C.

unit XIV about equally (apart from steric interactions between neighboring substituents of the polymer chain). The values of ΔH_e chosen in these instances correspond to those for aliphatic hydrocarbon monomers of the same degree of substitution. The heats of hydrogenation of these monomers have been measured directly by Kistiakowsky and co-workers.[42]

The heats of polymerization calculated for the monovinyl monomers listed in Table XXV, excepting ethylene and styrene, range from about 19 to 21 kcal. per mole. Those for the conjugated divinyl monomers are in the neighborhood of 17 to 18 kcal. per mole. Styrene falls between the two groups. The disappearance of resonance stabilization in the process of polymerization of conjugated monomers accounts for the lower heat evolved on polymerization by 2 or 3 kcal. per mole. Kistiakowsky[42] has pointed out that the resonance energy of the vinyl ether structure C=C—O—C is considerable but that it is negligible in the acrylate

structure $\overset{\displaystyle O}{\underset{\displaystyle \parallel}{\text{C}=\text{C}-\text{C}-\text{O}-}}$. Hence, the calculated heat of polymerization of vinyl ethyl ether places it with the conjugated diene monomers, while that of methyl methacrylate is nearly the same as for isobutylene.

In Table XXVI, observed heats of polymerization (column four) are compared with the values (column five) calculated as in Table XXV. Where the necessary data for calculating a heat of polymerization in the above manner are lacking, an estimate based on values for analogous compounds is given in parentheses in column five. Heats of polymerization may be obtained from the heats of combustion of monomer and polymer (or from the heats of formation calculated from heats of combustion), or they may be measured directly by observing calorimetrically the heat evolved. In place of the usual adiabatic methods involving observation of the temperature rise in a thermally isolated system of calibrated heat capacity, Tong and Kenyon[47] have introduced a convenient isothermal method in which the monomer and initiator contained in a sealed tube are immersed in a liquid such as carbon tetrachloride maintained at its boiling point in a vessel surrounded by refluxing vapors of the same liquid. The heat evolved from the polymerization causes the liquid to vaporize, and the amount so vaporized under the prevailing strictly isothermal conditions affords an accurate measure of the heat of polymerization.

The observed heats evolved $(-\Delta H_p)$ in polymerization are equal to or smaller than those calculated in all cases investigated. Their differences given in the last column of Table XXVI may be considered to represent the energy of steric repulsion between neighboring groups within the polymer chain. For mono-substituted units the difference is small; only for styrene does it definitely exceed 1 kcal. per mole. Heats of polymerization of disubstituted monomers are 3 to 9 kcal. per mole lower (in magnitude) than the calculated values, indicating severe compression of chain constituents. The external pressure required to produce a change in internal energy of the latter amount would exceed 25,000 atmospheres.

The pattern presented by heats of polymerization of various monomers appears to be very well explained by two dominant factors: elimination of resonance energy of conjugation, and steric interactions between substituents. The importance of the value of the heat of polymerization in determining the temperature above which reversal of the conversion of monomer to polymer occurs has been stressed by Dainton and Ivin.[54]

4. NONLINEARITY IN THE MACRO-STRUCTURE OF VINYL POLYMERS

The present chapter thus far has been concerned solely with the structural features of the chain unit and its steric relations with neighboring units. It remains to consider certain aspects of the polymer structure on a larger scale, i.e., to consider the over-all pattern of interconnection of the units.

While linear chain structure is the rule among vinyl polymers, the monotonous succession of bifunctional units may occasionally be interrupted by a unit of higher functionality. The frequency of incidence of such units along the chain depends of course on the monomer and on the conditions attending its polymerization. While they are usually in extreme minority, only a very small fraction of units of higher functionality—one in a polymer molecule—is sufficient to disqualify the polymer from the linear class and, of more importance, to alter its properties appreciably (Chap. IX). We shall be concerned here with two comparatively common sources of nonlinearity in vinyl polymerizations: *branching* resulting from chain transfer with monomer or polymer and *cross-linking* in polymerizations of diene or divinyl monomers. The present discussion is qualitative, the quantitative treatment of these deviations from linearity being necessarily reserved for Chapter IX, where the constitution of nonlinear polymers will be considered from a more general point of view.

4a. Branching in Vinyl Polymers.—It was pointed out in Chapter IV that a molecule of the monomer may sometimes perform the function of a chain transfer agent. If, as usually is the case, the chain transfer involves removal of an atom (often hydrogen) from the monomer, the ensuing growth from the transfer radical will produce a polymer molecule bearing a terminal unit which retains its vinyl group. Thus, on the presumption that the point of attack is in the substituent X, the sequence of reactions set off by chain transfer with the monomer may be represented as follows:

$$M_x\cdot + CH_2{=}\overset{\overset{\displaystyle X}{|}}{CH} \quad \rightarrow \quad M_x{-}H + CH_2{=}\overset{\overset{\displaystyle X\cdot}{|}}{CH} \xrightarrow{\ +\ M\ } \text{etc.}$$

$$\rightarrow \quad CH_2{=}CH{-}X{-}M_x \quad (XV)$$

Another chain radical may acquire the terminal unit of molecule XV in one of its propagation steps; i.e., the terminal unsaturated unit may be incorporated as one of the units combined in another growing chain. When this occurs a branched polymer molecule XVI is produced.[40]

$$\begin{array}{c} M_y \\ | \\ CH_2 \\ | \\ CH{-}X{-}M_x \\ | \\ M_z \end{array} \qquad (XVI)$$

The same result might follow from chain transfer by disproportionation between chain radical and monomer according to reaction (32) of Chapter IV.

The reaction of a chain radical with a unit of a previously formed polymer[40] represents an additional possible chain transfer process not previously considered in Chapter IV. The point of attack might again be located in the substituent X, or it might involve removal of the tertiary hydrogen on the substituted chain carbon. The following sequence of reactions, in which the latter alternative has arbitrarily been assumed, would then lead to a branched polymer molecule as indicated.[40,55]

$$M_x\cdot + H{-}\overset{\overset{\displaystyle |}{\underset{\displaystyle |}{C}}}{\underset{\displaystyle |}{C}}{-}X \ \rightarrow \ M_x{-}H + \cdot\overset{\overset{\displaystyle |}{\underset{\displaystyle |}{C}}}{\underset{\displaystyle |}{C}}{-}X \xrightarrow{\ +\ M\ } \text{etc.} \ \rightarrow \ \overset{\overset{\displaystyle |}{\underset{\displaystyle |}{C}}}{\underset{\displaystyle |}{C}}{-}X$$

(with CH₂ groups shown above each central carbon)

The immediate result of the intervention of the chain transfer process indicated in the first step is the termination of a growing chain and the reactivation of a polymer molecule, which then adds monomer to generate a branched molecule* containing a trifunctional unit, the structure being formally similar to XVI.   It is also apparent that a given polymer molecule may undergo repeated reactivation in the above manner, with growth of an additional branch at each recurrence.   Manifoldly branched molecules (XVII) may be produced in this way.   The previously considered mechanism

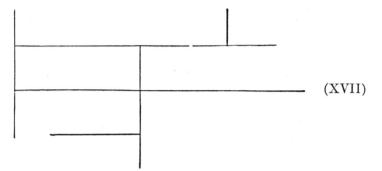

(XVII)

involving chain transfer with monomer may also yield multiply-branched molecules like XVII; and, of course, both mechanisms may contribute simultaneously (particularly, if the substituent X is vulnerable to attack both in the monomer and in the polymer).

According to either mechanism, the average degree of branching of the polymer molecules must increase with the conversion. In the one case the abundance of polymer molecules terminated with unsaturated units increases, relative to monomer, as the degree of conversion advances.   In the other, the incidence of chain transfer to polymer increases with the polymer-monomer ratio.   The effects of branching by transfer on the molecular constitution of a polymer will be considered further in Chapter IX.   There it will be shown that this branching process is not in itself capable of generating an infinite network structure, although a given molecule may be reactivated by chain transfer any (finite) number of times, with growth of an additional branch following each recurrence.   It is to be expected therefore that the branched polymers thus produced will remain soluble and fusible, in the absence of other processes.

Branching through chain transfer in vinyl polymerizations has been

---

* The term *branched* refers to the over-all pattern of the polymer molecule as depicted schematically above, rather than to the nature of the structural unit. The 1,2 unit of butadiene, for example, may be regarded as branched inasmuch as it possesses a pendant vinyl group, but the polymer should be considered linear nevertheless, provided that these units are connected in a single linear sequence.

the subject of much speculation supported by comparatively few experimental data. Chain transfer constants with monomers, or with substances which may be regarded as analogs of common polymer units such as styrene, methyl methacrylate, and methyl acrylate (see Tables XIII and XIV, pp. 143, 145) generally are less than about 10^{-4}. This would indicate that less than one trifunctional unit would be produced for each 10^4 monomers polymerized, unless the conversion is carried beyond about 75 percent. If the average degree of polymerization does not exceed about 10^4, therefore, most of the polymer molecules produced will be linear, and chain transfer with monomer and/or polymer is of little importance insofar as the macro-structure of the polymer chains is concerned.

Vinyl acetate is an exception to the above generalizations inasmuch as both the transfer constant with the monomer and that with the polymer unit are somewhat larger than usual.[56,57] The former is about 2×10^{-4} and the latter may be as great as 8×10^{-4} (at 60°C). The methyl group of the acetyl radical appears to be the principal site of transfer,

although in the polymer unit 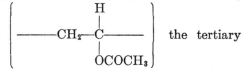 the tertiary

hydrogen atom may make a large contribution.[57] Nonlinearity in poly-(vinyl acetate)s having degrees of polymerization of several thousand or more may be by no means negligible, therefore.

Polyethylene, prepared commercially by polymerization of ethylene at high pressures and at temperatures exceeding 200°C, departs severely from linearity.[58,59] Although the average molecular weight may exceed 15,000, corresponding to approximately 1000 or more methylene groups per molecule, infrared absorption measurements[58,59] reveal the presence of as many as 5 methyl groups for each 100 methylene groups, corresponding to 50 per polymer molecule. The methyl group content decreases as the polymerization temperature is lowered, and may be reduced to a level of little more than 2 per molecule by polymerizing at sufficiently low temperatures.[59] The methyl-to-methylene ratios normally observed can be accounted for only by assuming deviation from the simple linear structure

$$(-CH_2-CH_2-)_x$$

The predominating type of nonlinearity in polyethylene appears to consist of short chain branches three or four chain atoms in length formed by intramolecular chain transfer as follows:[59]

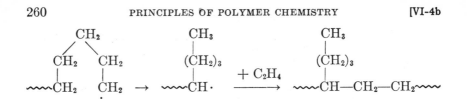

The methyl group content indicates approximately one such branch for each 20 to 100 carbon atoms in the main chain, the proportion of branches increasing with the temperature of polymerization. Intermolecular chain transfer also occurs with production of long chain branches in the manner previously discussed, but the number of such branches ordinarily does not exceed one-tenth of the short chain branches in polyethylene. There may be an average of several long chain branches per molecule, however, and this number is responsible for marked changes in the molecular weight distribution (Chapter IX) and in the flow properties of the molten polymer.[59] The susceptibility of ethylene polymerization to branching by chain transfer processes probably is related to the comparatively low reactivity of this monomer. Chain transfer consequently becomes competitive with chain growth. The high polymerization temperatures often used are also in part responsible for the relatively high incidence of chain transfer.

If a polymer of one kind is dissolved in the monomer of another and the mixture is then subjected to polymerizing conditions, chain transfer between growing chains of the monomer and the molecules of the polymer introduces the possibility of forming unique branched copolymers in which long chains of one type of unit are joined to a "backbone" chain consisting of the other unit. Thus Carlin and coworkers,[60] on polymerizing methyl acrylate monomer in the presence of poly-(p-chlorostyrene), obtained a product containing both units, which could not be resolved into the respective polymers by selective solvents. Other *graft copolymers* having interesting properties have been synthesized by Smets and Claesen,[61] who succeeded in "grafting" vinyl acetate, vinyl chloride, and styrene monomers to poly-(methyl methacrylate) chains, and methyl methacrylate monomer to polystyrene.

4b. Cross-Linking in Diene Polymers.—The residual double bond in the structural unit (either 1,4 or 1,2) of a diene polymer is potentially capable of reacting with an active chain radical in another way; it may add to the radical in place of a monomer.[55] Further addition of monomer to the active center continues the growth of the active chain, as is illustrated on the following page:

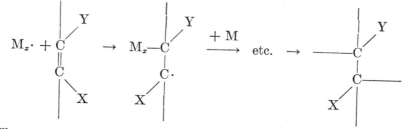

The net result of the intervention of the first step in the growth of the radical chain $M_x \cdot$ is the combination of two otherwise separate polymer molecules.   The unit through which they are joined constitutes a cross-linkage.   The pendant unsaturated group of a 1,2 diene unit may perform a similar function; it might in fact be expected to be somewhat more effective than an internal double bond.   In either case, the polymer unit should be less reactive than a diene monomer by several orders of magnitude; hence cross-linking should be rare in occurrence compared with the normal propagation step involving the diene monomer.   However, a single incidence of addition of a polymer unit suffices to unite two polymer molecules.

As in the transfer reactions leading to branching, it is possible for the same molecule to undergo repeated cross-linking.   Furthermore, the degree of cross-linking averaged over all molecules must increase as the extent of conversion advances.[55]   Unlike the branching reactions, however, cross-linking readily leads to the formation of infinite networks inasmuch as two polymer molecules are joined with each formation of an intermolecular cross-linkage.   Quantitative treatment of the consequences of cross-linking will be discussed in Chapter IX. It may suffice to point out here that only a minute degree of cross-linking—about one cross-linkage for four polymer molecules—is required to bring about the onset of infinite network formation, with the result that the polymer becomes partially insoluble.   If the degree of polymerization is large, the proportion of the units which must be cross-linked to produce gelation (*i.e.*, infinite network formation) is extremely small.

The polymerization of conjugated diolefins and their substitution products invariably tends to give insoluble, or at least partially insoluble, polymers owing to cross-linking reactions involving the residual double bond.   The degree of insolubility increases with the degree of conversion, as should be expected.   By limiting the degree of conversion and/or the chain length, it is usually possible nevertheless to prepare soluble diene polymers of moderately high molecular weight.

Divinyl monomers such as divinylbenzene, divinyl adipate, and ethylene dimethacrylate, in which the reaction of one double bond does not alter greatly the reactivity of the other, polymerize to highly cross-linked structures.

5. SUMMARY

Preferential addition to one end or the other of a vinyl (CH_2=CHX) or substituted vinyl (CH_2=CXY) monomer seems to be the rule to which exceptions are rare. This generalization appears to apply to ionic as well as to free radical polymerizations. The polymers of mono-unsaturated compounds consequently are characterized by a high degree of head-to-tail regularity in the arrangement of successive units. Little is known concerning the sequence of d and l configurations of the asym-

$$-CH_2-\underset{|}{\overset{\overset{\displaystyle X}{|}}{C}H-$$

metric atoms occurring in vinyl polymers ($-CH_2-\overset{X}{\underset{|}{C}}H-)_x$, but a near-random arrangement seems likely in most cases. A degree of order may prevail in cationically polymerized vinyl ethers prepared at low temperatures, however.

In the free radical polymerization of 1,3-dienes, 1,4 addition dominates 1,2 addition. The proportion of 1,2 (and 3,4)units decreases in passing from butadiene to its methyl and chlorine substitution products: isoprene, 2,3-dimethylbutadiene and chloroprene. The *trans* configuration of the 1,4 unit from butadiene is formed preferentially, the proportion of *trans* increasing rapidly with lowering of the polymerization temperature.

The 1,4 units constitute somewhat less than half of the total in polymerized dienes prepared by anionic polymerization at moderate temperature. Their proportion increases with the polymerization temperature.

The continuity of the polymer chain brings about a crowding of successive substituents of head-to-tail vinyl polymers which is not apparent in formulas as ordinarily written but is clearly shown by scale models. Steric interferences between substituents in polymers made up of disub-

stituted structural units of the type $-CH_2-\overset{X}{\underset{Y}{C}}-$ are so severe that

compression of the substituent groups ($-CH_3$, $-Cl$, $-\overset{O}{\overset{\|}{C}}OCH_3$, etc.) well below their normal van der Waals radii is required for the existence of the head-to-tail polymer chain structure in any configuration. These steric repulsions are reflected in the values of the heats of polymerization.   Heats of polymerization calculated from heats of formation of monomeric analogs of the structural units are close to -20 kcal. per mole for unconjugated monomers including both monosubstituted ($CH_2{=}CHX$) and disubstituted ($CH_2{=}CXY$) ethylenes. Whereas the observed heats for the former are in substantial agreement with the calculation, those for the latter are from 3 to 9 kcal. per mole lower in magnitude.   The discrepancy appears to represent the energy of repulsion between substituents of neighboring units.   The heat evolved in the polymerization of a conjugated monomer such as butadiene or styrene is about 2 to 3 kcal. per mole less than for an unconjugated monomer owing to loss of resonance energy in polymerization.

Nonlinear structures may arise in vinyl polymerizations through chain transfer with monomer or with previously formed polymer molecules, but such processes usually occur to an extent which is scarcely significant.   A more common source of nonlinearity in the polymerization of a 1,3-diene is the incorporation in a growing chain of one of the units of a previously formed polymer molecule.   The importance of both branching by chain transfer and cross-linking by addition of a polymer unit increases with the degree of conversion of monomer to polymer.

REFERENCES

1. F. R. Mayo and C. Walling, *Chem. Revs.*, **27**, 351 (1940).
2. E. T. Butler and M. Polanyi, *Trans. Faraday Soc.*, **39**, 19 (1943).
3. M. Szwarc, *J. Chem. Phys.*, **16**, 128 (1948); *Chem. Revs.*, **47**, 128–136 (1950).
4. P. J. Flory, *J. Polymer Sci.*, **2**, 36 (1947).
5. H. Staudinger and A. Steinhofer, *Ann.*, **517**, 35 (1935).
6. C. S. Marvel, J. H. Sample, and M. F. Roy, *J. Am. Chem. Soc.*, **61**, 3241 (1939).
7. P. J. Flory, *J. Am. Chem. Soc.*, **61**, 1518 (1939); F. T. Wall, *ibid.*, **62**, 803 (1940); **63**, 821 (1941).
8. C. S. Marvel and C. L. Levesque, *J. Am. Chem. Soc.*, **60**, 280 (1938).
9. P. J. Flory, *J. Am. Chem. Soc.*, **64**, 177 (1942); Wall, *ibid.*, **64**, 269 (1942).
10. C. S. Marvel, E. H. Riddle, and J. O. Corner, *J. Am. Chem. Soc.*, **64**, 92 (1942).
11. C. S. Marvel and C. E. Denoon, *J. Am. Chem. Soc.*, **60**, 1045 (1938).
12. P. J. Flory and F. S. Leutner, *J. Polymer Sci.*, **3**, 880 (1948); *ibid.*, **5**, 267 (1950).
13. C. E. Schildknecht *et al.*, *Ind. Eng. Chem.*, **40**, 2104 (1948).
14. I. M. Kolthoff and T. S. Lee, *J. Polymer Sci.*, **2**, 206 (1947); I. M. Kolthoff, T. S. Lee, and M. A. Mairs, *ibid.*, **2**, 220 (1947).

15. E. J. Hart and A. W. Meyer, *J. Am Chem. Soc.*, **71**, 1980 (1949); R. R. Hampton, *Anal. Chem.*, **21**, 923 (1949).
16. W. S. Richardson, private communication.
17. K. E. Beu, W. B. Reynolds, C. F. Fryling, and H. L. McMurry, *J. Polymer Sci.*, **3**, 465 (1948).
18. F. C. Foster and J. L. Binder, *J. Am. Chem. Soc.*, **75**, 2910, (1953).
19. G. B. B. M. Sutherland and A. V. Jones, *Faraday Soc. Discussions*, **9**, 281 (1950); J. E. Field, D. E. Woodward, and S. D. Gehman, *J. Applied Phys.*, **17**, 386 (1946).
20. W. S. Richardson and A. Sacher, *J. Polymer Sci.*, **10**, 353 (1953).
21. C. Harries, *Ber.*, **38**, 1195, 3985 (1905); **45**, 943 (1912); *Ann.*, **395**, 211 (1913).
22. R. Pummerer, G. Ebermayer, and K. Gerlach, *Ber.*, **64**, 809 (1931).
23. C. Harries, *Ann.*, **383**, 157 (1911); *Ber.*, **48**, 863 (1915).
24. C. Harries, *Ann.*, **395**, 264 (1913).
25. A. Klebanskii and K. Chevychalova, *J. Gen. Chem.* (U.S.S.R.), **17**, 941 (1947).
26. W. E. Mochel and J. B. Nichols, *Ind. Eng. Chem.*, **43**, 154 (1951).
27. R. Pummerer, *Kautschuk*, **10**, 149 (1934).
28. W. S. Richardson, private communication.
29. L. Pauling, *Nature of the Chemical Bond* (Cornell University Press, Ithaca, New York, 1945), Chap. V.
30. P. J. Flory, *J. Am. Chem. Soc.*, **65**, 372 (1943).
31. A. G. Evans, and M. Polanyi, *Nature*, **152**, 738 (1943).
32. A. G. Evans and E. Tyrrall, *J. Polymer Sci.*, **2**, 387 (1947).
33. C. S. Fuller, C. J. Frosch, and N. R. Pape, *J. Am. Chem. Soc.*, **62**, 1905 (1940).
34. R. C. Reinhardt, *Ind. Eng. Chem.*, **35**, 422 (1943).
35. M. L. Huggins, *J. Chem. Phys.*, **13**, 37 (1945).
36. E. J. Prosen and F. D. Rossini, *J. Research Nat. Bur. Standards*, **34**, 263 (1945).
37. E. J. Prosen and F. D. Rossini, *J. Research Nat. Bur. Standards*, **27**, 519 (1941); **34**, 163 (1945); **38**, 419 (1947).
38. G. S. Parks and J. R. Mosley, *J. Chem. Phys.*, **17**, 691 (1949).
39. F. D. Rossini, *Chem. Revs.*, **27**, 1 (1940); E. J. Prosen and F. D. Rossini, *J. Research Nat. Bur. Standards*, **27**, 519 (1941).
40. P. J. Flory, *J. Am. Chem. Soc.*, **59**, 241 (1937).
41. G. B. Kistiakowsky and co-workers, *J. Am. Chem. Soc.*, **57**, 65, 876 (1935); **58**, 137 (1936).
42. M. A. Dolliver, T. L. Gresham, G. B. Kistiakowsky, and W. E. Vaughan, *J. Am. Chem. Soc.*, **59**, 831 (1937); **60**, 440 (1938).
43. E. J. Prosen and F. D. Rossini, *J. Research Nat. Bur. Standards* **36**, 269 (1946).
44. E. J. Prosen, F. W. Maron, and F. D. Rossini, *J. Research Nat. Bur. Standards*, **46**, 106 (1951).
45. D. E. Roberts, *J. Research Nat. Bur. Standards*, **44**, 221 (1950).

46. J. E. Kilpatrick, C. W. Beckett, E. J. Prosen, K. S. Pitzer, and F. D. Rossini, *J. Research Nat. Bur. Standards*, **42**, 225 (1949).
47. L. K. J. Tong and W. O. Kenyon, *J. Am. Soc.*, **67**, 1278 (1945).
48. L. K. J. Tong and W. O. Kenyon, *J. Am. Chem. Soc.*, **69**, 1402 (1947).
49. D. E. Roberts, W. W. Walton, and R. S. Jessup, *J. Polymer Sci.*, **2,** 420 (1947).
50. L. K. J. Tong, and W. O. Kenyon, *J. Am. Chem. Soc.*, **69**, 2245 (1947).
51. L. K. J. Tong and W. O. Kenyon, *J. Am. Chem. Soc.*, **68**, 1355 (1946).
52. D. E. Roberts and R. S. Jessup, *J. Research Nat. Bur. Standards*, **46**, 11 (1951).
53. R. A. Nelson, R. S. Jessup, and D. E. Roberts, *J. Research Nat. Bur. Standards*, **48**, 275 (1952).
54. F. S. Dainton and K. J. Ivin, *Trans. Faraday Soc.*, **46**, 331 (1950).
55. P. J. Flory, *J. Am. Chem. Soc.*, **69**, 2893 (1947).
56. O. L. Wheeler, S. L. Ernst, and R. N. Crozier, *J. Polymer Sci.*, **8**, 409 (1952); O. L. Wheeler, E. Lavin, and R. N. Crozier, *ibid.*, **9**, 157 (1952).
57. W. H. Stockmayer, J. T. Clarke, and R. O. Howard, unpublished.
58. J. J. Fox and A. E. Martin, *Proc. Roy. Soc.* (London), **A175**, 226 (1940); H. W. Thompson and P. Torkington, *Trans. Faraday Soc.*, **41**, 248 (1945); W. M. D. Bryant, *J. Polymer Sci.*, **2**, 547 (1947).
59. M. J. Roedel, W. M. D. Bryant, F. W. Billmeyer, J. K. Beasley, C. A. Sperati, W. A. Franta, and H. W. Starkweather, Jr., to be published.
60. R. B. Carlin and D. L. Hufford, *J. Am. Chem. Soc.*, **72**, 4200 (1950); R. B. Carlin and N. Shakespeare, *ibid.*, **68**, 876 (1946).
61. G. Smets and M. Claesen, *J. Polymer Sci.*, **8**, 289 (1952)

CHAPTER VII

Determination of Molecular Weights

MOLECULAR weights of polymers may be established from appropriate physical measurements on very dilute solutions or, if circumstances permit, by chemical analysis for end groups. The former
methods, with which the present chapter is almost exclusively concerned, are applied with difficulty at molecular weights below 5,000 to
10,000, whereas the chemical methods generally are reliable only for
molecular weights below about 25,000. Hence neither physical nor
chemical methods alone are sufficient in all situations, but physical
methods are definitely indicated for the high molecular weight polymers
of general interest.

Application of a chemical method requires that the structure of the
polymer, as established from prior considerations, be such as to contain a known number of the chemically determinable functional groups
per molecule. This functional group invariably occurs as an end
group, and, with the exception of starch-like branched polymers (see
Chap. VIII), the polymers to which the chemical method has been
applied have been linear. If the polymer is precisely linear, quantitative determination of *all* end groups which may be present provides,
of course, a direct measure of the number of molecules present and
hence of the average molecular weight (number average, *see below*).
In the polyesters and polyamides formed by linear condensations
(see Chap. III, p. 92), the ratio of the reacting functional groups may
be controlled through the nature and stoichiometric proportions of the
reactants employed; for example, the two coreacting functional groups
may be introduced in equivalent quantities. To the extent that this
precise balance is not vitiated by impurities, by side reactions, or by
preferential loss of one ingredient, determination of one of the functional groups (e.g., COOH) suffices for the evaluation of the number
average molecular weight. The chemical method also may be applied
to vinyl polymers formed in the presence of a transfer agent, such as a

mercaptan or carbon tetrachloride, the elements of one molecule of which appear in the polymer for each incidence of transfer. Thus, if the polymer is formed under conditions such that its molecular weight is governed overwhelmingly by chain transfer (see Chap. IV), the number of polymer molecules may be computed from the number of transfer agent molecules incorporated in the polymer as determined by chemical analysis.[1,2] Similarly, the molecular weight of a vinyl polymer may sometimes be calculated from the number of initiator fragments occurring in the polymer (see Chap. IV); this method presupposes that the mechanisms of both initiation and termination are known and that chain transfer is unimportant under the conditions of polymerization.

Chemical methods for molecular weight determination become insufficiently sensitive when the molecular weight is large. Spurious sources of end groups not taken into account in the assumed reaction mechanism become increasingly consequential as the molecular weight increases, and the number of end groups eventually diminishes to the point where quantitative determination is impractical. For these reasons, the determination of molecular weights by chemical methods finds widespread use only for condensation polymers, which seldom have average molecular weights exceeding 25,000. End group determinations may also afford valuable information on the polymer structure and the mechanism of polymerization, as for example when used in conjunction with physically measured molecular weights to determine the number of chain ends of a given type per molecule.

The various physical methods in use at present involve measurements, respectively, of osmotic pressure, light scattering, sedimentation equilibrium, sedimentation velocity in conjunction with diffusion, or solution viscosity. All except the last mentioned are absolute methods. Each requires extrapolation to infinite dilution for rigorous fulfillment of the requirements of theory. These various physical methods depend basically on evaluation of the thermodynamic properties of the solution (i.e., the change in free energy due to the presence of polymer molecules) or of the kinetic behavior (i.e., frictional coefficient or viscosity increment), or of a combination of the two. Polymer solutions usually exhibit deviations from their limiting infinite dilution behavior at remarkably low concentrations. Hence one is obliged not only to conduct the experiments at low concentrations but also to extrapolate to infinite dilution from measurements made at the lowest experimentally feasible concentrations.

The physical reason for the necessity of working at high dilutions becomes evident from consideration of the nature of solutions of randomly coiled chain molecules of high molecular weight, as illustrated in Fig. 34. The spatial configurations of indiviual chain molecules will be discussed in Chapters X and XIV. Here it suffices to note that on the average a polymer molecule may be approximated by a spherically symmetric statistical distribution of chain elements about a center of gravity, and that the volume encompassed by this distribution may be many times the actual molecular volume. (This feature of polymer molecules is, in fact, of foremost importance with respect to all of those physical properties of polymers which distinguish them from other substances.) Thus the volume over which an individual polymer molecule exerts its influence may be as much as several hundred times its molecular volume; it will depend, of course, on the chain length, and also on the interaction between polymer elements and the medium in which the polymer is dissolved (see Chap. XIV).

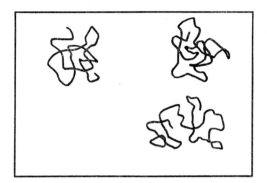

FIG. 34.—Schematic representation of polymer molecules in dilute solution.

All of the physical methods presently used for the determination of molecular weights of polymers require that the molecules contribute individually, i.e., additively, to the property being measured, contributions due to interactions between pairs (or clusters) of molecules being negligible. The light-scattering method requires that the total turbidity consist of the sum of the turbidities due to the individual molecules, the osmotic method that the decrease in activity of the solvent be proportional to the number of solute particles, and the viscosity method that the energy dissipated by each individual polymer molecule be uninfluenced by all others in the solution. Only if, as in Fig. 34, the solution is sufficiently dilute to permit the polymer molecules to occupy separate portions of the volume, without apppreciable overlapping by two or more molecules, will a satisfactory approach to these conditions be assured. Considering the voluminous character of the high polymer molecule, the necessity for measurements at concentrations below 1 percent, and in many cases down to several tenths of a percent, is apparent; even when measurements are made at such low concentrations, careful extrapolation to infinite dilution may yet be required. Furthermore, accurate measurement of the small physical effects observed at small concentrations of polymer calls for sensitive experimental methods (the nonabsolute viscosity method excepted). In view of these difficulties, the unfortunate lack of reliable molecular weight data in support of many otherwise excellent investigations is understandable.

An additional complication arises from the different types of molecular weight averages obtained by different methods. If the polymer comprises a broad range of species differing in chain length, as is usually the case (see Chap. VIII) unless a careful fractionation has been performed, different molecular weight methods will give substantially different values on this account.

1. OSMOTIC METHODS

1a. Theory.—As was pointed out in Chapter I, early investigators discounted the results of cryoscopic and osmotic molecular weight determinations on high polymers on the grounds that such methods are inapplicable to colloidal substances. In view of the peculiar character of high polymer solutions as set forth above, it is not surprising that the validity of the usual physicochemical methods in this area has often been questioned. It is appropriate therefore to reexamine the thermodynamical basis for the classical molecular weight determination methods.

The activities of the various components 1,2,3 . . . of an ideal solution are, according to the definition of an ideal solution, equal to their mole fractions N_1, N_2, ... The activity, for present purposes, may be taken as the ratio of the partial pressure P_i of the constituent in the solution to the vapor pressure P_i^0 of the pure constituent i in the liquid state at the same temperature.* Although few solutions conform even approximately to ideal behavior at all concentrations, it may be shown that the activity a_1 of the solvent must converge to its mole fraction N_1 as the concentration of the solute(s) is made sufficiently small. According to the most elementary considerations, at sufficiently high dilutions the activity a_2 of the solute must become *proportional* to its mole fraction, provided merely that it does not dissociate in solution. In other words, the escaping tendency of the solute must be proportional to the number of solute particles present in the solution, if the solution is sufficiently dilute. This assertion is equally plausible for monomeric and polymeric solutes, although the

---

* The vapor pressure of a polymer is, of course, far too small to measure. We may, nevertheless, insist that such a vapor pressure exists, however small it may be. Or we may resort to the use of the "escaping tendency," or fugacity, in place of the partial vapor pressure in the development given above, in accordance with usual thermodynamic procedures applied to the treatment of solutions. The treatment given here is in no way restricted to volatile solutes.

stipulation that the solution be sufficiently dilute will be considerably more stringent for the polymeric solute (see Chap. XII). Thus, at sufficiently low concentrations

$$a_2 = k_2 N_2 \tag{1}$$

where k_2 is a constant. (The range over which Eq. (1) is applicable in a given instance will depend on the solute and the solvent concerned; we need only observe that Eq. (1) must invariably hold if the solution is made sufficiently dilute.) According to the Gibbs-Duhem relation, which is an inexorable consequence of the nature of thermodynamic functions, for a binary solution

$$(\partial \ln a_1 / \partial \ln N_1)_{T,P} = (\partial \ln a_2 / \partial \ln N_2)_{T,P} \tag{2}$$

Substitution of Eq. (1) in (2) yields

$$(\partial \ln a_1 / \partial \ln N_1)_{T,P} = 1$$

from which it follows that

$$\ln a_1 = \ln N_1 + \text{const.}$$

Since $a_1 = 1$ when $N_1 = 1$, the constant must equal zero and

$$a_1 = N_1 \tag{3}$$

corresponding to ideality of the solvent over the concentration range within which Eq. (1) is applicable. In the dilute solution a_1 is very near unity, and it is therefore more illuminating to write

$$1 - a_1 = N_2 \tag{3'}$$

since $N_2 = 1 - N_1$. A measurement of the depression of the activity of the solvent by the presence of a sufficiently small proportion of the solute affords a measure of the mole fraction N_2 of the solute; the molecular weight of the solute may then be computed from its weight concentration. Colligative methods, all of which depend on the broad principles outlined above, must apply generally irrespective of the nature of the solute (provided it does not dissociate). It is essential, however, that the reduction of the activity of the solvent by the solute be evaluated within the range where a_2 remains proportional to the concentration of solute.

The depression of the activity may be measured in various ways. The most obvious would involve a measurement of the vapor pressure lowering, but this method is superseded by others both in accuracy and in simplicity of execution. The boiling point elevation and freezing point depression methods relegated vapor pressure measurement to obsolescence long ago.   In the former, instead of measuring the depression of the vapor pressure P of the solution below the value $P° = 760$ mm. at the boiling point of the pure solvent, the temperature increase ΔT_b required to restore P to 760 mm. is determined.   Then

$$(\Delta T_b/c)_0 = (RT^2/\rho l_v)/M \tag{4}$$

where R is the gas constant, T the absolute temperature, ρ the density of the solvent, l_v the latent heat of vaporization per gram, c the concentration in grams per ml., and M is the molecular weight of the solute.   The subscript zero is included on the left as a reminder that the relationship is exact only at infinite dilution.

In the freezing point depression method, one measures the temperature lowering ΔT_f required to render the activity of the solvent in the solution equal to that of the pure crystalline solvent (referred to the pure liquid as the standard state; see above).   Then

$$(\Delta T_f/c)_0 = (RT^2/\rho l_f)/M \tag{5}$$

where l_f is the latent heat of fusion per gram of solvent.

In the osmotic pressure method, the activity of the solvent in the dilute solution is restored to that of the pure solvent (i.e., unity) by applying a pressure π on the solution.   According to a well-known thermodynamic relationship, the change in activity with pressure is given by

$$(\partial \ln a_1/\partial P)_{T,N_1} = \bar{v}_1/RT$$

where $\bar{v}_1$ is the partial molar volume of the solvent, which for the dilute solution may be replaced by the molar volume v_1 of the pure solvent.   Thus, at osmotic equilibrium

$$\int_{a_1}^1 d \ln a_1 = \int_0^\pi (v_1/RT)dP$$

Since v_1 is practically independent of pressure for the minute pressures involved

$$- \ln a_1 = \pi v_1/RT \tag{6}$$

If the solution is sufficiently dilute, according to the considerations set forth above, $a_1 = N_1$, and since N_1 is very near unity

$$- \ln N_1 \cong 1 - N_1 = N_2 \cong cv_1/M \tag{7}$$

On substituting these expressions in the relation above

$$\pi/c \cong RT/M \tag{8}$$

which is van't Hoff's law. The approximations involved vanish at infinite dilution so that in analogy with Eqs. (4) and (5) we may write

$$(\pi/c)_0 = RT/M \qquad (8')$$

In Table XXVII $(\Delta T_b/c)_0$ and $(\Delta T_f/c)_0$ for benzene solutions of solutes having the molecular weights indicated are compared with $(\pi/c)_0$ for the same solutions, c being expressed in grams per 100 ml. The osmotic pressure in grams per cm.^{2} corresponds to the pressure head in cm. for a liquid of unit density. An accuracy of $\pm 0.001°$ seldom is surpassed in either the boiling point elevation or the freezing point lowering method. To be sure, certain other liquids exhibit larger freezing point depression constants (resulting from lower heats of fusion), but the advantage thus gained may be nullified by difficulties arising from enhanced supercooling. The rather large deviations from ideality occurring at the concentrations required for the desired precision of measurement, i.e., at concentrations of several grams per 100 ml. even for the lowest polymer molecular weight in Table XXVII, would in general require extrapolation of $\Delta T/c$ to $c=0$. It becomes clear from such considerations that these methods ordinarily are insufficiently accurate for molecular weights exceeding about 10,000.* At higher molecular weights deviations from ideality set in at even lower concentrations, while at the same time the magnitude of the ΔT at a given concentration is decreased.

TABLE XXVII.—COMPARISON OF CALCULATED BOILING POINT ELEVATION, FREEZING POINT DEPRESSION, AND OSMOTIC PRESSURE

M	$(\Delta T_b/c)_0$[a] in °C/(g./100 ml.)	$(\Delta T_f/c)_0$[a] in °C/(g./100 ml.)	$(\pi/c)_0$ (g./cm²)/(g./100 ml.)
10,000	0.0031	0.0058	25
50,000	.0006	.0012	5
100,000	.0003	.0006	2.5

[a] Values for benzene.

The figures in Table XXVII show that a 0.001°C depression in the melting temperature corresponds approximately to a 10-cm. change in hydrostatic pressure head in the osmotic pressure. With appropriate pains, osmotic pressures may be measured within ± 0.01 cm. of liquid

* With the aid of elaborate refinements, including a 20-junction thermocouple, Ray[3] was able to measure boiling point elevations with a precision of 0.0002°C and molecular weights of polyethylene up to 35,000 with an error not exceeding 10 percent.

level, corresponding to a superiority in sensitivity of about 10^3 over the cryoscopic and ebullioscopic methods. Molecular weights up to one or two million may be determined osmotically under favorable conditions.

1b. The Number Average Molecular Weight.—Inasmuch as polymer samples invariably consist of mixtures of polymer homologs, the particular average molecular weight obtained from a given type of measurement requires careful definition. All of the colligative methods discussed above depend on the determination of $1-a_1$ at concentrations so low that $a_1 = N_1$. For a heterogeneous solute such as a high polymer, we have $1-a_1 = N_2$ at sufficiently high dilution exactly as for a homogeneous solute except that N_2 must represent the *sum* of the mole fractions of all solute species. Thus

$$N_2 = \sum_i N_i$$

and for a dilute solution

$$N_2 \cong v_1 \sum_i c_i/M_i$$

where N_i is the mole fraction and c_i the concentration, in grams per ml. of solution, of polymer species i of molecular weight M_i, and v_1 is the molar volume of the solvent. The summations extend over all polymer (or solute) species. In analogy with Eq. (7), we may write

$$N_2 \cong cv_1/\overline{M}_n \tag{9}$$

where $c = \sum c_i$ is the combined concentration of all species of the polymer in the solution, and $\overline{M}_n$ is the *number average molecular weight* defined by

$$\overline{M}_n = c/\sum (c_i/M_i) \tag{10}$$

Employing Eq. (9) for N_2 in the derivation of Eqs. (8) and (8') we obtain

$$(\pi/c)_0 = RT/\overline{M}_n \tag{11}$$

which differs from (8') in that M is replaced by $\overline{M}_n$. The same replacement satisfies Eqs. (4) and (5) for the boiling point elevation and freezing point depression, respectively.

The number average molecular weight usually is defined by one or the other of the alternative expressions

$$\overline{M}_n = 1/\sum (w_i/M_i) = \sum N_i M_i/\sum N_i \tag{12}$$

where $w_i = c_i/c$ is the weight fraction of species i in the polymer and N_i

is the number of moles of species i. The right-hand expression in Eq. (12) also follows from (10) since there are $N_i = c_i/M_i$ moles of species i and $\sum N_i M_i = c$ grams of total polymer per unit volume of solution.

The number average molecular weight represents the total weight of a sample of the polymer divided by the total number of moles of molecules which it contains. Any of the colligative methods discussed above actually provides a count of the number of solute molecules present rather than a determination of their sizes. The "molecular weight" value obtained is merely the result of dividing this number into the total weight of solute (and multiplying by Avogadro's number). End group methods also yield what is primarily a count of the number of molecules present, and they likewise give the number average. The number average is characterized by its sensitivity to a small weight fraction of a constituent of low molecular weight and by its relative insensitivity to small proportions by weight of constituents

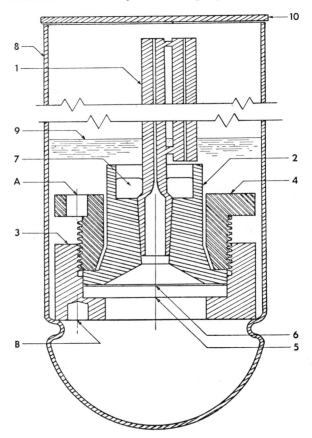

Fig. 35.—Assembled osmometer of Weissberg and Hanks.[4] (1) Solution and reference capillary; (2) solution cell; (3) osmometer base; (4) pressure ring; (5) perforated plate (perforations not shown); (6) semipermeable membrane; (7) mercury seal; (8) solvent container; (9) solvent level; (10) cover plate; (A) holes for wrench; (B) holes engaging pins in assembly bath.

much larger than the average in molecular size. These characteristics are easily verified through trial applications of either of the forms given in Eq. (12) to arbitrary distributions of the types indicated above.

1c. Experimental Methods.—One of the simpler types of osmometers, variants of which have been used by a number of authors,[4] is shown in Fig. 35. The membrane 6, usually consisting of a suitably swollen cellulose film, rests on the perforated metal plate 5, where it serves also as a gasket to seal the junction between the carefully machined lower edge of the metal cell 2 against plate 5. Pressure is applied by the large threaded ring 4. The glass capillary 1 of uniform bore connects to the cell through a carefully ground taper joint. The reference capillary is made of the same bore in order to eliminate correction for capillarity. (The assumption that the surface tension of the solution is the same as that of the solvent ordinarily is legitimate.) The solution to be measured is pipetted into the cell before inserting the capillary. By the application of suction as the capillary is inserted, the level of the solution in the capillary is adjusted close to the expected equilibrium value. The osmometer is then inserted in pure solvent in the container 8, and the whole is placed in a thermostat regulated within $\pm 0.005°C$ in order to eliminate the influence of thermal expansion and contraction of the solution on the capillary level. The difference in level compared with that in the reference capillary is observed over a period of time until it becomes satisfactorily constant. This may require from 6 hours to several days, depending on the permeability of the membrane and the precision required.

Osmometers of the type described offer the distinct advantage of simplicity both in construction and operation. Their major disadvantage results from the fact that the membrane is not firmly fixed over its entire area; hence its position relative to the lower plate changes with the osmotic pressure. This greatly reduces its sensitivity since much more solvent must be transported through the membrane to compensate for its change in position than is required for the change in height in the capillary.

Greater speed of attainment of equilibrium as well as greater precision are possible with a block-type osmometer[5,6,7,8] like the one shown schematically in Fig. 36. Osmometers of this type usually consist of a pair of matched, stainless-steel or brass blocks, in each of which is cut a shallow circular cell cavity. The membrane fits between the two blocks, preferably with a lead gasket on one side of the membrane. The blocks are firmly bolted together. Each cell may be emptied and refilled through a metal tube connected with the bottom of the cell and closed with a needle valve during operation. Various schemes have

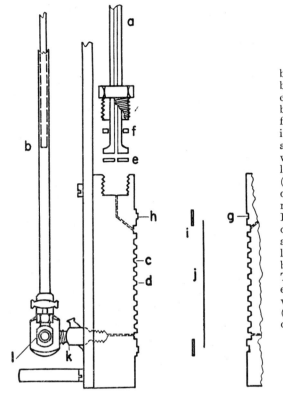

Fig. 36.—Diagram of the block-type osmometer used by Krigbaum[8] showing one cell in detail. Precision-bore 1-mm. capillary (a) is flanged and ground flat at its base. The packing nut, acting through the Teflon washer (f) seals the capillary foot against lead gasket (e). The cell consists of channels (c) with "island" membrane supports (d). Ridge (h) fits into the wider channel (g) when the blocks are bolted together using lead gasket (i) between left block and membrane (j). The cells may be filled and emptied through needle valve (k) and connection (l). The standpipe (b) facilitates changing the level.

been devised for supporting the membrane over the cell area. Meyer and collaborators[5] have used perforated metal disks which fit into the cell blocks on either side of the membrane. Others[6,7,8] have used systems of channels of one sort or another cut into the block; the residual intervening regions then fix the membrane firmly in place. Such a design[7,8] is shown in Fig. 37. It consists of two sets of shallow channels, semicircular in cross section, cut at right angles to each other over the face of each cell, with a circular channel running around the outer border of the cell. The "island points" remaining between the channels provide the necessary support for the membrane, yet cover only a small fraction of the surface.

Dynamic osmotic methods have been used occasionally,[6] the rate of permeation being measured as a function of the pressure difference. The interpolated pressure for zero rate should equal the osmotic pressure. In practice, however, more reliable results usually have been secured by establishing an equilibrium pressure which remains unchanged over an extended period of time.

Fig. 37.—Photograph of the block osmometer diagramed in Fig. 36. This design affords access of solution and solvent to maximum membrane area and rigid support of the membrane.[7,8]

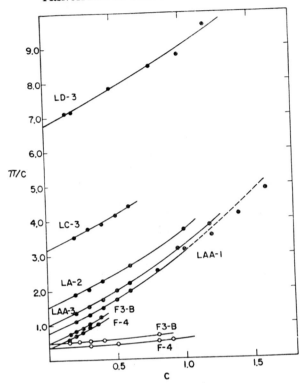

Fig. 38.—Plots of π/c against c for a series of polyisobutylene fractions ($M = 38,000$ to $720,000$) in cyclohexane ($\bullet$) and in benzene ($\bigcirc$), both at 30°C. The
osmotic pressure π is expressed in g./cm.^{2} and c in
g./100 cc. Curves have been calculated according to
Eq. (13). (Krigbaum.[8])

The membrane is critically important in osomometry. Selection
of a membrane involves reconciliation of high permeability toward the
solvent with virtual impermeability to the smallest polymer molecules
present in the sample. Membranes of cellulose are most widely used.
Commercially "regenerated" cellulose film is a common source. The
undried "gel" cellophane film is often preferred, but the dry film may
be swollen in water (or in aqueous solutions of caustic or zinc chloride[9])
to satisfactory porosity. Useful cellulose membranes may also be prepared by denitration of nitrocellulose films,[10] and special advantages
have been claimed for bacterial cellulose films.[11] The water in the
swollen membrane in any case may be replaced by a succession of
miscible organic solvents ending with the one in which osmotic measurements are to be made. Membranes of varying porosity may be

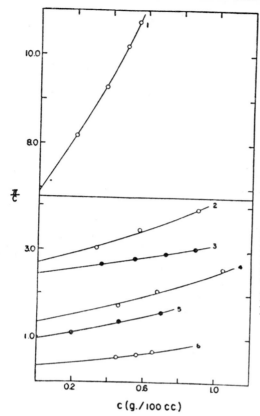

FIG. 39.—Plots of π/c against c from the data of Masson and Melville[13] for the following solvent-polymer pairs: curve 1, polyacrylonitrile in dimethylformamide at 13.5° C; curves 2 and 4, poly-(vinyl acetate)s in benzene at 20°C; curve 3, polyacenaphthylene in benzene at 25°C; curve 5, polyvinylxylene in benzene at 24°C; curve 6, poly-(methyl methacrylate) in benzene at 16°C. All curves have been calculated from Eq. (13). Units correspond to those in Fig. 38. (Fox, Flory, and Bueche.[14])

prepared depending on the swelling and solvent transfer treatment.[9,11] Drying out of the membrane at any stage destroys its permeability. "Fast" membranes may be permeable to polymers below about 50,000 in molecular weight; denser membranes retain polymers down to 5,000 in molecular weight, but at a considerable sacrifice in rate of permeation by solvent. Nitrocellulose membranes have been used successfully in hydrocarbon solvents, and suitably swollen poly-(vinyl alcohol) membranes are claimed to be especially adapted to the retention of polystyrene even at molecular weights of only 2,000.[12]

1d. **Treatment of Data.**—Typical osmotic data are shown in Figs. 38 and 39. Here the ratio (π/c) of the osmotic pressure to the concentration is plotted against the concentration. If the solutions behaved ideally, van't Hoff's law Eq. (11) would apply and π/c should be independent of c. Owing to the large effective size of the polymer molecules in solution (Fig. 34) and the interactions between them which consequently set in at low concentrations, π/c increases with c with a slope which depends on the interaction (see Chap. XII). The necessity for extrapolation to infinite dilution for the determination of molecular weights is evident from the data shown in Figs. 38 and 39. It is also apparent that some of these plots are appreciably curved, which complicates accurate extrapolation.

According to the theory of dilute polymer solutions, to be discussed in Chapter XII, the osmotic pressure may be expressed as a function of the concentration by an equation of the form[14]

$$\pi/c = (\pi/c)_0[1 + \Gamma_2 c + g\Gamma_2^2 c^2] \tag{13}$$

where $(\pi/c)_0$, representing the limiting value of π/c at infinite solution, equals $RT/\overline{M}_n$; Γ_2 is a parameter which depends on the polymer-solvent interaction, being greater the better the solvent; and g is a factor whose value depends on the polymer-solvent system, but which often may be approximated by 0.25 (see Chap. XII). The third term in Eq. (13) takes account of the curvature in the plot of π/c against c. The presence of Γ_2^2 in the coefficient of this term shows that its contribution decreases rapidly as the slope (Γ_2) decreases, i.e., as the solvent is made poorer. Owing to omission of higher terms, Eq. (13) is limited to low concentrations. It may be expected to hold in general for concentrations for which $(\pi/c) < 3(\pi/c)_0$, but it should not be applied much above this range.

If the value of g may be satisfactorily approximated by 0.25, and this usually is legitimate, then Eq. (13) may be written

$$(\pi/c)^{1/2} = (\pi/c)_0^{1/2}[1 + (\Gamma_2/2)c] \tag{13'}$$

and one may apply the graphical method of Berglund* consisting in plotting $(\pi/c)^{1/2}$ against c within the range for which π/c does not exceed $3(\pi/c)_0$. The intercept equals $(\pi/c)_0$ and the slope in conjunction with the intercept yields Γ_2. This method of treating osmotic data is illustrated in Fig. 40.

In instances where a value of g differing considerably from 0.25 is required, it is convenient to prepare a plot of the logarithm of the bracketed quantity in Eq. (13) against $\log(\Gamma_2 c)$.[14] The curve thus computed represents $\log(\pi/c) - \log(\pi/c)_0$ plotted against $\log c + \log \Gamma_2$ according to theory. Experimental values of $\log(\pi/c)$ for π/c not exceeding $3(\pi/c)_0$ are then plotted to the same scale against $\log c$ on transparent graph paper. The latter plot is superimposed on the theoretical curve and shifted rectilinearly to obtain the best fit. The vertical displacement of the ordinate scale of the experimental plot

* Private communication from Dr. U. Berglund-Larsson, Nobel Institute of Chemistry, Stockholm, Sweden.

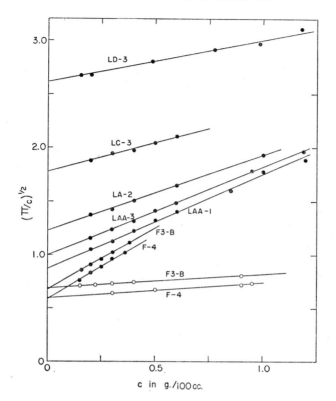

FIG. 40.—Plots of $\sqrt{\pi/c}$ against c according to the method of Berglund. The data are those of Krigbaum[8] for polyisobutylene fractions in cyclohexane (●) and in benzene (○) previously shown in Fig. 38. Units are the same as there given.

gives $-\log(\pi/c)_0$ from which $\overline{M}_n$ may be computed; the horizontal displacement gives $\log \Gamma_2$. The curves drawn in Figs. 38 and 39 have been calculated from Eq. (13) using values of the parameters $(\pi/c)_0$ and Γ_2 established in this manner.

It will be observed in Fig. 38 that the slope of the π/c plot for polyisobutylene in benzene at 30°C is small and that there is no evidence of curvature. This is in accord with Eq. (13), which requires the magnitude of the curvature to vary as the *square* of the slope. Of more immediate importance is the fact that the same limiting values, $(\pi/c)_0$, are obtained in different solvents, despite widely differing behavior at higher concentrations. Hence the observed molecular weights are independent of the solvent employed within experimental error. In those few instances where the molecular weight values obtained in different solvents do not agree,[15] chemical degradation, association due possibly to specific polar groups, or aggregation due to formation of crystallites is indicated.

The selection of a poor solvent for the polymer (such as benzene for polyisobutylene) offers the distinct advantage that the slope of π/c plotted against the concentration is small. Measurements may then be made at higher concentrations where π is larger, and therefore subject to a smaller percentage error, without exceeding the limit imposed on the extrapolation formula (13) by the stipulation that $(\pi/c) < 3(\pi/c)_0$. By judicious choice of solvent and temperature, the change in π/c with c may be made very small, and the extrapolation error is correspondingly reduced. In fact, at a uniquely defined temperature Θ for a given polymer-solvent system, namely at the critical consolute temperature for this system when the molecular weight is infinite (see Chap. XII), the slope is zero and π/c remains constant up to concentrations of several percent. This critical temperature Θ for polyisobutylene in benzene is 24.5°C. Experiments made precisely at the critical temperature may be uncomfortably close to the precipitation condition for the polymer (of finite M) being measured. Hence it is preferable to work at a temperature at least a few degrees above Θ and to extrapolate π/c to $c=0$, the relationship being very nearly linear and of low slope.

An alternative procedure for minimizing the change in π/c with c depends on the use of a mixture consisting of a solvent and a nonsolvent. In analogy with the situation for a single solvent component, reduction of the slope of π/c with c to zero requires close approach to precipitation conditions. A complication may arise, however, in osmotic experiments using mixed solvents owing to preferential absorption of one component by the polymer "particles," with the result that the ratio of solvent to nonsolvent in the two compartments of the osmometer will differ slightly at equilibrium. The time required for final equilibration might be prolonged as a result of the necessity for disproportionate transfer of the two solvent components through the membrane. Furthermore, if the solvent compartment is comparable in size with that for the solute (as in the osmometer shown in Fig. 36), the equilibrium composition in the solvent compartment may differ from that initially introduced, and it may vary with the polymer concentration in the solution compartment. The extrapolation procedure based on Eq. (13) may not apply exactly under these circumstances. In spite of these potential sources of difficulty, mixed solvents have been used successfully in the osmotic determination of polymer molecular weights.[16]

2. DETERMINATION OF MOLECULAR WEIGHTS AND POLYMER DIMENSIONS BY LIGHT SCATTERING

The scattering of light by the molecules of a gas (Rayleigh scattering) or by colloidal particles suspended in a liquid medium (Tyndall scattering) is well known. The intensity of scattered light depends on the polarizability of the particles (or molecules) compared with that of the medium in which they are suspended; it depends also on the size of the particles and, of course, on their concentration. If the solution is sufficiently dilute, the intensity of the scattered light is equal to the sum of contributions from the individual particles, each being unaffected by the others in the solution. The intensity of the light scattered in a given direction by a single particle will be proportional to the *square* of its size and independent of its shape if the particle is isotropic (i.e., if its polarizability is the same in different directions) and its dimensions are small compared with the wavelength of the light. The total light scattered by a solution containing particles at a specified concentration by weight consequently is greater the larger the individual particles. Hence the size of the particles may be deduced from the intensity of the light scattered by a dilute solution or suspension, assuming, of course, that the refractive indexes (or the polarizabilities) of the particles and of the medium are known. The irregularly coiled polymer particles in dilute polymer solutions fulfill the condition of isotropy, but if their molecular weights are very large, their mean dimensions in solution may not be much less than the wavelength of the light. Corrections must then be made for the *dissymmetry* of the scattered light in order to calculate the molecular weight. The intensity of the light scattered by a dilute polymer solution is small, and sensitive methods are required for its quantitative measurement.

The light-scattering method has been adapted in recent years, notably by Debye and collaborators,[17] to the determination of the molecular weights of polymers. The osmotic and light-scattering methods are the principal ones in use at the present time. Each offers certain advantages over the other, and in at least one respect they are supplementary. Thus the average molecular weight obtained by the light-scattering method is the *weight average* (see Sec. 2c), which for a heterogeneous polymer is greater than the *number average* obtained from osmotic measurements. The disparity between the two averages provides a measure of the degree of heterogeneity in the molecular weight distribution.

The determination of the turbidity of polymer solutions will be discussed briefly in the following pages, and the essentials involved in the deduction of molecular weights and molecular dimensions will be set forth.   For more detailed discussions, the reader is referred to papers and reviews by Debye,[17] Mark,[18] Oster,[19] and Doty and Edsall.[20]

2a. Experimental Methods.—A diagrammatic sketch of apparatus developed by Debye and co-workers[21] for measuring the intensity of light scattered by a high polymer solution is shown in Fig. 41.   Light from a medium-pressure mercury arc S (General Electric type AH-4) passes through an adjustable pinhole P and is collected by lens L, which is situated to transmit a nearly parallel beam.   The variable diaphragm D_1 restricts the diameter of the beam entering the cylindri-

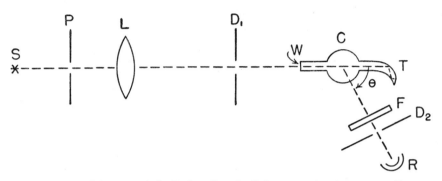

Fig. 41.—Diagram of the Debye-Bueche light-scattering instrument.[21]

cal Pyrex glass cell C, 40 mm. in diameter.   Two 12-mm. glass tubes are attached coaxially on either side of the cell as indicated in Fig. 41. To the one on the left, a plane glass window W is cemented; the one on the right, sealed off in a curved point, serves as a light trap for the transmitted beam.   The entire cell is painted black except for a strip around one side which is reserved for passage of the scattered beam. Light scattered at an angle θ passes through the filter F, which isolates the desired wavelength, and the fixed diaphragm D_2 collimates the beam before it enters the photomultiplier tube R.   The sensitivity of the photomultiplier tube can be varied by selecting different potentials across the stages of the tube.   The current through the tube, which is proportional to the light intensity, is measured with a galvanometer.

The parts of the apparatus except the source and the filter and phototube assembly (F, D_2, and R) are rigidly mounted within a box painted black inside.   The diaphragms and lens are mounted in metal panels which reach to the walls of the box for the purpose of minimizing stray light.   The scattered light is intercepted by a small reflecting prism (not shown) which passes the light downward through the filter F, diaphragm D_2, and photomultiplier tube R. These are placed perpendicular to the vertical beam, rather than as shown in Fig. 41, all three being mounted on a movable arm which may be rotated about a perpendicular axis through the center of the cell. The prism also is attached to this arm. The position of the latter, which defines the angle θ, can be adjusted externally and read on a dial. The cylindrical cell offers the advantage of equivalent optics for all angles θ (in contrast to rectangular cells); other features serve to minimize stray light —a major source of experimental errors.

No provision is made in the instrument described above for the measurement of the intensity of the incident beam. Hence deduction of the desired ratio of the intensity scattered in a given direction (θ) to the incident intensity requires the use of a reference standard of independently established scattering power. Debye and Bueche recommend a machined poly-(methyl methacrylate) block which may be inserted in place of the cell for this purpose. The scattering power of the block must be established independently as a function of angle. Recalibration at frequent intervals may be necessitated by changes in the optical properties of the block. A measurement of the galvanometer deflection for the solution of unknown scattering characteristics is preceded and followed by measurements at the same angle on the reference block. The influence of possible minor variations in the intensity of the source are minimized in this way. The ratio of the galvanometer deflection for the solution to the mean value for the two observations on the block, multiplied by the known scattering power of the block, yields the ratio (i_θ/I_0) of the intensity i_θ of light scattered per cc. of solution at the distance r from the cell in the direction θ to the intensity I_0 of the incident beam. The much smaller scattering of the pure solvent observed at the same angle is deducted from the scattering observed for the solution. The excess of the scattering of the solution over that of the solvent is the quantity of interest, and i_θ is assumed hereafter to have been so corrected.

A diagram of the experimental arrangement devised by Zimm[22] is shown in Fig. 42. Light from a 60-cycle a.c.-operated mercury arc S (AH-4) is focused by the pair of lenses L_1 on the diaphragm D_1 after passing through the filter F. The concave mirror M, consisting of a silvered watch glass with an unsilvered "hole" in the center, focuses a part of the beam on the photocell P_1 of low sensitivity. The remainder of the beam is focused by lens L_2 on the center of the small cell C containing the polymer solution. The cell is placed in the conical flask E, which is filled with a liquid of approximately the same refractive index as that of the solution being measured. The prism G tilts the

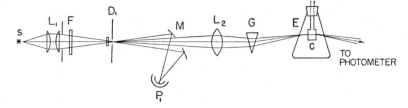

FIG. 42.—Diagram of the light-scattering apparatus designed by Zimm.[22]

beam upward to compensate for the refraction of the beam at the conical surface so that the beam traverses the flask and cell horizontally. With this arrangement, light reflected at the surfaces of the conical flask is dispersed in directions unfavorable for re-entry into the optical system. Temperature control of the flask E is provided. The projector system preceding the prism is placed in a dark box to further minimize stray light.

The photometer in Zimm's apparatus, not shown in Fig. 42, consists of a lens and diaphragm-collimating system which delivers light scattered at the angle θ from the center of the cell (where the image of the source is focused) to a sensitive photomultiplier tube. It is mounted on an arm which may be rotated about a vertical axis through the cell; a graduated scale indicates the scattering angle θ. The out-of-phase a.c. currents from the photomultiplier tube and the incident intensity-monitoring phototube P_1 are balanced with a precision potentiometer. The potentiometer gives the ratio of the two photocurrents, which is proportional to the ratio of the scattered to the incident intensities. Readings consequently are almost independent of fluctuations in the light source. The instrument measures only relative values of the ratio i_θ/I_0 of the scattered to the incident beam, and conversion to absolute values requires calibration. The angular dependence of i_θ from 25° to 150° can be measured accurately.

Absolute calibration is of critical importance for the determination of molecular weights by the light-scattering method. The scattered intensity is so low compared with the incident beam ($i_{90}°$ at a distance of a few cm. being of the order of $10^{-6} I_0$) that a reliable direct comparison of the two beams using the same photocell is impractical. Several methods have been used, with results which are not always comparable, however.

In one method[23] the intensity of the primary beam is diminished by several orders of magnitude through the use of a series of neutral filters, the percentage transmission of each at the wavelength λ having been accurately measured. Comparison of the intensity of light scattered by the solution with the intensity of the incident beam measured in this manner (applying appropriate correction for geometrical factors affecting the scattered beam) leads at once to the ratio i_θ/I_0. A related method [24] involves extracting a small fraction of the primary beam by several successive reflections from glass plates immersed in a liquid medium. The percentage reflection may be computed from the refractive indexes of the glass and liquid, which permits calculation of the ratio of the intensities of the multiply reflected and the primary beams. Comparison with the relative intensity of the light scattered by a solution, or by a reference standard, provides the required data for calculating i_θ/I_0.

If the turbidity is sufficient and a relatively long path is used, the fraction of light dissipated from the primary beam by the solute may be found by direct measurements of the relative intensity of the beam transmitted by the cell when filled with solution compared with that when filled with solvent.[23,25] High accuracy of measurement is of course required, since the turbidities measured are of the order of 10^{-4} cm.$^{-1}$. Division of the fractional decrease in light intensity by the path length yields the absolute turbidity τ, or relative decrease in intensity of the transmitted beam per cm. of path. Now the radiant energy scattered in all directions per cc. of solution must equal the product τI_0. If relative scattered intensities $i_\theta(\text{rel})$ are measured in enough directions, the total relative radiant energy scattered may be obtained by integrating the scattered radiation over all directions, i.e., from $\int_0^{180°} 2\pi r^2 i_\theta(\text{rel}) \sin \theta d\theta$. Comparison of the result with the absolute value τI_0 defines the factor for converting $i_\theta(\text{rel})$ to absolute values of i_θ/I_0.

Other methods of calibration are discussed by Zimm[25] and by Brice and co-workers.[23] Owing to difficulties of absolute calibration, a reference standard invariably is used. This may consist of a polymer solution of constant properties (i.e., not subject to degradation) or a bulk polymer, such as that used by Debye and Bueche.[21]

Careful filtration of the solvent and of the polymer solution is essential in all light-scattering work. Traces of dust or dirt may introduce serious errors.

2b. Theory for Particles Small Compared with the Wavelength of the Light.—Consider a wave traveling along the x-axis intercepted by an isotropically polarizable particle at the origin O of coordinates (Fig. 43), the dimensions of the particle being small compared with the wavelength of the radiation. Let the wave be plane-polarized in the zx plane. Its electric intensity may be written

$$E = E_{0z} \cos\left[(2\pi/\lambda)(\bar{c}t - \bar{n}_0 x)\right]$$

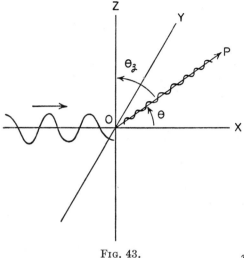

where E_{0z} is the amplitude of the incident plane-polarized wave, λ and $\tilde{c}$ are the wavelength and velocity of propagation in vacuum, respectively, t is the time, and $\tilde{n}_0$ is the refractive index of the medium. The wave will induce an electric moment p in the particle. If α represents the excess polarizability of the particle (which is assumed to be spherical) over that of the surrounding medium

Fig. 43.

$$p = \alpha E = \alpha E_{0z} \cos(2\pi\tilde{c}t/\lambda) \quad (14)$$

This oscillating dipole will radiate a secondary wave of the same frequency, which may be observed along the direction OP making an angle θ with the direction (x-axis) of the incident wave. The scattered wave will be polarized in the plane defined by P and the z-axis. The electric intensity E_{sec} of the scattered wave will depend on the acceleration of the induced electric moment, i.e., on d^2p/dt^2. Specifically the amplitude $E_{0,\text{sec}}$ of the wave scattered in the direction OP will depend on the amplitude of $(1/\tilde{c}^2)$ (d^2p/dt^2) calculated from Eq. (14), on the projection of the moment perpendicular to the direction P in the P,z plane, and inversely on the distance r from the scatterer. Thus,

$$E_{0,\text{sec}} = (\alpha E_{0z}/r)(2\pi/\lambda)^2 \sin\theta_z \quad (15)$$

where θ_z is the angle between OP and the z-axis. The intensity i_z' due to scattering by the single particle in the direction OP equals the square of the amplitude, or

$$i_z' = (16\pi^4/r^2\lambda^4)I_{0z}\alpha^2 \sin^2\theta_z \quad (16)$$

where $I_{0z} = E_{0z}^2$ is the intensity of the incident plane-polarized wave. Thus, the intensity of the scattered wave is a maximum in the x,y plane and zero along the z-axis; it is independent of the projection of the scattering direction in the x,y plane.

To calculate the intensity of scattered radiation along the direction P when the incident wave is unpolarized, let the incident wave of intensity I_0 be resolved into two components of equal intensity

($I_{0y} = I_{0z}$) polarized along the y and z directions, respectively. Then, taking $I_0 = I_{0y} + I_{0z}$ and $i' = i_y' + i_z'$, where i_y' is given by an expression exactly like (16) with z replaced by y, we obtain

$$i' = (8\pi^4 I_0 \alpha^2 / r^2 \lambda^4)(\sin^2 \theta_y + \sin^2 \theta_z)$$

The last factor is equal to $1 + \cos^2 \theta$. The combined intensity of scattered light at a distance r and in a direction specified by θ (Fig. 43) due to all particles in unit volume (N/V) of the very dilute solution becomes

$$i_\theta^0 = (I_0 N/V)(8\pi^4 \alpha^2 / \lambda^4)(1 + \cos^2 \theta)/r^2 \qquad (17)$$

The superscript ^{0} appearing above and in related symbols below will serve as a reminder that the particle dimensions have been assumed small compared to the wavelength of the light. Owing to failure to meet this condition it is often necessary to apply a correction (see Sec. 2d) to the observed intensity i_θ in order to obtain the intensity i_θ^0 applicable in this equation and subsequent relationships.

It is evident that the light scattered by the isotropic particle will be partially polarized in a manner depending on θ. For $\theta = 90°$, i.e., for transverse scattering, the light will be plane-polarized with the electric vector perpendicular to the incident beam. The polarization will vanish as θ approaches 0 or π. Also, according to Eq. (17), the intensity of scattered light will vary with the direction, being only half as great perpendicularly to the incident beam as it is in the forward or backward directions. It is important to note that the intensity is symmetrical about $\theta = 90°$; i.e., the intensity for $\theta = \pi/2 - \beta$ is the same as for $\pi/2 + \beta$.

The excess polarizability α is related to the (optical) dielectric constants ϵ_0 and ϵ of the solvent and of the solution, respectively, in the following manner:

$$(4\pi N/V)\alpha = \epsilon - \epsilon_0$$

The dielectric constants may be replaced by the squares of the respective refractive indexes of solution and solvent, which yields

$$\alpha = (V/4\pi N)(\tilde{n}^2 - \tilde{n}_0^2)$$

Since the solution is dilute

$$\alpha = (V/4\pi N)c(d\tilde{n}^2/dc)_0 = (\tilde{n}_0 c V/2\pi N)(d\tilde{n}/dc)_0$$

where c is the concentration in grams per cc. Substituting in Eq. (17) and noting that $c = NM/NV$ where N is Avogadro's number

$$i_\theta^0/I_0 = (2\pi^2/N\lambda^4 r^2)\tilde{n}_0^2(d\tilde{n}/dc)_0^2(1 + \cos^2\theta)Mc \tag{18}$$

The quantity $r^2 i_\theta^0/I_0$ is referred to as the Rayleigh ratio R_θ^0, which may be written

$$R_\theta^0 = K^*(1 + \cos^2\theta)Mc \tag{19}$$

or

$$R_\theta^0 = K^*(1 + \cos^2\theta)NM^2/VN \tag{19'}$$

where

$$K^* = 2\pi^2\tilde{n}_0^2(d\tilde{n}/dc)_0^2/N\lambda^4 \tag{20}$$

Eq. (18) expresses the intensity ratio for light scattered a distance r at an angle θ from an incident unpolarized beam by a solution of particles of molecular weight M at a (low) weight concentration c per unit volume. The more succinct Rayleigh ratio is given by Eq. (19). The parameter K^* depends on quantities which are readily ascertainable: the wavelength of the light, the refractive index of the medium, and the change in refractive index with concentration, $(d\tilde{n}/dc)_0$. Once K^* has been computed, the molecular weight of particles (molecules) much smaller than the wavelength of the light may be calculated according to Eq. (19) from the Rayleigh ratio R_θ measured at sufficiently low concentrations to justify the implicit assumption introduced above that the particles scatter independently of each other.

It is sometimes preferred to consider the fraction of the light scattered in all directions from the primary beam per cm. of path. Thus a beam of intensity I decreases, as a result of scattering, by the amount $\tau I dx$ in traveling the distance dx through the solution of *turbidity* τ.† After traversing a distance x, the incident beam will have been reduced from the initial intensity I_0 to I, where

$$I/I_0 = e^{-\tau x} \tag{21}$$

or, since the diminution in intensity will be extremely small

$$\tau x = (I_0 - I)/I_0 \tag{21'}$$

The turbidity may be obtained by integrating the total intensity of radiation scattered in all directions by the N/V particles per unit volume, i.e.

$$\tau = \int_0^\pi (i_\theta/I_0)2\pi r^2 \sin\theta d\theta$$

† Measured quantities i_θ, R_θ, and τ are expressed without the superscript ^{0}.

If the actual intensity i_θ is replaced by the corrected intensity i_θ^0, which would be observed if intraparticle interference (see below) were of negligible consequence, then on substituting from Eq. (18) for i_θ^0/I_0 and performing the integration we obtain for the similarly corrected turbidity

$$\tau^0 = HcM \tag{22}$$

where

$$H = (32\pi^3/3\lambda^4 N)\tilde{n}_0^2(d\tilde{n}/dc)_0^2 \tag{23}$$

Comparing with Eqs. (19) and (20)

$$\tau^0 = (16\pi/3)R_\theta^0/(1 + \cos^2\theta) \tag{24}$$

Thus the turbidity may be calculated directly from the Rayleigh ratio. Both of these quantities enter into the subsequent discussion, where it is to be borne in mind that either one is readily calculated from the other.

As already pointed out, direct measurement of the turbidity from the diminution in intensity of the primary beam per cm. of path generally is impractical owing to its low magnitude. The relative intensity i_θ/I_0 of the scattered light is measured at a given angle, usually 90°, therefore, and the result converted to the Rayleigh ratio, or to the turbidity using Eq. (24). The molecular weight may then be calculated from Eq. (19) or (22). In practice, matters are not so simple owing to the necessity of taking into account two sources of complications not dealt with thus far. One of these has to do with extrapolating the ratio of τ/c to its value at infinite dilution, in analogy with the extrapolation of osmotic data. The other arises because of the fact that polymer molecules in solution often assume dimensions which are not small compared with the wavelength of the light, as has been assumed above. When the dimension of the particle exceeds about $\lambda/20$, the angular dependence of the intensity of the scattered light deviates appreciably from Eq. (18); hence the values observed for $R_\theta/(1+\cos^2\theta)$ will depend on the angle θ, and the (average) molecular weight cannot be calculated unambiguously using the foregoing equations and the intensity measured at an arbitrary angle. Before entering into a discussion of methods for introducing appropriate corrections, however, the type of average molecular weight obtained by the light-scattering method will be considered.

2c. The Weight Average Molecular Weight.—The amplitude of the secondary radiation scattered by a particle, or molecule, depends on the polarization induced in it by the incident wave; i.e., it depends on α as shown by Eq. (15). But for particles composed of matter having given optical characteristics, α is proportional to the size of the particle. Hence the *amplitude* of the radiation scattered by one of these particles will be proportional to its size.   The *intensity* of the light scattered (see Eq. 16) by one particle is proportional to the *square* of its polarizability α, and hence to the *square* of its size.   This is the basis for the appearance of the product of the number of particles and the square of the molecular weight in the alternative Eq. (19′).   Indeed, it is the dependence of scattered intensity per particle on a power of M greater than the first which permits determination of the molecular weight.   If, for example, we were dealing with a property such as the heat capacity, the refractive index, or the specific volume, to each of which the polymer particle makes a contribution proportional to the first power of its mass, the value observed would depend only on the total *quantity* of polymer in a given volume of solution, and (to a very close approximation) would not depend at all on the number of molecules into which the given amount of solute is divided.   Molecular weight determination from the measurement of a property of this nature is impossible.

If the solute consists of polymer homologs differing in molecular weight, the larger particles will make a greater contribution to the scattering, or the turbidity, than an equal weight of smaller ones.   This is evident from Eqs. (19) and (22).   The situation contrasts with that for colligative properties, which depend only on the number of particles present, provided the solution is sufficiently dilute.   For the purpose of expressing quantitatively the total turbidity of a heterogeneous polymer in dilute solution, we may write

$$\tau^0 = \sum_i \tau_i^0$$

where τ_i^0 is the turbidity due to molecules of molecular weight M_i. Since $d\tilde{n}/dc$ will be practically identical for a series of polymer homologs, a single value of H applies to all of them and we obtain in analogy with Eq. (22)

$$\tau^0 = H \sum c_i M_i$$

The forbidding appearance of the summation sign may be avoided merely by writing

$$\tau^0 = Hc\overline{M}_w \qquad (25)$$

Where $\overline{M}_w$ is the *weight average* molecular weight defined by any of the following alternate expressions:

$$\overline{M}_w = \sum c_i M_i / c = \sum w_i M_i = \sum N_i M_i^2 / \sum N_i M_i \qquad (26)$$

Here w_i is the weight fraction and $c = \sum c_i$. The light-scattering method must invariably yield the weight average molecular weight irrespective of the nature of the molecular weight distribution. Eq. (25) replaces Eq. (22), in which molecular weight heterogeneity was disregarded. Similar considerations obviously apply to the Rayleigh ratio; instead of Eq. (19), we should use

$$R_\theta^0 = K^*(1 + \cos^2 \theta)\overline{M}_w c \qquad (27)$$

for a heterogeneous polymer.

The weight average molecular weight is important elsewhere in the interpretation of polymer properties; hence its relationship to other averages deserves careful inspection. The mean-square molecular weight is, of course,

$$\overline{M^2} = \sum N_i M_i^2 / \sum N_i \qquad (28)$$

Combining this relation with Eqs. (12) and (26)

$$\overline{M}_w = \overline{M^2}/\overline{M}_n \qquad (29)$$

i.e., $\overline{M}_w$ could be defined as the ratio of the mean square to the number average molecular weight. If $\overline{M}_{rms}$ is written for the root-mean-square molecular weight

$$\overline{M}_w/\overline{M}_{rms} = \overline{M}_{rms}/\overline{M}_n \qquad (30)$$

or, $\overline{M}_{rms}$ is the geometric mean of $\overline{M}_n$ and $\overline{M}_w$. From another viewpoint, suppose that a structural unit is selected at random from the mass of heterogeneous polymer. It may be shown that the weight average molecular weight represents the statistically *expected* size of the molecule of which this unit is a part. Of foremost importance, the weight average is particularly sensitive to the presence of larger species, whereas the number average is sensitive to the proportion by weight of smaller molecules. Thus, equal *parts by weight* of molecules with $M = 10,000$ and $100,000$ give a number average of $18,200$, whereas the weight average is $55,000$. Again, if equal *numbers* of molecules having $M = 10,000$ and $100,000$ are mixed, $\overline{M}_n = 55,000$ and $\overline{M}_w = 92,000$ for the mixture.

2d. Light Scattering by Polymer Particles with Dimensions Approaching the Wavelength of the Light.—The large volume engulfed by a polymer molecule in solution has been emphasized earlier in the present chapter. If the molecular weight exceeds several hundred thousand, then the breadth of the polymer coil may reach several hundred Ångstrom units or more. When the dimensions of the scattering particle exceed about one-twentieth of the wavelength of the light, it is no longer permissible to consider the particle as a simple source of scattered radiation; induced dipoles in widely separated portions of the particle will emit radiation for which the phase difference is appreciable. This is illustrated in Fig. 44. The length of the path traversed by a portion of the incident wave impinging on element B and then scattered toward P_1 differs from the length of path for a portion of the wave impinging on element A and scattered in the same direction. Moreover, the path difference is greater for the portions of the wave scattered from A and from B in the backward direction P_2 than for scattering from the same two points in the forward direction

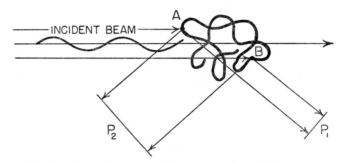

Fig. 44.—Interference of light scattered by different parts of a polymer chain comparable in dimension with the wavelength.

P_1. Destructive interference owing to phase differences between rays scattered by different elements of the molecule will therefore be greater the larger the scattering angle θ; it will vanish as θ approaches zero. Owing to interference effects of this nature, which obviously assume importance as the linear dimensions approach the wavelength $\lambda' = \lambda/\tilde{n}_0$ of the light in the medium, the scattering will not then be symmetrical about 90° as prescribed by Eq. (18); it will be less in the direction for which $\theta = 90° + \beta$ than for $\theta = 90° - \beta$. The observed ratio of these intensities is customarily referred to as the dissymmetry coefficient designated by z_β, i.e.,

$$z_\beta = i_{90°-\beta}/i_{90°+\beta}$$

The root-mean-square distance $\sqrt{\overline{r^2}}$ separating the ends of the polymer chain is a convenient measure of its linear dimensions. The dissymmetry coefficient z_β will be unity for $(\sqrt{\overline{r^2}}/\lambda') \ll 1$ and will increase as this ratio increases.

If the scattering intensity can be measured with sufficient accuracy down to very small angles, the Rayleigh ratio R_0 at zero angle may be secured by extrapolation. Since intraparticle interference must vanish as θ goes to zero (Fig. 44), the scattering at zero angle is unperturbed. The molecular weight may therefore be calculated from R_0 using Eq. (27), which reduces at $\theta = 0$ to

$$R_0 = 2K^*\overline{M}_w c \tag{27'}$$

Alternatively, the hypothetical turbidity which would obtain in the absence of interference may be calculated from Eq. (24), which at $\theta = 0$ reduces to

$$\tau^0 = (8\pi/3)R_0 \tag{24'}$$

The molecular weight may then be obtained from Eq. (25) using this hypothetical turbidity.

Measurements at low angles are subject to considerable error, and for this reason it is often preferred to apply appropriate corrections to scattering intensities measured at larger angles. The observed intensity i_θ in a direction θ will be reduced on account of intraparticle interference by a factor cusomarily designated by $P(\theta)$, which depends on the size and shape of the particle as well as on the angle θ. Thus, by definition

$$P(\theta) = i_\theta/i_\theta^0 = R_\theta/R_\theta^0$$

where i_θ^0 and R_θ^0 are the intensity and Rayleigh ratio, respectively, which would be observed at the angle θ in the absence of interference. Debye has shown[17,18,19,20] that for randomly coiled polymer molecules each of which may be represented approximately as a Gaussian distribution of chain elements about its center of gravity (see Chap. X)

$$P(\theta) = (2/v^2)\left[e^{-v} - (1 - v)\right] \tag{31}$$

where

$$v = (2/3)(\overline{r^2}/\lambda'^2)\left[2\pi \sin(\theta/2)\right]^2 \tag{32}*$$

The lowermost curve in Fig. 45 represents $P(\theta)$ plotted against $v^{1/2}$ according to Eq. (31) for random coil molecules. The results of similar calculations for spherical and for rod-shaped particles of uniform density are shown also. The curve for the former of these is not very different from that for randomly coiled polymers at corresponding values of the abscissas; the factor $P(\theta)$ for rods differs appreciably, however.

* The dimension which enters into the actual calculation is the root-mean-square distance $\sqrt{\overline{s^2}}$ (or "radius of gyration") of an element from the center of gravity. However, it may be shown[26] (see p. 430) that $\overline{s^2} = \overline{r^2}/6$, which has been substituted to replace $\sqrt{\overline{s^2}}$ with $\sqrt{\overline{r^2}}$ in the expression leading to Eq. (32).

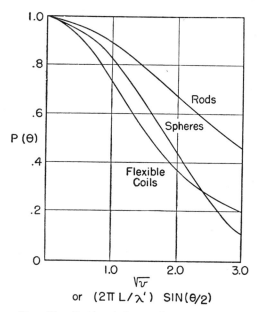

FIG. 45.—Ratio of observed to corrected intensity as a function of $\sqrt{v} = \sqrt{2/3}(\sqrt{\overline{r^2}}/\lambda')2\pi$ sin $(\theta/2)$ for flexible coils and $(2\pi L/\lambda')$ sin $(\theta/2)$ for spheres of diameter L or rods of length L. (Doty and Edsall.[20])

In all three cases $P(\theta)$ differs insignificantly from unity if the particle dimension is less than about $\lambda'/20$, i.e., if the root-mean-square end-to-end distance $\sqrt{\overline{r^2}}$ of the coil, the diameter of the sphere, or the length of the rod does not exceed about $\lambda'/20$. Then the scattering intensity should vary with angle as $(1+\cos^2 \theta)$, and the previous relations between the Rayleigh ratio and the molecular weight apply without correction. For larger particles (i.e., for larger ratios $\sqrt{\overline{r^2}}/\lambda'$), the angular dependence of the scattering should be expressed as $P(\theta)(1+\cos^2\theta)$.

For the purpose of calculating the weight average molecular weight we require the Rayleigh ratio R_θ^0 (or the turbidity τ^0) which would obtain in the absence of intraparticle interference. This ratio is given by the observed scattering ratio divided by $P(\theta)$; i.e., by $R_\theta/P(\theta)$. It is necessary therefore to evaluate $P(\theta)$ experimentally. This may be accomplished conveniently by measuring the scattering (i_θ) at two angles symmetrically disposed with respect to 90°, e.g., at 45° and 135°. The dissymmetry coefficient z_β thus observed may be compared with that calculated from Eqs. (31) and (32) as a function of $\sqrt{\overline{r^2}}/\lambda'$ as shown in Fig. 46. Having found $\sqrt{\overline{r^2}}/\lambda'$ in this manner, v may be computed from Eq. (32) for any angle θ, and the correction factor $1/P(\theta)$ for the observed intensity ratio at that angle is obtained from Eq. (31), or from Fig. 45.

It should be noted that relative intensity measurements suffice for the determination of z_β. The absolute value of i_θ/I_θ required for calculation of the molecular weight usually is determined at 90° and the appropriate correction factor $1/P(90°)$, obtained from the dissymmetry as described above, is applied to obtain $R_{90°}^0$. The factor $P(\theta)$ is strictly applicable only at infinite dilution, and it is therefore impor-

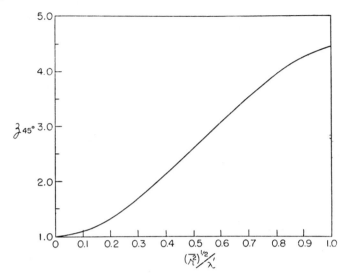

Fig. 46.—Dissymmetry ratio for light scattered at 45° and 135° as a function of $\sqrt{\overline{r^2}}/\lambda'$ for random coil polymer chains.[19]

tant that the dissymmetry measurements be made at the lowest concentration consistent with adequate precision.

Apart from their utility in determining the correction factor $1/P(\theta)$, light-scattering dissymmetry measurements afford a measure of the dimensions of the randomly coiled polymer molecule in dilute solution. Thus the above analysis of measurements made at different angles yields the important ratio $\sqrt{\overline{r^2}}/\lambda'$, from which the root-mean-square end-to-end length of the chain is at once obtained. Alternatively, one may for this purpose measure the scattering for a series of wavelengths at the same angle θ, e.g., at 90°. If there were no dissymmetry, R_θ would vary as λ^{-4} according to Eqs. (19) and (20); $R_\theta\lambda^4$ would remain constant. Changes observed in this product reflect changes in $P(\theta)$ with wavelength, from which $\sqrt{\overline{r^2}}$ may be determined through the use of Eqs. (31) and (32). Light-scattering measurements offer the only *absolute* experimental method at present available for determining the dimensions of randomly coiled polymer molecules. This method is applicable, however, only within the somewhat restricted range $1/10 < \sqrt{\overline{r^2}}/\lambda < 1$.

2e. Dependence of Scattering on Concentration.—In order to treat light scattering by nonideal solutions at finite concentrations, it is necessary to consider the problem from the more general point of view developed by Smoluchowski[27] and Einstein.[28] They considered fluctuations of refractive index within arbitrarily chosen volume elements

δV small compared to λ'. These fluctuations originate from two sources: variations in density and in concentration. The density fluctuations are responsible for the scattering by the pure solvent and are of no concern here; their contribution is eliminated by deducting the scattering intensity observed for the pure liquid from that for the solution. We shall confine the following discussion, therefore, to a consideration of the turbidity, or the scattering, arising from concentration fluctuations.

The particles of the preceding treatment are here replaced by small elements of volume δV of the solution. The excess polarizability of one of these volume elements due to the deviation of its concentration from the average may be written

$$\Delta\alpha = \Delta\epsilon \, \delta V/4\pi$$

where $\Delta\epsilon$ is the difference in optical dielectric constant for the volume element compared with the average for the entire solution. The scattered intensity depends on the average square of $\Delta\alpha$ for all volume elements, $\overline{(\Delta\alpha)^2}$, which replaces α^2 in Eq. (17). Noting further that the number of particles per unit volume N/V is to be replaced by the number of volume elements per unit volume given by $1/\delta V$, we have from Eq. (17)

$$i^0 = I_0\pi^2\overline{(\Delta\epsilon)^2}\delta V(1 + \cos^2\theta)/2\lambda^4r^2 \qquad (33)$$

Here $\overline{(\Delta\epsilon)^2}$ is the average square of the fluctuation of the dielectric constant in a volume element. It may be written

$$\overline{(\Delta\epsilon)^2} = (\partial\epsilon/\partial c)^2\overline{(\Delta c)^2} = [2\tilde{n}_0(d\tilde{n}/dc)_0]^2\overline{(\Delta c)^2}$$

According to the thermodynamic theory of fluctuations, the mean-square concentration fluctuation is given[19,28] by

$$\overline{(\Delta c)^2} = kT/(\partial^2 F/\partial c^2) = kTv_1c/\delta V(-\partial\mu_1/\partial c)$$

where F is the free energy of the entire solution and c is the concentration in a volume element δV; k is Boltzmann's constant, μ_1 is the chemical potential of solvent in the solution, and v_1 is its molar volume. Substituting these results in Eq. (33) and converting to the Rayleigh ratio, we obtain

$$R_\theta^0 = K^*RTv_1c(1 + \cos^2\theta)/(-\partial\mu_1/\partial c) \qquad (34)$$

where R is the gas constant. Or, since $\mu_1 - \mu_1^0 = -\pi v_1$ where π is the osmotic pressure,

$$(-\partial\mu_1/\partial c) = v_1(\partial\pi/\partial c)$$

and

$$R_\theta^0/(1 + \cos^2\theta) = K^*cRT/(\partial\pi/\partial c) \tag{35}$$

Similarly for the turbidity

$$\tau^0 = HcRT/(\partial\pi/\partial c) \tag{36}$$

Introducing Eq. (13) for the osmotic pressure as a function of concentration and replacing $(\pi/c)_0$ with RT/M, we obtain

$$Hc/\tau^0 = (1/M)[1 + 2\Gamma_2 c + 3g\Gamma_2^2 c^2] \tag{37}$$

which obviously reduces to the previous equation (22) at infinite dilution. Extension of the consideration of fluctuations to a solution of a heterogeneous polymer again shows[29] that the molecular weight M is to be replaced by the weight average $\overline{M}_w$. The confusion resulting from molecular weight heterogeneity is avoided if we adopt the form

$$c/\tau^0 = (c/\tau^0)_0[1 + 2\Gamma_2 c + 3g\Gamma_2^2 c^2] \tag{37'}$$

which is the analog of Eq. (13) for the osmotic pressure. (For practical purposes g may again be assigned the fixed value 1/4.) The limiting value $(c/\tau^0)_0$ of the concentration-turbidity ratio leads at once to the weight average molecular weight according to Eq. (25), which is more appropriately expressed as follows:

$$(c/\tau^0)_0 = 1/H\overline{M}_w \tag{25'}$$

Equivalent expressions may be written for the Rayleigh ratio R_θ; in particular, we may replace c/τ^0 and $(c/\tau^0)_0$ in Eq. (37') with c/R^0 and $(c/R_\theta^0)_0$, respectively.

Equation (37), or (37'), may be expected to represent the turbidity as a function of concentration over the range in which $(c/\tau^0)/(c/\tau^0)_0 < 4$ with adequate accuracy. Molecular weight heterogeneity may affect the coefficients Γ_2 entering the osmotic and light-scattering expression somewhat differently[8,30] with the result that the values obtained for Γ_2 by the two methods may not necessarily be identical if unfractionated samples are used.

2f. Treatment of Data.—The sensitivity of the light-scattering method depends on the magnitude of the constant K^* (or H), which in turn depends on the square of the difference between the refractive indexes of the solute and of the solvent, as represented by the square of $(d\tilde{n}/dc)_0$ in Eq. (24). Ordinarily a solvent should be chosen which will give a value of at least 0.02 in (g./ml.)$^{-1}$, and preferably in excess of 0.05, for this coefficient. It may be established by differential refractivity measurements on solvent and solution of known concentration.[21] Since K^* depends on the inverse fourth power of the wavelength λ, monochromatic radiation of short wavelength may be preferred. However, the advantage gained from a larger K^* may be offset to some extent by the increased dissymmetry at the shorter wavelength.

Before scattering intensity measurements can be converted to molecular weights, the two corrections previously discussed—the dissymmetry correction for intraparticle interference and the extrapolation to zero concentration—must be introduced, or established to be negligible. The relationships given in the preceding sections unfortunately account rigorously for either only in the absence of the other. The theory of the concentration dependence of the scattered intensity applies to the turbidity corrected for dissymmetry, and the treatment of dissymmetry is strictly valid only at zero concentration (where interference of radiation scattered by different polymer molecules vanishes).

The dissymmetry, if appreciable, invariably depends on the concentration, and hence should be measured as a function of the concentration. If, however, the dissymmetry coefficient, $z_{45°}$ for example, is not much greater than unity (see Fig. 46), its change with concentration is scarcely significant in respect to molecular weight determination. The correction factor $P(\theta)$ may then be determined in the manner described above from the dissymmetry observed at finite concentrations without extrapolating the dissymmetry coefficient to infinite dilution. If the dissymmetry is large, as will ordinarily be true when the molecular weight is large, the observed R_θ's should be extrapolated graphically to zero angle to obtain R_0. Both the angular and concentration expolation may be performed graphically on the same plot by the method of Zimm.[22,31] Accurate measurements down to small angles are required, however, and for this reason it may be preferable to carry out the concentration extrapolation first. The observed ratios c/R_θ may be extrapolated using a relationship of the form of Eq. (37'), for example.* The dissymmetry z_β at infinite dilution (sometimes referred to as the intrinsic dissymmetry $[z]$) may then be used to calculate both the chain dimension $\sqrt{\overline{r^2}}$ and the correction factor $P(\theta)$ applicable at infinite dilution.

The dissymmetry to be expected for the polymer in a given solvent may be estimated with the aid of the following semiempirical formula, provided the *intrinsic viscosity* $[\eta]$ (see Sec. 4b) in the same solvent and the approximate molecular weight of the polymer are known. The

---

* The empirical values of the parameter Γ_2 found using uncorrected Rayleigh ratios or turbidities should be angle-dependent and do not necessarily represent the customary thermodynamic parameter Γ_2.

root-mean-square end-to-end length of the chain in Ångstrom units is
expressed approximately by

$$\sqrt{\overline{r^2}} \cong 8(M[\eta])^{1/3} \tag{38}$$

The approximate dissymmetry correction may then be computed from
$\sqrt{\overline{r^2}}/\lambda'$ using Eqs. (31) and (32). Corrections so calculated are not
regarded as acceptable substitutes for actual dissymmetry measure-
ments, but they may prove highly useful as preliminary estimates of
the correction.

Typical results of light-
scattering measurements
are shown in Figs. 47 and
48 for polystyrene fractions
of medium[14] and of very
high molecular weight,[22]
respectively. The small
dissymmetry correction for
the former was calculated
from the measured dissym-
metry coefficient $z_{45°}(\cong 1.2)$
using Eqs. (31) and (32).
For the high molecular

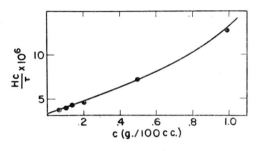

FIG. 47.—Hc/τ vs. c for polystyrene of
medium molecular weight (2.9×10^5) in benzene. (Fox, Flory, and Bueche.[14])

weight polymer (Fig. 48) c/τ^0 values were obtained at each concentration by extrapolating observed intensities to $\theta = 0$. The curves drawn
through the points in these figures have been calculated from Eq. (37′)

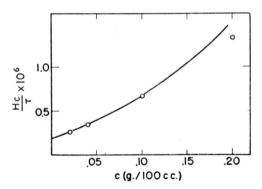

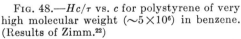

FIG. 48.—Hc/τ vs. c for polystyrene of very
high molecular weight ($\sim 5 \times 10^6$) in benzene.
(Results of Zimm.[22])

using values of Γ_2 affording
optimum agreement with
the data. The extrapola-
tion may be conducted
graphically after the manner described above for extrapolating π/c. That is,
an experimental plot of
$\log(c/\tau)$ against $\log c$ may
be fitted to the graph of
$\log[(c/\tau)/(c/\tau)_0]$ vs. $\log(\Gamma_2 c)$
calculated from Eq. (37′)
using an appropriate value
of g. Alternatively, if
$g \sim 1/3$,

$$(c/\tau^0)^{1/2} \cong (c/\tau^0)_0^{1/2}(1 + \Gamma_2 c) \tag{37″}$$

which suggests a plot of $(c/\tau^0)^{1/2}$ against c.

The similarity between the plots of c/τ vs. c shown in Figs. 47 and 48 and those for π/c vs. c shown in Figs. 38 and 39 is apparent. Deviations from ideality (i.e., the changes in π/c and in c/τ with c) have the same origin for both types of measurements. As with the osmotic pressure-concentration ratio, the change of c/τ with c may be reduced by choosing a poor solvent. A further advantage of a poor solvent enters because of the smaller size assumed by the polymer molecule in a poor solvent environment, which reduces the dissymmetry correction.

TABLE XXVIII.—COMPARISON OF LIGHT SCATTERING WITH OSMOTIC MOLECULAR
WEIGHTS FOR POLYSTYRENE FRACTIONS
(Brice, Halwer, and Speiser[23])

Fraction	$\overline{M}_w \times 10^{-3}$ light scat. at $\lambda = 4360$ Å	$\overline{M}_w \times 10^{-3}$ light scat. at $\lambda = 5460$ Å	$\overline{M}_n \times 10^{-3}$ osmotic
1	116	115	101
2	180	178	151
3	270	268	238

Molecular weights of polystyrene fractions determined by light scattering at two wavelengths are compared with the osmotic values in Table XXVIII. The corrections for dissymmetry amount to only a few percent in this molecular weight range, and the uncertainty in the extrapolation to zero concentration is small. The somewhat higher values obtained by the light-scattering method compared with the osmotic results may reasonably be attributed to heterogeneity of the fractions which causes $\overline{M}_w$ to exceed $\overline{M}_n$.

Mixed solvents are generally unsatisfactory for use in the determination of polymer molecular weights owing to the likelihood of selective absorption of one of the solvent components by the polymer coil.[32] The excess of polarizability of the polymer particle (polymer plus occluded solvent) is not then equal to the difference between the polarizabilities of the polymer and the solvent mixture. For this reason the refractive increment $d\tilde{n}/dc$ which would be required for calculation of K^*, or of H, cannot be assumed to equal the observed change in refractive index of the medium as a whole when polymer is added to it, unless the refractive indexes of the solvent components happen to be the same. The "size" $\sqrt{\overline{r^2}}$ may, however, be measured in a mixed solvent, since only the dissymmetry ratio is required for this purpose.

In conclusion, the light-scattering method offers the advantage that the property being measured increases with increase in molecular weight, whereas the reverse is true for the osmotic pressure. This apparent advantage cannot be fully exploited owing to the complications arising from dissymmetry corrections, which increase with molecular weight. Neither method offers distinct advantages over the other, and a choice between them should depend on the use to be made of the molecular weight measurements and on circumstances peculiar to the polymer being investigated. The two methods actually are supplementary to a considerable extent, owing to the fact that different averages are obtained. Comparison of $\overline{M}_w$ with $\overline{M}_n$ affords an index of the heterogeneity of the polymer. Light-scattering dissymmetry measurements are unique in providing an absolute measure of the dimensions of a random coil polymer molecule in solution. The importance of dissymmetry measurements for this purpose, beyond their use in correcting turbidities for the deduction of molecular weight, should not be overlooked.

3. DETERMINATION OF MOLECULAR WEIGHTS WITH THE ULTRACENTRIFUGE[33,34]

Ultracentrifugation techniques, developed largely by Svedberg and his collaborators in Uppsala over the past thirty years,[33] are the most intricate of the existing methods for the determination of the molecular weights of macromolecular substances. The ultracentrifuge consists of a steel rotor several inches in diameter which is spun at high speed in a vacuum (in order to minimize thermal effects). The rotor may be driven electrically at constant speed, or, in older models, by an oil or an air turbine. The small cylindrical cell containing the solution being centrifuged is held within the rotor near its periphery. Concentration gradients in the cell are determined optically by using a beam of light traveling in a direction parallel to the axis of rotation and perpendicular to the cell, the beam being intercepted by the cell at each rotation. Refraction of rays of the beam passing through different portions of the cell may be determined either by the Lamm scale method[33] or by use of schlieren optics.[33] In either case, the measured refractions are converted to concentration gradients along the direction of sedimentation within the cell. The progress of the solute under the influence of the centrifugal field is followed in this manner in determinations of the sedimentation velocity, and the ultimate distribution of solute in a lower centrifugal field is similarly deduced in the equilibrium method. The fields employed for sedimentation velocity measurements may be made as great as several hundred thousand times the acceleration of gravity.

3a. Molecular Weights from the Sedimentation and Diffusion Constants.—At a point in the cell a distance x from the axis of rotation the force acting on a particle of mass m and (partial) specific volume $\bar{v}$ immersed in a medium of density ρ is given by $m(1-\bar{v}\rho)x\omega^2$, where ω is the angular velocity of rotation. Equating this to the frictional force, obtained by multiplying the velocity dx/dt by the *frictional coefficient* f, leads to the primary sedimentation velocity relationship

$$(1/\omega^2 x)(dx/dt) = m(1 - \bar{v}\rho)/f \tag{39}$$

The quantity on the left, which is determined by measuring the rate of movement of the sedimentation boundary in the ultracentrifuge operated at the constant angular velocity ω, is called the *sedimentation constant* s. Thus

$$s = m(1 - \bar{v}\rho)/f \tag{40}$$

The frictional coefficient varies with concentration, but at infinite dilution it reduces to the coefficient (f_0) for an isolated polymer molecule moving through the surrounding fluid unperturbed by movements of other polymer molecules (see Chap. XIV). At finite concentrations, however, the motion of the solvent in the vicinity of a given polymer molecule is affected by others nearby; binary encounters (as well as ones of higher order) between polymer molecules contribute also to the observed frictional effects. The influence of these interactions will persist to very low concentrations owing to the relatively large effective volume of a polymer molecule, to which attention has been directed repeatedly in this chapter. Since the sedimentation "constant" depends inversely on the frictional coefficient, s must also depend on concentration.

The mass of the sedimenting particle could be deduced from its rate of sedimentation at high dilution in a given field, i.e., from its sedimentation constant, if the frictional coefficient f could be determined independently. Rates of diffusion may be utilized to secure this necessary supplementary information, since the diffusion constant D depends also on the frictional coefficient. Thus[35]

$$D = kT(1 + d \ln \gamma/d \ln c)/f \tag{41}$$

where γ is the activity coefficient of the solute. This reduces at infinite dilution to

$$D = kT/f_0 \tag{42}$$

which in combination with Eq. (40) at infinite dilution (multiplied by Avogadro's number) gives

$$D_0/s_0 = RT/M(1 - \bar{v}\rho) \tag{43}$$

Hence the molecular weight may be calculated from the ratio of the limiting values of D and s at infinite dilution. The problem is to secure accurate values for D_0 and s_0.

The change in the sedimentation constant with concentration enters solely from the change in $1/f$, and it is customary therefore to extrapolate a plot of $1/s$ against c to infinite dilution. The results of sedimentation studies by Newman and Eirich[36] on several polystyrene

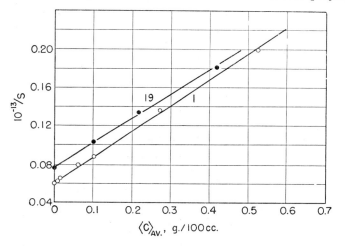

FIG. 49.—Reciprocal of the sedimentation constant s plotted against concentration for two polystyrene fractions in chloroform. Molecular weights 13×10^5 and 5.5×10^5 for fractions 1 and 19, respectively. (Newman and Eirich.[36])

fractions in chloroform are shown in Fig. 49. The change of D with concentration is not so simple, for it depends not only on $1/f$ but also on the thermodynamic nonideality as embodied in the term $d\ln\gamma/d\ln c$ of Eq. (41). Application of the thermodynamic theory from which Eq. (13) for the osmotic pressure was derived yields

$$D = kT[1 + 2\Gamma_2 c + 3g\Gamma_2^2 c^2]/f \tag{44}$$

Thus both the numerator and denominator terms in Eq. (41), or in Eq. (44), depend on the concentration.[37] Because of this situation empirical extrapolation of D is particularly hazardous (for random coiling polymers). If Γ_2 is known from osmotic or light-scattering measurements at a series of concentrations, extrapolation according to Eq. (44) will be facilitated. (If such measurements have been carried out, however, the molecular weight also will have been determined.)

A preferable procedure would involve elimination of the frictional coefficient from Eqs. (40) and (44) to obtain

$$D/s = (RT/M)\left[1 + 2\Gamma_2 c + 3g\Gamma_2^2 c^2\right]/(1-\bar{v}\rho) \qquad (45)$$

A single extrapolation of the ratio D/s measured at a series of very low concentrations would then suffice. The graphical method advocated above for handling osmotic and turbidity data could be applied here also for the purpose of obtaining $(D/s)_0 = RT/M$. Again, if a poor solvent is used, so that Γ_2 is very small, D/s should be nearly independent of c over the dilute range.

The results of sedimentation and diffusion measurements carried out by Meyerhoff and Schulz[38] on semifractionated poly-(methyl methacrylate) samples in acetone at 20°C are summarized in Table XXIX. The measurements on each sample were extended to sufficiently low concentrations for reliable extrapolation of both s and D to their limiting values. The molecular weights calculated from Eq. (43) are given in the fourth column. These are weight averages according to the method used by Meyerhoff and Schulz to assign the positions of the mean boundaries in the sedimentation and diffusion runs. (Various averages may be obtained[34] depending on the basis used for selecting a mean position in the ill-defined sedimentation and diffusion boundaries which are broadened by molecular heterogeneity.) The osmotic molecular weights given in the last column are considerably lower, owing no doubt to the heterogeneity of the fractions,[38] which introduces a rather large disparity between the weight and number averages. Light-scattering molecular weights[39,40] (weight averages)

TABLE XXIX.—MOLECULAR WEIGHTS OF POLY-(METHYL METHACRYLATE)
FRACTIONS FROM SEDIMENTATION AND DIFFUSION MEASUREMENTS[38]

Polymer	$s_0 \times 10^{13}$ sec./g.	$D_0 \times 10^{-7}$ cm.²/sec.	$\overline{M}_w \times 10^{-3}$ from D_0/s_0[a]	$\overline{M}_n \times 10^{-3}$ osmotic
A	107.0	0.95	7,440	
B	82.0	1.18	4,590	
C	69.0	1.42	3,210	
D	59.5	1.95	2,020	
E	48.5	2.25	1,420	1,000[b]
F	36.5	3.95	611	410[b]
G	25.2	5.45	306	210
H	18.8	8.30	148	116
J	14.1	12.05	77.2	58

[a] $\bar{v}\rho = 0.631$ in acetone at 20°C.

[b] Calculated from intrinsic viscosities using an empirical relationship based on osmotic measurements.

are in good agreement with the results of sedimentation and diffusion, however.

3b. Sedimentation Equilibrium.—By the use of a lower speed of rotation and operation of the centrifuge for a long period of time under constant conditions with precautions to avoid disturbing influences such as convection within the cell, a state of equilibrium may be reached in which the polymer is distributed over the length of the cell in a manner depending on its molecular weight and molecular weight distribution. If the solution is so dilute that ideal thermodynamic behavior is closely approached and if the polymer is homogeneous, the molecular weight is related to the ratio of the concentrations c_2/c_1 at two points x_2 and x_1 in the cell, according to the equation

$$M = \frac{2RT \ln(c_2/c_1)}{(1 - \bar{v}\rho)\omega^2(x_2 - x_1^2)} \tag{46}$$

The centrifugal field in the sedimentation equilibrium experiment is the analog of the membrane in an osmometer.

If the polymer is heterogeneous, then from detailed measurements of the concentration as a function of x throughout the cell it is possible to construct higher molecular weight averages[33,41] starting with the weight average $\overline{M}_w$ and including the so-called z and $z+1$ averages defined by

$$\overline{M}_z = \sum N_i M_i^3 / \sum N_i M_i^2 = \sum w_i M_i^2 / \sum w_i M_i \tag{47}$$

and

$$\overline{M}_{z+1} = \sum N_i M_i^4 / \sum N_i M_i^3 = \sum w_i M_i^3 / \sum w_i M_i^2 \tag{48}$$

It is essential that the solution be sufficiently dilute to behave ideally. a condition which is difficult to meet in practice. Ordinarily the dilutions required are beyond those at which the concentration gradient measurement by the refractive index method may be applied with accuracy. Corrections for nonideality are particularly difficult to introduce in a satisfactory manner owing to the fact that nonideality terms depend on the molecular weight distribution, and the molecular weight distribution (as well as the concentration) varies over the length of the cell. Largely as a consequence of this circumstance, the sedimentation equilibrium method has been far less successful in application to random-coil polymers than to the comparatively compact proteins, for which deviations from ideality are much less severe.

The difficulties arising from nonideality could again be minimized through the selection of a poor solvent. Indeed, in the case of sedimentation equilibrium such a choice would seem to be imperative, for by selecting a solvent and temperature (Θ, see Chap. XII) for which the nonideality parameter Γ_2 is zero, the relatively simple relationships for ideal solutions may be applied, enabling $\overline{M}_w$, $\overline{M}_z$, and $\overline{M}_{z+1}$ to be calculated unequivocally. The possibility of determining from a single set of measurements a series of averages with which to characterize the distribution is especially attractive. Lest all this should appear too easily achieved in practice, attention should be directed to the further requirements in the selection of a solvent: $1 - \bar{v}\rho$ of Eq. (46) should not be too small, although a large value is not required, and the refractive indexes of polymer and solvent must differ sufficiently to permit application of the refractive index gradient method.

4. INTRINSIC VISCOSITIES IN RELATION TO MOLECULAR WEIGHTS OF HIGH POLYMERS

Staudinger[42] many years ago called attention to the utility of viscosity measurements on dilute polymer solutions as a means of characterization. High polymer molecules possess the unique capacity to greatly increase the viscosity of the liquid in which they are dissolved, even when present at concentrations which are quite low. This is, of course, another manifestation of the voluminous character of randomly coiled long chain molecules, the influence of the polymer on the viscosity of the medium being closely related to the frictional effects encountered in sedimentation and diffusion (see Chap. XIV). The higher the molecular weight within a given series of linear polymer homologs, the greater the increase in viscosity produced by a given weight concentration of polymer. In other words, the *intrinsic viscosity*, which represents the capacity of a polymer to enhance the viscosity, increases with M; hence, viscosity measurements afford a measure of molecular weight. An absolute value for the molecular weight of a polymer cannot be derived from solution viscosity measurements, however. The dependence of the intrinsic viscosity on molecular weight must be established more or less empirically in each individual case by comparison with molecular weights determined by one of the absolute methods discussed above. Nevertheless, dilute solution viscosity measurements are extremely valuable for characterization of high polymers, as attested by their widespread use. The measurement of viscosity is so much simpler than execution of any of the absolute methods that the establishment of the intrinsic viscocity-molecular weight relationship is almost always a primary objective in a molecular weight investigation on a given polymer series.

Viscosities at higher concentrations (5 to 10 percent) are frequently used to advantage as empirical measures of the molecular weight.

The much higher viscosities prevailing at the higher concentrations are more sensitive to changes in molecular weight. They are, however, less amenable at present to theoretical interpretation. If the average molecular weights are not too great ($<100,000$), bulk viscosity measurements on the undiluted molten polymer may be used for the same purpose. The relationship[43,44]

$$\log \eta = A + B\overline{M}_w^{1/2}$$

where η is the melt viscosity and A and B are constants at a given temperature, holds with remarkable accuracy for a number of linear condensation polymers, although in other cases it is inapplicable.[45]

4a. Evaluation of the Intrinsic Viscosity.—Viscosities of dilute polymer solutions are conveniently measured in capillary viscometers of the Ostwald, Fenske, or Ubbelohde types. In order to assure the required accuracy, measurements should be conducted in a constant temperature bath regulated within $\pm 0.02°C$. The measured time of flow t should exceed 100 seconds. The viscosity η is calculated from an equation of the form

$$\eta = \alpha\rho(t - \beta/\alpha t) \qquad (49)$$

where ρ is the density of the solvent (or solution) and α and β are calibration constants, the latter of which takes account of the small correction for kinetic energy. The viscosity of the solution is divided by the viscosity of the solvent to obtain the relative viscosity η_r. The specific viscosity, $\eta_{sp} = \eta_r - 1$, expresses the incremental viscosity attributable to the polymeric solute. The ratio η_{sp}/c is a measure of the specific capacity of the polymer to increase the relative viscosity. The limiting value of this ratio at infinite dilution is called the intrinsic viscosity,[46] which is designated by $[\eta]$; i.e.

$$[\eta] = (\eta_{sp}/c)_{c\to 0} \equiv [(\eta_r - 1)/c]_{c\to 0} \qquad (50)$$

The concentration c is customarily expressed in grams per 100 ml. of solution, the intrinsic viscosity being given in the reciprocal of this unit, i.e., in deciliters per gram. The kinetic energy correction in Eq. (49) should not be ignored, for to do so would introduce an error into the intrinsic viscosity approximately proportional to $\beta/\alpha t^2$.[47]

Again it is necessary to extrapolate to infinite dilution, but the procedure for doing so is relatively simple. Plots of η_{sp}/c against c usually are very nearly linear for $\eta_r < 2$, and it has been pointed out[48] that the slopes of these plots for a given polymer-solvent system vary approximately as the square of the intrinsic viscosity. Thus the equation proposed by Huggins[48] is

$$\eta_{sp}/c = [\eta] + k'[\eta]^2 c \tag{51}$$

where k' is approximately constant for a series of polymer homologs in a given solvent. Usually (but not always) k' is in the range from 0.35 to 0.40.*

The intrinsic viscosity may also be defined[46] as follows

$$[\eta] = [(\ln \eta_r)/c]_{c \to 0} \tag{50'}$$

Series expansion of the natural logarithm demonstrates the equivalence of this definition to Eq. (50). It follows also from Eq. (51) that

$$(\ln \eta_r)/c = [\eta] + k''[\eta]^2 c \tag{51'}$$

where $k'' = k' - 1/2$. Generally, k'' is negative and smaller in magnitude than k'; hence $(\ln \eta_r)/c$ changes less rapidly with concentration than does η_{sp}/c. For this reason extrapolation of $(\ln \eta_r)/c$ is somewhat preferred over extrapolation of η_{sp}/c.

In practice it is customary to measure the relative viscosity at two or more concentrations so chosen as to give relative viscosities in the range from about 1.10 to 1.50. Either η_{sp}/c or $(\ln \eta_r)/c$ (or both) is extrapolated graphically to $c = 0$.

If the intrinsic viscosity is large (i.e., greater than about 4 deciliters per gram), the viscosity is likely to be appreciably dependent on the rate of shear in the range of operation of the usual capillary viscometer. Measurements in a viscometer permitting operation at a series of rates of shear extending to very low rates are then required in order to extrapolate η_{sp}/c to its limiting value at a shear rate of zero.[49] Extrapolation to infinite dilution does not eliminate the effect on this ratio of a dependence on shear rate.

4b. Intrinsic Viscosity–Molecular Weight Relationship.—When the logarithms of the intrinsic viscosities of a series of fractionated linear polymer homologs are plotted against the logarithms of their molecular weights, relationships which are linear within experimental error are obtained. Typical results are shown in Fig. 50 for polyisobutylene in diisobutylene and in cyclohexane. Molecular weights of the fractions were determined osmotically. The indicated linear relationships may be expressed by simple equations of the form

$$[\eta] = K' M^a \tag{52}$$

where K' and a are constants determined, respectively, by the intercept and the slope of a plot of the type shown in Fig. 50.

---

* Polymers bearing electric charges, i.e., polyelectrolytes, exhibit a very different behavior. See Chapter XIV.

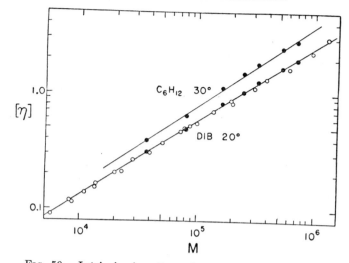

FIG. 50.—Intrinsic viscosity-molecular weight relationship for polyisobutylene in diisobutylene (DIB) at 20° and in cyclohexane at 30°C. Open circles from Ref. 7; filled circles, Ref. 8.

Values obtained for K' and a for a number of polymer-solvent pairs are given in Table XXX. It will be observed that the exponent a varies with both the polymer and the solvent. It does not fall below 0.50 in any case and seldom exceeds about 0.80.* Once K' and a have been established for a given polymer series in a given solvent at a specified temperature, molecular weights may be computed from intrinsic viscosities of subsequent samples without recourse to a more laborious absolute method.

It should be emphasized that Eq. (52) is empirical in origin. However, the more complicated theoretical expressions to be discussed in Chapter XIV can be approximated quite closely by this simple equation over ranges of as much as a hundredfold in M. The convenience of application of the empirical relationship assures its continued use for correlating intrinsic viscosities and molecular weights.

4c. The Viscosity Average Molecular Weight and the Influence of Molecular Heterogeneity.—For the purpose of investigating the type of average molecular weight obtained from viscosity measurements

---

* Exceptions occur in the case of polyelectrolytes where a, in the absence of added salts, may approach two. The difference arises here because at infinite dilution the charges cause the polyelectrolyte chain to become nearly fully extended.

TABLE XXX.—DEPENDENCE OF INTRINSIC VISCOSITY ON MOLECULAR WEIGHT
FOR VARIOUS POLYMER-SOLVENT SYSTEMS

Polymer	Solvent	T°C	Mol. wt. range $\times 10^{-3}$	$K' \times 10^4$	a	Method[a] and reference
Polystyrene	Benzene	25	32–1,300	1.03	0.74	Os, 50, 51
Polystyrene	Methyl ethyl-ketone	25	2.5–1,700	3.9	.58	LS, 31
Polyisobutylene	Diisobutyl-ene	20	6–1,300	3.6	.64	Os, 7, 52
Polyisobutylene	Cyclohexane	30	6–3,150	2.6	.70	Os, 8, 52
Polyisobutylene	Benzene	24	1–3,150	8.3	.50	Os, 8
Polyisobutylene	Benzene	60	6–1,300	2.6	.66	Os, 52
Natural rubber	Toluene	25	40–1,500	5.0	.67	Os, 53
Poly-(methyl methacrylate)	Benzene	20	77–7,440	0.84	.73	SD, 38
Poly-(methyl methacrylate)	Benzene	25		.57	.76	LS, 40
Poly (vinyl acetate)	Acetone	25	65–1,500	1.76	.68	Os, 54
Poly-(ε-capro-amide)	Sulfuric acid	25	4– 37 (wt. av.)	2.9	.78	E, 43

[a] Methods used for molecular weight measurements are as follows: osmotic, Os;
light-scattering, LS; end group, E; SD, sedimentation-diffusion.

using the empirical equation (52), consider a heterogeneous polymer
in a solution so dilute that individual molecules can be assumed to
contribute to the viscosity independently of one another. Then

$$\eta_{sp} = \sum_i (\eta_{sp})_i \tag{53}$$

where $(\eta_{sp})_i$ is the contribution due to species of size i. According
to Eq. (52), $(\eta_{sp})_i = K' M_i^a c_i$, where c_i and M_i are the concentration
and molecular weight of the same species. Introducing this expres-
sion into Eq. (53)

$$\eta_{sp} = K' \sum_i M_i^a c_i$$

and since the solution is extremely dilute

$$[\eta] = \eta_{sp}/c = K' \sum_i M_i^a c_i/c$$

where $c = \sum_i c_i$ is the total concentration of all polymer species. If we define the viscosity average molecular weight $\overline{M}_v$ as follows:

$$\overline{M}_v = \left[\sum w_i M_i^a \right]^{1/a} = \left[\sum N_i M_i^{1+a} / \sum N_i M_i \right]^{1/a} \tag{54}$$

where $w_i = c_i/c$ is the weight fraction of species i in the whole polymer and N_i is the number of molecules of this species, then

$$[\eta] = K' \overline{M}_v^a \tag{55}$$

For molecularly heterogeneous polymers this equation must be used in place of (52).

The viscosity average molecular weight depends on the nature of the intrinsic viscosity–molecular weight relationship in each particular case, as represented by the exponent a of the empirical relationship (52), or (55). However, it is not very sensitive to the value of a over the range of concern. For polymers having the "most probable" distribution to be discussed in the next chapter, it may be shown,[43] for example, that

$$\overline{M}_n : \overline{M}_v : \overline{M}_w :: 1 : [(1 + a)\Gamma(1 + a)]^{1/a} : 2 \tag{56}$$

where $\Gamma(1+a)$ is the gamma function of $1+a$. As a varies from 0.50 to 1.00, $\overline{M}_v/\overline{M}_n$ for this particular molecular weight distribution increases from 1.67 to 2.00. When $a = 1$, the viscosity average is, of course, identical with the weight average. The results just cited for the most probable distribution can be extended to the broader conclusion that the viscosity average molecular weight will always be considerably closer to the weight average than to the number average for any distribution likely to be encountered in a high polymer.

The establishment of an empirical relationship between intrinsic viscosity and molecular weight involves comparison of a property, the intrinsic viscosity, which depends on the peculiar viscosity average molecular weight with another average obtained by an absolute method. If the osmotic method is used for absolute molecular weight determination, the latter will be the number average. The relationship established obviously will be in error to the extent that the two averages differ for the samples used. If the ratio of $\overline{M}_v$ to $\overline{M}_n$ is about the same for all of the samples, the error will reside in the value found for K'; if this ratio varies irregularly for the different samples, no consistent relationship will be found between the property (i.e., the intrinsic viscosity) and the measured molecular weight.

The disparity between different average molecular weights may be made small by careful fractionation. Complete molecular homogeneity is by no means required, however; it is necessary only that the different modes of the distribution be brought to close proximity, which is a far less stringent requirement.   The intrinsic viscosity-molecular weight relationship may be established without serious error from this source by carrying out the comparison using well-fractionated samples.   A point in favor of the light-scattering method in preference to the osmotic method for this purpose is the fact that the former gives the weight average, which is closer to the viscosity average; hence the efficiency of fractionation is less important.   The values of K' and a for a given polymer-solvent combination having once been established, the *viscosity average* molecular weight may be computed from the intrinsic viscosity of a polymer sample of any degree of heterogeneity through use of Eq. (55).   It is necessary merely that the molecular species included fall within the range over which the empirical equation applies with sufficient accuracy.

REFERENCES

1. R. A. Gregg and F. R. Mayo, *J. Am. Chem. Soc.*, **70**, 2373 (1948).
2. C. H. Bamford and M. J. S. Dewar, *Proc. Roy. Soc.* (London), **A192**, 329 (1948).
3. N. H. Ray, *Trans. Faraday Soc.*, **48**, 809 (1952).
4. G. V. Schulz, *Z. physik. Chem.*, **A176**, 317 (1936); R. H. Wagner, *Ind. Eng. Chem.*, Anal. Ed., **16**, 520 (1944); G. D. Sands and B. L. Johnson, *Anal. Chem.*, **19**, 261 (1947); S. G. Weissberg, and G. A. Hanks, private communication.
5. K. H. Meyer, E. Wolff, and C. G. Boissonnas, *Helv. Chim. Acta*, **23**, 430 (1940).
6. R. M. Fuoss and D. J. Mead, *J. Phys. Chem.*, **47**, 59 (1943); K. B. Goldblum, *ibid.*, **51**, 474 (1947); A. F. Sirianni, L. M. Wise, and R. L. McIntosh, *Can. J. Research*, **25B**, 301 (1947).
7. P. J. Flory, *J. Am. Chem. Soc.*, **65**, 372 (1943).
8. W. R. Krigbaum and P. J. Flory, *J. Am. Chem. Soc.*, **75**, 1775 (1953).
9. S. R. Carter and B. R. Record, *J. Chem. Soc.*, **1939**, 660, 664.
10. R. E. Montonna and L. T. Jilk, *J. Phys. Chem.*, **45**, 1374 (1941).
11. C. R. Masson and H. W. Melville, *J. Polymer Sci.*, **4**, 323 (1949).
12. H. T. Hookway and R. Townsend, *J. Chem. Soc.*, **1952**, 3190.
13. C. R. Masson and H. W. Melville, *J. Polymer Sci.*, **4**, 337 (1949).
14. T. G. Fox, P. J. Flory, and A. M. Bueche, *J. Am. Chem. Soc.*, **73**, 285 (1951); W. R. Krigbaum and P. J. Flory, *J. Polymer Sci.*, **9**, 503 (1952); see also Ref. 8.
15. P. M. Doty, H. L. Wagner, and S. Singer, *J. Phys. Colloid Chem.*, **51**, 32 (1947).
16. G. Gee, *Trans. Faraday Soc.*, **36**, 1171 (1940); S. R. Palit, G. Colombo, and H. Mark, *J. Polymer Sci.*, **6**, 295 (1951).
17. P. Debye, *J. Applied Phys.*, **15**, 338 (1944); *J. Phys. & Colloid Chem.*, **51**, 18 (1947).

18. H. Mark, *Chemical Architecture*, ed. by R. E. Burk and O. Grummitt (Interscience Publishers; New York, 1948), pp. 121–173).

19. G. Oster, *Chem. Revs.*, **43**, 319 (1948).

20. P. Doty and J. T. Edsall *Advances in Protein Chemistry* Vol. VI (Academic Press, New York, 1951), pp. 35–121.

21. P. Debye and A. M. Bueche, Report to Office of Rubber Reserve, 1949. See also P. P. Debye, *J. Applied Phys.*, **17**, 392 (1946).

22. B. H. Zimm, *J. Chem. Phys.*, **16**, 1099 (1948).

23. B. A. Brice, M. Halwer, and R. Speiser, *J. Opt. Soc. Am.*, **40**, 768 (1950).

24. F. W. Billmeyer, Jr., Reports to the Office of Rubber Reserve, Jan. 15, and March 1, 1945.

25. C. I. Carr and B. H. Zimm, *J. Chem. Phys.*, **18**, 1616 (1950).

26. P. Debye, *J. Chem. Phys.*, **14**, 636 (1946).

27. M. Smoluchowski, *Ann. Physik*, **25**, 205 (1908); *Phil. Mag.*, **23**, 165 (1912).

28. A. Einstein, *Ann. Physik*, **33**, 1275 (1910).

29. H. C. Brinkman and J. J. Hermans, *J. Chem. Phys.*, **17**, 574 (1949); J. G. Kirkwood and R. J. Goldberg, *ibid.*, **18**, 54 (1950); W. H. Stockmayer, *ibid.*, **18**, 58 (1950).

30. P. J. Flory and W. R. Krigbaum, *J. Chem. Phys.*, **18**, 1086 (1950).

31. P. Outer, C. I. Carr, and B. H. Zimm, *J. Chem. Phys.*, **18**, 830 (1950).

32. R. H. Ewart, C. P. Roe, P. Debye, and J. R. McCartney, *J. Chem. Phys.*, **14**, 687 (1946).

33. T. Svedberg and K. O. Pedersen, *The Ultracentrifuge* (Clarendon Press, Oxford, 1940).

34. P.-O. Kinell and B. G. Rånby, *Advances in Colloid Science*; Vol. III, ed. by H. Mark and E. J. W. Verwey (Interscience Publishers, New York, 1950), pp. 161–215.

35. L. Onsager and R. M. Fuoss, *J. Phys. Chem.*, **36**, 2689 (1932).

36. S. Newman and F. Eirich, *J. Colloid Sci.*, **5**, 541 (1950).

37. L. Mandelkern and P. J. Flory, *J. Chem. Phys.*, **19**, 984 (1951).

38. G. Meyerhoff and G. V. Schulz, *Makromol. Chem.*, **7**, 294 (1952).

39. J. Bischoff and V. Desreux, *Bull. soc. chim. Belges*, **61**, 10 (1952).

40. T. G. Fox, D. R. Conlon, H. F. Mason, E. M. Schuele, and J. F. Woodman, forthcoming publication.

41. M. Wales, F. T. Adler, and K. E. VanHolde, *J. Phys. Chem.*, **55**, 145 (1951); M. Wales, *ibid.*, **55**, 282 (1951).

42. H. Staudinger and W. Heuer, *Ber.*, **63**, 222 (1930); H. Staudinger and R. Nodzu, *ibid.*, **63**, 721 (1930).

43. J. R. Schaefgen and P. J. Flory, *J. Am. Chem. Soc.*, **70**, 2709 (1948).

44. P. J. Flory, *J. Am. Chem. Soc.*, **62**, 1057 (1940).

45. T. G. Fox and P. J. Flory, *J. Am. Chem. Soc.*, **70**, 2384 (1948).

46. E. O. Kraemer, *Ind. Eng. Chem.*, **30**, 1200 (1938).

47. G. V. Schulz, *Z. Elektrochem.*, **43**, 479 (1937).

48. D. J. Mead and R. M. Fuoss, *J. Am. Chem. Soc.*, **64**, 277 (1942); M. L. Huggins, *ibid.*, **64**, 2716 (1942).

49. H. T. Hall and R. M. Fuoss, *J. Am. Chem. Soc.*, **73**, 265 (1951); T. G. Fox, J. C. Fox, and P. J. Flory, *ibid.*, **73**, 1901 (1951).

50. R. H. Ewart and H. C. Tingey, unpublished; C. E. H. Bawn, R. F. J. Freeman, and A. R. Kamaliddin, *Trans. Faraday Soc.*, **46**, 1107 (1950).

51. W. R. Krigbaum and P. J. Flory, *J. Polymer Sci.*, **11**, 37 (1953).

52. T. G. Fox and P. J. Flory, *J. Phys. and Colloid Chem.*, **53**, 197 (1949).

53. W. C. Carter, R. L. Scott, and M. Magat, *J. Am. Chem. Soc.*, **68**, 1480 (1946).

54. R. H. Wagner, *J. Polymer Sci.*, **2**, 21 (1947).

CHAPTER VIII

Molecular Weight Distributions in

Linear Polymers

ALL synthetic polymers, and most natural ones too, consist of a multitude of molecular species covering a broad range of sizes. As was pointed out in Chapters I and II, high polymers are molecular mixtures rather than chemical individuals in the customary sense. No one average molecular weight, whether it is the number average, the weight average, or some other, is fully satisfactory for the characterization of such a mixture. Determination of two different average molecular weights, the number and weight averages for example, furnishes some measure of the breadth of the molecular weight distribution, but it does not provide an altogether satisfactory substitute for complete resolution of the molecular composition of the polymer.

Ideally, perhaps, interpretation of a given property of a polymer should proceed from consideration of the contribution of each molecular species to that property, the final result being obtained by summing the contributions from all species present. Knowledge of the mole or weight fraction of each species, i.e., of the complete molecular size distribution, would assure that realization of such an objective is possible, in principle at least. On the other hand, it often happens that the given property depends uniquely on a particular average molecular weight, so that determination of this average alone suffices for the interpretation of the given property. Such is the case for colligative properties (number average), the turbidity (weight average), and the intrinsic viscosity (viscosity average), as was shown in the preceding chapter. Nevertheless, for the purpose of correlating properties which depend on different averages, for comparing polymers prepared under circumstances leading to different types of molecular size distributions, and, most important, for the mere satisfaction to be gained from an understanding of the molecular composition of polymers, a grasp of the molecular weight distribution is of fundamental importance.

The experimental approach to the problem of molecular weight distribution, involving separation of the polymer by fractional precipitation into fractions covering *comparatively* narrow molecular weight ranges followed by reconstruction of the distribution curve from the amount and average molecular weight of each fraction, is laborious to say the least, and the results which have been obtained in this manner are only semiquantitative. Sharp separations are never achieved, and reliable results require time-consuming fractionation and refractionation into a large number of cuts. Several more elegant physical methods—sedimentation equilibrium and sedimentation velocity measurements in particular—for ascertaining the molecular weight distribution in some detail have been much discussed, but the paucity of concrete results obtained by such methods up to the present time is not encouraging.

The most significant progress toward the elucidation of molecular weight distributions has been achieved through the application of theory. By consideration of the circumstances which control molecular growth under given conditions of polymerization in conjunction with the well-established propensity of chemical reactions to proceed in a random manner, molecular weight distibutions usually may be deduced from simple statistical considerations. Moreover, analysis of the factors governing molecular weight distributions leads at once to a system of classification whereby interrelationships between various polymer types are clarified. The discussion of molecular weight distributions presented in this chapter and the next is developed, therefore, primarily from the statistical point of view, which in fact prevails throughout most of the remainder of the book.

1. CONDENSATION POLYMERS

1a. Linear Chain Molecules; the Most Probable Distribution.—The principle of equal reactivity of all functional groups during condensation polymerization, as established in Chapter III, is the foundation for theoretically derived molecular size distribution relationships. No other postulates or assumptions are required. According to the principle of equal reactivity, at every stage of the polymerization process an equal opportunity for reaction is available to each functional group of a given chemical type, irrespective of the size of the molecule to which it is attached. (The chemical reactivity is not, however, asserted to remain constant throughout the course of the polymerization.) The probability that a given functional group has reacted will then be equal to the fraction p of all functional groups of the same type which have condensed. The significance of this statement lies in the fact that if a given unit is known to be attached through one of its functional groups to a sequence of x consecutive units combined in a linear polymer chain, the probability that the other functional group of the unit also has reacted is still exactly equal to p and is independent of the length of the chain indicated by x.

Consider a linear condensation polymer of type i formed from a monomer of the A——B type (see Chap. III). Polymerization of an ω-hydroxy acid will serve as an example. Suppose that the terminal hydroxyl group of a molecule has been selected at random and that we wish to know the probability that this molecule is composed of exactly x units.

$$\text{H—ORCO—ORCO—}\cdots\cdots\text{—ORCO—OH}$$
$$\quad\quad 1 \quad\quad\quad 2 \quad\quad\quad x-1 \quad\quad x$$

The probability that the carboxyl group of the first unit is esterified is equal to p. The probability that the carboxyl of the second unit is esterified, this probability being independent of whether or not linkage 1 has been formed, is likewise equal to p. The probability that this sequence continues for $x-1$ linkages is the product of these separate probabilities, or p^{x-1}. This is the probability that the molecule contains at least $x-1$ ester groups, or at least x units. The probability that the x-th carboxyl group is unreacted, thus limiting the chain to exactly x units, is $1-p$. Hence the probability that the molecule in question is composed of exactly x units is given[1] by

$$N_x = p^{x-1}(1-p) \tag{1}$$

If none but linear open chain polymer molecules are present, which for the present we assume to be the case, then obviously the probability, N_x, that any molecule selected at random is composed of x units must equal the mole fraction of x-mers.

The total number of x-mers is given by

$$N_x = N(1-p)p^{x-1}$$

where N is the total number of molecules of all sizes. If we let N_0 represent the total number of units, then as was shown in Chapter III (see Eq. III-9)

$$N = N_0(1-p)$$

which gives

$$N_x = N_0(1-p)^2 p^{x-1} \tag{2}$$

If the added weight of the end groups (equal to $H+OH$ for each molecule) is neglected, the molecular weight of each species is directly proportional to x. Hence the weight fraction can be written

$$w_x = xN_x/N_0$$

The error introduced by this approximation will be significant only at very low molecular weights. Substituting from Eq. (2)

$$w_x = x(1 - p)^2 p^{x-1} \tag{3}$$

The same derivation holds for type ii polymers formed from precisely equivalent proportions of A——A and B——B reactants. Here x represents the combined number of both types of units in the polymer chain. Eq. (3) applies also to polymers "stabilized" (see Chap. III) with small amounts of monofunctional units, although here it becomes necessary to replace the extent of reaction p with another quantity, namely, the probability that a given functional group has reacted with a *bifunctional* monomer. Type ii polymers stabilized with an excess of one or the other ingredient will be discussed later.

The susceptibility of polyesters and polyamides to interchange reactions, such as may occur in the former between a terminal hydroxyl of one molecule and an interunit ester group of another, was discussed in Chapter III. These interchange processes do not decrease the number of molecules, and hence do not affect $\overline{M}_n$, but they might permit some molecular species to be formed in preference to others. In other words, they may conceivably bring about an alteration of the molecular size distribution.

If interchange reactions proceed under conditions which do not permit further polymerization (the total number of molecules remaining constant), ultimately each species will be formed at a rate equal to its destruction. The size distribution then existing will represent a state of dynamic equilibrium. It is easy to show that the resulting distribution will be identical with the one described by Eqs. (1) and (3).[2,3] These relationships have been derived above from the principle of equal reactivity in the *kinetic* sense that all functional groups of the same chemical type are equally susceptible to condensation. Here we introduce the equally plausible postulate of equal reactivity from the point of view of *thermodynamic equilibrium* in the sense that every functional group shares an equal probability of existing at any instant in the reacted condition, regardless of the size of the molecule to which it belongs. This probability again may be designated by p, the fraction of all groups which exist in the condensed state at any instant, and the previous derivation applies without revisions.

Hence interchange equilibration and random condensation lead to the same size distribution, and occurrence of interchange during condensation polymerization normally will have no effect. Whether or not the interunit linkages are reversibly or irreversibly formed in condensation polymerization is of no importance insofar as the molecular weight distribution is concerned. The preceding size distribution relations apply also to polymers formed by random scission (degradation) of the interunit bonds of substantially infinitely long polymer molecules,[4] as must be obvious without further elaboration. Because of its occur-

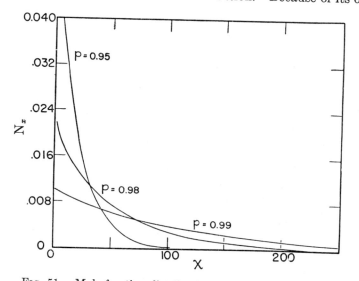

FIG. 51.—Mole fraction distribution[1] of chain molecules in a linear condensation polymer for several extents of reaction p.

rence under a variety of controlling circumstances, the distribution described by these relations is appropriately designated as the "most probable distribution."

Mole fraction (or number) distribution curves calculated from Eq. (1) are shown in Fig. 51 for polymers having average degrees of polymerization $\bar{x}_n$ of 20, 50, and 100. The corresponding weight fraction distributions calculated from Eq. (3) are shown in Fig. 52. As is evident from the equations also, monomers are present in greater numbers than any other single species, and this is true at all stages of the condensation. The number curves exhibit monotonic decreases as the number of units is increased. On a weight basis, however, the proportion of very low polymers is small and is diminished as the average molecular weight is increased. Maxima in the weight-distribution

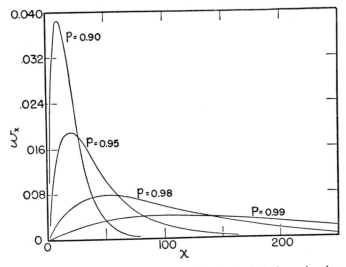

Fig. 52.—Weight fraction distributions[1] of chain molecules in linear condensation polymers for several extents of reaction p.

curves occur very near the number average value of x, i.e., very near $x = \bar{x}_n = 1/(1-p)$ (see Eq. III-9).

In type ii condensations three types of molecular species are present:

$$(A{-}{-}AB{-}{-}B)_{x/2}$$
Type iiAB

$$(A{-}{-}AB{-}{-}B)_{(x-1)/2}A{-}{-}A$$
Type iiAA

and

$$B{-}{-}B(A{-}{-}AB{-}{-}B)_{(x-1)/2}$$
Type iiBB

All molecules containing an *even* number of units necessarily will be of type iiAB, while those containing an *odd* number of units will be of type iiAA or iiBB. If the numbers of A——A and B——B units in the reaction mixture are exactly equal, there will be as many *even* as *odd* molecules, and the *odd* molecules will be equally divided between types iiAA and iiBB. The molecular weight distribution given above applies here also; only this alternation in end groups will occur between successive even and odd values of x.

If an excess of one component is employed in the reaction, the relative numbers of these three types of species will be altered; in particular, there will be more *odd-x* molecules (including both *odd* subtypes together) than *even* molecules. This is readily evident from the fact that there are more B than A end groups when the B——B unit is the one present in excess. At completion of the reaction ($p=1$) only type iiBB molecules will remain. The weight fraction distribution at completion of the condensation is given[1] by

$$w_x = xr^{(x-1)/2}(1 - r)^2/(1 + r) \qquad (4)$$

where r is the ratio all A to all B groups ($r<1$), and x is restricted to odd integral values. Eq. (4) resembles Eq. (3) and describes a similar distribution curve.

In the more general case of incomplete reaction[1] ($p\neq1$) and non-equivalence of the reactants ($r\neq1$), all three of the above species are present. The number average degree of polymerization is given by Eq. (12) of Chapter III. The weight distribution for each type resembles Eq. (3). The curves are parallel, with maxima occurring at the same value of x for each curve; the heights of the curves decrease in the order: BB, AB, AA if $r<1$. It is impossible to represent the distribution by a single curve since the value of the weight fraction alternates between *odd* (AA+BB types) and *even* (AB type) members, the former exceeding the latter if $r\neq1$. The average curve, however, corresponds very nearly to that given by Eq. (3) with p replaced by $pr^{1/2}$. This generalization includes Eq. (4), which applies when $p=1$, if account is taken of the absence of *even* x-mers in this case.

It follows from considerations of this nature that the form of the molecular size distribution obtained in bifunctional condensation is unaffected by such variations as portion-wise addition of one reactant to the other, polymerization followed by partial degradation, and so forth. Virtually all conceivable polymerization schemes involving reactions

$$x\text{-mer} + y\text{-mer} = (x + y)\text{-mer}$$

where x and y may assume the unrestricted range of integral values lead inevitably to the same distribution at the same average molecular weight.

1b. Experimental Confirmation of the Most Probable Size Distribution.—The only direct experimental verification of Eqs. (1) and (3) as applied to linear condensation polymers is due to Taylor,[5] who succeeded in separating poly-(hexamethyleneadipamide) ("nylon 66") into 46 fractions by a series of successive fractional precipitations executed with exceptional care by the addition of increments of water (precipitant) to phenol solutions of the polymer at 70°C. The differential distribution obtained from the data is shown by the points in Fig. 53. Curves calculated from Eq. (3) for two values of p are shown for comparison. The agreement with theory is within the experimental error of even this unusually precise fractionation.

Further confirmation of the theoretical size distribution equations comes from the various successful applications of these relationships to the interpretation of melt viscosities[6] and to the equilibration of the size distribution in mixtures of condensation polymers.[2] In these applications the average molecular weight (e.g., the weight average) on which the property being observed depends directly is calculated from another average (e.g., the number average) which is measured or controlled; the calculation is performed with the aid of the size distribu-

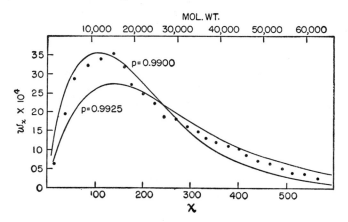

Fig. 53.—Molecular weight distribution in poly-(hexamethylene adipamide) as obtained by fractionation (points) compared with curves calculated from Eq. (3) for two values of p. (Taylor.[5])

tion relationships. The magnitude of whatever disparity may exist between the actual and the theoretical distribution must certainly vary with the average degree of polymerization. If the disparity is appreciable, then its variation with $\bar{x}_n$ should have manifested itself in a lack of agreement between theory and experiment in the applications referred to. The observed agreement actually is very good. The severity of this test of the size distribution equations exceeds that provided by fractionation.

The size distribution equations rest squarely on the validity of the principle of equal reactivity. Should they be proved appreciably in error in a given instance, this principle would have to be modified, or abandoned, for the polymerization process concerned. Since the principle seems well established, the likelihood of need for such qualifications seems remote.

1c. Average Molecular Weights for the Most Probable Distribution.—The number average degree of polymerization, $\bar{x}_n$, is given according to definition by the following expression, which may be obtained by dividing Eq. (VII-12) by the molecular weight M_0 of the structural unit:

$$\bar{x}_n = \sum xN_x / \sum N_x \tag{5}$$

or

$$\bar{x}_n = \sum xN_x \tag{5'}$$

For a polymer possessing the most probable distribution, the mole fraction N_x is given by Eq. (1); hence

$$\bar{x}_n = \sum xp^{x-1}(1-p)$$

Evaluation of the summation* gives

$$\bar{x}_n = 1/(1-p) \tag{6}$$

which corresponds to the result previously obtained in a more direct manner in Chapter III, Eq. (9).   The number average molecular weight is $\overline{M}_n = M_0\bar{x}_n$, neglecting the weight of terminal groups of the polymer chain, where M_0 is the molecular weight of the structural unit.   Eq. (6) may be regarded as an alternate definition of the parameter p in terms of the number average degree of polymerization.

The weight average degree of polymerization (see Eq. VII-27) is similarly obtained:

$$\bar{x}_w = \sum_1^\infty xw_x = \sum x^2 p^{x-1}(1-p)^2 \tag{7}$$

$$\bar{x}_w = (1+p)/(1-p) \tag{8}$$

The root-mean-square degree of polymerization for the most probable distribution is found to be

$$\bar{x}_{rms} = (1+p)^{1/2}/(1-p) \tag{9}$$

Thus

$$\overline{M}_n : \overline{M}_{rms} : \overline{M}_w : : 1 : (1+p)^{1/2} : (1+p) \tag{10}$$

* The values of this and related summations are as follows:

$$\sum_1^\infty xp^{x-1} = 1/(1-p)^2$$

$$\sum_1^\infty x^2 p^{x-1} = (1+p)/(1-p)^3$$

$$\sum_1^\infty x^3 p^{x-1} = (1+4p+p^2)/(1-p)^4$$

When the molecular weight is large, p may be replaced by unity with small error, and the above ratios become $1:\sqrt{2}:2$, i.e., the ratios of the various averages are practically independent of the degree of polymerization.

1d. Cyclic Condensation Polymers.—The foregoing discussion has proceeded under the assumption that the only products of bifunctional condensation are open chain polymer molecules—an assumption which obviously will not be exactly valid since cyclic polymers must always occur to some extent.   The nature of the error introduced by this assumption will be examined in the course of the following discussion of cyclic polymer components.

The formation of cyclic monomers when the repeating unit is of a size which permits formation of a ring of five, six, or seven members has been discussed in Chapter III.   The presence of such a ring, which in many instances constitutes the major product, is easily disposed of insofar as the subject matter of the present chapter is concerned by the mere assertion that it is not really polymeric and hence need not be regarded as a component of the polymer distribution.   Larger rings composed of a number of repeating units are genuine polymeric constituents, however. Unlike their monomeric antecedent, they may be discussed in the light of the theory developed by Jacobson and Stockmayer,[7] which should be generally applicable to bifunctionally condensing systems.

The proportion of rings of a given size must depend on the probability that the ends of a chain polymer of the same size meet in juxtaposition compared with the probability that one or the other of the ends occurs next to a functional group of another molecule with which it may react. If the chain contains more than about 12 or 15 single bonds, the former probability may be estimated from the statistical theory of chain configuration discussed in Chapter X; the latter probability depends directly on the concentration.   From these premises, Jacobson and Stockmayer[7] have shown that the weight fraction of cyclic polymers composed of x structural units in a type i polymer *at equilibrium*, which may be established through interchange processes (or reversible polymerization-depolymerization), should be given by

$$w_{rx} = (2B'M_0/c)(p')^x/x^{3/2} \tag{11}$$

where

$$B' = (3/2\pi\xi)^{3/2}/2l^3N \tag{12}$$

and c is the concentration in grams per cc. and M_0 is the molecular weight per structural unit, ξ is the number of chain atoms per unit, l is the "effective" bond length as corrected for bond angles and hindrances to free rotation (see Chap. X), N is Avogadro's number, and p' is a revised extent of reaction defined by

$$(1 - p') = (1 - p)/(1 - w_r) \tag{13}$$

Here w_r is the combined weight fraction of the cyclic polymers in the system, and p is the fractional degree of condensation of the functional groups for the system as a whole, as would be determined by chemical analysis, in accordance with its earlier definition. Noting that $1 - w_r$ is the weight fraction of open chain polymers and that cyclic polymers possess no unreacted groups, we observe from Eq. (13) that $1 - p'$ represents the fraction of the functional groups contained in open chain polymer molecules which are unreacted; i.e., p' is the extent of reaction in the chain polymer portion. The weight fraction w_r of all cyclic polymers (including the calculated fraction of cyclic monomer for which the statistical method may be quite unreliable on account of the small number of bonds) is given by

$$w_r = \sum_{x=1}^{\infty} w_{rx} = (2B'M_0/c)\phi(p', 3/2) \tag{14}$$

where

$$\phi(p', 3/2) = \sum_{x=1}^{\infty} (p')^x x^{-3/2} \tag{15}$$

Values for this function are listed by Jacobson and Stockmayer. The same equations apply to a polymer prepared from a single monomer A——A, except that the numerical factor 2 is to be omitted from Eq. (11).

A type ii polymer may contain rings made up of an even number x of *structural* units only. If equivalent proportions of the reactants A——A and B——B are used, then

$$w_{rx} = (2B'M_0/c)(p')^x/(x/2)^{3/2} \tag{16}$$

B' is again defined by Eq. (12), but with ξ equal to the number of chain atoms per *repeating* unit (i.e., per pair of the alternating structural units); M_0 is the mean molecular weight per *structural* unit. It follows that

$$w_r = (2B'M_0/c) \sum_{x/2=1}^{\infty} (p')^x/(x/2)^{3/2} \tag{17}$$

$$= (2B'M_0/c)\phi(p'^2, 3/2) \tag{18}$$

The weight fraction of rings w_r in a type ii polymer is plotted against the extent of reaction in Fig. 54 for several values of $B'M_0/c$. These

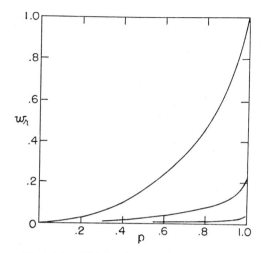

w_r

p

FIG. 54.—Weight fraction w_r of rings vs. the extent of reaction p for a type ii polymer for $B'M_0/c=0.005$ (lowest curve), 0.05 (middle curve), and 0.5 (uppermost curve). The curves correspond to successively increasing dilutions. (Jacobson and Stockmayer.[7])

curves were calculated by Jacobson and Stockmayer from Eqs. (13) and (18) with the aid of tabulated values for the function ϕ. Plausible values of $B'M_0$ for a polyester, or polyamide, formed from a polymethylene dibasic acid and a polymethylene glycol, or diamine, would be in the neighborhood of 0.01 in units of (g./cc.)$^{-1}$ for $\xi=10$ to 20 (taking $M_0=16\ \xi/2$ and $l\cong4\times10^{-8}$ cm., a likely value as indicated by dilute solution viscosity measurements on polymer molecules; see Chap. XIV). Hence the curves for an undiluted polymer should lie somewhat above the lowest one shown in Fig. 54. The middle curve should apply to a system diluted severalfold and the uppermost to a highly diluted system. When $B'M_0/c$ exceeds the critical value 0.19, the curve intersects the ordinate axis on the right at $w_r=1$, which signifies that the final product at complete conversion will then consist exclusively of ring structures![7]

Of greatest interest is the range of high extents of reaction—$p>0.95$ perhaps—where the average degree of polymerization (for chains) is not uninterestingly low. If the weight fraction of rings w_r is not too large, $1-p'$ is not very different from $1-p$ according to Eq. (13); hence p' also is near unity. The function $\phi(p'^2,3/2)$ increases only from 1.6 to 2.6 as p'^2 increases from 0.90 to 1.00. The increase in the weight fraction of rings as the condensation nears completion is correspondingly small, since according to Eq. (18) w_r is equal to the product of $B'M_0/c$ and ϕ. For the value of $B'M_0$ given above, w_r at $p=p'$ $=1$ should amount to only about 2.5 percent at $c=1$ g./cc. Thus the work of Jacobson and Stockmayer predicts that the proportion of cyclic polymers will be small but not negligible in undiluted systems, and that this proportion should increase approximately as the dilution. In type i polymers w_r should be about twice as great, for the same values of ξ and other parameters. It should be understood that the relationships under discussion apply only to equilibrated polymer distributions; they may not necessarily hold for primary products of condensation where the proportions of open chain and of cyclic products may depend on kinetic factors.

Molecular weight distributions for cyclic constituents are shown by the solid curves in Fig. 55 calculated from Eq. (16) for a type ii polymer where $B'M_0/c = 0.010$ and $p' = 0.95$ and 1.00; the combined weight fractions w_r of ring constituents are 0.0324 and 0.0522, respectively, and the corresponding values of the over-all extent of reaction p are 0.9516 and 1.00. Unlike the distribution for open chain polymers, these curves decrease monotonically with increase in size. Of even more significance is the virtual confinement of the distributions to the very low range, and the comparative lack of dependence on the extent of reaction except in the range of large size. The amount of a cyclic polymer of low degree of polymerization greatly exceeds that of the open chain polymer of the same size (compare broken curve). The primary effect of moderate dilution at fixed extent of reaction is to increase the weight fractions of each species proportionately. However, according to Eq. (13) p' must decrease somewhat with dilution at

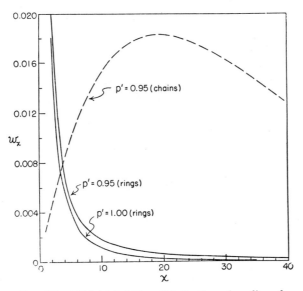

FIG. 55.—Weight fraction distribution of cyclic polymers for a type ii polymer with $B'M_0/c = 0.01$ (g./cc.)$^{-1}$ as calculated from Eq. (16) for $p' = 0.95$ and 1.00 (solid curves); only even integral values of x apply. The chain distribution for $p' = 0.95$ is shown for comparison by the broken curve calculated from Eq. (3'), p. 330.

fixed p, and if the dilution is great the distribution of ring components will be shifted in favor of lower species on this account (see Eq. 16).

The number average degree of polymerization $\bar{x}_{rn}$ of the cyclic species under no circumstances exceeds four for type ii, or two for type i, regardless of the extent of reaction.[7,*]   The weight average $\bar{x}_{rw}$ may be shown[7] to increase without limit as p (and hence p') proceeds to unity. However, it is always very much smaller than the weight average for the open chain species.

For complete rigor the distribution previously presented (Eqs. 1, 2, and 3 and Figs. 51 and 52) for linear polymer chains should be modified through substitution of p' for the over-all extent of reaction p.   Thus, for example, Eq. (3) for the weight fraction should be replaced with

$$w_{cx} = (1 - w_r)(1 - p')^2 x (p')^{x-1} \tag{3'}$$

where w_{cx} is the weight fraction of chain x-mer referred to the combined mixture of chain and ring species.   Likewise, p in other formulas given in the preceding sections of the present chapter should be replaced by p', and it should be stipulated that the formulas apply to open chain species exclusively.   For polymers formed (or equilibrated) in the absence of diluent, p' will be only very slightly less than p; hence the revision usually is scarcely significant.   In particular, the distribution (3') for chain species will be shifted only slightly toward smaller sizes as compared with that previously given by Eq. (3). The area under the curve, however, will be a little less than unity, being equal to $1 - w_r$, of course.   The dashed curve shown in Fig. 55 has been calculated from Eq. (3').   In order to obtain the complete size distribution including both chain and ring species, the curve for the latter (Fig. 55) should be added to the dashed curve.   This revision would greatly increase the proportion of very low polymers over the amounts previously considered (Fig. 52), and it would cause the combined distribution for chains and rings to pass through a minimum at a low value of x.   With dilution (at fixed p), not only do cyclic polymers increase at the expense of the open chain species, but the disparity between p' and p increases, with the result that the average degree of polymerization of both ring and open chain polymers decreases.

Jacobson, Beckmann, and Stockmayer[8] have taken advantage of this latter prediction to compare the theory with experiments on the molecular weights of decamethylene adipate ($\xi = 18$) polyesters equili-

---

* This calculation includes the lowest member of the ring series, the cyclic monomer, for which the theory is unreliable.   In any event, the conclusion that $\bar{x}_{rn}$ always is very small certainly must hold.

brated at a series of dilutions. Weight average molecular weights
were estimated empirically from intrinsic viscosities. Dilutions of
the polymer up to thirtyfold decreased $\bar{x}_w$ by a factor of about three,
the dependence of $\bar{x}_w$ on dilution being in accordance with the theory.
The value of $B'M_0$ calculated from their experimental data is 0.011
(g./cc.)$^{-1}$, which is reasonable.

Although the proportion of rings at equilibrium in an undiluted
polymer may be scarcely significant on a weight basis, they may make
a relatively large contribution to the total *number* of molecules and,
hence, markedly lower the over-all number average degree of polymerization. This will be particularly true if p is large and the number
of chain polymer molecules consequently is small. Cyclic polymers
would be likely to escape recognition in osmotic molecular weight methods, however, for they occur in the size range to which osmotic membranes ordinarily are permeable.

1e. Multichain Condensation Polymers.[9]—From the course of the
preceding discussion it might appear that bifunctional condensation
must inevitably yield the most probable distribution, for the open chain
species at equilibrium at any rate. Such a sweeping assertion would indeed invite citation of exceptions. Partly for the purpose of pointing
out how the monotony of this distribution may be avoided, we discuss
here polymers formed by condensing a type i bifunctional monomer
(A——B) together with a small proportion of an f-functional substance
which may be designated R—A$_f$. Polymers thus formed will consist
of f chains attached to the central unit R, and the type formula may be
written

$$R[—A(B——A)_y]_f \qquad (19)$$

where it is to be understood that y, the number of units in a chain, may
differ for each of the f chains. For example, ϵ-aminocaproic acid may
be condensed with a small proportion of tetrabasic acid.[9] If the
condensation is carried close to completion, linear species
HO[—CO(CH$_2$)$_6$NH—]$_x$H virtually disappear in favor of multichain
polymer molecules, R[—CO(—NH(CH$_2$)$_6$CO—)$_y$OH]$_4$, the number of
these being equal to the number of tetrabasic acid units in the system.
(Molecules possessing cyclic structures are disregarded here.) The
multichain polymers (except for $f=2$) are not linear; hence a discussion
of them in this chapter may seem out of place. However, these distributions are more closely related to those of linear polymers than to
the nonlinear polymers to be discussed in the following chapter.

Each chain of a molecule conforming to the formula (19) is subject
to the same statistical opportunities for development as a linear molecule in ordinary bifunctional condensation. The difference lies in the

fact that the total number x of units in the entire molecule depends on the sum of the y values for each of its chains. A size x much larger, or much smaller, than the average is likely to occur only if several (or all) of the chains in the molecule are abnormally large, or small, and such cooperation among statistically uncoordinated components will be comparatively rare. Hence the distribution will be narrower than for the ordinary *monochain* polymers.

If p is chosen to represent the probability that a given A group (COOH) has reacted, the probability that a particular chain contains y units is $p^y(1-p)$, in analogy with Eq. (1). The probability that the f chains have lengths y_1, y_2, $y_3 \cdots y_f$, respectively, is

$$p^{y_1}p^{y_2} \cdots p^{y_f}(1 - p)^f$$

Since

$$y_1 + y_2 + \cdots + y_f = x - 1 \qquad (20)$$

counting the central unit as one of the total (x), this may be written

$$p^{x-1}(1 - p)^f$$

The probability of each combination of the y's for which the total number of A——B units equals $x-1$ is given by this same expression. Hence the probability that the molecule in question contains exactly x units distributed over the f chains *in any manner whatever* must equal the product of this expression and the total number of combinations of the y's which fulfill Eq. (20). Noting that zero is a permissible value for the individual y's, the number of such combinations is[*] $(x+f-2)!/(f-1)!(x-1)!$, and the probability, or mole fraction, of x-mers becomes[9]

$$N_{x,f} = [(x + f - 2)!/(f - 1)!(x - 1)!]p^{x-1}(1 - p)^f \qquad (21)$$

The weight distribution may be shown[9] to be

$$w_{x,f} = \left[\frac{x(x + f - 2)!}{(f - 1)!(x - 1)!}\right]\left[\frac{(1 - p)^{f+1}}{fp + 1 - p}\right]p^{x-1} \qquad (22)$$

which for values of p near unity is approximated satisfactorily by[10]

---

[*] This result may be ascertained by considering the $x-1$ A——B monomers to be arranged in a linear succession with $f-1$ spacers distributed between them at random. The number of monomers in each of the resulting f sequences of monomer units shall be considered to correspond to the number of monomer units in each of the f chains of the molecule. The total number of ways of distributing the $x-1$ monomers over the f chains must correspond to the number of ways of arranging the $f-1$ spacers among the $x-1$ monomer units, i.e., to the number of combinations of $x-1+f-1$ elements taken $f-1$ at a time, which gives at once the result stated in the text.

$$w_{x,f} \cong (x^f/f!)(- \ln p)^{f+1}p^x \qquad (22')$$

If the molar ratio of the multifunctional unit to the bifunctional unit is Q/f, then at completion of the condensation Q moles of end groups A will remain unreacted per mole of A——B units, and $p = 1/(1+Q)$. The number of moles of polymer will equal Q/f per mole of bifunctional units in the system; hence

$$\bar{x}_n = (1 + Q/f)/(Q/f) = (Q + f)/Q$$
$$= (fp + 1 - p)/(1 - p) \qquad (23)$$

where p is the fraction of the A groups reacted at completion. It may be shown further[9] that

$$\bar{x}_w/\bar{x}_n \cong 1 + 1/f \qquad (24)$$

indicating that the distribution becomes sharper as f increases.

Distribution curves calculated for several values of f are shown in Fig. 56. Values of p have been adjusted to give the same number average (see Eq. 23), which also locates the maxima in the curves very nearly at the same abscissa value. The sharpening of the curves with increase in f is evident. The curve for $f=1$, corresponding to the most probable distribution, is included for comparison. Even for $f=2$, which represents the *linear* polymer prepared by condensing to completion an A——B monomer with a small amount of A——A (e.g., a dibasic acid), the alteration of the distribution is marked. The proportion of very low species is greatly reduced, the curve passes through an inflection preceding the maximum, and w_x falls more rapid-

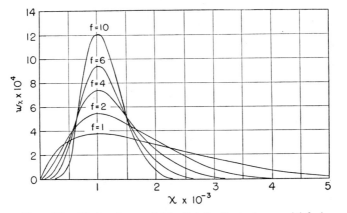

FIG. 56.—Molecular weight distributions for multichain polymers for the several values of f indicated, as calculated from Eq. (22'). Number averages for the different distributions are identical. (Schulz.[10])

ly toward zero as x increases beyond the maximum.   This narrowing of the distribution causes $\bar{x}_w/\bar{x}_n$ to approximate three-halves contrasted with two for the most probable distribution.   Thus, by combining two statistically independent chains into one molecule, a linear polymer having an equilibrium distribution much narrower than the most probable one may be obtained.

The stipulation that monomers of the A——B type be used in the preparation of multichain polymers conforming to formula (19) stems from the necessity for avoiding condensations between multichain molecules.   Owing to the use of an A——B monomer exclusively, all terminal groups on multichain molecules consist of A's, and these are presumed to be of such a character that condensation with one another does not occur.   Incorporation of B——B monomers in systems where the functionality f of the unit R—A_f exceeds two permits formation of network structures; experimentally, gelation is observed to occur.

2. ADDITION POLYMERS

2a. Vinyl Polymers.—The molecular weight distributions occurring in vinyl polymers follow no simple pattern such as holds for linear condensation polymers, and generalizations must be hedged with annoying qualifications.   Consider first the relatively simple case in which nearly all molecules are terminated by chain transfer.   Then, in accordance with the notation used in Chapter IV, the rates of propagation and chain transfer may be written $k_p[M][M\cdot]$ and $k_{tr}[S][M\cdot]$, respectively, where $[M]$, $[S]$, and $[M\cdot]$ are the concentrations of monomer, transfer agent, and radicals, respectively; k_p and k_{tr} are rate constants for propagation and transfer, and their ratio $C_S = k_{tr}/k_p$ is the transfer constant introduced in Chapter IV.   These velocity constants should be independent of the size of the growing radical $M\cdot$ according to the principle of equal reactivity.   Hence the probability that a radical containing i units undergoes transfer rather than addition of another monomer is given by

$$1 - p = k_{tr}[S]/\{k_p[M] + k_{tr}[S]\}$$
$$= \{1 + /C_S)[M]/[S]\}^{-1} \qquad (26)$$

the value of p being independent of the radical size i.   The number average degree of polymerization, equal to the number of units (including the transfer radical as one unit) polymerized per transfer, is

$$\bar{x}_n = 1/(1 - p) \qquad (27)$$

The parameter p is the intermediary connecting the reaction kinetics and the molecular distribution.   It expresses the probability that a growing chain of i units adds at least one more monomer; hence it is precisely equivalent in this sense to the p used in the treatment of condensation polymers where it was equal to the extent of reaction. Equations (1) and (3), and (5) through (10), inclusive, for the most probable distribution are immediately applicable, therefore.* However, the ratio of $[S]$ to $[M]$ ordinarily will change during the course of the polymerization and p will change accordingly. The increment of polymer formed within a small range of conversion will be characterized by a certain value of p, which defines the average molecular weight and distribution for this increment. The aggregate of polymer formed over a wide conversion range may be regarded as the summation of such increments, each having a most probable distribution about an average which shifts progressively for successive increments. The aggregate distribution will be broader than the most probable distribution to an extent depending on the variation of p, and hence of $[S]/[M]$, with conversion. Only if $C_S = 1$ (i.e., $k_{tr} = k_p$) so that this concentration ratio remains constant with conversion (see Eq. IV–44), or if the polymerization is confined to a low conversion so that the change in $[S]/[M]$ is in any case small, will the product conform to the most probable distribution.

To consider another extreme, suppose that transfer contributes negligibly to the cessation of chain growth and, further, that the chains are initiated by monoradicals and terminated by coupling. Each polymer molecule consists of two sequences of units which grew independently and were joined together at their mutual termination. They are analogous to the dichain ($f = 2$) condensation polymers discussed above. Further advantage may be taken of this analogy by defining p as the probability of continuation of either chain from one of its units to the next. Then $1 - p$ is the probability that the i-th unit in the growth of the chain reacts by termination; it is therefore equal to the ratio of terminated to total units. Since two units are involved in each termination step

$$1 - p = 2k_t[M\cdot]/\{k_p[M] + 2k_t[M\cdot]\} \tag{28}$$

where k_t is the rate constant for termination. Then, since each molecule consists of two chains

$$\bar{x}_n = 2/(1 - p) = 2 + k_p[M]/k_t[M\cdot] \tag{29}$$

---

* The size distributions in polymers may be derived in a more elegantly impressive manner through the use of differential rate equations for each species. After introduction of the inevitable simplifying assumptions, the final results are the same as those found above.

The weight fraction distribution[10] is readily shown to be

$$w_x = x(x - 1)p^{x-2}(1 - p)^3/2 \tag{30}$$

where each initiator radical is counted as a unit.   This expression corresponds very nearly to Eq. (22) when applied to dichain polymers ($f=2$).   Minor differences occur because the latter contains a single central unit, whereas two must be counted here.

As previously pointed out, this distribution is much sharper than the most probable one (see Fig. 56).   Unless the initiation mechanism is of a sort which depends on the square of the monomer concentration (e.g., thermal initiation; see Chap. IV), $[M\cdot]/[M]$ will vary with conversion, causing the distribution for the aggregate of polymer formed to broaden with conversion.

In the more general case of joint control of molecular weight by both transfer and radical termination, it is appropriate to consider that two distributions are formed simultaneously.   One of these distributions consists of molecules terminated by chain transfer; the other of pairs of chains joined by the combination of radicals.   For any conversion increment, the two coexisting distributions will depend on the same parameter p representing the probability of continuation of the growth of any chain, i.e.

$$p = k_p[M]/\{k_p[M] + k_{tr}[S] + 2k_t[M\cdot]\} \tag{31}$$

The portion of the polymer consisting of molecules terminated by transfer will conform to the most probable distribution, its average degree of polymerization being

$$\bar{x}_n = 1/(1 - p)$$

The distribution of the coupling-terminated molecules will be represented by the sharper dichain distribution discussed above, its degree of polymerization being given by

$$\bar{x}_n = 2/(1 - p)$$

The proportion by weight of the former will, of course, equal $k_{tr}[S]/\{k_{tr}[S]+2k_t[M\cdot]\}$.   These remarks apply only to increments of polymer; for wide conversion ranges, the broadening effects due to changes in p and in this ratio must be considered.

2b. Polymers Formed by Monomer Addition without Termination.— Polymerization of cyclic monomers such as ethylene oxide, as was pointed out in Chapter II, may be induced by small amounts of substances capable of cleaving the ring with generation of a hydroxyl group, which may add another monomer, and so forth as indicated by the following scheme:

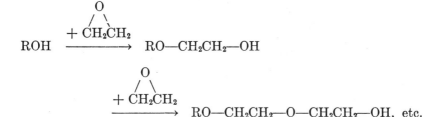

Other cyclic compounds such as the N-carboxyanhydrides of α-amino acids,[11] and lactams may be polymerized similarly with regeneration of an amino group at each step. According to the mechanism postulated, the number of polymer molecules formed should equal the number of initiator molecules (e.g., ROH) introduced, and the average number of monomers per polymer molecule should equal the ratio of monomer consumed to initiator.

The distinguishing feature of such a mechanism occurs in the fact that the growth of all polymer molecules proceeds simultaneously under conditions affording equal opportunities for all. (This will hold provided the addition of monomer to the initiator is not much slower than succeeding additions.) These circumstances are unique in providing conditions necessary for the formation of a remarkably narrow molecular weight distribution—much narrower than may be obtained by polymer fractionation, for example. Specifically, they are the conditions which lead to a Poisson distribution[12] of the number and mole fraction, i.e.[13]

$$N_x = e^{-\nu}\nu^{x-1}/(x-1)! \tag{32}*$$

where ν represents the number of monomers reacted per polymer molecule (or per initiator); in defining x, the initiator is counted as one unit. The weight fraction distribution is given by [13] the slightly modified Poisson distribution

$$w_x = [\nu/(\nu+1)]xe^{-\nu}\nu^{x-2}/(x-1)! \tag{33}$$

The number average degree of polymerization is, of course

$$\bar{x}_n = \nu + 1 \tag{34}$$

and the weight to number average ratio can be shown to be given by

$$\bar{x}_w/\bar{x}_n = 1 + \nu/(\nu+1)^2 \tag{35}$$

Weight fraction distributions according to Eq. (33) for three values

* Eq. (32) may be deduced from kinetic arguments.[13] This is unnecessary, however, since the conventional derivation of the Poisson distribution proceeds from equivalent statistical conditions.[12]

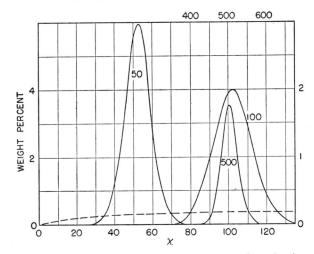

Fig. 57.—Weight percent ($100\,w_x$) vs. number of units for polymers formed by successive addition of monomers to a fixed number of active centers, as calculated from Eq. (33) for the values of ν indicated.[13] The $\nu = 500$ curve is drawn to the scales along the upper and right-hand margins; scales for the other curves are given along the lower and left-hand margins. The broken curve represents a most probable distribution, $\bar{x}_n = 101$, shown for comparison.

of ν are shown in Fig. 57. The relative breadth of the distribution decreases as the average degree of polymerization increases. This is shown also by the decrease of the $\bar{x}_w/\bar{x}_n$ ratio, given by Eq. (35), as ν increases. The portion of the most probable distribution curve for $\bar{x}_n = 101$, shown for comparison, illustrates the marked difference in character of the two distributions.

The conditions essential for the formation of this exceptionally sharp distribution are the following: (1) growth of each polymer molecule must proceed exclusively by consecutive addition of *monomers* to an active terminal group, (2) all of these active termini, one for each molecule, must be equally susceptible to reaction with monomer, and this condition must prevail throughout the polymerization, and (3) all active centers must be introduced at the outset of the polymerization and there must be no chain transfer or termination (or interchange). If new active centers are introduced over the course of the polymerization, a much broader distribution will be produced for the obvious reason that those introduced late in the process will enjoy a shorter period in which to grow.[14] If the chains suffer transfer, or if termination occurs with constant replenishment of the active centers by one means or another, the polymerization reduces formally to a vinyl polymerization type.

If a polymer formed initially by the addition of monomers to a fixed number of centers is subjected to conditions permitting interchange processes to occur, either during polymerization or subsequent thereto, the distribution will broaden. Polymerization of a lactone, for example, according to the mechanism

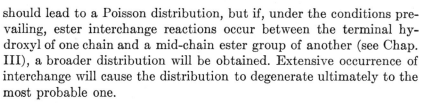

should lead to a Poisson distribution, but if, under the conditions prevailing, ester interchange reactions occur between the terminal hydroxyl of one chain and a mid-chain ester group of another (see Chap. III), a broader distribution will be obtained. Extensive occurrence of interchange will cause the distribution to degenerate ultimately to the most probable one.

The addition mechanism under discussion appears to have been definitely established for the polymerization of the N-carboxyanhydrides of α-amino acids,[11,15] and there are indications that it applies, approximately at least under appropriate conditions, to others of the examples cited above. It is important principally because of the possibilities it suggests for the preparation of polymers having exceptionally narrow molecular weight distributions. It offers a possible explanation for the formation of naturally occurring polymers, proteins in particular, reputed to be homogeneous. It is doubtful if the degree of heterogeneity associated with a Poisson distribution at a high degree of polymerization could be distinguished from absolute homogeneity by any presently known physical method.

3. POLYMER FRACTIONATION

Procedures for separating high polymers into fractions of restricted molecular weight range vary, but all of them depend on the very small decrease in solubility with increase in molecular weight in a given poly-

mer series.[16] Fractional precipitation from a dilute solution is most
widely used. A suitable combination of a solvent and a precipitant
miscible with one another is selected, and the latter is added to a dilute
solution of the polymer in the former until a slight turbidity develops
at the temperature chosen for the fractionation. For the purpose of
assuring establishment of equilibrium between the two phases which
form, it is usually considered desirable to warm the solution to homogeneity and then to allow precipitation to take place during gradual
cooling to the temperature of the fractionation bath, which should be
controlled within $\pm 0.05°C$. The solution is stirred slowly during precipitation. Thereafter the precipitated phase is allowed to settle.
After it has formed a coherent layer, which may require from 2 to 24
hours, the supernatant phase is removed. Polymer in the precipitated
phase, representing the first fraction, is recovered by drying or by co-
agulation in a large excess of precipitant. A judiciously chosen incre
ment of precipitant is added to the supernatant portion from the first
separation, and the operation is repeated to obtain the second fraction.

The precipitation process is more appropriately regarded as a separation of the system into two liquid phases. The one relatively rich in
polymer is called, somewhat inaccurately, the precipitate and the other
the supernatant phase. The precipitated phase actually contains much
more solvent than polymer; the concentration of polymer in it may be
only about 10 percent. Each polymer species is partitioned between
these two phases. It may be shown that every polymer species, including both the large and the small, is more soluble in the more concentrated, or precipitated, phase (see Chap. XIII). However, the
smaller species will be distributed at more nearly equal *concentrations*
in the two phases. This is merely a result of the fact that they possess
fewer units per molecule to interact with the less favorable environment of the supernatant phase; they consequently show less discrimination in choosing their surroundings. To put the matter another way,
for a given *amount* of polymer transported from the precipitated to the
supernatant phase, the entropy gain arising from the larger volume of
the supernatant phase will depend directly on the number of particles
transported. Lower polymer molecules are better able to take advantage of the larger volume of the supernatant phase because of their
greater number in a given amount of polymer.

The theory of fractionation will be discussed in Chapter XIII. It is
sufficient to observe here that polymer fractionation depends on the
tenuous differentiation indicated above. In order to make the most of
it, the ratio of the volumes of the supernatant and precipitated phases
should be as large as possible; the ratio should be at least ten and

preferably much larger. Since the precipitated phase is highly swollen, this objective can be achieved only by using a very dilute polymer solution. The required dilution increases with the molecular weight being fractionated. As a useful guide, the polymer concentration, expressed in volume fraction, immediately preceding phase separation should not exceed about one-fourth of the square root of the ratio of molar volumes of solvent and of polymer of the molecular weight being fractionated. Even lower concentrations are preferable, but the difficulties of handling larger volumes of solution weigh heavily against further dilution. The importance of high dilution frequently has not been properly appreciated, with the result that the volume ratio between the two phases has been inadequate to assure satisfactory separation. Stepwise precipitation of the polymer all too often has been carried out without achieving substantial fractionation, particularly in the higher molecular weight range. Molecular weight distribution curves constructed from results so obtained may show fictitious "second peaks" in the higher range.

Successive refractionation has been popular as a means to achieve better separations. Further analysis shows, however, that a single fractional separation at high dilution may be as efficient as the laborious separation of the unwieldy array of overlapping fractions obtained by successive refractionations. The large volume of solution required to secure the required dilution while at the same time maintaining an adequate sample size poses a serious difficulty. It may be alleviated to a considerable extent by twice precipitating each fraction according to the following procedure.[17] From the main solution at a moderate dilution, e.g., ca. 0.5 percent for $\overline{M}_w = 10^6$, a somewhat larger fraction than desired is precipitated. The comparatively small amount of polymer thus precipitated may be redissolved and diluted to about 0.1 percent without reaching a volume which is excessive. A more precise fractional separation is then achieved from this more dilute solution. This second precipitate is isolated as a final fraction, and the polymer recovered from the dilute phase by evaporation is returned to the main solution. The operation is repeated to obtain the next fraction, etc.

The successive fractions may be obtained by lowering the temperature in suitable increments rather than by varying the solvent composition isothermally. If a poor solvent is found from which the polymer precipitates in a convenient temperature range, it may be used alone without addition of another component.

Fractional extraction occasionally is employed. The polymers of lowest molecular weight are extracted first from the sample, which is

highly swollen by the solvent medium.   The solvent power of the
medium is varied by suitable changes in composition or in temperature
in order to remove fractions of successively increasing molecular
weight.   The principles involved are the same as for fractional precipitation.   This scheme lends itself to continuous operation using a flowing, or circulating, solvent medium.[18]   The polymer must be retained
in a finely divided form or as a thin film in order to assure efficient transfer between the polymer and the solvent medium.

The average molecular weights of the fractions are determined either
by an absolute method or by measuring the intrinsic viscosity and applying a previously established relationship with molecular weight.
The type of average molecular weight is of minor importance; if the
fractionation is adequate for satisfactory determination of the molecular weight distribution curve, the differences between the different average molecular weights for a given fraction must be fairly
small (i.e., less than 10 percent between $\overline{M}_w$ and $\overline{M}_n$).

Having determined the amount and molecular weight of each fraction, it is customary to plot the cumulative, or integral, distribution
curve, expressing the combined weight fraction of all fractions having
molecular weights up to and including M.   Cumulative distribution
curves obtained by Baxendale, Bywater, and Evans[19] for poly-(methyl
methacrylate) polymerized by hydroxyl radicals in aqueous emulsion
are shown in Fig. 58.   The differential distribution, expressed by w_x
above, is obtained by taking slopes from the integral distribution curve
plotted against the degree of polymerization.   Thus, if W_x represents

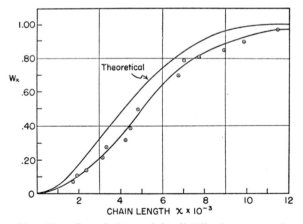

Fig. 58.—Cumulative weight distribution curves for
poly-(methyl methacrylate) as established by fractionation compared with theoretical curve. (Results of
Baxendale, Bywater, and Evans.[19])

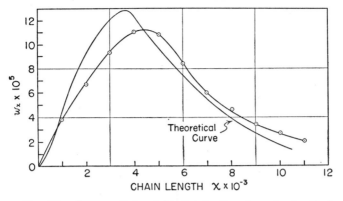

FIG. 59.—Differential weight distribution for poly-(methyl methacrylate), as deduced from curve in Fig. 58, in comparison with theoretically calculated curve. (Baxendale, Bywater, and Evans.[19])

the integral curve, then $w_x = dW_x/dx$. The differential distribution curve calculated in this way is shown in Fig. 59. The theoretical curves shown in Figs. 58 and 59 have been calculated from kinetic considerations assuming termination of growth by coupling without appreciable chain transfer, the influence of change with conversion on the ratio $[M \cdot]/[M]$ (see Eq. 28) having been taken into account.

Probable errors in assigning the integral distribution curve, as indicated by scatter of the points in Fig. 58, are magnified in the process of taking the slope for the deduction of the differential distribution. Only the approximate location of the maximum and breadth of the latter are experimentally significant.

In Fig. 60 are shown the results obtained by Merz and Raetz[20] from a large-scale fractionation of a polystyrene prepared by thermal polymerization to 15 percent conversion. Thirty-nine fractions were obtained. The integral plot is first expressed in a stepwise manner, with the vertical displacement of each step corresponding to the size of the fraction having the molecular weight at the position of the step. The slope of the smoothed curve through the stepwise plot gives the differential distribution shown in the same figure. Fractional precipitation was conducted from solutions approximating 1.5 percent in concentration, which is too high for satisfactory resolution of the higher molecular weight components. The sudden dip of the differential distribution curve toward zero at higher molecular weights may therefore be artificial.

Complete fractionation of a high polymer with sufficient precision and enough fractions for even approximate definition of the differential

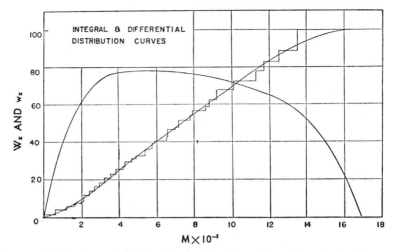

Fig. 60.—Molecular weight distribution for thermally polymerized polystyrene as established by fractionation. (Results of Merz and Raetz.[20])

distribution curve is an arduous undertaking. Fractionation serves another very useful function however—that of providing samples of sufficient homogeneity to minimize the disturbing effects of heterogeneity on the interrelationship of various physical measurements. This usually means that the disparity between $\overline{M}_w$ and $\overline{M}_n$ should be reduced to about 10 percent, which is a less drastic requirement than is needed (though seldom attained) for accurate assignment of the distribution curve.

In the case of a polymer which may occur in the crystalline form, it is desirable to adopt conditions for fractionation which allow the precipitate to separate as a totally amorphous liquid without formation of polymer in the crystalline condition. This may be difficult or even impossible to achieve if the melting point of the polymer is high. Whereas liquid-liquid separation from a dilute solution is rapidly reversible, separation of crystalline polymer usually occurs as a poorly reversible process often with large supercooling effects. The rate of crystallization consequently plays a role comparable in importance to equilibrium solubility in the selection of molecules for the newly formed phase. Although the larger polymer molecules are less soluble, they will be slower to undergo the configurational changes required for crystallization. Whereas equilibrium factors would cause the higher species to precipitate (in crystalline form) at a slightly higher temperature, their slower rates may prevent them from doing so.

These opposing tendencies may defeat the purpose of the fractional precipitation process. The fractional precipitation of crystalline polymers such as nitrocellulose, cellulose acetate, high-melting polyamides, and polyvinylidene chloride consequently is notoriously inefficient, unless conditions are so chosen as to avoid the separation of the polymer in semicrystalline form. Intermediate fractions removed in the course of fractional precipitation may even exceed in molecular weight those removed earlier.[21] Separation by fractional extraction should be more appropriate for crystalline polymers inasmuch as both equilibrium solubility and rate of solution favor dissolution of the components of lowest molecular weight remaining in the sample.

From a study of the phase diagram, it is usually not difficult to differentiate crystalline from liquid precipitation. The curve for liquid-liquid equilibrium representing the miscibility temperature plotted against the concentration of polymer in a solvent consisting of a single component shows a characteristic maximum at a low concentration. The liquid-crystalline phase equilibrium curve, on the other hand, increases monotonically with polymer concentration. (See Figs. 133 and 134, pp. 574, 575.) In order to avoid the latter, it may be necessary to conduct the fractionation at an elevated temperature not too far below the melting point of the pure polymer, using a sufficiently poor solvent to obtain liquid-liquid separation under these conditions.

Characteristic features of the molecular weight distribution may be determined by turbidimetric titration of a very dilute (0.01 g./100 cc. or less) solution of the polymer.[22,23] A precipitant is added slowly to the agitated solution at constant temperature, and the turbidity due to precipitated polymer is measured by recording the decrease in the intensity of a transmitted beam of light. (This turbidity is of course very much greater than that due to the dissolved molecules in solution, which is observed in the light-scattering method for measuring the molecular weight.) An empirical correlation must first be established between the concentration of the precipitant and the molecular weight of the polymer precipitated at a specified polymer concentration. The experimental curve expressing the turbidity as a function of the concentration can then be converted to an integral molecular weight distribution curve, from which an approximate differential curve can be derived. This method, developed principally by Morey and Tamblyn,[22] has been applied successfully to poly-(vinyl acetate),[23] nitrocellulose,[24] and poly-(methyl methacrylate).[25] A mixture of two fractions of the latter polymer differing fivefold in average molecular weight was easily resolved by this method.[25]

REFERENCES

1. P. J. Flory, *J. Am. Chem. Soc.*, **58**, 1877 (1936); *Chem. Revs.*, **39**, 137 (1946).
2. P. J. Flory, *J. Am. Chem. Soc.*, **64**, 2205 (1942).
3. P. J. Flory, *J. Chem. Phys.*, **12**, 425 (1944).
4. W. Kuhn, *Ber.*, **63**, 1503 (1930).
5. G. B. Taylor, *J. Am. Chem. Soc.*, **69**, 638 (1947).
6. P. J. Flory, *J. Am. Chem. Soc.*, **62**, 1057 (1940).
7. H. Jacobson and W. H. Stockmayer, *J. Chem. Phys.*, **18**, 1600 (1950).
8. H. Jacobson, C. O. Beckmann, and W. H. Stockmayer, *J. Chem. Phys.*, **18**, 1607 (1950).
9. J. R. Schaefgen and P. J. Flory, *J. Am. Chem. Soc.*, **70**, 2709 (1948).
10. G. V. Schulz, *Z. physik. Chem.*, **B43**, 25 (1939).
11. S. G. Waley and J. Watson, *Proc. Roy. Soc.* (London), **A199**, 499 (1949).
12. See for example, W. Feller, *Probability Theory and Its Applications* (John Wiley and Sons, New York, 1950), I, 115–123.
13. P. J. Flory, *J. Am. Chem. Soc.*, **62**, 1561 (1940).
14. H. Dostal and H. Mark, *Z. physik. Chem.*, **B29**, 299 (1935).
15. J. H. Fessler and A. G. Ogston, *Trans. Faraday Soc.*, **47**, 667 (1951).
16. L. H. Cragg and H. Hammerschlag, *Chem. Revs.*, **39**, 79 (1946).
17. P. J. Flory, *J. Am. Chem. Soc.*, **65**, 372 (1943).
18. V. Desreux, *Rec. trav. chim.*, **68**, 789 (1949); V. Desreux and M. C. Spiegels, *Bull. soc. chim. Belges*, **59**, 476 (1950). O. Fuchs, *Makromol. chem.*, **5**, 245 (1950); *ibid.*, **7**, 259 (1952).
19. J H. Baxendale, S. Bywater, and M. G. Evans, *Trans. Faraday Soc.*, **42**, 675 (1946).
20. E. H. Merz and R. W. Raetz, *J. Polymer Sci.*, **5**, 587 (1950).
21. D. R. Morey and J. W. Tamblyn, *J. Phys. Chem.*, **50**, 12 (1946).
22. D. R. Morey and J. W. Tamblyn, *J. Applied Phys.*, **16**, 419 (1945).
23. D. R. Morey, E. W. Taylor, and G. P. Waugh, *J. Colloid Sci.*, **6**, 470 (1951).
24. A. Oth, *Bull. soc. chim. Belges*, **58**, 285 (1949).
25. I. Harris and R. G. J. Miller, *J. Polymer Sci.*, **7**, 377 (1951).

CHAPTER IX

Molecular Weight Distributions in Nonlinear Polymers and the Theory of Gelation

THE various molecular species occurring in linear polymers formed by condensation of bifunctional units differ only in one parameter, the chain length, apart from possible differences in end groups. If some of the units possess a higher functionality, i.e., if some of the units combine with more than two other units, nonlinear structures develop and the complete description of the molecular constitution assumes a higher order of complexity. Mere designation of the size of a molecule, or of the total number of units it contains, fails to define its structure; the degree of branching, or number of units of higher functionality in the molecule, requires specification also. Moreover, an enormous array of isomeric structures may be derived from given numbers of bifunctional and of polyfunctional units $(f > 2)$. While these ramifications of composition and structure for each polymer species of specified degree of polymerization must be kept clearly in mind, one is obliged to be content for the most part with description of the *size* distribution as an objective, without attempting a more detailed quantitative account of the constitution of a given nonlinear polymeric substance. The size distribution may be markedly distorted by even a very small proportion of polyfunctional units—so much so that little resemblance to the distribution for a linear polymer may remain.

The presence of polyfunctional units nearly always presents the possibility of forming chemical structures of macroscopic dimensions, to which the term *infinite network* is appropriately applied (Chap. II). It is at this point that the concept of molecules as the primary chemical entities must be abandoned, for the infinite network reaches the macro-

347

scopic dimensions of the sample itself. On the ordinary molecular level, it is unbounded in extent, hence the term infinite network. The presence of these *space* (or *three-dimensional*) networks dominates in importance other aspects of molecular constitution in most nonlinear polymer systems. They are responsible for the interesting phenomenon of gelation, already described briefly in Chapter II, which is widely observed to occur in nonlinear polymerizations, e.g., in vinyl-divinyl copolymerizations, in the vulcanization of rubber, in the moulding of thermosetting resins such as those formed from phenol and formaldehyde, and in the drying of the oils used in protective coatings.

The critical conditions for the formation of infinite networks will be discussed at the outset of the present chapter. Molecular weight distributions for various nonlinear polymers will then be derived. Experimental data bearing on the validity of the theory will be cited also.

1. CRITICAL CONDITIONS FOR THE FORMATION OF INFINITE NETWORKS

1a. Polyfunctional Condensation Polymerization.[1,2]—Consider the polymerization of a bifunctional unit A——A, a trifunctional unit

A——$\langle$, and a bifunctional unit of opposite character B——B, where

condensation occurs exclusively between A and B. Polymers having structures such as the one shown in Fig. 61 will be formed. Let all functional groups of each kind, A and B, be chemically equivalent and hence equally reactive; the principle of equal reactivity is assumed to hold throughout the condensation so that the reactivity of a given A, or B, group is independent of the size or structure of the molecule (or network) to which it is attached. The further assumption that reactions between A and B groups on the same molecule are forbidden, or that their occurrence is negligible, introduces an error which frequently is appreciable in magnitude. The introduction of this assumption is dictated by expediency; treatment of nonlinear polymerizations without it appears hopeless at present.

To examine the significance of this approximation further, it should be noted that a highly branched condensation polymer molecule, such as the one shown in Fig. 61, retains many unreacted functional groups which offer a number of opportunities for reaction between pairs on the same molecule. That intramolecular reaction between them proceeds to an appreciable degree in competition with intermolecular condensa-

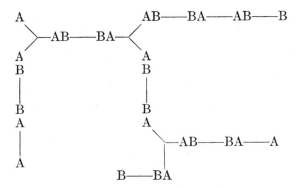

Fig. 61.—Trifunctionally branched polymer composed of A——A and B——B
bifunctional units, and of A—$\big\langle$ $\genfrac{}{}{0pt}{}{A}{A}$ trifunctional units. AB or BA is the product of
the condensation of two functional groups.

tion is shown by experiments of Kienle and co-workers[3] on the condensation of glycerol with dibasic acids. If esterification were exclusively intermolecular, the number of molecules present should decrease by one for each ester group formed. Kienle and co-workers observed that the cryoscopically determined number of particles decreased less than the increase in the extent of esterification; the discrepancy amounted to about 5 percent of the total esterification. With less highly branched molecules the discrepancy presumably should be smaller. It is convenient to consider that the assumption of an exclusively intermolecular condensation fails to eliminate from consideration the fraction of the condensation which is "wasted" on the formation of intramolecular connections which do not increase the molecular weight or, rather, do not decrease the number of molecules. The error so introduced generally is small. To take account of the intramolecular condensations would complicate the theory without adding materially to its content beyond improvement in numerical agreement with experiment.

The purpose of the following treatment is to define the conditions under which indefinitely large chemical structures, or infinite networks, will occur. To this end we seek the answer to the question: Under what conditions is there a finite probability that an element of the structure selected at random occurs as part of an infinite network? In order to simplify the problem, any given molecule such as the one shown in Fig. 61 may be regarded as an assemblage of *chains* connected together through polyfunctional, or branch, units (trifunctional in

Fig. 61). Specifically, a chain is defined as that portion of a molecule between two branch units, or between a branch unit and a terminal unreacted functional group (e.g., OH or COOH). The lengths of the chains will vary, but for the present this variation is unimportant.

First of all it is necessary to determine the branching coefficient α, which is defined as the probability that a given functional group of a branch unit leads via a chain of bifunctional units to another branch unit. In a polymer of the type shown in Fig. 61, α is the probability that an A group selected at random from one of the trifunctional units is connected to a chain the far end of which connects to another trifunctional unit. As will be shown later, both the location of the gel point and the course of the subsequent conversion of sol to gel are directly related to α.

The formation of chains, as defined above, can be represented by the equation

$$A\text{---}A + A\text{---}\Big\langle\begin{matrix}A\\[4pt]A\end{matrix} + B\text{---}B$$

$$\rightarrow\ \Big\rangle\text{---}A[B\text{---}BA\text{---}A]_iB\text{---}BA\text{---}\Big\langle$$

where i may have any value from 0 to ∞. Other chains will be formed, one end, or both ends, of which consist of unreacted terminal groups. Under the assumption of equal reactivities of all A and of all B functional groups, the probability that the first A group of the chain shown on the right has reacted is given by p_A, the fraction of all A groups which have reacted; similarly, the probability that the B group on the right of the first B---B unit has reacted is given by p_B. Let ρ represent the ratio of A's (reacted and unreacted) belonging to branch units to the total number of A's in the mixture. Then, the probability that a B group has reacted with a branch unit is $p_B\rho$; the probability that it is connected to a bifunctional A---A unit is $p_B(1-\rho)$. Hence the probability that the A group of a branch unit is connected to the sequence of units shown in the preceding formula is given by

$$p_A[p_B(1-\rho)p_A]^i p_B\rho$$

The probability, α, that the chain ends in a branch unit regardless of the number, i, of pairs of bifunctional units is given by the sum of such expressions having $i = 0, 1, 2, \cdots$ etc., respectively. That is

$$\alpha = \sum_{i=0}^{\infty} [p_A p_B(1-\rho)]^i p_A p_B\rho$$

Evaluation of this summation (see p. 325, footnote), yields

$$\alpha = p_A p_B \rho / \left[1 - p_A p_B (1 - \rho) \right] \tag{1}$$

If we let the ratio of A to B groups initially present be represented by r, as in the case of type ii linear polymers, then

$$p_B = r p_A$$

Substituting in Eq. (1) to eliminate either p_B or p_A

$$\alpha = r p_A^2 \rho / \left[1 - r p_A^2 (1 - \rho) \right] \tag{2}$$

or

$$\alpha = p_B^2 \rho / \left[r - p_B^2 (1 - \rho) \right] \tag{3}$$

In the application of these equations, ordinarily r and ρ will be determined by the proportions of the initial ingredients employed, and either the unreacted A or B group will be determined analytically at various stages of the reaction. Then α can be calculated employing either Eq. (2) or Eq. (3), depending on which group is determined directly. Hence α is readily calculable from experimentally observed and controlled quantities.

Several special cases are of particular interest. When there are no A——A units, $\rho = 1$ and

$$\alpha = r p_A^2 = p_B^2 / r \tag{4}$$

When A and B groups are present in equivalent quantities, $r = 1$, $p_A = p_B = p$, and

$$\alpha = p^2 \rho / \left[1 - p^2 (1 - \rho) \right] \tag{5}$$

In a system consisting of bifunctional A——A units and f-functional units R—A$_f$, where A may condense with A

$$\alpha = p \rho / \left[1 - p (1 - \rho) \right] \tag{6}$$

If the branch unit is other than trifunctional, e.g., if it is tetrafunctional, the same equations for the calculation of α can be employed; r and ρ have been so defined as to preserve these equations independent of the functionality of the branching unit. This scheme is not completely general, however. For example, two multifunctional units, one bearing A and the other B groups, may be present. Or, the multifunctional unit may possess both A and B groups. Other variations are possible also. In general, an α can be calculated from the proportions of reactants and the extent of reaction by a procedure resembling that given above but adapted to the particular type of reaction involved.

A somewhat more difficult case is encountered when the functional groups of the polyfunctional monomer differ in reactivity. Glycerol is such an example, the secondary hydroxyl being less reactive than either of the two primary groups. If the difference in reactivity were known from kinetic studies, extents of reaction p' and p'', for the respective types of hydroxyl groups, conceivably could be calculated from the average extent of reaction determined by analysis. Calculation of α from p' and p'' might then be carried out. A more serious problem would arise if the reactivity of one hydroxyl group were altered by the condensation of one of its neighbors, owing to their close proximity in a molecule like glycerol. Nevertheless, if the magnitude of such an effect can be established, calculation of α is possible. The point to be made is that the branching probability α is calculable, in principle at least, for virtually any polyfunctional system from suitable analytical measurements of the extent of reaction supplemented by appropriate information on the reaction kinetics. An excessive amount of information may be required in complex cases, however.

The critical value of α at which the formation of an infinite network becomes possible can be deduced as follows: If the branching unit is trifunctional, as in Fig. 61, each chain which terminates in a branch unit is succeeded by two more chains. If both of these terminate in branch units, four more chains are reproduced, and so on. If $\alpha < 1/2$, there is less than an even chance that each chain will lead to a branch unit and thus to two more chains; there is a greater than even chance that it will end at an unreacted functional group. Under these circumstances the network cannot possibly continue indefinitely. Eventually termination of chains must outweigh continuation of the network through branching. Consequently, when $\alpha < 1/2$ all molecular structures must be limited, i.e., finite, in size.

When $\alpha > 1/2$, each chain has better than an even chance of reproducing two new chains. Two such chains will on the average reproduce 4α new chains, and so on; n chains can be expected to lead to $2n\alpha$ new chains, which is greater than n when $\alpha > 1/2$. The *expected** number of chains in each succeeding generation of chains is greater than the number of chains in the preceding generation. Under these circumstances, branching of successive chains may continue the structure indefinitely. Unlimited structures, or what we have called infinite networks, are then possible. Hence $\alpha = 1/2$ represents the critical condition for incipient formation of infinite networks in the trifunctionally branched system depicted in Fig. 61.

* The expected number is the average number which would be observed from many trials made under equivalent circumstances.

It is important to note, however, that beyond $\alpha = 1/2$ by no means all of the material will be combined into infinite molecules. For example, in spite of the favorable probability of branching, a chain selected at random may be terminated at both ends by unreacted functional groups. Or it may possess a branch at only one end, and both of the succeeding two chains may lead to unreacted "dead ends." These and other finite species will coexist with infinite networks as long as $1/2 < \alpha < 1$. The relative amounts of sol and gel portions will be discussed later.

The preceding treatment can be generalized to include other cases, such as that in which only one type of functional group $(A = B)$ is present, and these groups are capable of condensing with one another. Polymerizations in which the branched unit is of higher functionality than three, or in which more than one type of branched unit is present, also may be included. A general statement of the critical condition for formation of infinite networks is the following: Infinite network formation becomes possible when the expected number of chains (or elements) which will succeed n chains (or elements), through branching of some of them, exceeds n. That is, if f is the functionality of the branching unit (i.e., the unit having a functionality greater than two), gelation will occur when $\alpha(f-1)$ exceeds unity. The critical value of α is, therefore,

$$\alpha_c = 1/(f - 1) \tag{7}$$

If more than one type of branching unit is present, $(f-1)$ must be replaced by the appropriate average, weighted according to the numbers of functional groups attached to the various branched units and the molar amount of each present. The critical condition can be expressed in various ways; Eq. (7) is a particularly convenient form for application to condensation polymers.

A close analogy exists between these three-dimensional polymerizations and gas-phase chain reactions which may undergo branching. If the probability of termination of the kinetic chain $(1 - \alpha)$ in the gas-phase reaction exceeds the probability of branching (α) with reproduction of two chain carriers, the chains are of finite length, and the rate of reaction attains a finite steady-state value. If the probability of branching exceeds the probability of termination, the reaction accelerates without limit and an explosion is observed. The formation of infinite networks, which as will be shown is concomitant with gelation, is analogous to explosion in chain reactions. A very slight alteration of temperature or pressure is sufficient to cause a gas-phase chain reaction accompanied by branching to accelerate from a moderate (or even negligible) rate to explosion. Similarly, a very small change, brought about by one means or another, in the total number of intermolecular linkages may convert a nonlinear polymer from a moderately viscous liquid to a gelled material having infinite viscosity.

Reference to the infinite structures as networks would seem inconsistent with our assumption, introduced as an approximation, that no intramolecular reactions occur. A randomly branched structure devoid of intramolecular linkages could hardly be called a network; the latter term conveys the notion of circuitous interconnections within the structure. Actually, as will appear later, the assumption referred to need only be applied to the finite molecular species; its extension to the infinite structure is superfluous. Certainly it will contain an abundance of intramolecular connections, which, in fact, is an essential feature of the gel structure.

1b. Experimentally Observed Gel Points in Polyfunctional Condensations.—If gelation in polyfunctional polymerizations is a manifestation of the formation of an infinite network as has been postulated above, the observed gel point should coincide with the point at which α reaches its critical value as specified by Eq. (7). In the reactions between glycerol and equivalent amounts of various dibasic acids, Kienle and Petke[3] observed that gelation occurred at 76.5 ± 1 percent esterification; i.e., at $p = 0.765$. If the difference in reactivities of the primary and secondary hydroxyl groups in glycerol is neglected, the branching coefficient α should be given by $\alpha = p^2$ according to Eq. (5) with $\rho = 1$. The observed critical extent of reaction corresponds, therefore, to $\alpha_c = 0.58$, compared with the calculated $\alpha_c = 0.50$. Correction for the lower reactivity of the secondary hydroxyl group would lower the observed α_c somewhat but not enough to eliminate the discrepancy.

The gel points in reactions of diethylene glycol with succinic or adipic acid and varying proportions of the tribasic acid, tricarballylic, have been investigated by the author.[1] The extents of reaction (p), determined by titration and plotted against time as in Fig. 62, were accurately extrapolated to the gel point, which was precisely observed by the sudden loss in fluidity as demonstrated by the failure of bubbles to rise in the mixture; α was calculated from the proportions of ingredients $(r$ and $\rho)$ and from the extent of esterification of the carboxyl groups, p_A, using Eq. (2). The number average degree of polymerization $\bar{x}_n$, calculated from the relationship

$$\bar{x}_n = \frac{\text{Number of units}}{\text{Number of molecules}} = \frac{f(1 - \rho + 1/r) + 2\rho}{f(1 - \rho + 1/r - 2p_A) + 2\rho} \tag{8}$$

is not very large nor is it increasing rapidly at the gel point. This

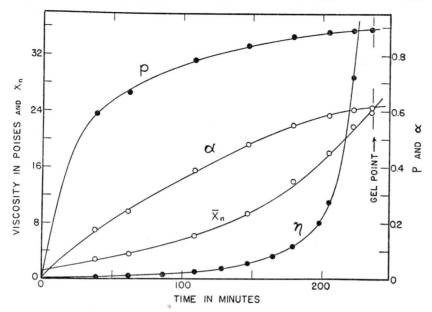

Fig. 62.—The course of a typical three-dimensional polyesterification.[1] The results shown are those for the third experiment reported in Table XXXI.

merely means that at the gel point many molecules are still present; it does not preclude the formation of a fractional amount of indefinitely large structures beyond the gel point.

TABLE XXXI.—GEL POINTS FOR POLYMERS CONTAINING TRICARBALLYLIC ACID[1,2]

Additional ingredients, diethylene glycol and	$r =$ COOH/OH	ρ[a]	p at gel point		$\alpha_{obs.}$ at gel point
			Obs.	Calcd.[b]	
Adipic acid	1.000	0.293	0.911	0.879	0.59
Succinic acid	1.000	.194	.939	.916	.59
Succinic acid	1.002	.404	.894	.843	.62
Adipic acid	0.800	.375	.9907	.955	.58

[a] $\rho = 3 \times$ [tricarballylic acid]/[total carboxyl groups].
[b] Calculated from Eq. (2) when $\alpha = 1/2$.

The results of the experiment depicted in Fig. 62 and others similarly obtained are summarized in Table XXXI. In every case the observed gel point is reached at higher than the theoretical extent of reaction. The discrepancies between the observed and calculated α_c's appear to be due to the failure of the theory to take into account a minor degree of intramolecular condensation. Since some of the interunit linkages are wasted in forming these intramolecular connections, the reaction must be carried somewhat farther to reach the critical point.

The most convincing confirmation of the gelation theory as applied to condensation polymerizations has been obtained by Stockmayer and Weil,[4] who investigated gelation in the condensation of pentaerythritol ($f=4$) with adipic acid. Here $\alpha=p^2$ and α_c (calc'd) $=1/3$, or $p_c=0.577$, compared with p_c (obs.) $=0.63$. Since intramolecular condensation should increase with dilution, they reasoned that in a hypothetical reaction conducted at infinite concentration intramolecular condensation would be suppressed entirely. They therefore measured p_c in the presence of varying amounts of diluent and extrapolated the results plotted against $1/c$ to deduce the intercept at $1/c=0$. The value of p_c thus found was $0.578 \pm .005$, in remarkable agreement with the calculated value. This result justifies the conclusion that disregard of intramolecular condensation is responsible for the discrepancies to which attention has been directed.

The identification of the gel point with the stage in the polymerization at which infinite networks make their appearance is confirmed by the results cited, and the extension of the assumption of random reaction to polyfunctional systems appears to be warranted.

1c. Cross-Linking of Polymer Chains.—Formation of chemical bonds between linear polymer molecules, commonly referred to as cross-linking, also may lead to the formation of infinite networks. Vulcanization of rubber is the most prominent example of a process of this sort. Through the action of sulfur, accelerators, and other ingredients present in the vulcanization recipe, sulfide cross-linkages are created by a mechanism not fully understood (see Chap. XI). Vulcanized rubbers, being typical network structures, are insoluble in all solvents which do not disrupt the chemical structure, and they do not undergo appreciable plastic, or viscous, flow.

To mention other instances in which gels have been prepared by cross-linking linear polymers, Signer and Tavel[5] converted methyl cellulose to a gel through the action of oxalyl chloride on the residual hydroxyl groups in the cellulose derivative. Jullander and Blom-Sallin[6] introduced cross-linkages into cellulose nitrate through the action of the tetrachlorides of titanium or silicon, which join hydroxyl groups intermolecularly as follows:

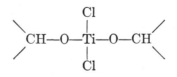

Hydroxyl containing polymers may be cross-linked with diisocyanates. Fordyce and Ferry[7] cross-linked styrene–maleic anhydride copolymers through the action of glycols. The copolymerization of divinyl with vinyl monomers may be looked upon as a method of cross-linking chain polymers. The cross-linkages are introduced simultaneously with the growth of the linear polymer chains, rather than afterwards, but this difference is secondary.

Let us consider a linear polymer consisting of molecules each of which possesses exactly y-units. We suppose that a certain number $\nu/2$ of cross-linkages are introduced which connect pairs of chain units at random. (The polymer may be a copolymer in which only one of the types of units is susceptible to the cross-linking process. Fulfillment of the condition of random cross-linking merely requires that the susceptible units be distributed at random over the molecules, so that established incidence of cross-linking with one unit of a chain does not affect the expectancy of cross-linking among its neighbors or elsewhere in the chain.) It is assumed, as an apparently unavoidable approximation, that cross-linking is exclusively intermolecular among finite species, each cross-linkage decreasing the number of molecules by one. The cross-linking density ρ will be defined as the fraction of the units which are cross-linked. There being two cross-linked units for each cross-linkage

$$\rho = \nu/N_0 \tag{9}$$

where N_0 is the total number of units contained in the polymers. The quantity y will be referred to as the *primary degree of polymerization* and yM_0, where M_0 is the (mean) molecular weight per unit, as the *primary molecular weight*. These terms are convenient, for it will be necessary to distinguish the actual molecular weight (of whatever average) of the cross-linked polymer from that which prevailed prior to the introduction of cross-linkages, or that which would result if all cross-linkages were severed.

Suppose a primary molecule is selected at random from the partially cross-linked mixture. Let this primary molecule be the one designated A in the structure shown schematically in Fig. 63. Two of its units happen to be cross-linked, and these two cross-linkages lead to the two primary molecules B_1 and B_2, which constitute the "generation" immediately succeeding A. (The allusion must not be overdrawn; progeny are indistinguishable from progenitors in the situation under discussion.) The former of these bears two cross-linkages, the latter none; and so forth to a complete description of the molecule indicated. We wish again to establish the conditions under which a primary molecule selected at random might possibly belong to an infinite network. The answer lies in the expected number of cross-linked units in any one of the progeny of the primary molecule selected initially. Thus, in close analogy with the argument used in the discussion of polyfunctional condensations, the question must be asked: In passing via the cross-linkage from one primary molecule to another (e.g., A to B_1, or from B_1 to C_2), what is the expected number of cross-linked units to be found among all the other $y-1$ units of the latter? Or, what is the

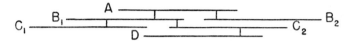

Fig. 63.—Diagramatic representation of a structure formed by cross-linking linear polymer molecules of uniform length.

expected number of additional cross-linked units to be found in the as yet unexplored chain B_1? If the cross-linking process is random, each of these expectancies ϵ will be the same, and equal to the probability ρ that any one unit is cross-linked multiplied by the total number $(y-1)$ of units in question; i.e.

$$\epsilon = \rho(y - 1) \tag{10}$$

Clearly, if $\epsilon < 1$, each "generation" will tend to be smaller than the one preceding, and indefinite continuation of the structure without limit is impossible. Only if $\epsilon > 1$ may indefinite continuation sometimes occur. Hence the critical condition for incipient formation of infinite structures is $\epsilon_c = 1$, or [8]

$$\rho_c = 1/(y - 1) \cong 1/y \tag{11}$$

If the primary molecules are large, the proportion of cross-linkages required for gelation becomes remarkably small; a ratio of one cross-linkage for two primary molecules is sufficient for onset of formation of infinite structures (gel) according to Eq. (11).

The same pattern may be followed in treating the cross-linking of primary molecules of any arbitrary distribution of sizes.[9] Referring again to Fig. 63, our question as to the number of additional cross-linked units to be found in a primary molecule reached by passage via a cross-linkage from a preceding primary molecule rests on two statistically governed contingencies: the number of units in this primary molecule and the proportion of them which happen to be cross-linked. Suppose the cross-linkage in question is the one leading from A to B_1. All that is known about B_1 is that one (at least) of its units is cross-linked, namely, the one attached to the cross-linkage. The probability that this unit is part of a primary molecule composed of y units is, of course, equal to the fraction of all units occurring in y-mers. This is the weight fraction w_y of y-mers. The expected number of additional cross-linked units in the y-mer is $\rho(y-1)$. It follows that the expected number of additional cross-linked units in the primary molecule (B_1) of which our cross-linked unit is a part is given by the sum of products $\rho w_y(y-1)$ over all values of y. Thus

$$\epsilon = \rho \sum_{y=1}^{\infty} w_y(y-1) = \rho(\bar{y}_w - 1) \tag{12}$$

where $\bar{y}_w$ is the weight average degree of polymerization of the primary molecules. The critical value of ϵ is again unity, giving[9]

$$\rho_c = 1/(\bar{y}_w - 1) \cong 1/\bar{y}_w \tag{13}$$

The relation (13) should hold regardless of the primary molecular weight distribution, provided only that the cross-linking proceeds between units at random. The number of cross-linked units per primary molecule, which has been called the *cross-linking index*,[7] equals $\rho \bar{y}_n$. For a homogeneous primary polymer $\bar{y}_w = \bar{y}_n = y$, and the critical value of the cross-linking index is unity. For a heterogeneous primary polymer $\bar{y}_w > \bar{y}_n$; hence the critical value of this index falls below unity; fewer than one cross-linked unit per molecule is required. For primary molecules having the most probable distribution $\bar{y}_w = 2\bar{y}_n$; hence the critical cross-linking index is one-half; a ratio of one *cross-linkage* to four molecules is sufficient for the onset of gelation. Thus, comparing polymers having the same $\bar{y}_n$, i.e., the same number of primary molecules per unit weight, fewer cross-linkages are required to produce gelation the broader the molecular weight distribution. This is understandable because the larger molecules included in the more heterogeneous polymer will acquire proportionately more intermolecular connections.

A pair of cross-linked chain units may be regarded as a single tetrafunctional unit according to the scheme applied to polyfunctional condensation. In order to demonstrate the correspondence between the two treatments, consider the condensation of tetrafunctional units

$$A\!-\!\!\underset{\displaystyle A}{\overset{\displaystyle A}{|}}\!\!-\!A,$$ or RA$_4$, with bifunctional units A——A where A may condense

with A. Observe that the definition of ρ used in the preceding section, where ρ is taken to be the ratio of A's on tetrafunctional units to the total number of A's, and the present definition (Eq. 9), according to

which A——|——A should be regarded as a pair of cross-linked units, are
    A

identical in this case.   Considered as a condensation, the critical value
of α is $1/3$; hence according to Eq. (6) the critical p for a given ratio
ρ is

$$p_c = 1/(1 + 2\rho) \tag{14}$$

If we regard it as a cross-linked polymer, we require $\bar{y}_w$, the weight
average degree of polymerization if all cross-linkages were severed.
This operation would amount to replacing each tetrafunctional unit,
or pair of cross-linked units, with two bifunctional units.   The extent
of reaction p would not be affected.   Hence, according to Eq. (VIII–8)

$$\bar{y}_w = (1 + p)/(1 - p)$$

At the gel point, $(\bar{y}_w - 1) = 1/\rho$, which with the foregoing expression
gives Eq. (14), thus establishing equivalence of the two procedures.
The primary molecules in a condensation polymer must almost invariably conform to a most probable distribution (see Chap. VIII).   The
random cross-linking of primary molecules otherwise distributed in size
has no counterpart in polyfunctional condensation, therefore.

The correspondence between condensation of A————A, B————B, and
RA$_4$ molecules and random cross-linking of primary molecules having
the most probable distribution can be demonstrated also.   In this case,
however, it is desirable to define $\bar{y}_w$ in terms of the *repeating* unit
A————AB————B, instead of the structural units, in order to preserve
the correspondence of the ρ's used in each treatment.

To extend the analogy in another way, the cross-linking treatment
given in this section may be adapted readily to interunit connections
of other *even* functionalities.   If, for example, three units are connected
by a cross linkage of higher order (to yield what could be regarded as
a hexafunctional unit), the only modification required is the substitu-
tion of the critical condition $\epsilon_c = 1/2$; or, in general $\epsilon_c = 2/(f-2)$.   The
analog of a cross-linking process which produces polyfunctional units
of *odd* functionality is a little hard to imagine; the specification of
primary molecules is even more awkward, but not impossible.

Copolymers prepared from a vinyl monomer and a small proportion
of a divinyl monomer are closely related to cross-linked polymers.
They will be considered in the final section of the present chapter,
where the application of the critical conditions set forth above will be
discussed in greater detail.

1d. General Conditions for the Formation of Infinite Networks.[1,9.2]—
Infinite structures are possible only in systems possessing structural

units some of which are capable of joining with more than two other units. This is by no means a sufficient condition, however; the multichain polymers discussed in Chapter VIII possess units of higher functionality, but they are incapable of forming infinite structures (without the assistance of reactions not considered in the formal scheme). This leads to a second, more far-reaching assertion, namely, that complex molecules must somehow be generated having the capacity to combine with one another; this capacity must increase with the size and complexity of the polymer species. It is totally lacking in the multichain polymers; in linear condensation polymers, each molecule regardless of size possesses just two terminal groups capable of combining with other molecules. In ordinary polyfunctional condensations, on the other hand, the number of functional groups potentially capable of reacting with another molecule increases with each coalescence of two molecules into a single, more complex structure. In a cross-linking process, the larger the molecule becomes, the greater its expectancy for participation in further cross-linking by virtue of the increased number of units it contains.

The two conditions stated above do not assure the occurrence of gelation. The final and sufficient condition may be expressed in several ways not unrelated to one another. First, let structural *elements* be defined in an appropriate manner. These elements may consist of primary molecules or of chains as defined above; or they may consist of the structural units themselves. The necessary and sufficient condition for infinite network formation may then be stated as follows: The expected number of elements united to a given element selected at random must exceed two. Stated alternatively in a manner which recalls the method used in deriving the critical conditions expressed by Eqs. (7) and (11), the expected number of additional connections for an element known to be joined to a previously established sequence of elements must exceed unity. However the condition is stated, the issue is decided by the frequency of occurrence and functionality of branching units (i.e., units which are joined to more than two other units) in the system, on the one hand, as against terminal chain units (joined to only one unit), on the other.

2. MOLECULAR DISTRIBUTIONS IN POLYFUNCTIONAL CONDENSATIONS

2a. Special Case of Random Branching without Network Formation.—Consider a monomer of the type A-$<^{B}_{B}$ where A may condense

with B, but reactions between like functional groups are forbidden. The polymers will have structures as indicated in Fig. 64.   Each x-meric species will contain $x+1$ unreacted B groups and only one unreacted A, intramolecular condensation being neglected.   Analogous branched polymers would be obtained from monomers of high functionality of the type A—R—B$_{f-1}$.   The number of unreacted A's per molecule remains at one; the number of unreacted B's will be

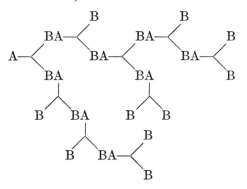

FIG. 64.—Randomly branched molecule.

$(f-2)x+1$.   As a further modification, bifunctional units of the type A——B may be copolymerized with the A—R—B$_{f-1}$ units without altering the essential character of the molecular plan.   This has the effect of merely inserting sequences of A——B units, varying in length, between the branch units.   Inclusion of A——A units, or of A——A and B——B units, would at once introduce the possibility of infinite structure formation.

The branching probability α equals p_B, the fraction of B groups reacted.   Since $p_B = p/(f-1)$ where p is written for p_A, the fraction of the A groups which have reacted

$$\alpha = p/(f - 1) \qquad (15)$$

The total number of molecules is $N_0(1-p)$ where N_0 is the total number of units.   Hence

$$\bar{x}_n = 1/(1 - p) = 1/[1 - \alpha(f - 1)] \qquad (16)$$

The critical value of α being $1/(f-1)$ according to Eq. (7), it is apparent from Eq. (15) that α may not reach α_c, since the extent of reaction p may approach but never reach unity.   These polymers comply with the first two conditions mentioned at the close of the preceding section, i.e., they may acquire a multitude of polyfunctional units and the number of reactive groups increases with each condensation of a pair of molecules, but the third condition cannot be met. Their inability to meet this condition is linked with the limitation of the number of unreacted A groups to one per molecule.

Examples of branched polymers of the type under consideration are few. Polymers prepared by Friedel-Crafts condensation of benzyl halides[10] doubtless are properly described as such

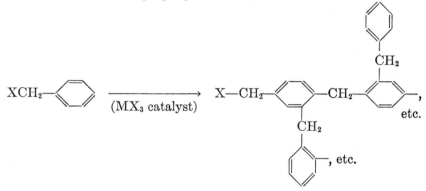

where X, corresponding to A above, may be chlorine or fluorine. The polymers are noncrystalline (indicative of irregular structure), and they remain soluble and fusible provided reaction conditions are sufficiently mild to avoid excessive rearrangement in the presence of the Friedel-Crafts catalyst. The little-investigated polymers formed by the rapid elimination of metal halide from alkali metal salts of trihalophenols[11]

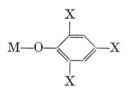

appear also to be of the type represented in Fig. 64.

Pacsu and Mora[12] have shown more recently that D-glucose undergoes intermolecular etherification in the presence of dilute acids, giving a soluble polyglucose. Every condensation occurs between a 1-hydroxyl group of one unit and a 2-, 3-, 4-, or 6-hydroxyl of the other. The 6 position may be favored, under some experimental conditions at least, thereby yielding a preponderance of 1,6 units along with a variety of trifunctional units, 1,2,6 or 1,4,6, etc., and some of higher functionality also. Every molecule should retain one unreacted 1-hydroxyl in conformity with the foregoing scheme.

Starch and glycogen are naturally occurring analogs. In the amylopectin fraction of starch, for example, most of the units are bifunctional 1,4-α-anhydroglucose

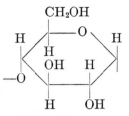

but about one unit in 15 to 20 consists of a trifunctional 1,4,6-α-glucose branched unit.

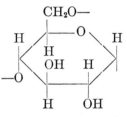

The structure of amylopectin according to Meyer[13] is shown in Fig. 65. The single reducing end group is indicated by A; all other terminal units are attached at the 1-position only.

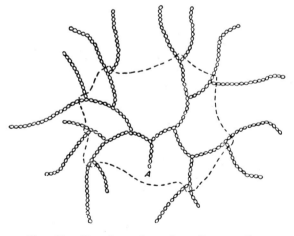

FIG. 65.—Structure of amylopectin according to Meyer.[13]

Branched polymers derived from monomers of the A—B$_{f-1}$ type are discussed here despite the dearth of prominent examples primarily because they occupy a unique intermediate position between linear polymers, on the one hand, and network-forming polyfunctional types, on the other. Furthermore, the molecular weight distribution relations for these branched polymers are easily extended to the more common polyfunctional types.

2b. Molecular Size Distribution Obtained by Condensing A—R—B$_{f-1}$ Monomers.[14]—For convenience in enumeration only, the $f-1$ B groups of a unit are considered to be distinguishable from one another, although identical in reactivity. Whether or not an actual basis for differentiation is provided by asymmetry of the monomer is immaterial; this assumption is introduced merely as an artifice to simplify enumeration of configurations and has no effect on the final results. Thus the simple dimeric structures

will be regarded for the present as different, the possibility that B^1 and B^2 may be physically indistinguishable notwithstanding. Any given molecular structure may be specified by stipulating which of the B groups of each successive unit have reacted, starting with the one bearing the unreacted A. The probability that an unreacted A group subjoins an x-mer of the structure thus specified equals the probability that the particular sequence of $x-1$ B groups have reacted while the remaining $fx-2x+1$ have not. This specific probability

$$\alpha^{x-1}(1 - \alpha)^{fx-2x+1}$$

is the same for each x-meric configuration. Hence the probability that any given unreacted A group is attached to an x-mer molecule of any structural configuration whatever is

$$N_x = \omega_x \alpha^{x-1}(1 - \alpha)^{fx-2x+1} \tag{17}$$

where ω_x is the total number of configurations. Since, according to the assumption that intramolecular condensation may be neglected, each molecule bears a single unreacted A group, N_x is also the mole fraction of x-mer.

In order to evaluate ω_x, it will be assumed tentatively that the individual monomer molecules are distinguishable from one another. Since the B's within a monomer must be considered distinguishable (in order to validate the procedure adopted above for enumerating configurations), then every B is distinguishable from every other in the system. The x monomers from which an x-mer is to be constructed having been chosen, an arbitrary set of $x-1$ B groups are selected for reaction out of the total of $(f-1)x$ B's. The total number of sets which may be selected is simply the number of combinations of $fx-x$ things taken $x-1$ at a time, or

$$(fx - x)!/(fx - 2x + 1)!(x - 1)!$$

Identical polymer configurations cannot be constructed using different sets of B's, but many configurations may be had from each set by combining them with the A's in different ways. The number of different ways in which the $x-1$ B's may be paired with $x-1$ A's, avoiding always a combination between a B and the A of the same molecule, is $(x-1)!$. On multiplying the above expression by this quantity, we obtain for the total number of arrangements, under the assumption that the individual monomers are distinguishable

$$(fx - x)!/(fx - 2x + 1)!$$

This physically unrealistic assumption may be abolished by dividing by $x!$, the number of permutations of the monomer units for a given configuration as originally defined. Hence

$$\omega_x = (fx - x)!/(fx - 2x + 1)!x! \tag{18}$$

Equation (17) may for convenience be written

$$N_x = [(1 - \alpha)/\alpha]\omega_x\beta^x \tag{19}$$

where

$$\beta = \alpha(1 - \alpha)^{f-2} \tag{20}$$

It remains to evaluate the summations required for the calculation of the weight fraction w_x of x-mer and the weight average degree of polymerization $\bar{x}_w$. These are of the type

$$S_m = \sum^{\infty} (fx - x)!x^m\beta^x/(fx - 2x + 1)!x! \tag{21}*$$

or

$$S_m = [\alpha/(1 - \alpha)]\sum_1^{\infty} x^m N_x \tag{21'}$$

* The radius of convergence of S_m for any finite m is $\beta_c = (f-2)^{f-2}/(f-1)^{f-1}$, which corresponds to $\alpha = 1/(f-2) = \alpha_c$. We have pointed out that in the systems under consideration $\alpha < \alpha_c$; hence $\beta < \beta_c$ and the sums are always convergent. At $\beta = \beta_c$ only S_0 would converge.

Differentiating the expression for S_0 and S_1 as given by Eq. (21)

$$S_1 = \beta(dS_0/d\beta) = \beta(dS_0/d\alpha)(d\alpha/d\beta) \tag{22}$$

$$S_2 = \beta(dS_1/d\beta) = \beta(dS_1/d\alpha)(d\alpha/d\beta) \tag{23}$$

Since $\sum_{1}^{\infty} N_x = 1$, according to Eq. (21')

$$S_0 = \alpha/(1 - \alpha) \tag{24}$$

Substitution of $d\alpha/d\beta$ from Eq. (20) and $dS_0/d\alpha$ from (24) in Eq. (22) yields

$$S_1 = \alpha/(1 - \alpha)[1 - \alpha(f - 1)] \tag{25}$$

Similarly, from Eq. (23)

$$S_2 = [\alpha/(1 - \alpha)][1 - \alpha^2(f - 1)]/[1 - \alpha(f - 1)]^3 \tag{26}$$

With the aid of these results, the following are readily obtained. For the weight fraction distribution:

$$w_x = x N_x / \sum x N_x = [\alpha/(1 - \alpha)] x N_x / S_1$$
$$= [(1 - \alpha)/\alpha][1 - \alpha(f - 1)] x \omega_x \beta^x \tag{27}$$

Substituting Eq. (25) for S_1 in the expression for the number average degree of polymerization

$$\bar{x}_n = \sum x N_x = [(1 - \alpha)/\alpha] S_1$$

gives Eq. (16), which was previously derived directly.   Similarly

$$\bar{x}_w = \sum x^2 N_x / \sum x N_x = S_2/S_1$$
$$= [1 - \alpha^2(f - 1)]/[1 - \alpha(f - 1)]^2 \tag{28}$$

which could not have been deduced from stoichiometric considerations alone.   Finally

$$\bar{x}_w/\bar{x}_n = [1 - \alpha^2(f - 1)]/[1 - \alpha(f - 1)] \tag{29}$$

It will be observed that both of the distribution equations, (19) and (27) for the mole and weight fraction distributions, respectively, contain factors ω_x and β^x.   Since α is limited to values less than $\alpha_c = 1/(f - 1)$, β is always much less than unity (for $f = 3$ the maximum value of β is $\beta_c = 1/4$), and the factors ω_x and β^x change in opposite directions as x increases.   The decrease of the latter outweighs the increase of the former for all permissible values of β ($\beta < \beta_c$; see p. 366, footnote); hence both N_x and w_x decrease monotonically with x, the latter less rapidly than the former.   The weight fraction distribution

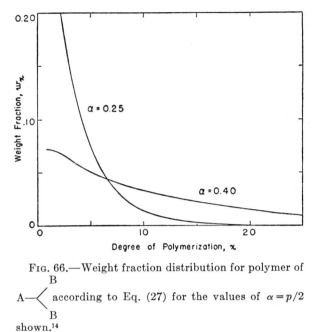

FIG. 66.—Weight fraction distribution for polymer of

A—⟨ $\begin{matrix}B\\\\B\end{matrix}$ according to Eq. (27) for the values of $\alpha = p/2$

shown.[14]

according to Eq. (27) is plotted in Fig. 66 for several extents of reaction α when $f = 3$. Broadening of the distribution with increase in α is evident. As α increases to its limit, one-half for $f = 3$, the distribution curve merges with the x-axis; according to Eq. (27) all weight fractions are zero at $\alpha = \alpha_c = 1/2$. But the total area under the distribution curve must remain equal to unity! It follows that at $\alpha = \alpha_c$ the distribution is infinitely wide and infinitesimally high.

The broadening of the distribution as the reaction proceeds toward completion is shown also by the expressions for the number and weight average degrees of polymerization. Both go to infinity as $\alpha \rightarrow \alpha_c$, of course, but $\bar{x}_w$ gets there faster—so much so that the ratio of $\bar{x}_w/\bar{x}_n$ goes to infinity also (see Eq. 29).

The preceding equations will, of course, be somewhat in error owing to the neglect of intramolecular condensations. Very large species will be suppressed relatively more on this account. All conceivable errors can do no more, however, than to effect a distortion of the quantitative features of the predictions, which will be small in comparison with the vast difference between the branched polymer distribution and that usually prevailing in linear polymers. From this point of view, the statistical theory given offers a useful description of the state of affairs.

If an A——B co-monomer is present, the structural pattern is as

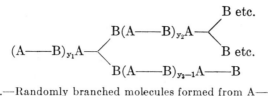

FIG. 67.—Randomly branched molecules formed from A——B and

$$A \underset{B}{\overset{B}{\big<}} \text{ monomers.}$$

shown in Fig. 67, which, incidentally, resembles that of the naturally occurring polymers amylopectin and glycogen.[13]   The weight fraction of species comprised of n f-functional and l bifunctional units, derived by an extension of the above treatment to this case, is given by the following expression:[14]

$$w_{n,l} = [(1 - p_B)^2/p_B - (f - 2)(1 - p_B)\rho](n + l)\zeta^n\eta^l\omega_{n,l} \quad (30)$$

where ρ is the fraction of A's belonging to branch units (i.e., ρ is the mole fraction of branch units), and

$$\zeta = p_B\rho(1 - p_B)^{f-2}$$
$$\eta = (1 - \rho)p_B$$
$$\omega_{n,l} = (l + fn - n)!/l!n!(fn - 2n + 1)!$$

Eq. (30) and its accessory expressions are too unwieldy to be of much use.   This is inherent in the presence of two parameters, n and l. For many purposes the distribution according to n only, which may be obtained, for example, by summing Eq. (30) over l, is more informative.   We then obtain the weight fraction of all species containing n branch units irrespective of the value of l.   This will be referred to as the *complexity distribution*.   For larger $n(>2$ or $3)$ it parallels the actual size distribution in $x = n + l$, and hence may be used as a substitute for the latter.   This follows because there are in an "n-mer" $(f-1)n+1$ chains, each of which competes statistically for its share of all of the bifunctional units in the system.   A molecule of large complexity n therefore is almost certainly one of high molecular weight.

It is possible to show[14] that the complexity distribution retains all of the principal characteristics of the distribution for the condensation of A—R—B$_{f-1}$ alone (Fig. 66).   Indeed, these are already present in Eq. (30).   The major difference has to do with the presence of linear molecules (A——B)$_l$ in the copolymers containing bifunctional units. If the proportion of these units is large, the mole fraction of linear molecules remains correspondingly large under all conditions.   (For example, for $f=3$, and $\rho \ll 1$, their mole fraction falls only to one-half in the limit of complete condensation.) Their weight fraction, however, becomes vanishingly small as the condensation proceeds to completion.

The somewhat analogous complexity distribution for systems containing R—A$_f$ branch units is discussed in the Appendixes to this chapter.

2c. The Distribution in Ordinary Polyfunctional Condensations.[14,15,16]—The simplest case of a polyfunctional condensation is represented by the self-condensation of an f-functional monomer, R—A$_f$, giving polymer structures as indicated in Fig. 68 for $f=3$. The branching coefficient $\alpha = p_A = p$ may acquire any value from zero to unity; it is not restricted to values less than α_c as in the preceding case. Herein lies the basis for the sharp contrast in the behavior of the two systems, as will be evident from the discussion which follows. Similar circumstances prevail in condensations between equivalent quantities

of A—$\big\langle \begin{smallmatrix} A \\ A \end{smallmatrix}$ and B—$\big\langle \begin{smallmatrix} B \\ B \end{smallmatrix}$, or of B——B and A—$\big\langle \begin{smallmatrix} A \\ A \end{smallmatrix}$. In the latter case

$\alpha = p^2$; if non-equivalents are used $p_A \neq p_B$, with the result that $\alpha = p_A p_B$. The following treatment applies with but minor modifications (see Appendix A) to these and other variants of polyfunctional condensation.

Returning to consider a simple f-functional condensation such as that represented in Fig. 68, we note in passing that the total number of molecules is

$$N = N_0(1 - \alpha f/2) \tag{31}$$

and that

$$\bar{x}_n = 1/(1 - \alpha f/2) \tag{32}$$

neglecting intramolecular condensation. This approximation deprives these relations of all meaning beyond the point where the number of interunit linkages equals the number of units, whereupon all would be combined into a single molecule if the condensation were exclusively intramolecular at this stage. This point would be reached at $\alpha = 2/f$, which lies beyond $\alpha_c = 1/(f-1)$. Fortunately, there will be no need to apply Eqs. (31) and (32) beyond $\alpha = \alpha_c$; hence the approximation will be no more serious here than in the problems considered previously.

Exploration of the molecular size distribution is conveniently carried out by selecting an unreacted A group at random, and then proceeding in a manner which parallels the procedure applied above

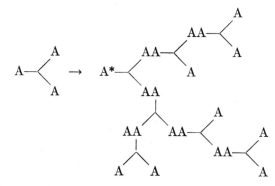

FIG. 68.—Polymer from a simple trifunctional monomer.

to polymers of A—R—B_{f-1}.   Let the A group marked with an asterisk in Fig. 68 be selected for example.   The probability that $x-1$ additional identical units are attached according to a specific pattern to the unit bearing this A group is

$$\alpha^{x-1}(1 - \alpha)^{(f-2)x+1}$$

there being $x-1$ linkages and $(f-2)x+1$ *additional* unreacted groups, precisely as in the preceding problem.   Unlike the branched polymer of A—R—B_{f-1} units (or B—R—A_{f-1} units), some of the specific configurations laid out in this fashion about the unit bearing the chosen unreacted A group will differ from each other only in the location of this group.   The correspondence to the preceding situation holds *only* after the unreacted A group has been selected.   Each of the specific configurations built about the unit bearing this unreacted group, once it has been selected, corresponds exactly to one, and to only one, of the configurations for the x-meric polymer of the A—R—B_{f-1} monomer. Hence the total number of specific configurations here involved is again equal to ω_x as given by Eq. (18).   Consequently, the probability P_x *that the unreacted A group selected at random is part of an x-meric structure of any configuration whatever* must be given by

$$P_x = \omega_x \alpha^{x-1}(1 - \alpha)^{fx-2x+1}$$

which looks exactly like Eq. (17) for the mole fraction N_x in the previous problem.   However, P_x is *not* the mole fraction of x-mer here. To obtain it we write, in accordance with the definition of P_x given in italics above,

$$P_x = \frac{\text{Number of unreacted A's on } x\text{-mers}}{\text{Total unreacted A's}}$$

The number of unreacted A's in an x-mer molecule is $(f-2)x+2$; hence

$$P_x = [(f-2)x+2]N_x/N_0f(1-\alpha)$$

where N_x is the *number* of x-mer molecules, and N_0 is the total number of units. Eliminating P_x from these two expressions and substituting from Eq. (18) for ω_x

$$N_x = N_0[(1-\alpha)^2/\alpha]\omega'_x\beta^x \tag{33}$$

where

$$\omega'_x = f(fx-x)!/(fx-2x+2)!x! \tag{34}$$

and β is defined again by Eq. (20).

The mole fraction, obtained by dividing Eq. (33) by the total number of molecules as given by Eq. (31), is [15]

$$N_x = [(1-\alpha)^2/\alpha(1-\alpha f/2)]\omega'_x\beta^x \tag{35}$$

and the weight fraction[15] by multiplying N_x by x/N_0

$$w_x = [(1-\alpha)^2/\alpha]x\omega'_x\beta^x \tag{36}$$

As a consequence of having used Eq. (31) to obtain Eq. (35) for the mole fraction, this expression should not be applied beyond the gel point. The same limitation does not apply to the number and the weight fraction formulas, (33) and (36), which should in fact be just as valid at high α's (corresponding to high extents of reaction) as elsewhere. The necessity for abandoning Eq. (35) in this range is of no concern inasmuch as its proper substitute at high α's is easily found (see Sec. 2d).

The weight average degree of polymerization $\bar{x}_w$ may be derived as follows, where it is assumed that $\alpha \leqq \alpha_c$. We begin by defining

$$S'_m = \sum_{x=1}^{\infty} x^m\omega'_x\beta^x \tag{37}$$

From Eq. (35), since $\sum N_x = 1$

$$S'_0 = \alpha(1-\alpha f/2)/(1-\alpha)^2 \tag{38}$$

and from Eq. (36), since $\sum w_x = 1$

$$S'_1 = \alpha/(1-\alpha)^2 \tag{39}$$

Applying the procedure previously used to obtain S_2 (see Eq. 26), we find

$$S'_2 = \alpha(1+\alpha)/(1-\alpha)^2(1-\alpha f+\alpha) \tag{40}$$

Recalling the derivation of Eq. (28), we obtain at once

$$\bar{x}_w = S_2'/S_1' = (1 + \alpha)/[1 - \alpha(f - 1)] \tag{41}$$

which, incidentally, reduces to Eq. (VIII–8) if $f=2$. From Eq. (32)

$$\bar{x}_w/\bar{x}_n = (1 + \alpha)(1 - \alpha f/2)/[1 - \alpha(f - 1)] \tag{42}$$

At the critical point $\alpha_c = (f-1)^{-1}$, $\bar{x}_n$ is finite, which signifies merely that the number of molecules is by no means zero; the weight average, however, is infinite at α_c; and, of course, the ratio of $\bar{x}_w$ to $\bar{x}_n$ is infinite, indicating extreme heterogeneity.

The weight fraction of x-mer calculated according to Eq. (36) for a trifunctional monomer is shown in Fig. 69 for several degrees of reaction. The curves resemble those for the polymers of an A—R—B$_2$ monomer shown in Fig. 66, with the important difference that they do not vanish into the axis at $\alpha = \alpha_c$. This is simply a consequence of the fact that condensation is far short of completion at the critical point.

Although the theory is developed above for a polyfunctional condensation of the utmost simplicity, exploration of variants of this case indi-

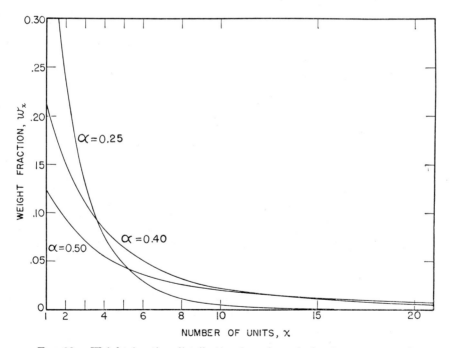

FIG. 69.—Weight fraction distribution for a branched polymer prepared from a simple trifunctional monomer at the α's indicated.[2]

cates only minor changes in the essential features. Usually the principal alteration occurs in the relation of α to reaction parameters, these being the proportion ρ of branching units of functionality f, the ratio r of A to B groups, and the extent of reaction p.

If the polymer contains a preponderance of bifunctional units A——A (or of both A——A and B——B) in addition to the branch units R—A$_f$, the general characteristics of the distribution—its trend toward infinite breadth in particular—are retained.* The size of a given molecule depends then not only on the number n of branch units but also on the number l of bifunctional units it contains. (This case is discussed in Appendix A to this chapter.) Just as in the condensation of A—R—B$_{f-1}$ with A——B, it is expedient to adopt the simpler complexity distribution in n only. The molecular constitution of products of condensation of R—A$_f$ with bifunctional units may be described jointly by the complexity distribution, and by the average number of bifunctional units per chain. As the condensation reactions proceed, the branching probability α increases, and species of higher complexity become more prominent in the distribution. Simultaneously, the average length of the chains increases, which also contributes to the advancement of the size distribution.

2d. Post-gelation Relationships.[2,16]—If we undertake to sum either Eq. (33) for the number of x-mers or Eq. (36) for their weight fraction over all species $x = 1$ to ∞, a superficially anomalous result appears, the importance of which warrants careful attention. The latter sum, for example, may be written

$$\sum_1^\infty w_x = \frac{(1-\alpha)^2}{\alpha} \sum_1^\infty \frac{f(fx-x)!\beta^x}{(x-1)!(fx-2x+2)!} \tag{43}$$

The summation on the right, which is identical with S_1' as defined by Eq. (37), depends only on the value of β. But β according to Eq. (20) increases with α to a maximum occurring at $\alpha = \alpha_c$, then decreases as α increases further. Hence for each permissible value of β there are two roots α in the physically accessible range $0 < \alpha < 1$ (except of course at the maximum, where the roots are identical). When $f = 3$, for example, $\beta = \alpha(1-\alpha)$ reaches its maximum value, $1/4$, at $\alpha_c = 1/2$; for each value of β the roots are α and $\alpha' = 1-\alpha$. Clearly, the summation occurring on the right side of Eq. (43), since it depends only on β,

---

* The complete distribution functions for the case of N_{A_1}, N_{A_2}, $\cdots$ N_{A_i} moles of reactants of functionalities f_1, f_2, $\cdots$ f_i in A groups which are allowed to react with N_{B_1}, N_{B_2}, $\cdots$ N_{B_j} moles of reactants of functionalities g_1, g_2, $\cdots$ g_j in B groups, where condensation occurs between A and B groups, have been derived by Stockmayer, *J. Polymer Sci.*, **9**, 69 (1952).

must have the same value for either root, α or α'. The factor $(1-\alpha)^2/\alpha$ does *not*, however, equal $(1-\alpha')^2/\alpha'$. Consequently, $\sum_1^\infty w_x$ cannot retain the same value (unity) beyond the gel point as it possessed prior to gelation. We are forced to face the disturbing conclusion that Eq. (36) (and Eq. 33 likewise) must fail to comply with the standard requirement $\sum_1^\infty w_x = 1$ for a proper weight fraction when $\alpha > \alpha_c$. On the other hand, no condition has been introduced in deriving Eq. (36) which should limit it to $\alpha < \alpha_c$.*

The key to the resolution of the apparent contradiction becomes evident upon re-examining the initial derivation which proceeds from Fig. 68. Finite, or bounded, molecular species are implied in the expression for the probability of a specific x-mer configuration; thus $fx - 2x + 1$ unreacted ends in addition to the one selected at random are prescribed. An infinite network, on the other hand, is terminated only partially by unreacted end groups; the walls of the macroscopic container place the ultimate limitation on its extent. Hence the network fraction is implicitly excluded from consideration, with the result that the distribution functions given above are oblivious of it. Failure of $\sum_1^\infty w_x$ to retain the same value throughout the range in α is a consequence of omission of the weight fraction of infinite network.

Re-examination shows that $\sum_1^\infty w_x$, where w_x is given by Eq. (36), does indeed equal unity when $\alpha \leq \alpha_c$. It follows that S_1' is correctly given by Eq. (39) provided $\alpha \leq \alpha_c$; the same applies to S_0' and S_2' as given by Eqs. (38) and (40), respectively. Mathematically speaking, there should be appended to Eqs. (38), (39), and (40) the stipulation that α is to be taken as the lowest real root of Eq. (20) for the given value of β.

Suppose that the condensation has been carried beyond the gel point so that $\alpha > \alpha_c$. Then the sum of the weight fractions of everything other than the network is, according to Eq. (36)

$$\sum_{\text{all finite species}} w_x = [(1 - \alpha)^2/\alpha]S_1'$$

In the light of the preceding discussion this summation represents the weight fraction w_s of sol. According to Eq. (39), $S_1' = \alpha'/(1-\alpha')^2$ where α' designates the lowest root of Eq. (20). Hence

$$w_s = (1 - \alpha)^2\alpha'/(1 - \alpha')^2\alpha \qquad (44)$$

* The validity of the weight fractions given by Eq. (36) for the lowest species throughout the range $\alpha = 0$ to 1 may be reaffirmed directly. Thus, the probability that any arbitrarily selected unit exists as an unreacted monomer is obviously $w_1 = (1-p)^f = (1-\alpha)^f$; similarly $w_2 = f\alpha(1-\alpha)^{2f-2}$ may be verified directly, and so forth.

The weight fraction of gel obtained by difference is

$$w_g = 1 - (1 - \alpha)^2 \alpha' / (1 - \alpha')^2 \alpha \qquad (45)$$

With somewhat greater generality we may let α represent the actual branching probability determined by the extent of condensation irrespective of whether or not the critical point has been exceeded, while α' is the lowest root of Eq. (20) for the same value of β. If $\alpha \leqq \alpha_c$, then $\alpha = \alpha'$ and according to Eqs. (44) and (45) $w_s = 1$ and $w_g = 0$. If $\alpha > \alpha_c$, then $\alpha > \alpha'$ and $w_s < 1$. In the trifunctional case, for example, $\alpha' = 1 - \alpha$ when $\alpha > \alpha_c$, giving

$$w_s = (1 - \alpha)^3 / \alpha^3$$

The weight fraction of gel calculated for this case from Eq. (45) is plotted against α in Fig. 70. It will be observed that the formation of the infinite network is indicated to commence suddenly at the critical point. This prediction of the theory is abundantly confirmed by the characteristic abruptness with which gel appears in polyfunctional condensations. Also shown in Fig. 70 are the weight fractions w_x of various species as functions of α, calculated according to Eq. (36).

Let w_x' represent the weight fraction of the sol which consists of x-mer; i.e., $w_x' = w_x / w_s$. From Eqs. (36) and (44)

$$w_x' = [(1 - \alpha')^2 / \alpha'] x \omega_x' \beta^x \qquad (46)$$

which is identical with Eq. (36) except for the replacement of α with the corresponding lower root α'. In a simple f-functional condensation, therefore, the molecular distribution within the sol fraction at $\alpha > \alpha_c$ is identical with that which prevailed for the entire polymer when the branching probability was equal to α'. Thus, the curves of Fig. 69 for $\alpha = 0.25$ and 0.40 apply also to the *sol fraction* at $\alpha = 0.75$ and at 0.60, respectively.

The change in molecular constitution over the entire course of an f-functional condensation may be described on the basis of these considerations. As condensation progresses, larger species are formed gradually at the expense of smaller ones; however, monomer is always more abundant than any other finite species. The distribution broadens, as shown in Fig. 69, maximum heterogeneity being reached at the gel point, but even here the proportion of highly branched molecules is small. At the gel point the generation of the infinite network commences abruptly, and the proportion of gel increases rapidly with further condensation. The sol which remains decreases in average molecular weight owing to the preferential conversion of the larger, more complex species to gel. The molecular size distribution within

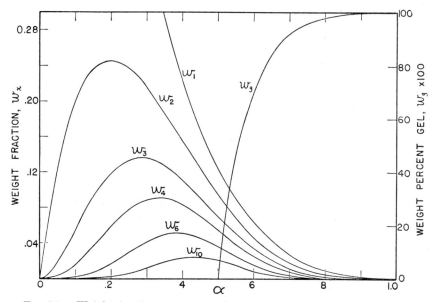

FIG. 70.—Weight fractions of various finite species and of gel in a simple trifunctional condensation as a function of α, which in this case equals the extent of reaction p. Curves have been calculated from Eqs. (36) and (45).[2]

the sol, and the averages ($\bar{x}_n$ and $\bar{x}_w$) characterizing it, undergo retroversion over the exact course followed up to the gel point.

It now becomes clear that the approximation of neglecting intramolecular condensations need only be applied to finite species. It need not apply to the gel, since the theory does not deal directly with gel, its amount being obtained only by difference. As already emphasized, the gel must surely possess an abundance of intramolecular connections. These are an essential part of its network character. Errors arising from the aforementioned approximation will be greatest among the more complex (finite) species. Since these are present in maximum abundance at the gel point, the error should be greatest here and should diminish in *either direction* from the gel point. The error may be imagined to introduce a distortion of the abscissa scale in Fig. 70, this distortion being greatest at $\alpha = \alpha_c$.

That the distinction between sol and gel is by no means an arbitrary one is shown by the character of the distribution as depicted in Figs. 69 and 70. The distribution curves in the former figure always are asymptotically zero at large x, even at the gel point. Extremely large (i.e., "almost infinite") molecules, which might be regarded as intermediate between sol and gel, never account for more than an extremely small fraction of the total. The structural distinction between sol and gel is as well defined as the difference in physical state of a liquid and its vapor.   The sol and gel do not, however, represent physically separate phases, as this otherwise close analogy might suggest; the former occurs interspersed throughout the latter.

Other polyfunctional condensations which proceed to gelation and beyond are subject to similar interpretation (see Appendix B).   Differences occur only in quantitative details.

3. MOLECULAR DISTRIBUTIONS IN CROSS-LINKED SYSTEMS

3a. Molecular Size Distributions.—General relationships for calculating the distribution obtained by cross-linking primary molecules of any arbitrary distribution have been derived by Stockmayer.[9] Tractable expressions are obtained, of course, only for certain specified primary distributions.   Two special cases will be discussed briefly.

If the units of primary molecules of *uniform length* are subjected to cross-linking at random, the weight fraction of polymer molecules comprised of z primary molecules is[8,9]

$$w_z = (z^{z-1}/\gamma z!)(\beta/e)^z \tag{47}$$

where γ is the cross-linking index, or number of cross-linked units per primary molecule in the system as a whole,* and

$$\beta = \gamma e^{1-\gamma} \tag{48}$$

This β, like the previous one, possesses in the physically significant region $(\gamma \geqq 0)$ a single maximum. This occurs at $\gamma = 1$, which is the gel point (see Eq. 11).   The summation $\sum_1^\infty w_z = 1$ for $\gamma \leqq 1$.   For $\gamma > 1$, the weight fraction of sol is

$$w_s = \sum_1^\infty w_z = \gamma'/\gamma \tag{49}$$

where γ' is the lower root of Eq. (48.)   This result is readily apparent from Eq. (47) in the light of the previous derivation of the weight fraction of sol for polyfunctional condensations.   Combining Eqs. (48) and (49), we obtain the relation[17]

$$-\ln w_s/(1 - w_s) = \gamma \tag{50}$$

The sol fractions occurring in a series of cross-linked "butyl rubber" copolymers[17] consisting of a small proportion of isoprene with isobutylene units have been found to vary with γ in accordance with Eq. (50). The unfortunately large experimental inaccuracy precluded a precise test, however.

* $\gamma = \rho y$ where y is the number of units per primary molecule.

The other case which we consider is that of a most probable primary distribution. The molecular size distribution after random cross-linking must correspond exactly to that which would be obtained by random condensation of a mixture of bifunctional and tetrafunctional units. This follows as an extension of the correspondence between these two cases considered in the discussion of the critical condition given in the preceding section. The equations developed there are applicable to this case.

3b. The Partitioning of Components between Sol and Gel.[18]—Having established as a principle of broad generality the sharp demarcation between sol and gel, we may consider the probability ϕ_s that a *non-cross-linked* polymer unit selected at random belongs to the sol fraction in a randomly cross-linked system of any arbitrary primary molecular weight distribution. An arbitrarily selected *cross-linked* unit will belong to the sol fraction only if *both* the unit itself and its cross-linked partner would belong to the sol in the absence of the cross-linkage in question. Hence the probability that a cross-linked unit is part of the sol must equal ϕ_s^2. The probability that a unit selected at random is cross-linked and a part of the sol fraction is $\rho\phi_s^2$; the probability that it is not cross-linked and a part of the sol is $(1-\rho)\phi_s$. Hence the weight fraction of sol is

$$w_s = (1 - \rho)\phi_s + \rho\phi_s^2$$

or

$$w_s/\phi_s = 1 - \rho(1 - \phi_s) \qquad (51)$$

If the degree of cross-linking ρ is small (and the primary chain length therefore great), $\phi_s \cong w_s$, giving

$$w_s/\phi_s \cong 1 - \rho(1 - w_s) \qquad (51')$$

the utility of which will appear below.

The probability that a primary molecule composed of y units possesses i cross-linked units is

$$P_y(i) = [y!/(y - i)!i!]\rho^i(1 - \rho)^{y-i}$$

(assuming that each unit may be cross-linked only once), and the probability that none of the partners of these cross-linked units would belong to an infinite network in the absence of these i cross-linkages is ϕ_s^i. Hence the probability s_y that a y-mer primary molecule selected at random is a part of the sol fraction is given by

$$s_y = \sum_{i=0}^{y} P_y(i)\phi_s^i$$

which on substitution of the expression given above for $P_y(i)$ reduces to

$$s_y = [1 - \rho(1 - \phi_s)]^y = (w_s/\phi_s)^y \tag{52}$$

It follows that

$$w_s = \sum_{y=1}^{\infty} w_y s_y$$

$$= \sum_{1}^{\infty} w_y(w_s/\phi_s) \tag{53}$$

where w_y is the weight fraction of primary y-mer molecules. Given the primary molecular weight distribution (the w_y) and ρ, Eqs. (53) and (51) may in principle be solved for the weight fraction of sol. Or, if ρ is small, the approximate Eq. (51') may be substituted in Eq. (53) to give

$$w_s = \sum_{y=1}^{\infty} w_y[1 - \rho(1 - w_s)]^y \tag{53'}$$

which, though not explicitly soluble, may be solved by substituting trial values of w_s in the right-hand side and evaluating the summation graphically. In any event, there is here provided a relatively simple procedure for computing the weight fraction of sol remaining at a degree of cross-linking ρ for *any* primary distribution.

Bardwell and Winkler[19] applied the converse of this procedure to the determination of the degree of cross-linking of butadiene-styrene copolymer latices by persulfate ions. They first calculated the primary molecular weight distribution from the consumption of the mercaptan "modifier" (chain transfer agent) during the polymerization. Assuming random incidence of transfer, each increment of polymer formed should possess a most probable (primary) distribution. The value of $\bar{y}_w$ for the increment was computed from the amount of modifier disappearing in the interval, one polymer molecule being created for each molecule of mercaptan consumed. Summation over all increments gave the cumulative distribution, i.e., the w_y. These were inserted in Eq. (53) to establish by trial the value of w_s/ϕ_s which would yield the experimentally observed weight fraction of sol w_s in the persulfate-treated polymer; ρ was then calculated from Eq. (51). The values of ρ thus deduced for various concentrations of persulfate and for various times of reaction were consistent with a simple kinetic interpretation of the cross-linking reaction.

The relative abundances of cross-linkages and of various primary species in the sol and in the gel fractions may be obtained by extension of this procedure.[18]   Since ϕ_s^2 equals the fraction of the cross-linkages (or of the cross-linked units) occurring in the sol fraction, we can write

$$\phi_s^2 = \rho' w_s / \rho$$

where ρ' represents the cross-linking density in the sol.   This equality may be employed to define ρ', i.e.,

$$\rho' = \rho \phi_s^2 / w_s$$

Substituting for $(\phi_s/w_s)^2$ from Eq. (51')

$$\rho' \cong \rho w_s / [1 - \rho(1 - w_s)]^2$$
$$\cong \rho w_s [1 + 2\rho(1 - w_s)] \tag{54}$$

or to a further approximation

$$\rho' \cong \rho w_s \tag{54'}$$

Similarly, the fraction ρ'' of the units in the gel which are cross-linked is

$$\rho'' = \rho(1 - \phi_s^2)/(1 - w_s) \tag{55}$$

which, on introducing approximations corresponding to those used above, reduces to

$$\rho'' \cong \rho(1 + w_s - 2\rho w_s^2) \tag{55'}$$

or

$$\rho'' \cong \rho(1 + w_s) \tag{55''}$$

These equations show that the densities of cross-linkages in the sol and in the gel vary linearly with w_s, in first approximation.   At incipient gelation $\rho'' = 2\rho' = 2\rho$; the cross-linking density of the first increment of gel is twice that of the sol, regardless of the primary molecular weight distribution.   This method was successfully used by Bardwell and Winkler to calculate the degree of cross-linking ρ'' in gels obtained in the investigation to which reference has been made.[19]

The weight fraction of the sol consisting of y-mer primary molecules is

$$w_y' = w_y s_y / w_s = w_y (w_s/\phi_s)^y / w_s \tag{56}$$

The number and weight average values of y in the sol fraction are, respectively,[18]

$$\bar{y}'_n = 1/\sum (w'_y/y) = w_s/\sum [(w_y/y)(w_s/\phi_s)^y] \qquad (57)$$

and

$$\bar{y}'_w = \sum yw'_y = (1/w_s)\sum [yw_y(w_s/\phi_s)^y] \qquad (58)$$

The previous expressions for w_s/ϕ_s may be used in these equations.

The total number of molecules in the sol fraction, neglecting intramolecular cross-linking, will equal the number of primary molecules minus the number of cross-linkages in the sol. Expressing these as numbers of moles per equivalent of structural units, we have $N' = 1/\bar{y}'_n$ primary molecules and $\rho'/2$ cross-linkages in the sol. The number average degree of polymerization in the sol is then

$$\bar{x}'_n = 1/(N' - \rho'/2)$$
$$\bar{x}'_n = w_s/\left\{\sum [(w_y/y)(w_s/\phi_s)^y] - \rho\phi_s^2/2\right\} \qquad (59)$$

Its weight average degree of polymerization may be shown[9] to be

$$\bar{x}'_w = \bar{y}'_w(1 + \rho')/[1 - \rho'(\bar{y}'_w - 1)] \qquad (60)$$

The results of the application of these relationships to the cross-linking of a rectangular primary distribution consisting of equal parts by weight of every species $y = 1$ to 1000 (i.e., $w_y = 0.001$ for $y = 1$ to 1000 and $w_y = 0$ for $y > 1000$) are shown in Figs. 71 and 72. The de-

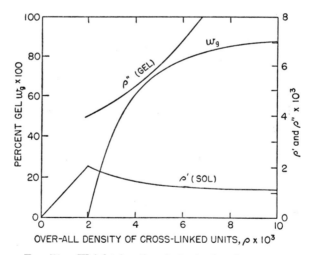

Fig. 71.—Weight fraction (w_g) of gel and concentrations of cross-linked units in the sol (ρ') and in the gel (ρ'') vs. the over-all degree of cross-linking (ρ) for the rectangular primary distribution described in the text.[16]

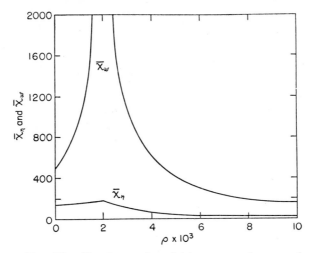

Fig. 72.—Number and weight average degrees of polymerization prior to gelation and for the sol after gelation vs. the over-all degree of cross-linking (ρ) for the rectangular distribution.[18]

crease of ρ' and the increase in ρ'' as the over-all degree of cross-linking ρ advances is completely general.   In Fig. 72 $\bar{x}_n$ is observed to increase only slightly up to the gel point (at $\rho = 0.002$), whereas the weight average degree of polymerization reaches infinity at the critical point. Beyond the gel point $\bar{x}'_w$, the weight average degree of polymerization of the sol, returns to finite values, and both $\bar{x}'_n$ and $\bar{x}'_w$ decrease with further cross-linking.   This again is a consequence of the preferential acquisition of larger molecules (and larger primary molecules as well) by the gel.

The properties of nonlinear polymer molecules have intrigued many investigators, who all too often have undertaken to prepare polymers of nonlinear structure for their experiments by the simple expedient of incorporating a small proportion of a polyfunctional unit or cross-linking agent; e.g., a small amount of divinyl monomer added to its monovinyl analog often has seemed attractive.   The fallacy of this procedure is manifest in the foregoing analysis.   If enough poly-functional unit, or cross-linking agent, is used to introduce a really significant departure from linearity, gelation intervenes.   The sol extracted from the gel offers no advantage, for it has suffered deple-tion of cross-linkages by the gel.   Separation of the small amount of highly branched soluble polymers by fractionation is unattractive. Choice of some other procedure by which to introduce nonlinearity without incidence of gelation is clearly indicated for this purpose.

4. BRANCHING AND CROSS-LINKING IN VINYL ADDITION POLYMERS

4a. Branching by Chain Transfer.—We have pointed out in Chapter VI how, during the course of a vinyl polymerization propagated by free radicals, a growing chain may occasionally enter into a chain transfer reaction with a polymer molecule. The growing chain is terminated and the reactivated polymer molecule proceeds to grow a new chain attached to the unit involved in the chain transfer (see p. 257). The resulting branched polymer molecule may subsequently undergo activation at another of its units, and another chain is added to the molecule at this point. Repetition of this process a number of times may yield highly branched molecules having a structural pattern like that indicated below:

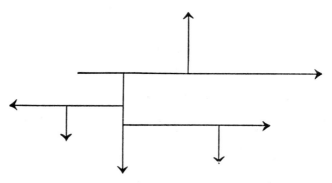

The direction of each chain's growth is indicated by an arrow, which also marks the point at which transfer terminates its growth. Instances in which branching by transfer may assume importance were cited in Chapter VI.

The structure shown above bears an obvious resemblance to that of the A—R—B$_{f-1}$-type condensation polymers shown in Fig. 64, or, rather, of a copolymer of A——B units with these polyfunctional units. It follows by extension of this analogy that branching by chain transfer cannot alone produce an infinite network.[20] The conclusion can be derived independently by computation of the branching probability factor α. Each chain transfer generates one branch and one chain end; hence α may never exceed the critical value, one-half, for a trifunctionally branched polymer. To look at the problem another way, chain transfer merely adds one more chain to the existing structure without providing any mechanism for joining together two complex molecules. Furthermore, while some of the molecules may become very large if chain transfer occurs extensively, the mechanism does not yield circuitous connections within the structure, which are essential for the formation of a network.   On the other hand, the complex structures generated by extensive chain transfer may offer favorable circumstances for the formation of a network through the assistance of independently occurring intermolecular reactions of one sort or another (e.g., cross-linking).*

The degree of branching by transfer with polymer obviously will increase with the conversion since the relative incidence of branching must depend on the ratio of polymer to monomer in the system.   To examine the matter from the point of view of reaction rates, let θ represent the fraction of monomer molecules which have polymerized out of a total of N_0 in the system, and let ν represent the total number of branches.   (At variance with the definition used elsewhere, N_0 is the total number of units polymerized and *unpolymerized as well.*)   The rates of generation of branches and of polymerization can then be written

$$d\nu/dt = k_{tr,P}[M \cdot]\theta N_0$$

$$d\theta/dt = k_p[M \cdot](1 - \theta)$$

giving

$$d\nu/d\theta = C_P N_0 \theta/(1 - \theta) \tag{61}$$

where $C_P = k_{tr,P}/k_p$ is the transfer constant (Chap. IV).   Integrating from 0 to θ, with $\nu = 0$ at $\theta = 0$, and introducing the branching density $\rho = \nu/N_0\theta$

$$\rho = - C_P[1 + (1/\theta) \ln(1 - \theta)] \tag{62}$$

The equivalent function of the degree of conversion is encountered in the cross-linking of diene polymers discussed below.   It is plotted in Fig. 73 in relation to the latter problem.   For present purposes it is necessary merely to replace the ordinate in Fig. 73 with ρ/C_P.   Regardless of the absolute magnitude of the branching transfer constant, the *relative* amount of branching must increase rapidly with conversion.

A molecule whose first primary chain was formed early in the polymerization will, on the average, acquire more branches than a molecule originated later; even among those originating in the same time inter-

* If chain transfer of the radical center to a previously formed polymer molecule is followed ultimately by termination through coupling with another similarly transferred center, the net result of these two processes is the combination of a pair of previously independent polymer molecules.   T. G. Fox (private communication of results as yet unpublished) has suggested this mechanism as one which may give rise to network structures in the polymerization of monovinyl compounds.   His preliminary analysis of kinetic data indicates that proliferous polymerization of methyl acrylate may be triggered by networks thus generated.

val, the final degree of branching will vary statistically. It is evident therefore that the molecular weight distribution will be broadened as a result of the wide variability in the branching expectancy in different portions of the population. Bamford and Tompa* have developed methods for quantitative treatment of the molecular distribution in cases such as this one, and their results reveal the expected broadening of the distribution brought about by chain transfer to the polymer. If the degree of branching is such as to amount to an average of several branches per molecule, the broadening is marked.

Trifunctional branches may be introduced also as a result of chain transfer with monomer (see p. 257). In the subsequent growth of a chain from the monomer activated by chain transfer, the vinyl group of this terminal monomer remains unimpaired; hence it is available for addition in a subsequent chain. When the latter event occurs, a trifunctional branch is formed. The highly branched molecules which could conceivably develop in this way resemble those formed by chain transfer with polymer. The last chain introduced into a molecule according to the present scheme corresponds to the first in the former; the order of development is reversed in the sense that branching sites are created in advance of chain growth. Details of the kinetic scheme differ, but similar conclusions hold. The average degree of branching increases with conversion, but formation of infinite networks is impossible (without the assistance of other processes). The molecular weight distribution is broadened. Chain transfer with a monomer whose vinyl group may subsequently polymerize is actually somewhat less efficient in approaching gelation conditions than is chain transfer with polymer; each chain transfer with a monomer molecule terminates the growth of a chain, but only those "transfer" monomers whose vinyl groups subsequently polymerize yield branches.

4b. Cross-Linking in the Polymerization of Dienes.—The unsaturated structural units into which conjugated dienes are converted on polymerization are potentially capable of copolymerizing with monomer in subsequent stages of the polymerization (see Chap. VI). A polymer radical may conceivably add either to the 1,4 or to the 1,2 unit as indicated below; the greater susceptibility to be expected for the latter may be compensated to some extent by the usually smaller proportion of 1,2 units.

* C. H. Bamford and H. Tompa, *J. Polymer Sci.*, **10**, 345 (1953), first derive the moments of the distribution for the case of chain transfer to polymer. They then obtain the molecular weight distribution from these moments by appropriate mathematical methods. Their procedure should be applicable to a wide variety of polymerization mechanisms.

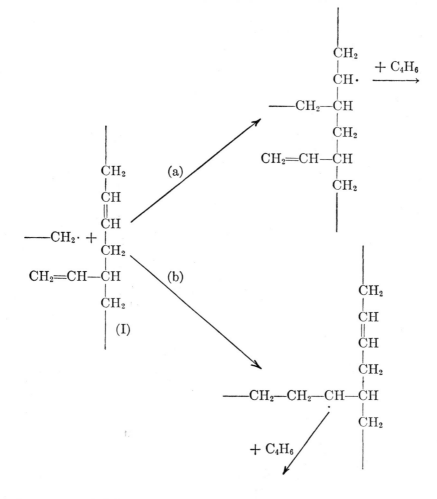

Occurrence of either step causes the polymer molecule I to be incorporated in the growing chain.   The unit of the polymer to which the radical adds becomes a tetrafunctional unit, which is equivalent to a pair of cross-linked units, i.e., to a cross-linkage.

The kinetic treatment corresponds to that given above for the case of chain transfer with polymer.   We have merely to write for the rate of generation of cross-linked units

$$d\nu/dt = 2k_{pP}[M\cdot]\theta N_0$$

where k_{pP} represents the effective mean rate constant for the addition of a polymer unit according to reactions (a) and (b) above, weighted according to the relative proportions of 1,4 and 1,2 units present. The factor 2 enters because two cross-linked units are involved in each cross-linkage. Then, in place of Eqs. (61) and (62)

$$d\nu/d\theta = (2k_{pP}/k_p)\theta N_0/(1 - \theta) \tag{63}$$

$$\rho = \nu/\theta N_0 = -(2k_{pP}/k_p)[1 + (1/\theta)\ln(1 - \theta)] \tag{64}$$

The latter equation, shown graphically in Fig. 73, expresses quantitatively the expected rapid increase in cross-linking with conversion. The rate constant ratio k_{pP}/k_p corresponds to the inverse of the reactivity ratio used in copolymerization theory (Chap. V).

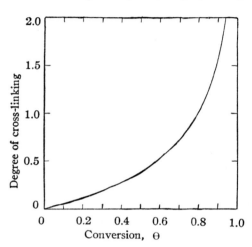

FIG. 73.—Degree of cross-linking expressed in arbitrary units $(\rho k_p/2k_{pP})$ vs. conversion as calculated from Eq. (64.)[20]

The commonly observed, and technically troublesome, formation of gel in the polymerization of dienes testifies to the occurrence of cross-linking processes during their polymerization. According to Eq. (13), gel should begin to form when ρ reaches the critical value $\rho_c = 1/\bar{y}_w$, where $\bar{y}_w$ is the weight average degree of polymerization of the primary molecules. Thus

$$1/\bar{y}_w = -(2k_{pP}/k_p)[1 + (1/\theta_c)\ln(1 - \theta_c)] \tag{65}$$

where θ_c is the extent of polymerization at incipient gelation. The definition of primary molecules as those which would exist if all cross-linkages were severed presents a minor difficulty here, since the method by which the severing process is to be executed is not altogether obvious. This is particularly true in the case of addition to the 1,4 unit according to process (a) above. All hazards of ambiguity are avoided, however, if the polymer unit entering the growing chain by either of the processes (a) or (b) is abstracted therefrom, and succeeding monomer units are united directly with one another. One then has the primary molecules which would have existed if all processes (a) and (b) had been prevented from participating in the polymerization. If $\bar{y}_w$ for the primary molecules so defined were known and if

k_{pP}/k_p could be established independently, the degree of conversion θ_c at which gel formation should commence could be computed from Eq. (65).   More important, the converse of this procedure may be used to obtain k_{pP}/k_p, which would indeed be difficult by any other method.   That is, given $\bar{y}_w$ (the evaluation of which will be discussed below) and θ_c at incipient gelation, k_{pP}/k_p can be calculated.

Before pursuing further this application of cross-linking theory, it should be recalled that in deriving Eq. (13), as elsewhere, the structural units are assumed to be cross-linked at random.   The present system does not conform exactly to this condition.   Polymer molecules formed at the very beginning of the polymerization will be quite linear, but they are exposed to attack by radicals of other growing chains over a large span of conversion and, hence, have the maximum opportunity to acquire cross-linking connections *subsequent* to their formation. Those formed late in the process will, on the average, contain more cross-linked units initially but will have little opportunity to acquire more.   The counterbalancing tendencies do not equalize; a simple analysis[20] of the situation shows that the cross-linking density is somewhat greater among primary molecules formed late in the polymerization.

Under these circumstances the probability that a given structural unit is cross-linked is not entirely independent of the status of other units in the same primary molecule.   If an abnormally large fraction of some of the units of a given primary molecule are found to be cross-linked, the likelihood that it was formed toward the end of the polymerization process is enhanced; hence the probability that one of its other units is cross-linked will be greater than the over-all ρ for the system.   Calculations indicate that the magnitude of the non-randomness is not excessive below about 70 percent conversion.   For most purposes its effect probably may be ignored without serious error, thus obviating a more elaborate theory which would take into account non-randomness of this nature.

Morton and Salatiello[21] have deduced the ratio k_{pP}/k_p for radical polymerization of butadiene by applying the above described procedure, appropriately modified for the emulsion system they used.   The primary molecular weight was controlled by a mercaptan acting as chain transfer agent, as in the experiments of Bardwell and Winkler[19] cited above.   Measurement of the mercaptan concentration over the course of the reaction provided the necessary information for calculating $\bar{y}_w$ at any stage of the process, and in particular at the critical conversion θ_c for the initial appearance of ¦gel.   The velocity constant ratios which they obtained[21] from their results through the use of Eq.

TABLE XXXII.—THE CROSS-LINKING RATE CONSTANT FOR POLYBUTADIENE[21]

$T°C$	$(k_{pP}/k_p) \times 10^4$	$k_{pP} \times 10^2$ in $(moles/l.)^{-1}sec.^{-1}$ [a]
40	1.02	0.42
50	1.36	.87
60	1.98	1.98

$$E_{pP} - E_p = 7.5 \pm 0.6 \text{ kcal. per mole.}$$
$$E_{pP} = 17 \pm 1 \text{ kcal. per mole.}$$

[a] Obtained from the independently established value of k_p. See Table XVII (p. 158) and Ref. 90, p. 177.

(65) are given in Table XXXII. Taking the value at 60°, for example, and considering a bulk polymerization (unlike emulsion polymerization where monomer is continually fed to the polymerizing particle) carried to 60 percent conversion, we find from Fig. 73 that $\rho = 2 \times 10^{-4}$, or one unit in about 5000 should be cross-linked. Even this very small proportion may induce the profound effects of gelation if the (primary) molecular weight is large. The above values of k_{pP} refer, of course, to the effective mean for the various units (1,2, and 1,4-*cis* and -*trans*) present in the polymer formed under the conditions of the experiment.

Chain transfer with previously formed polymer also may occur in diene polymerizations. Its extent has not been established, however. The consequences of cross-linking through copolymerization of previously polymerized diene units contrast sharply with the less sweeping effects of branching by transfer. It has become customary to differentiate these sources of nonlinearity in diene polymers by the terms *branching* and *cross-linking* and to point out that, whereas infinite network formation is a direct consequence of cross-linking, branching yields only finite molecules. Taken out of its context, this assertion may be dangerously misleading. The most obvious difference between the structures produced resides in the functionalities of the polyfunctional units; the one is trifunctional, the other tetrafunctional. Hence one is tempted to make the erroneous generalization that trifunctional units in any polymer system do not generate infinite networks, whereas tetrafunctional units may do so. The functionality of the polyfunctional unit is not the critical factor involved (apart from the necessity that it exceed two). Whether or not a space network of unlimited extent is generated depends on the circumstances controlling the generation of polyfunctional units in relation to chain ends. To avoid the confusion related above, reservation of *branching* as a generic term, which would include any type of nonlinearity, would be preferable. *Cross-linking* would then refer to a specific type of *branching*.

4c. Copolymerization of Divinyl with Monovinyl Monomers.—The now classical copolymerization of a small amount of divinylbenzene with styrene[22] offers the possibility of introducing cross-linkages in a manner superficially analogous to that involved in the polymerization of a conjugated diene, but more amenable to control. Here the ethylenic group remaining in the divinyl unit after one of its vinyl groups has polymerized is approximately as susceptible to polymerization as it was initially. A better system for quantitative study than the one mentioned consists of ethylene dimethacrylate-methyl methacrylate[23]

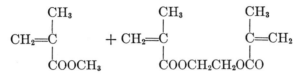

where the unsaturated groups of the direactive monomer may be assumed with confidence to be equal in reactivity with the identical group of the methyl methacrylate monomer; nor should the reactivity of one group in the ethylene dimethacrylate depend perceptibly on the status of its partner. (Divinylbenzene, by contrast, is known to be appreciably more reactive than styrene.) Each twice-reacted ethylene dimethacrylate molecule contributes one cross-linkage. If the fraction of methacrylate groups occurring in dimethacrylate molecules in the initial mixture is represented by ρ_0 and all of the groups in the system are assumed to be equally reactive, the number ν of cross-linked units will be $\theta^2 N_0 \rho_0$ where N_0 is the total number of methacrylate groups in the system. The density of cross-linked units in the polymer is

$$\rho = \nu/\theta N_0 = \theta \rho_0 \qquad (66)$$

At incipient gelation, therefore

$$\theta_c = 1/\rho_0 \bar{y}_w \qquad (67)$$

Gel points in the above system containing varying proportions of the dimethacrylate, and in vinyl acetate-divinyl adipate mixtures also, were observed by Walling,[23] who compared his results with an equation equivalent to, but less general than, Eq. (67). His results indicate approximate agreement with theory, provided the proportion of direactive monomer is low enough to make $\theta_c > 0.10$. For $\bar{y}_w = 5000$ to 10,000, the approximate range for Walling's experiments, the value of ρ_0 giving θ_c (calc'd) $= 0.10$ is 0.001 to 0.002. The values used by

Walling for $\bar{y}_w$ unfortunately are unreliable. Hence no precise test of the theory is afforded by his results.

When larger proportions of the direactive monomer were used, thereby reducing θ_c well below 0.10, unmistakably large deviations from theory were observed; the value of θ_c observed exceeded θ_c, calculated, by severalfold or more. The explanation for this deviation, as was pointed out by Walling, undoubtedly lies in the excessive occurrence of intramolecular cross-linking at the high dilutions of polymer with monomer prevailing up to the critical point. Reactions involving two polymer molecules, as required for cross-linking, are thereby suppressed to the relative advantage of intramolecular cross-linking. The wastage of cross-linkages in this manner, neglected by theory, may account for the discrepancy in a satisfactory manner.

5. SUMMARY

The molecular distributions occurring in nonlinear polymers assume a higher order of complexity than do linear polymer distributions. The formation of infinite networks whose presence is manifested by gelation usually, although not always, occurs in the course of polymerizations leading to nonlinear structures, or during cross-linking processes which convert linear polymers to nonlinear ones. The phenomenon of gelation dominates other aspects of nonlinear polymerizations.

Critical conditions for the formation of an infinite network are readily calculated statistically by considering the probability of continuation of the structure along any arbitrary path, or chain. If branched or cross-linked units are sufficiently numerous compared with chain ends to cause the proliferation of chains to dominate chain ending, an infinite structure will have been formed. A very small proportion of polyfunctional units ordinarily suffices to bring about gelation; the individual polymer molecules existing up to the gel point, or beyond for that matter, are not highly branched, therefore, according to the theoretical distributions. Comparison of experimentally observed gel points with those calculated reveals a minor discrepancy owing to neglect of intramolecular connections among finite molecular species. The error thus introduced into the molecular distribution theory is greatest at the gel point.

The condensation of A and B groups in a system consisting of an A—R—B_{f-1} monomer, and possibly also an A——B co-monomer, represents an intermediate case in which gelation is impossible according to the condition stated above. Other features of the molecular distribution resemble those of the more commonly encountered polyfunctional condensations which are unrestricted in their capacity to generate infinite networks. Specifically, the molecular size distribution is very broad, and many isomeric arrangements of the units are possible among species possessing a number of branch units.

The molecular distributions for polymers formed by condensations involving polyfunctional units of the type R—A_f resemble those for the branched polymers mentioned above, except for the important modification introduced by the incidence of gelation. The generation of an infinite network commences abruptly at the gel point, and the amount of this gel component increases progressively with further condensation. Meanwhile, the larger, more complex, species of the sol are selectively combined with the gel fraction, with the result that the sol fraction decreases in average molecular complexity as well as in amount. It is important to observe that the distinction between soluble finite species on the one hand and infinite network on the other invariably is sharp and by no means arbitrary.

A similar description applies to the cross-linking of linear polymers. The degree of cross-linking in the sol fraction reaches a maximum at the gel point, which occurs when an average of one unit is cross-linked out of the number of units occurring in a primary molecule having the weight average degree of polymerization. The partitioning of the primary molecules and of the cross-linkages between the sol and the gel may be calculated, provided the primary molecular weight distribution is known and the cross-linkages are distributed at random.

The cross-linking occurring during the polymerization of a monovinyl with a small proportion of a divinyl (or a diene) monomer may be treated from this point of view. An error, which usually is small, enters in this case owing to the fact that the cross-linkages thus formed are not distributed at random among the primary molecules generated at different stages of the polymerization.

APPENDIX A

Derivation of the Molecular Distribution for the Random Condensation of Bifunctional with Polyfunctional Monomers.[15]

—Let a bifunctional monomer A——A condense with a polyfunctional monomer R—A_f. If $f=3$, structures such as the following will form:

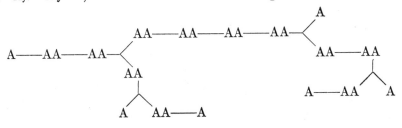

Let ρ represent the ratio of A groups belonging to branch units to the total number of A's, i.e.

$$\rho = fN_0/(fN_0 + 2L_0) \tag{A-1}$$

where N_0 and L_0 are the numbers of polyfunctional and bifunctional units, respectively. If p is the extent of reaction as used above, the probability that a given A has condensed with an A of a branch unit is $p\rho$, the probability that it has condensed with a bifunctional unit is $p(1-\rho)$, and the probability that it remains unreacted is $1-p$. Proceeding according to the derivation given in the text of this chapter for the distribution in a simple condensation of R—A_f units, we select an unreacted A group at random. The probability that it is joined to a specific configuration (as previously defined) comprising n polyfunctional and l bifunctional units is readily found to be

$$\rho(p\rho)^{n-1}[(1 - \rho)p]^l(1 - p)^{fn-2n+1}$$

To find the number of specific configurations for an n,l-mer, each of which may occur with a probability given by the above expression, we note that the *branching* pattern may assume arrangements which correspond exactly to those for the simple f-functional case. Thus, the factor ω_n, as given by Eq. (18) with n replacing x, will be involved. In addition, the l bifunctional units may be partitioned among the $(f-1)n+1$ chains of each branching pattern. Each chain may have zero or more of them. The number of these arrangements is given by the combinatory factor

$$(l + fn - n)!/l!(fn - n)!$$

Hence the probability that the arbitrarily selected unreacted A group belongs to an n,l-mer of any configuration whatever may be written

$$P_{n,l} = [(1 - p)/p]\zeta^n\eta^l\omega_{n,l} \tag{A-2}$$

where

$$\zeta = p\rho(1 - p)^{f-2} \tag{A-3}$$

$$\eta = (1 - \rho)p \tag{A-4}$$

$$\omega_{n,l} = (l + fn - n)!/l!n!(fn - 2n + 1)! \tag{A-5}$$

At extents of reaction which are not too large (i.e., up to the gel point; see text), the total number N of molecules may be taken as the number of units less the number of linkages formed, intramolecular condensation being neglected as previously. Thus

$$N = (N_0 + L_0) - (fN_0 + 2L_0)p/2$$
$$= N_0[(1 - f/2) + (f/2\rho)(1 - p)] \tag{A-6}$$

The total number of unreacted A groups in the system is $(fN_0+2L_0)(1-p)$, and the number of unreacted A groups attached to an n,l-mer is $fn-2n+2$; hence, recalling the derivation of Eq. (33)

$$P_{n,l} = N_{n,l}(fn - 2n + 2)/(fN_0 + 2L_0)(1 - p)$$
$$= \rho N_{n,l}(fn - 2n + 2)/fN_0(1 - p) \tag{A-7}$$

where $N_{n,l}$ is the number of n,l-mers. Substituting from Eq. (A-2) for $P_{n,l}$ and solving for $N_{n,l}$

$$N_{n,l} = N_0[(1 - p)^2/p\rho]\zeta^n\eta^l\omega'_{n,} \tag{A-8}$$

where

$$\omega'_{n,} = f(l + fn - n)!/l!n!(fn - 2n + 2)! \tag{A-9}$$

These equations were first derived by Stockmayer using another method.[15]

Utilizing Eq. (A-6) for the total number of molecules, we obtain for the mole fraction

$$N_{n,l} = \{(1 - p)^2/p\rho[1 - f/2 + (f/2\rho)(1 - p)]\}\zeta^n\eta^l\omega'_{n,l} \tag{A-10}$$

If the molecular weight of each species can be taken to be proportional to its total number of units, the weight fraction of n,l-mers is

$$w_{n,l} = (n + l)N_{n,l}/(N_0 + L_0)$$
$$= \{(1 - p)^2/p[\rho + (f/2)(1 - \rho)]\}\zeta^n\eta^l(n + l)\omega'_{n,l} \tag{A-11}$$

If $\rho=1$, then $l=0$, $\zeta=\beta$, $\alpha=p$, $x=n$, and Eq. (A-11) reduces to Eq. (36). If $\rho=0$, Eq. (A-11) reduces to Eq. (VIII–3) for the most probable linear distribution.

APPENDIX B

The Complexity Distribution.[2,8,15]—The number of molecules composed of n branch units irrespective of the number l of bifunctional units, which we have called the complexity distribution, may be obtained by summing over all l in Eq. (A-8). A more intuitive approach to the same result will be followed here. We consider the probability α that an A group of one branch leads via a sequence of zero or more A——A units to another branch.

$$\alpha = \sum_{i=0}^{\infty} [p(1 - \rho)]^i p\rho = p\rho/[1 - p(1 - \rho)] \tag{B-1}$$

The probability α^* that an *unreacted* A group selected at random leads to a branch unit is found in a like manner to be given by

$$\alpha^* = \alpha/p \tag{B-2}$$

From here on the deduction of the probability P_n that the arbitrarily selected unreacted A group is part of a molecule containing n branch units proceeds identically with the deduction of P_x for the simple f-functional distribution.   Hence

$$\left. \begin{array}{l} P_n = (\alpha^*/\alpha)(1 - \alpha)\beta^n\omega_n \qquad n > 0 \\[2mm] P_0 = 1 - \alpha^* \end{array} \right\} \tag{B-3}$$

where ω_n is defined by Eq. (18) and β by Eq. (20).   The number N_n of molecules containing n branch units may be found by repetition of earlier procedures as follows:

$$N_n = P_n(fN_0/\rho)(1 - p)/(fn - 2n + 2)$$

Then from Eqs. (B-1), (B-2), and (B-3)

$$N_n = N_0[(1 - \alpha)^2/\alpha]\beta^n\omega_n' \qquad n > 0 \tag{B-4}$$

where ω_n' is defined by Eq. (34).   Similarly

$$N_{n=0} = N_0[(1 - \alpha)(1 - \alpha^*)/\alpha^*](f/2) \tag{B-4'}$$

where the symbol $N_{n=0}$ representing the number of linear molecules is so written in order to avoid confusion with N_0, the number of f-functional units.   Expressions for the mole fractions may be obtained by dividing Eqs. (B-4) and (B-4') by the total number of molecules as given in Eq. (6).

For the purpose of deriving the weight fraction distribution, attention is directed to the fact that a molecule containing n f-functional branches is composed of $fn - n + 1$ chains.   The average size of a chain being independent of the location of the chain in a branched structure, the quantity $fn - n + 1$ may be taken as a measure of the average weight of an n-chain polymer.   It follows that

$$w_n = (fn - n + 1)N_n/(\text{Total number of chains})$$

The total number of chains, being by definition equal to half the sum of the unreacted functional groups and the functional groups attached to branch units, is

$$[(fN_0/\rho)(1 - p) + fN_0]/2$$

which reduces with the aid of Eqs. (B-1) and (B-2) to

$$fN_0[1 + (1 - \alpha)/\alpha^*]/2$$

Hence, for $n > 0$

$$w_n = \{(1 - \alpha)^2/[\alpha + (1 - \alpha)(\alpha/\alpha^*)]\}\beta^n 2\omega_n'(fn - n + 1)/f \quad \text{(B-5)}$$

Similarly, it is found that

$$w_{n=0} = (1 - \alpha^*)(1 - \alpha)/(1 - \alpha + \alpha^*) \quad \text{(B-5')}$$

This complexity distribution resembles the size distribution derived in the text of this chapter for the simple f-functional case. Their characteristics are so similar as to obviate a separate discussion here. The only difference of real significance arises from the presence in the copolymer of a linear component with $n = 0$, which has no counterpart in the simple f-functional case.

If the proportion of f-functional units is small ($\rho \ll 1$), the condensation presumably will be carried relatively close to completion in order to achieve a substantial average molecular weight. Then p will be near unity and it is permissible, consequently, to take $\alpha^* \cong \alpha$. Eqs. (B-5) and (B-5') simplify to

$$w_n \cong (1 - \alpha)^2[2(fn - n + 1)/f]\omega_n'\beta^n \quad n \geqq 0 \quad \text{(B-6)}$$

This expression differs from Eq. (36) for the weight fraction of x-mer in a simple f-functional condensation chiefly through the replacement of x in the latter by its analog $2(fn - n + 1)/f$ for the branched copolymer.

Summation of the weight fractions of all finite species according to Eq. (B-6) gives unity when $\alpha < \alpha_c$. When $\alpha > \alpha_c$, there is obtained for the weight fraction of sol

$$w_s = \sum_{\text{all finite species}} w_n = (1 - \alpha)^2/(1 - \alpha')^2 \quad \text{(B-7)}$$

where α' is the lowest root of Eq. (20) as discussed in the text of this chapter. The weight fraction of gel, therefore, is

$$w_g = 1 - (1 - \alpha)^2/(1 - \alpha')^2 \quad \text{(B-8)}$$

The complexity distribution of the sol undergoes retroversion beyond the gel point in the same manner as the molecular distribution in the simple f-functional condensation. Whereas the complexity distribution in the sol at α is the same as that for the polymer as a whole at α' (α and α' being roots of Eq. 20 for the same value of β, with $\alpha' < \alpha$), its average molecular weight is greater owing to the greater lengths, at higher degrees of condensation, of the linear chains reaching between the branch points of the structures.

REFERENCES

1. P. J. Flory, *J. Am. Chem. Soc.*, **63**, 3083 (1941).
2. P. J. Flory, *Chem. Revs.*, **39**, 137 (1946).
3. R. H. Kienle, P. A. van der Meulen, and F. E. Petke, *J. Am. Chem. Soc.*, **61**, 2258, 2268 (1939); R. H. Kienle and F. E. Petke, *ibid.*, **62**, 1053 (1940); **63**, 481 (1941).
4. W. H. Stockmayer and L. L. Weil, results quoted in *Advancing Fronts in Chemistry*, ed. by S. B. Twiss (Reinhold Publishing Corp., New York 1945), Chap. 6, by W. H. Stockmayer.
5. R. Signer and P. Tavel, *Helv. Chim. Acta*, **26**, 1972 (1943).
6. I. Jullander and B. Blom-Sallin, *J. Polymer Sci.*, **3**, 804 (1948).
7. D. B. Fordyce and J. D. Ferry, *J. Am. Chem. Soc.*, **73**, 62 (1951).
8. P. J. Flory, *J. Am. Chem. Soc.*, **63**, 3097 (1941).
9. W. H. Stockmayer, *J. Chem. Phys.*, **12**, 125 (1944).
10. R. A. Jacobson, *J. Am. Chem. Soc.*, **54**, 1513 (1932); S. Bezzi, *Gazz. chim. ital.*, **66**, 491 (1936).
11. W. H. Hunter and G. H. Woollett, *J. Am. Chem. Soc.*, **43**, 135 (1921).
12. E. Pacsu and P. T. Mora, *J. Am. Chem. Soc.*, **72**, 1045 (1950).
13. K. H. Meyer, *Natural and Synthetic Polymers*, 2d ed. (Interscience Publishers, New York-London, 1950), pp. 456 ff.
14. P. J. Flory, *J. Am. Chem. Soc.*, **74**, 2718 (1952).
15. W. H. Stockmayer, *J. Chem. Phys.*, **11**, 45 (1943).
16. P. J. Flory, *J. Am. Chem. Soc.*, **63**, 3091 (1941).
17. P. J. Flory, *Ind. Eng. Chem.*, **38**, 417 (1946).
18. P. J. Flory, *J. Am. Chem. Soc.*, **69**, 30 (1947).
19. J. Bardwell and C. A. Winkler, *Can. J. Research*, **B27**, 116, 128, 139 (1949).
20. P. J. Flory, *J. Am. Chem. Soc.*, **69**, 2893 (1947). See also *ibid.*, **59**, 241 (1937).
21. M. Morton and P. P. Salatiello, *J. Polymer Sci.*, **6**, 225 (1951); M. Morton, P. P. Salatiello and H. Landfield, *ibid.*, **8**, 215 (1952).
22. H. Staudinger and W. Heuer, *Ber.*, **67**, 1164 (1935); H. Staudinger and E. Husemann, *ibid.*, **68**, 1618 (1935).
23. C Walling. *J. Am. Chem. Soc.*, **67**, 441 (1945).

CHAPTER X

Configuration of Polymer Chains

IN THE usual chemical formulas written for chain polymers the successive units are projected as a co-linear sequence on the surface of the sheet of paper. This form of representation fails to convey what is perhaps the most significant structural characteristic of a long polymer chain, namely, its capacity to assume an enormous array of configurations. This configurational versatility is a consequence of the considerable degree of rotational freedom about single bonds of the chain. In the simple polymethylene chain, for example, the conventional formula

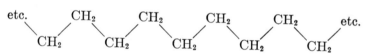

expresses only one configuration, the fully extended one, out of the very great number which are possible. An almost limitless variety of irregular shapes may be realized by performing rotations about the bonds of the chain. Molecular models display this feature most clearly.

To an approximation which is adequate for all ordinary purposes, it is permissible to consider the lengths (l) of the valence bonds in the chain and the valence angle (θ) between successive bonds to be fixed quantities for a given structure. A given configuration, obtained from the fully extended form by a series of arbitrary rotations about single bonds, may be specified by the angle (ϕ) of rotation for each bond. To visualize this more clearly, consider the portion of a singly bonded carbon chain skeleton represented in Fig. 74. Carbon atoms 1, 2, and 3 define a plane. With reference to these, atom 4 may occur any place on the circle which is the base of the cone described by the rotation of bond 3 (between atoms 3 and 4). The position of atom 4 is specified by the angle of rotation ϕ_2 about bond 2. C_4' and C_4'', which represent positions assumed by atom 4 for $\phi_2=0$ and π, respectively, are in the plane defined by bonds 1 and 2. The former is the planar zigzag form, corresponding to the fully extended configuration. The valence angle θ is expressed without subscript, inasmuch as it is assumed to be constant and identical for all pairs of successive bonds. Similarly, bonds 2 and 3 define a plane, and the direction of bond 4

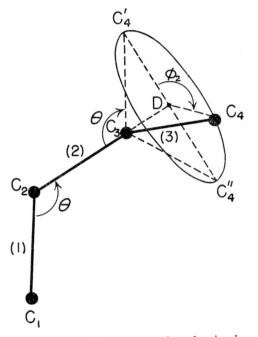

FIG. 74.—Spatial representation of a simple
singly bonded carbon chain.

(not shown in Fig. 74) is correspondingly specified by an angle ϕ_3 of rotation about bond 3; the position of atom 5 relative to atoms 2, 3, and 4 is thus defined. In this manner the configuration of a molecule of n bonds may be specified by $n-2$ angles of rotation about single bonds. We may speak of a given *configuration* as one in which each angle ϕ is specified within an arbitrary small range $\delta\phi$. If the repeating unit possesses double bonds, rings, or more than one type of single bond (or bond angle), the analysis of configurations is correspondingly more complicated but may be described adequately nevertheless by assigning a value to the rotation angle about each single bond.

Hindrances to free rotation of a steric or any other nature may of course restrict the values which the rotation angles ϕ may assume.

The foregoing description is in no way inconsistent with this; we consider merely that values of ϕ are permitted over some finite range for each bond. The total number of configurations which the polymer molecule may assume will in any event increase as an exponential function of the number of single bonds, becoming incomprehensibly large for values of n in the usual range of 100 to 10,000. Detailed analysis of the molecular configurations would indeed be quite hopeless. On the other hand, the overwhelming complexities of the situation provide ideal circumstances for the application of statistical methods. As we shall be interested only in average properties, the sacrifice of more detailed information is of no real concern. Owing to the large number of statistical elements, or of bonds, in a polymer chain, statistical formulas may be reduced in nearly all cases to asymptotic expressions which are of the same general form irrespective of the particular structure.

A physical property which depends on the configuration of the polymer molecule ordinarily can be expressed as a function of an average dimension of one sort or another. The dimension of a polymer most widely used to characterize its spatial, or configurational, character is the distance r from one end group to the other of the chain molecule. This quantity is sometimes designated the *displacement length* of the chain. In order to avoid all possibility of confusion, the length of the fully extended chain may be referred to as the *contour length*. An average value of r obviously is required, the usually appropriate average being the root mean square, $\sqrt{\overline{r^2}}$. Another important measure of the effective size of a polymer molecule is the root-mean-square distance of the elements of the chain from its center of gravity. This quantity, which we shall designate by $\sqrt{\overline{s^2}}$, is often referred to as the radius of gyration of the molecule. For linear chain polymers $\sqrt{\overline{s^2}}$ is easily obtained from $\sqrt{\overline{r^2}}$; hence these quantities need not be discussed separately.

The statistical distribution of r values for long polymer chains and the influence of chain structure and hindrance to rotation about chain bonds on its root-mean-square value will be the topics of primary concern in the present chapter. We thus enter upon the second major application of statistical methods to polymer problems, the first of these having been discussed in the two chapters preceding. Quite apart from whatever intrinsic interest may be attached to the polymer chain configuration problem, its analysis is essential for the interpretation of rubberlike elasticity and of dilute solution properties, both hydrodynamic and thermodynamic, of polymers. These problems will be dealt with in following chapters. The content of the present chapter is preparatory to discussion of the physical properties and behavior of polymer systems from the point of view of their constitution.

1. STATISTICAL DISTRIBUTION OF END-TO-END DIMENSIONS

It is appropriate to consider first a hypothetical chain consisting of linkages of length l joined in linear sequence without any restrictions whatever on the angles between successive bonds. Thus, for the case to be discussed θ may assume all values from 0 to π with equal probability and ϕ is likewise unrestricted. Only the bond length is fixed. Of

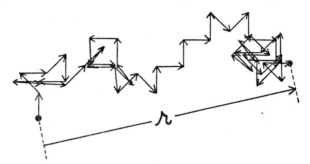

Fig. 75.—Vectorial representation in two dimensions of a freely jointed chain. A random walk of fifty steps.

course, no polymer chain is known which conforms even approximately to this model, but, as will be pointed out later, the statistical properties of a real chain for which the angle θ is virtually fixed and ϕ is restricted to some extent will not be different in character from those of the *freely jointed* chain. Such a chain may be considered to consist of a set of n vectors of equal magnitude but unrestricted in direction. The vector r from beginning to end of the chain is, of course, the vector sum of the bond vectors as shown in Fig. 75. The chain configuration problem is reduced in this manner to that of *random flight*, which has been treated by Rayleigh[1] and others.[2] The configuration of the freely jointed polymer chain resembles the path described by a diffusing particle such as a gas molecule. A free path between collisions of the latter corresponds to a bond vector; the correspondence would be exact if the length of the free path between collisions were always the same.

1a. The Freely Jointed Chain in One Dimension.—To proceed with analysis of the configurations of the freely jointed polymer chain, it is convenient to consider first the projection of the configuration on one coordinate axis, say the x-axis. Since each bond is assumed to be capable of choosing any direction with equal probability, it must follow that the average x-component $\bar{l}_x$ is zero, since positive and negative values of l_x occur with equal probability. In mathematical language

$$\bar{l}_x = \int_0^l l_x p(l_x) dl_x \tag{1}$$

where $p(l_x)dl_x$ is the probability that the projection lies between l_x and $l_x + dl_x$. If polar coordinates are chosen with ψ representing the angle between the bond and the x-direction, then $l_x = l \cos \psi$. All directions being equally probable, the probability $p(l_x)dl_x$ must be proportional to the size of the solid angle of bond directions over which l_x lies in the range specified. This solid angle is $2\pi \sin \psi d\psi$, and the total solid angle for all directions is 4π. Hence $p(l_x)dl_x = (1/2) \sin \psi d\psi$, which on substitution in Eq. (1) yields

$$\bar{l}_x = (1/2) \int_0^\pi l \sin \psi \cos \psi d\psi = 0 \tag{2}$$

The *algebraic* average of the component of $\boldsymbol{r}$ in any arbitrarily specified direction must likewise be zero.

The *average square* of the projection of a bond vector on the x-axis is not zero. It may be calculated in a similar manner as follows:

$$\bar{l_x^2} = \int_0^l l_x^2 p(l_x) dl_x$$

$$= (1/2) \int_0^\pi l^2 \sin \psi \cos^2 \psi d\psi$$

$$= l^2/3 \tag{3}$$

Thus the *root-mean-square* value of the projection is given by

$$\sqrt{\bar{l_x^2}} = l/\sqrt{3} \tag{4}$$

Let the components of the end-to-end vector $\boldsymbol{r}$ along rectangular axes be expressed by x, y, and z. For the solution of the problem in one dimension, we require the probability of a given value of the component of r in the chosen direction. Letting this direction coincide with the x-axis, we require the probability $W(x)dx$ that x assumes a value between x and $x + dx$. Under the assumption that the number n of bonds in the chain is large, it is permissible to consider that each bond makes a contribution along the x-axis equal in magnitude to the root-mean-square projection $\sqrt{\bar{l_x^2}}$, positive and negative values occurring with equal probability. If we let n_+ and n_- represent the numbers of bonds making positive and negative contributions, respectively, it is at once evident that

$$x = (n_+ - n_-)\sqrt{\overline{l_x^2}} = (n_+ - n_-)l/\sqrt{3}$$

and the probability of a given value of x reduces to the calculation of the probability of the corresponding excess of heads in a series of coin tosses. The well-known solution (see Appendix A) of this problem, for $|n_+ - n_-| \ll n$, i.e., for x values much smaller than would correspond to full extension of the chain, is expressed by the Gaussian distribution

$$W(x) \, dx = (\beta/\pi^{1/2})e^{-\beta^2 x^2}dx \qquad (5)$$

where

$$\beta = (1/2n\overline{l_x^2})^{1/2} = \sqrt{3/2}/n^{1/2}l \qquad (6)$$

According to Eq. (5) the most probable value of x occurs at $x=0$. As x increases in magnitude, $W(x)$ decreases monotonically from its maximum at $x=0$, the decrease being the more rapid the smaller n. Eq. (5) obviously becomes invalid at values of x approaching nl; $W(x)$ remains finite, though very small, according to Eq. (5) even for $x > nl$, where it necessarily is zero.

The foregoing derivation may appear artificial in view of the assumptions involved. The contribution of a given bond to x is by no means restricted to the two unique values, $\pm \sqrt{\overline{l_x^2}}$, as has been assumed. On the contrary, one may show that all values of l_x from 0 to l occur with equal probability for freely jointed connections between links. A more detailed study of the problem shows that the final result is unaffected by this assumption so long as n is large. The freely jointed chain model under consideration is an artifice also, but the *form* of the results obtained will be shown to apply also to real polymer chains.

1b. The Freely Jointed Chain in Three Dimensions.—Equivalent expressions may be employed for the probability distributions, $W(y)$ and $W(z)$, for the components y and z. Furthermore, it is possible to show[1] that $W(y)$ is independent of the value previously assigned to x (aside from effects of higher order) provided that n is large and x is much smaller than the length nl for full extension of the chain. Similarly, $W(z)$ is independent of x and y when n is large and both of the conditions $x \ll nl$ and $y \ll nl$ are fulfilled. Thus, for small extensions of a long chain, it is permissible to consider that $W(x)$ depends only on x, $W(y)$ only on y, and so forth. The probability $W(x, y, z)dxdydz$ that the components of $\mathbf{r}$ are x to $x+dx$, and so on, may therefore be taken as the product of the separate probabilities under the conditions assumed. Thus

$$W(x, y, z)dxdydz = W(x)W(y)W(z)dxdydz$$
$$= (\beta/\pi^{1/2})^3 e^{-\beta^2 r^2} dxdydz \qquad (7)$$

where r is the magnitude of the chain displacement vector $\mathbf{r}$, i.e.

$$r^2 = x^2 + y^2 + z^2$$

If one end of the freely jointed polymer chain is placed at the origin of a coordinate system, as represented in Fig. 76 and the chain is per-

mitted to assume any configuration at random (being subject to no externally imposed restraints), the probability that the other end occurs in the volume element of size $dxdydz$ located at x, y, z is given by Eq. (7). Otherwise stated, $W(x, y, z)dxdydz$ expresses the *probability* that the chain displacement vector $\mathbf{r}$ reaches from the origin to the volume element specified. If the chain displacement vectors for each of a large number of identical polymer molecules are plotted from the same origin, then $W(x, y, z)$ represents also the *density of distribu-*

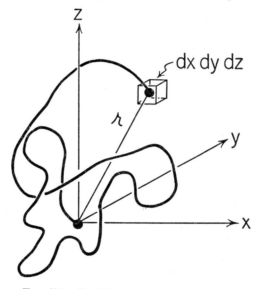

Fig. 76.—Spatial configuration of a polymer chain taking the origin of coordinates at one end of the chain.

tion of the end points of these vectors. These alternate interpretations of the distribution function may be used interchangeably. The probability, or density, depends only on the magnitude of $\mathbf{r}$ and not on its direction.

The Gaussian function (7) is shown graphically in Fig. 77. The most probable location of one end of the chain relative to the other is at coincidence, i.e., at $r=0$. The density, or probability, decreases monotonically with $\mathbf{r}$, exactly as noted above for the one-dimensional case. Equation (7) likewise is unsatisfactory for values of r not much less than the full extension length nl. The extent of this limitation will be discussed presently.

The probability that r assumes a certain magnitude *irrespective of*

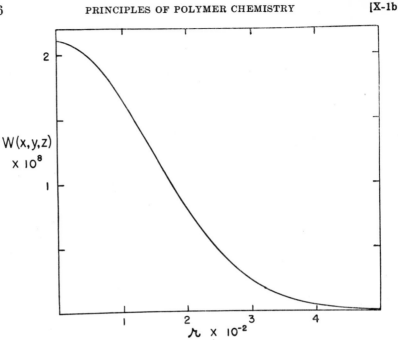

Fig. 77.—Gaussian density distribution of the chain displacement vectors for chain molecules consisting of 10^4 freely jointed segments, each of length $l = 2.5$ Å. The end-to-end length r is in Ångstrom units and $W(x, y, z)$ is expressed in Å^{-3}.

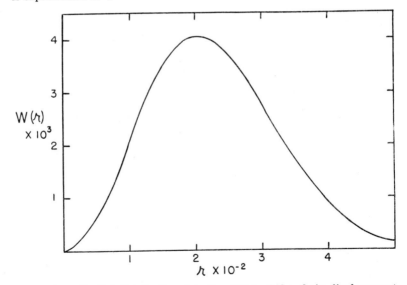

Fig. 78.—Radial distribution function $W(r)$ of the chain displacement vectors for the same polymer chains as in Fig. 77. $W(r)$ is expressed in Å^{-1}.

the direction of the vector r will be given by $W(x, y, z)$ multiplied by the total volume of *all volume elements* at a distance r from the origin. This total volume is, of course, $4\pi r^2 dr$. Hence the probability that the chain displacement length has a value in the range r to $r+dr$, *irrespective of direction*, is

$$W(r)dr = (\beta/\pi^{1/2})^3 e^{-\beta^2 r^2} 4\pi r^2 dr \tag{8}$$

The radial distribution function $W(r)$, which is shown in Fig. 78, exhibits a maximum at a value of r greater than zero. This maximum occurs at

$$r = 1/\beta \tag{9}$$

which therefore represents the most probable value of r. This result may appear inconsistent with the previous conclusion that the probability density function $W(x, y, z)$ possesses a maximum at $r=0$. The confusion which might arise from this apparent paradox may be avoided though proper cognizance of the significance of the two functions. If the origin of coordinates is taken at one end of the molecule in either case, $W(x, y, z)$ represents the probability *per unit volume* of finding the other end of the chain at x, y, z. For a collection of many chains, ν in number, $W(x, y, z)$ represents the *density* distribution of end points, i.e., $\nu W(x, y, z)$ represents the density, or number per unit volume, at x, y, z. $W(r)$, on the other hand, represents the probability *per unit in r* irrespective of direction. $W(r)$ is obtained from $W(x, y, z)$ by multiplying by the size of the volume element throughout which r is constant. The size of this volume element, $4\pi r^2 dr$, increases as r^2, and this fact alone is responsible for the difference between the two functions. Thus, although the density distribution always diminishes as r is increased, the total volume of all volume elements a distance r from the origin increases with r^2.

We turn now to the consideration of average values of r. The average of r, unlike that of one of its components, e.g., x, exceeds zero inasmuch as r is restricted to values equal to or greater than zero. We have

$$\bar{r} = \int_0^\infty rW(r)dr \bigg/ \int_0^\infty W(r)dr$$

The integral in the denominator equals unity; i.e., $W(r)$ is normalized. Substituting for $W(r)dr$ from Eq. (8) and evaluating the numerator integral

$$\bar{r} = 2/\pi^{1/2}\beta \tag{10}$$

Similarly, the average square end-to-end length is

$$\overline{r^2} = \int_0^\infty r^2 W(r) dr$$

which gives

$$\sqrt{\overline{r^2}} = \sqrt{3/2\,\beta^2} \tag{11}$$

Equations (10) and (11) are characteristic of the Gaussian distribution, Eq. (8), irrespective of the relationship of β to chain dimensions in any given instance. In the particular case of the freely jointed chain β assumes the value given by Eq. (6). Substituting Eq. (6) in Eqs. (10) and (11) yields

$$\bar{r} = \sqrt{8/3\pi}\; ln^{1/2} \tag{12}$$

$$\sqrt{\overline{r^2}} = ln^{1/2} \tag{13}$$

The latter relation is particularly noteworthy. It recalls the previously indicated analogy between the configuration of the freely jointed chain and the path described by a diffusing particle. According to diffusion theory, the mean-square displacement of a diffusing particle is proportional to the time. Inasmuch as the particle moves with a fixed average velocity at a given temperature, it is evident that the time corresponds to the chain length of the polymer. Thus Eq. (13) may be looked upon as a necessary consequence of the analogy existing between the two phenomena.

The end-to-end distance, or displacement length, of the chain enters directly into the treatment of rubber elasticity and related problems, but it is the root-mean-square distance $\sqrt{\overline{s^2}}$ of an element from the center of gravity which bears directly on dilute solution properties such as viscosity and the dissymmetry of scattered light (see Chap. VII). If the distribution of the distance r_{ij} between any two elements i and j of the chain can be expressed as a Gaussian distribution $W(r_{ij})$ with the parameter β_{ij} varying as the inverse square root of the number n_{ij} of elements between i and j (i.e., if each such distribution is the same as it would be in the absence of elements preceding i and succeeding j in the chain), then it is possible to show[3] that (see Appendix C)

$$\sqrt{\overline{s^2}} = \sqrt{\overline{r^2}/6} \tag{14}$$

or

$$\sqrt{\overline{s^2}} = 1/2\beta \tag{14'}$$

where $\overline{r^2}$ is, as above, the mean-square end-to-end distance for the entire chain. Owing to the existence of this relationship between the two quantities, either may be used as a measure of the size of the polymer chain. In spite of the direct dependence of the properties often measured on $\sqrt{\overline{s^2}}$, the root-mean-square end-to-end distance has come to be the quantity more widely used. It is important to observe, however, that Eq. (14) holds only for *linear* polymers of random configuration such as to permit representation of the end-to-end distribution by a Gaussian function. The average dimensions of nonlinear polymers will be considered later (see p. 422).

1c. The Distribution at High Extensions.—It has been noted that the Gaussian distribution is unsatisfactory at high extensions and that it is totally unacceptable for displacement lengths approaching full extension of the chain. This was indeed to be expected in view of the mathematical approximations introduced in the derivation (see Appendix A). Partly for the purpose of exploring the extent of the resulting limitations on the convenient Gaussian function, there is considered here a more accurate treatment of the problem, which is valid even at high extensions.

The distribution of the displacement length r for the freely jointed chain may be treated[4,5] according to the well-known methods applied to the orientation of magnetic or electric dipoles by an external field of sufficient strength to produce effects approaching saturation (i.e., complete orientation). The results, though cumbersome, are accurate for all extensions up to the maximum $r_m = nl$. This treatment, which is given briefly in Appendix B of this chapter, was adapted to the polymer configuration problem by Kuhn and Grün[4] and by James and Guth.[5] These authors show that Eq. (8) should be replaced by

$$W(r)dr = \text{Const} \cdot \exp\left[-\int_0^r \mathcal{L}^*(r/r_m)dr/l\right]4\pi r^2 dr \qquad (15)$$

where $r_m = nl$, the maximum length corresponding to full extension of the chain, and $\mathcal{L}^*$ is the inverse Langevin function. The Langevin function $\mathcal{L}$ is defined by

$$\mathcal{L}(u) = \coth(u) - 1/u$$

Letting $\mathcal{L}(u) = v$, we have

$$\mathcal{L}^*(v) = u$$

Series expansion of Eq. (15) gives

$$W(r)dr = \text{Const} \cdot \exp\{-n[(3/2)(r/r_m)^2 + (9/20)(r/r_m)^4 \\ + (99/350)(r/r_m)^6 + \cdots]\}4\pi r^2 dr \qquad (16)$$

Inasmuch as $3n/2r_m^2 = \beta^2$, the distribution function may also be written

$$W(r)dr = \text{Const} \cdot \exp\{- \beta^2 r^2[1 + (3/10)(r/r_m)^2$$
$$+ (33/175)(r/r_m)^4 + \cdots]\}4\pi r^2 dr \quad (16')$$

Eq. (16') reverts to (8) for sufficiently small values of the ratio, r/r_m, of the displacement length to the maximum length. The bracketed quantity in the exponential of Eq. (16') may be looked upon as a correction factor on the value of r^2 which should have been used in the simpler Eq. (8). For r/r_m less than about one-half, this correction is negligible.* The ratio of the root-mean-square length to the maximum length varies inversely as the square root of the maximum length, according to Eq. (13), i.e., $\sqrt{\overline{r^2}}/r_m = 1/n^{1/2}$, from which it follows that the correction factor for a freely jointed chain of 100 links is negligible up to extensions (r/r_m) of five times the root-mean-square extension. For a chain of 10^4 units it may be disregarded up to 50 times the root-mean-square extension. Thus the approximation of the actual distribution by the Gaussian function, Eq. (8), should be satisfactory throughout the range ordinarily of interest. Only under conditions leading to extension of the polymer chains far beyond their average configurations is it necessary to resort to the more accurate formulas. The Gaussian approximation is acceptable for the treatment of rubber elasiticity, swelling, and the configuration of polymer molecules in solution, except for polyelectrolytes in media of very low ionic strength (Chap. XIV). In the latter case, the influence of the charges attached to the polyelectrolyte chain may be sufficient to cause it to extend almost to full length. Except where specified otherwise, the Gaussian approximation is regarded as adequate throughout the remainder of this chapter and those which follow.

1d. Influence of Bond Angle Restrictions.—In all real polymer chains the direction assumed by a given bond is strongly influenced by the direction of its predecessor in the chain. The orientation of other nearby bonds (second, third, and possibly fourth neighbors) may also exert an appreciable influence, but the orientation of the immediate predecessor usually is of greatest importance. The exact nature of these restrictions on the direction assumed by a given bond

---

* On the other hand, the correction factor by which $W(r)$ is altered through this refined treatment, namely, $\exp[-(9n/20)(r/r_m)^4 \cdots]$ from Eq. (16), depends both on n and on r/r_m. If the distance of separation of the ends of the chain lies in the vicinity of its root-mean-square value, i.e., if $r \sim \sqrt{\overline{r^2}} = n^{1/2}l$, then $(r/r_m) \sim n^{-1/2}$, and this correction factor is negligible for any value of n large enough to validate statistical treatment. The correction factor applying to r, and considered in the text, is more significant from the standpoint of applications to rubber elasticity and other problems.

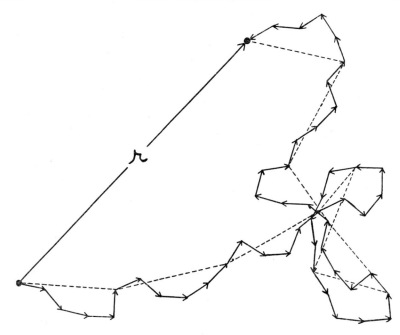

FIG. 79.—Representation of a hindered chain in two dimensions. A random walk of fifty steps with angles between successive bonds limited to the range $-\pi/2$ to $\pi/2$. The scale is identical with that in Fig. 75 for an unrestricted random walk of the same number of steps.

depends specifically on the structure of the chain unit. Invariably, their resultant effect is to expand the configuration in comparison with that for the freely jointed chain of the same contour length. This is illustrated in Fig. 79 where a configuration for a restricted "random walk" in two dimensions is shown; each step is subject to the restriction that the change in direction compared with the preceding step may not exceed 90°. The influence of this restriction is apparent by comparison with Fig. 75 (p. 402) for a freely jointed random walk of the same number of steps.

Kuhn[6] has shown how a real polymer chain may be approximated by an *equivalent* freely jointed chain. Instead of taking the individual bonds as statistical elements, one may for this purpose choose sequences of m bonds each. In Fig. 79, arbitrarily chosen statistical elements consisting of five bonds are indicated, the displacement vectors for these elements being shown by the dashed lines. The direction assumed by a statistical element will be nearly independent of the direction of the preceding element, provided the number m of bonds per element is sufficiently large.   The lengths of the elements are variable. However, Kuhn[6] showed that the correct statistical distribution of displacement lengths for the chain as a whole will be obtained if each element is replaced by a hypothetical element of length l' equal to the root-mean-square length of the actual elements consisting of m bonds.   With this substitution, the statistical elements are both random in orientation and fixed in length, and the configuration problem reduces to that of a freely jointed chain consisting of $n' = n/m$ elements each of length l'.   We are thus led to the important conclusion that the *form* of the statistical distribution of chain displacement lengths for a real chain will be the same (i.e., approximately Gaussian) as for a freely jointed chain.   The characteristic parameter β will differ however; i.e., Eq. (6) for the freely jointed chain must be replaced by

$$\beta = (3/2n'l'^2)^{1/2} \tag{17}$$

For a given chain the parameter β will, of course, have a definite value, but the individual values of n' and l' depend on the choice of the size m of the chain element; the length l' will depend also on the specific nature of the restriction on the bond angles θ and ϕ.   It is required merely that m be sufficiently large to render negligible the influence of one element on the direction of the next; at the same time the size of the element must not be chosen so large as to reduce the number $n' = n/m$ of elements below that required (about ten) for the Gaussian approximation to hold.   Within the limits of these diffuse requirements, the choice of m remains arbitrary.   Fulfillment of both of them is possible, of course, only if the number n of actual bonds is large; the lower limit on n is greater the more hindered the chain.

The introduction of vectors of constant displacement length to represent the individual elements, which actually vary in length, is rendered more plausible by inquiry into the effect of incorporating this artifice in the treatment of the freely jointed chain.   In this case $l' = m^{1/2}l$. Upon substitution of this expression together with $n' = n/m$ in Eq. (17), the previous expression for β, Eq. (6), is recovered.   Hence the calculated distribution is unaffected by an arbitrary subdivision of the chain in this manner.   We conclude that the value chosen for m in the reduction of the *real* chain to an equivalent freely jointed chain likewise is inconsequential (within the limits on m stated above).

The foregoing discussion of equivalent chains requires merely that its root-mean-square end-to-end distance shall equal that of the real chain.   In order to define completely the equivalent chain, its contour lengths may also be required to coincide with that of the real chain.

Thus the fully extended length of the equivalent chain will be set equal to that of the real chain with all valence angle (θ and ϕ) restrictions removed. Then n' and l' (and therefore m) for the equivalent chain are completely defined by

$$nl = n'l'$$
$$\sqrt{\overline{r^2}} = l'(n')^{1/2} \tag{18}$$

This hypothetical equivalent chain should resemble in statistical behavior the actual chain of given size n and mean extension $\sqrt{\overline{r^2}}$.

According to the preceding analysis, the introduction of bond angle restrictions will affect the configuration distribution function $W(r)$ in a manner which may be taken into account by suitably altering the parameter β without changing the form of the distribution. Average dimensions are correspondingly altered, as is clear from Eqs. (10), (11), and (14'). The complications inherent in the nature of steric repulsions precludes calculation of these dimensions, and of the parameter β also, from structural data on the actual polymer chain unit. It is this feature of the theory of freely jointed chains, for which β is related in a simple way to n and l, that is sacrificed in adapting the treatment to actual chains. The proportionality of β to $n^{-1/2}$ remains (Eq. 17), but the coefficient assumes a value depending on the specific structure of the given polymer. Thus we may write in place of Eqs. (6) and (17)

$$\beta = \sqrt{3/2}/Cn^{1/2} \tag{6'}$$

and

$$\sqrt{\overline{r^2}} = Cn^{1/2} \tag{11'}$$

where C is a constant characteristic of the given chain structure. For a polymer chain composed of a sequence of identical bonds, C will, of course, be proportional to the length l of the bond. If $\overline{r^2}$ can be determined either by calculation or by experiment for a polymer of given length, C is at once established. In the following section the calculation of $\overline{r^2}/n$, and hence of C and $\beta n^{1/2}$, for various structures will be attempted. In anticipation of the conclusion to be reached, it may be stated that reliable values cannot as yet be obtained from theory; hence $\beta n^{1/2}$ and $(\overline{r^2}/n)^{1/2}$ (and $(\overline{s^2}/n)^{1/2}$ also) are to be regarded as empirical parameters for a given series of linear polymer homologs.

2. CALCULATION OF AVERAGE DIMENSIONS FOR VARIOUS POLYMER CHAIN STRUCTURES

2a. The Freely Rotating Polymethylene Chain.—Consider a polymer chain consisting of n bonds of identical length l joined at fixed valence angles θ. The polymethylene chain is a typical example. Vinyl

$$\text{X} \\ | \\ \text{polymers consisting of} \quad —CH_2—C— \quad \text{units likewise conform to these} \\ | \\ \text{Y}$$

stipulations to the extent that the bond angles (θ) at the alternating chain carbons are the same, i.e., to the extent that the bond angle is unaffected by substituents. Let us assume first that the rotation angle ϕ is unrestricted, an admittedly erroneous assumption. The vector $\boldsymbol{r}$ may be expressed as the sum of the n bond vectors $\boldsymbol{l}_i$

$$\boldsymbol{r} = \sum_{i=1}^{n} \boldsymbol{l}_i$$

Then

$$r^2 = \boldsymbol{r} \cdot \boldsymbol{r} = \sum_{i=1}^{n} \sum_{j=1}^{n} \boldsymbol{l}_i \cdot \boldsymbol{l}_j$$

The average square of r may be written

$$\overline{r^2} = \sum_{i=1}^{n} \sum_{j=1}^{n} \overline{\boldsymbol{l}_i \cdot \boldsymbol{l}_i} \tag{19}$$

the bar indicating that an average value is to be taken. If the chain were freely jointed (i.e., if θ, as well as ϕ, were unrestricted), the average projection of any bond vector on any other would always be zero. **i.e.**

$$\overline{\boldsymbol{l}_i \cdot \boldsymbol{l}_j} = 0 \qquad\qquad \text{when } i \neq$$

There would be left then only product terms for which $i = j$; hence

$$\overline{r^2} = \sum_{1}^{n} l_i^2 = nl^2$$

which was obtained previously (Eq. 13).

The solution of the present problem[7] is made clearer by expanding the summations in Eq. (19) into a square array, thus

$$\overline{r^2} = \begin{vmatrix} \overline{l_1 \cdot l_1} + \overline{l_1 \cdot l_2} + \overline{l_1 \cdot l_3} + \cdots \overline{l_1 \cdot l_n} \\ + \overline{l_2 \cdot l_1} + \overline{l_2 \cdot l_2} + \overline{l_2 \cdot l_3} + \cdots \overline{l_2 \cdot l_n} \\ + \overline{l_3 \cdot l_1} + \overline{l_3 \cdot l_2} + \overline{l_3 \cdot l_3} + \cdots \overline{l_3 \cdot l_n} \\ + \cdots \\ + \cdots \end{vmatrix} \qquad (20)$$

Each of the n terms $\overline{l_i \cdot l_i}$ occurring on the principal diagonal is equal to l^2. Terms $\overline{l_i \cdot l_{i\pm1}}$ on the immediately adjacent diagonals involve the projection of a bond on its immediate neighbor. As is obvious from Fig. 74 (p. 400), this projection is $-l \cos \theta$, and the value of each term on either diagonal adjacent to the principal diagonal is $-l^2 \cos \theta$. Terms $\overline{l_i \cdot l_{i\pm2}}$ involve projections on the second preceding, or the second succeeding, bond; for example, the projection of bond 3 on bond 1 of Fig. 74. Since angle ϕ_2 in Fig. 74 assumes all values with equal probability, the average projection of bond 3 perpendicular to bond 2 is zero, and its entire average contribution consists of $C_3D = -l \cos \theta$ along bond 2. The projection of this on bond 1 is $l \cos^2 \theta$. Hence, $\overline{l_i \cdot l_{i\pm2}} = l^2 \cos^2 \theta$. In general, $\overline{l_i \cdot l_{i\pm m}} = l^2(-\cos \theta)^m$. Introducing these values into Eq. (20) and combining equal terms along diagonals of the array

$$\overline{r^2} = l^2[n + 2(n-1)(-\cos \theta) + 2(n-2)(-\cos \theta)^2 + \cdots \\ + 2(-\cos \theta)^{n-1}]$$

which may be rearranged to

$$\overline{r^2} = nl^2 \left\{ 1 + 2(z - z^n)/(1 - z) - (2z/n)\frac{d}{dz}[(z - z^n)/(1 - z)] \right\} \qquad (21)$$

where $z = -\cos \theta$. For large n

$$\overline{r^2} \cong nl^2(1 - \cos \theta)/(1 + \cos \theta) \qquad (22)$$

to an adequate approximation. In the particular case of a tetrahedrally bonded chain, $\theta = 109.5°$ and $\cos \theta = -\frac{1}{3}$, giving

$$\overline{r^2} = 2nl^2$$

or just twice the value for the freely jointed chain.

2b. Hindered Polymer Chains.—Molecular dimensions predicted by calculations made under the assumption of free rotation are found to be appreciably in error. Although the freely rotating chain of fixed valence angle θ conforms more nearly to reality than the freely jointed chain, it is nevertheless quantitatively unsatisfactory. The reason for this failure is immediately evident on examination of a scale model of a polymer chain.   The range of values accessible to each of the rotation angles ϕ is severely restricted by steric interferences between successive units of the chain; hence the assumption of free rotation obviously is far from realistic.   Restrictions of this sort are operative even in a simple polymethylene chain having no bulky substituents.   The influence on chain dimensions of limitations on ϕ, due to steric or other effects, is far more difficult to treat than the fixed valence angle problem, and no general solution can be given.   Inquiry into the origin of these

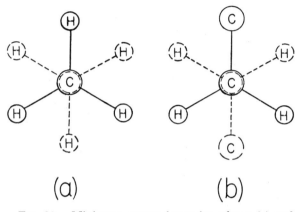

(a)   (b)

Fig. 80.—Minimum energy forms for ethane (a) and for a section of a polymethylene chain (b) as viewed along the axis of the C—C bond about which hindrance to rotation is under consideration.   Dotted portions in each figure represent atoms or substituents attached to the lower carbon of the bond.

hindrances and evaluation of the effects of the simplest types of hindrance, namely, those which act symmetrically about $\phi=0$, nevertheless are illuminating.

Hindrance to rotation in its most primitive form is encountered in simple singly bonded molecules such as ethane.   Owing to repulsions between nonbonded atoms, the staggered configuration shown in Fig. 80,a corresponds to an energy minimum.   Rotation of either methyl group through 60° places the hydrogen atoms directly above one another, and the potential energy associated with rotation reaches its maximum value.   As a consequence of the symmetry of the molecule, the potential energy must be periodic in $2\pi/3$; i.e., the potential energy $V(\phi)$ will pass through three equal maxima and three equal minima for each complete rotation 2π.   As a first approximation, the potential energy may be expressed as follows:

$$V(\phi) = (V_0/2)(1 - \cos \phi/3) \tag{23}$$

where V_0 is the energy difference between maxima and minima. This function is shown in Fig. 81,a. Comparison of the entropy of ethane calculated from the third law of thermodynamics with that computed from molecular parameters and vibration frequencies obtained spectroscopically leads to the conclusion that V_0 is about 2800 cal. per mole for ethane.[8]

The threefold symmetry of rotation about the C—C bond of ethane disappears when substituents are introduced on both of the carbon

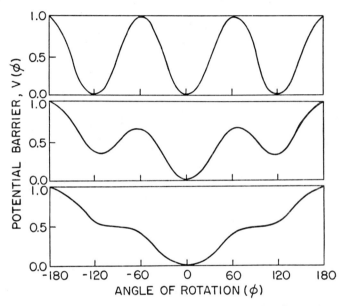

Fig. 81.—Potential energy associated with bond rotation as a function of angle. (a) Symmetrical potential according to Eq. (23); (b) and (c) potential energy functions with lowest minimum at $\phi = 0$ corresponding to the planar zigzag form of a polymethylene chain. These curves were calculated by Taylor.[10]

atoms, as in butane or higher homologs[9] including the polymethylene chain. Three energetically favored positions corresponding to those of ethane persist, but the energy for the one of them in which the substituents (methylene groups) are in opposite (*trans*) positions, as shown in Fig. 80, b, will be lower than for the other two.* In either of the latter two positions (not shown) the attached groups are sufficiently

* An excellent discussion of the influence of hindrance to rotation on the configuration of polymethylene chains has been given by Taylor.[10]

close together to cause appreciable repulsion between them. On this basis, potential energies of the sort shown in Fig. 81, curves b and c, should be expected. In Fig. 81,b shallow minima remain at positions corresponding to rotations of $\pm 120°$ from the configuration shown in Fig. 80,b, but the most stable configuration corresponds to the planar zigzag with $\phi = 0$. Fig. 81,c represents a case of stronger repulsions between the substituents, with the result that only the single minimum at $\phi = 0$ remains.

This analysis of hindrances to rotation is oversimplified even for a polymethylene chain. However, with the tacit assumption that only interactions between atoms or groups directly connected to the C—C bond under consideration affect the value of ϕ, the hindrance potential should be symmetrical about $\phi = 0$ regardless of the precise nature of the interactions. That is to say, the repulsion for an angle ϕ should be the same as for $-\phi$. When this is true, the resultant of bond (3) in Fig. 74 (p. 400) averaged over all angles ϕ_2, each angle occurring with a probability proportional to $e^{-V(\phi)/kT}$, can be resolved into two components both of which lie in the plane of bonds (1) and (2); the component perpendicular to this plane must be zero by the conditions of symmetry requiring a given value ϕ and $-\phi$ to occur with equal probability. One of the two components is, of course, $-l \cos \theta$ along bond (2); the other perpendicular to it must lie along the direction $\overline{C_4'C_4''}$. This component is $\sin \theta \overline{\cos \phi}$, where $\overline{\cos \phi}$ designates the average value of $\cos \phi$. By an elaboration of the method used to derive Eq. (22), it may be shown[10,11] that

$$\overline{r^2} = nl^2 \left(\frac{1 - \cos \theta}{1 + \cos \theta}\right) \left(\frac{1 + \overline{\cos \phi}}{1 - \overline{\cos \phi}}\right) \tag{24}$$

to an approximation which is excellent when n is large and $\overline{\cos \phi}$ is not too near unity. If the rotation is unrestrained, $\overline{\cos \phi} = 0$ and Eq. (24) reverts to (22). Furthermore, if the restricting potential is of the form given by Eq. (23) and shown graphically in Fig. 81,a, the average value of $\cos \phi$ is again zero, and the dimensions of the chain are the same as if the rotation were free. The value of $\cos \phi$ will be positive, and $\overline{r^2}$ will exceed its value for free rotation, if the hindrance potential is such as to favor values of $|\phi|$ less than 90° (Fig. 81, b and c). Steric interferences between successive chain atoms and their laterally bonded members virtually assure that this will be true for any actual polymer chain.

Failure to take account of interactions with elements of the chain other than those bonded directly to the atoms of the bond under consideration constitutes an approximation which is not justified physically. Thus the hindrance potential associated with the rotation angle ϕ_i must depend on the value of ϕ_{i-1}, and to some extent also on ϕ_{i-2}, and so forth. Introduction of substituents such as the phenyl groups of polystyrene, or the (two) methyl groups of polyisobutylene on alternate carbon atoms, so complicates the steric interactions as to preclude meaningful interpretation in terms of a hindrance potential characteristic of one chain bond without regard for the rotation angles of its neighbors in the chain. The methyl substituents of polyisobutylene interfere severely in all possible configurations (see Chap. VI). As inspection of models shows, these interferences are by no means at a minimum for $\phi=0$. The optimum configuration is a spiraled one. The ϕ values may be assumed to fluctuate about those for the spiral, and these fluctuations introduce irregularity into the over-all configuration.

In the planar zigzag form of polystyrene

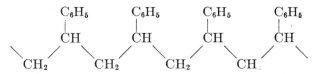

the phenyl groups are more closely crowded together than is apparent from this structural formula. A fortuitous arrangement of the asymmetric carbon atoms in successively alternating configurations so that the phenyl groups fall alternately above and below the plane of the zigzag would alleviate the crowding to some extent, without eliminating it, however. Experimental evidence bearing on the structure is provided by measurements of Debye and Bueche[12] on the dielectric polarization in dilute solutions of poly-(p-chlorostyrene). The observed dipole moment cannot be reconciled with a model in which the hindrance to rotation is a symmetrical function of the ϕ_i defined as above with reference to the planar zigzag structure. The results suggest rather that the phenyl groups tend to occur on alternate sides of the chain. Hence the structure of *minimum energy* for polystyrene (or its substitution product) probably is

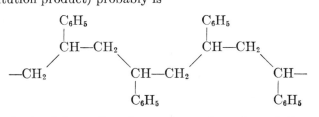

which is obtained from the planar zigzag form by rotations of 180°

about *every other* bond. This configuration is described by alternate
ϕ's of 0 and 180° in the previous notation. To be sure, wide deviations
from this regular structure must occur owing to the considerable latitude of rotational freedom which exists. The point to be stressed,
however, is that a grave error would be introduced if the hindrance
potential were assumed to be the same function of ϕ for every bond in a
chain of this sort.

The relationship of the configuration of a polymer chain to the structure of its units is an obviously complex problem. Hindrances to free
rotation about a given bond as it occurs in monomeric analogs, though
important, fail to represent the situation in its entirety. In more
complicated chains the mean thickness of the chain unit in relation to
the distance between singly bonded chain atoms may turn out to be
a more reliable criterion of its tortuosity and flexibility. Owing to
the complexities involved, mere comparison of experimentally determined average polymer dimensions with those computed, assuming
free rotation about all single bonds, represents the present level of
accomplishment in this direction. It is appropriate, therefore, to
discuss briefly the dimensions calculated on the assumption of free
rotation in chains other than the polymethylene chain and its substitution products such as the vinyl polymers.

2c. The Silicone Chain.—The silicone chain

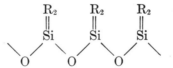

consists of identical bonds Si—O, but the bond angles differ at the
alternating silicon and oxygen atoms along the chain. It may be
shown[13] by application of the method given above for the polymethylene chain that

$$\overline{r^2} = nl^2(1 - \cos \theta_1)(1 - \cos \theta_2)/(1 - \cos \theta_1 \cos \theta_2) \qquad (25)$$

where θ_1 and θ_2 are the angles at the silicon and at the oxygen atoms,
respectively, and free rotation about the chain bonds is assumed.
This expression reduces to Eq. (22) when $\theta_1 = \theta_2$. The length l of the
silicon-oxygen bond is 1.65 Å.[14,15] The silicon angle θ_1 must be very
nearly tetrahedral (ca. 110°); the oxygen angle in the silicones is in
doubt; it probably is greater than 130° but less than 160°.[14,15,16]
If a value $\theta_2 = 145°$ is assumed, then $(\overline{r^2}/n)^{1/2} \times 10^{10} = 300$ cm. Through
the use of the factor 10^{10}, the figure on the right of this equality is

rendered equal to the root-mean-square length in Ångstrom units for a chain of 10^4 atoms.

2d. Cellulosic Chains.—Cellulose and its derivatives consist of six-membered rings connected through ether oxygens with one another in the manner indicated schematically in Fig. 82. The shaded bar represents the ring, whose carbon atoms 1 and 4 are connected by single bonds to the oxygen atoms. By projecting the oxygen-carbon bonds to the points C_1' and C_4', located so that the angles at C_1' and C_4' are right angles, the structure can be considered to consist of a bond of

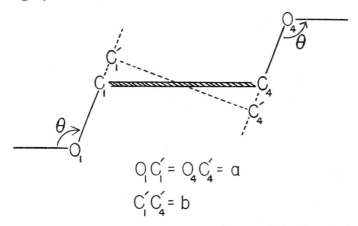

$$O_1 C_1' = O_4 C_4' = a$$
$$C_1' C_4' = b$$

Fig. 82.—Geometry of the cellulose chain unit.   (After Benoit.[17])

length a connected at an angle of 90° to one of length b, about which rotation cannot occur, and this bond connects at an angle of 90° to another bond of length a. The latter bond joins at an angle θ (the oxygen angle) with an "a" bond of the next unit, and so forth. Employing this model, Benoit[17] has shown that the mean-square displacement length under the assumption of free rotation about all "a" bonds is given by

$$\overline{r^2} = x\left[b^2 + (2a)^2(1 - \cos\theta)/(1 + \cos\theta)\right] \tag{26}$$

for large chain lengths, where x is the number of units. The results of X-ray diffraction measurements on crystalline cellulose give $a = 2.715$ Å, $b = 1.452$ Å, and $\theta = 110°$, from which is obtained

$$(\overline{r^2}/x)^{1/2} \times 10^{10} = 790 \text{ cm.}$$

2e. Natural Rubber and Gutta-Percha.—These *cis* and *trans* forms of 1,4-polyisoprene may be similarly reduced to structures amenable to calculation.[17,18] Assuming the usual bond dimensions and angles,

Wall[18] has shown that

$$(\overline{r^2}/x)^{1/2} \times 10^{10} = 402 \text{ cm.}$$

for the *cis* (natural rubber) form, and

$$(\overline{r^2}/x)^{1/2} \times 10^{10} = 580 \text{ cm.}$$

for the *trans* (gutta-percha) form.

In view of the previous generalization concerning the applicability of a Gaussian distribution to any flexible chain regardless of its valence angles, this distribution function may be assumed to apply to each of the preceding chains provided it contains many units and that configurations approaching full extension are excluded from consideration. The calculations here presented refer, of course, to the chains with unhindered rotations about all single bonds. The value of β corresponding to this approximation may be obtained from $\overline{r^2}$ through the use of Eq. (11).

2f. Nonlinear Polymer Molecules.—The root-mean-square end-to-end distance $\sqrt{\overline{r^2}}$ used above to characterize the size of a linear polymer obviously would be ambiguous if applied to nonlinear polymers, which possess a multiplicity of ends depending on the degree and functionality of branching. We resort, therefore, to $\sqrt{\overline{s^2}}$, the radius of gyration or root-mean-square distance of an element from the center of gravity of the nonlinear molecule, for the purpose of characterizing dimensions of such molecules. Recalling that $\overline{s^2} = \overline{r^2}/6$ for a random linear polymer chain (see Appendix C), it follows that for a freely jointed linear chain

$$\overline{s^2} = (l^2/6)n$$

The corresponding relation for a real linear chain may be expressed in terms of the equivalent freely jointed chain comprising n' bonds each of length l' as follows:

$$\overline{s^2} = (l'^2/6)n'$$

As has been pointed out on page 409, $\sqrt{\overline{s^2}}$ is a satisfactory alternative for $\sqrt{\overline{r^2}}$ as a measure of the effective size of a linear polymer molecule.

A branched molecule obviously will extend over a smaller volume than would a linear one of the same molecular weight, or number of units. In other words, $\overline{s^2}$ will be smaller for the nonlinear polymer of the same number n' of equivalent elements. Hence it is convenient to write for a nonlinear polymer[19]

$$\overline{s^2} = g(l'^2/6)n' \tag{27}$$

where g, representing the ratio of the mean-square distance of an element from the center of gravity of a nonlinear polymer to that of the corresponding linear polymer of the same degree of polymerization, will be less than unity.

Statistical calculations by Zimm and Stockmayer[19] for various forms of branching show that g decreases rather slowly with increase in the number of branches in the molecule. For random branching, for example, five trifunctional units per molecule decrease g only to about 0.70. Branches of higher functionality are not markedly more effective in reducing g. Inasmuch as a much smaller degree of nonlinearity (less than one cross-linkage per primary molecule, for example) normally is sufficient to induce gelation and the accompanying extreme distortion of the molecular weight distribution, it is apparent that the effect of nonlinearity on the molecular dimensions will in general be obscured by its effect on the molecular weight distribution. Thus, in seeking to observe the influence on a given physical property of the introduction of polyfunctional units into an otherwise linear polymer, the resulting broadening of the distribution is likely to be of much greater consequence than the alteration of the average dimensions of molecules of a given degree of polymerization.

3. CONFIGURATION OF POLYMER MOLECULES IN DILUTE SOLUTION

3a. Long-Range Intramolecular Interactions.[20]—A somewhat subtle approximation not mentioned heretofore persists throughout the foregoing account of polymer chain configurations. This has to do with the assumption that the disposition of a given bond, though it may depend on the orientations of its immediate predecessors in the chain, is independent of the coordinates describing all other bonds. Inasmuch as two elements of the molecule, possibly remote from one another in sequence along the chain, cannot occupy the same space, certain arrangements of the set of n bonds will be forbidden. Hence the assumption just mentioned cannot hold rigorously.

It is at this point that the analogy between the path described by a diffusing particle and the configuration of a linear polymer chain breaks down. The former may return to a position in space which it occupied previously without violating any physical restriction, whereas a corresponding configuration of the polymer molecule is forbidden. In the two-dimensional analogs depicted in Figs. 75 and 79 (pp. 402, 411), a configuration involving one or more "crossovers" is forbidden; hence both of the two-dimensional configurations shown in these figures are impossible. In three dimensions, this requirement becomes less restrictive.   A sequence of vectors, being mathematical lines, will never intersect exactly in space.   If, however, allowance is made for the finite cross section of the chain, *interferences* violating the necessity of volume exclusion will, for chains of many units, rule out the majority of configurations which may be generated by random assignment of bond directions taking into account only the previously considered restrictions on bond angles.   In other words, of the various diffusion paths that are acceptable for a particle which at each encounter with another particle changes direction in a way corresponding exactly to the valence angle restrictions for a given polymer structure, only a minute fraction, when the chain is very long, fulfill the requirement that no two portions of the chain shall overlap in space.   Those diffusion paths which happen to be expanded over a greater volume will be more likely to fulfill this condition.   Hence the actual chain will tend on this account to be more extended than is implied by the preceding calculations, which actually are strictly applicable to the diffusion analog rather than to a real polymer chain.   Average dimensions will be correspondingly greater and the entire distribution curve will be shifted toward larger values of r.

The configuration of the polymer molecule must depend also on its environment.   In a good solvent, where the energy of interaction between a polymer element and a solvent molecule adjacent to it exceeds the mean of the energies of interaction between the polymer-polymer and solvent-solvent pairs, the molecule will tend to expand further so as to reduce the frequency of contacts between pairs of polymer elements.   In a poor solvent, on the other hand, where the energy of interaction is unfavorable (endothermic), smaller configurations in which polymer-polymer contacts occur more frequently will be favored.

The intramolecular interference effect and the influence of the energy of mixing may be treated simultaneously from the broader point of view of intramolecular thermodynamic interaction, as will be discussed fully in Chapter XIV.   It suffices to observe here that the problem of polymer configuration is twofold.   It depends in the first place on the bond dimensions and angles as discussed above; these are the *short-range* effects having to do with relationships between units of the chain which are very near one another in sequence.   Secondly, the configuration is influenced also by thermodynamic interactions (including the interference effect) between the polymer elements and their environment.   Since the latter problem reduces to that of encounters between pairs of polymer elements which in general are remote from each other in the sequence along the chain, it is referred to as a *long-range* effect.   It depends on the polymer molecule and on its environment, whereas the first factor depends on parameters of the polymer molecule alone. In general, the effects of the second factor are by no means negligible. The not uncommon attempts to draw conclusions regarding hindrances to rotation from a direct comparison of an average dimension (e.g., $\sqrt{\overline{r^2}}$ obtained experimentally from light-scattering dissymmetry or from the intrinsic viscosity) with theoretical formulas based on the bond dimensions and angles characterizing the chain structure are therefore unsound. Perturbations due to the long-range interactions should be taken into account in one way or another before such correlations are attempted.

3b. The Unperturbed Polymer Chain.—If the solvent medium is sufficiently poor, i.e., if the interaction energy with the polymer is sufficiently positive, the energy of interaction may compensate exactly the influence of volume exclusion. When this condition is achieved,[20] the polymer chain will assume its so-called random flight configuration, its over-all dimensions then being determined solely by bond lengths and angles in accordance with the premises of the principal developments set forth in this chapter. This state of affairs will prevail in a poor solvent for the given polymer at a unique temperature, which can be deduced by relatively simple experiments (see Chap. XIII). Physical measurements made under these conditions will reflect the characteristics of the unperturbed molecule.

In view of the complications introduced by intramolecular interactions, revisions in notation are in order. Hereafter, in all circumstances where the perturbed and unperturbed dimensions, or parameters, are to be distinguished, the latter will be subscripted thus: $\sqrt{\overline{r_0^2}}$, $\sqrt{\overline{s_0^2}}$, and β_0. These subscripts would be appropriate throughout the equations given in this chapter, since they refer to unperturbed dimensions. The perturbed dimensions, represented generally without subscript, will differ from these by the average expansion α of the molecule arising from the long-range effects. Thus, we may write

$$\sqrt{\overline{r^2}} = \alpha\sqrt{\overline{r_0^2}}$$

$$\sqrt{\overline{s^2}} = \alpha\sqrt{\overline{s_0^2}} \tag{28}$$

$$\beta = \beta_0/\alpha$$

The quantitative treatment of α will be given in Chapter XIV. We note here merely that it often exceeds unity appreciably. Hence the differentiation is important.

Perturbed and unperturbed polymer dimensions deduced from intrinsic viscosity measurements, according to procedures which will be discussed later. are given in Table XXXIX of Chapter XIV. The unperturbed dimensions obtained experimentally are compared with those calculated, assuming free rotation, from the appropriate formula given previously. The rather large disparity serves to emphasize the importance of hindrances to free rotation, as discussed previously. Further discussion of the determination of polymer chain dimensions will be postponed.

As regards the distribution of configurations, the previous derivation obviously applies to the unperturbed molecule, and the symbols r_0 and β_0 should hereafter be used to represent the chain displacement length and Gaussian parameter, respectively. To an approximation which ordinarily is trivial, though perhaps not rigorously justified, the distribution $W(r)$ for the perturbed molecule may be taken to be Gaussian with a parameter β related to β_0 according to the last of the equations (28) above.

It may be shown that when the polymer concentration is large, the perturbation tends to be less. In particular, in a bulk polymer containing no diluent $\alpha = 1$ for the molecules of the polymer. Thus the distortion of the molecular configuration by intramolecular interactions is a problem which is of concern primarily in dilute solutions. In the treatment of rubber elasticity—predominantly a bulk polymer problem—given in the following chapter, therefore, the subscripts may be omitted without ambiguity.

APPENDIX A

Derivation of the Gaussian Distribution for a Random Chain in One Dimension.—We derive here the probability that the vector connecting the ends of a chain comprising n freely jointed bonds has a component x along an arbitrary direction chosen as the x-axis. As has been pointed out in the text of this chapter, the problem can be reduced to the calculation of the probability of a displacement of x in a random walk of n steps in one dimension, each step consisting of a displacement equal in magnitude to the root-mean-square projection $l/\sqrt{3}$ of a bond on the x-axis. Then

$$x = (n_+ - n_-)l/\sqrt{3}$$

where n_+ and n_- are the numbers of positive and negative projections, respectively, with $n_+ + n_- = n$. The problem becomes that of computing the probability of n_+ heads out of n tosses of a coin, which is

$$W(n_+, n_-) = (1/2)^n n!/n_+!n_-! \tag{A-1}$$

Letting $n_+ - n_- = m$, this may be written

$$W(n, m) = (1/2)^n n!/[(n + m)/2]![(n - m)/2]! \tag{A-2}$$

Introducing Stirling's approximations, $n! \cong \sqrt{2\pi} n^{n+1/2}/e^n$, for the factorials, which is legitimate provided both n and $n-m$ are large, and consolidating the resulting expression, we obtain

$$W(n, m) = (2/\pi n)^{1/2}[1 - (m/n)^2]^{-(n+1)/2}[(1-m/n)/(1 + m/n)]^{m/2} \quad \text{(A-3)}$$

If $|m/n| \ll 1$, corresponding to a small extension, the last factor in brackets may be replaced by $(1 - m/n)^m$, which may be written $[(1-m/n)^{n/m}]^{m^2/n}$. The quantity in brackets approaches e^{-1} when n/m becomes large. With this substitution and the similar one

$$[1 - (m/n)^2]^{-(n/m)^2} \cong e$$

for large $(n/m)^2$,

$$W(n, m) \cong (2/\pi n)^{1/2}e^{-m^2/2n} \quad \text{(A-4)}$$

The approximations introduced above are mutually compensating to some extent; hence the limitation on the size of m/n is not as stringent as otherwise might be inferred.

In this random walk analog with steps of fixed magnitude, m is permitted to assume only every other integral value; hence x must change in steps of $\pm 2l/\sqrt{3}$. To establish the connection with the distribution $W(x)$ for the molecule, which is continuous in x, we note that $W(m,n)$ must equal $W(x)\Delta x$, where Δx corresponds to $\Delta m = 2$. Hence

$$W(x) = W(m, n)/\Delta x$$

By substitution of $m = \sqrt{3}x/l$ and $\Delta x = 2l/\sqrt{3}$

$$W(x)dx = (3/2\pi)^{1/2}(1/n^{1/2}l)e^{-3x^2/2nl^2}dx$$

which is equivalent to Eq. (5).

APPENDIX B

Exact Treatment for the Freely Jointed Chain (or Equivalent Chain).[4,5]—Consider one of the bonds of a freely jointed chain acted upon by a tensile force τ in the x direction. Letting ψ_i represent the angle between the bond and the x-axis, its component on the x axis is $x_i = l \cos \psi_i$. The orientation energy of the bond is $-\tau x_i$, and the probability that its x component has a value between x_i and $x_i + dx_i$ therefore is proportional to

$$\exp(\tau x_i/kT)2\pi \sin \psi_i d\psi_i$$

or to

$$\exp(\tau x_i/kT)dx_i$$

The average value of x_i under the influence of the force τ must therefore be

$$\bar{x}_i = \frac{\displaystyle\int_{-l}^{l} x_i \exp(\tau x_i/kT)dx_i}{\displaystyle\int_{-l}^{l} \exp(\tau x_i/kT)dx_i}$$

$$= l[\coth(\tau l/kT) - kT/\tau l] \tag{B-1}$$

The quantity in brackets is the *Langevin function* of $\tau l/kT$, which may be written $\mathcal{L}(\tau l/kT)$.   Hence

$$\bar{x}_i = l\mathcal{L}(\tau l/kT) \tag{B-2}$$

Now the average of the algebraic sum of the average projections of each of the n bonds on the x-axis is

$$\bar{x} = \sum_{i=1}^{n} \bar{x}_i = nl\mathcal{L}(\tau l/kT) \tag{B-3}$$

Solving for the force required to maintain an average projection $\bar{x}$, we have

$$\tau = (kT/l)\mathcal{L}^*(\bar{x}/nl) \tag{B-4}$$

where $\mathcal{L}^*$ is the *inverse Langevin function*.   Conversely, τ represents the average force for an extension $\bar{x}$, and

$$\int_0^r \tau dr = (kT/l) \int_0^r \mathcal{L}^*(r/nl)dr$$

represents the work required to extend the chain to a length r.   It follows that the probability of an end-to-end length between r and $r+dr$ is

$$W(r)dr = \text{Const}\cdot\exp\left[-\left(\int_0^r \tau dr\right)\Big/ kT\right] 4\pi r^2 dr$$

$$= \text{Const}\cdot\exp\left[-l^{-1}\int_0^r \mathcal{L}^*(r/nl)dr\right] 4\pi r^2 dr \tag{B-5}$$

which is equivalent to Eq. (15) of the text.

APPENDIX C

The Mean-Square Distance of a Chain Element from the Center of Mass.[3,19]—In this appendix there is presented a derivation of the important relationship, expressed by Eq. (14) and frequently used in this chapter and elsewhere, between the mean-square distance $\overline{s^2}$ of a chain element (or segment) from the center of gravity of the random coiling chain and its average square end-to-end length $\overline{r^2}$.

As an extension of previous notation we let r_i represent the vector leading from one end of the chain to its i-th element in a given configuration. Similarly, s_i will be the vector from the center of mass to the same element. If Z is chosen to represent the vector from the end of the chain to the center of mass (i.e., $Z \equiv -s_1$), then

$$s_i = r_i - Z \qquad \text{(C-1)}$$

which must hold for all i. The definition of the center of mass requires that

$$\sum s_i = 0 \qquad \text{(C-2)}$$

or, from Eq. (C-1),

$$Z = (1/n) \sum r_i \qquad \text{(C-3)}$$

hence

$$Z^2 = (1/n^2) \sum_i \sum_j r_i \cdot r_j \qquad \text{(C-4)}$$

where n is the total number of elements in the chain, and Z is, of course, the magnitude of the vector Z.

In order to arrive at the desired relationship between the average of $\overline{s_i^2}$ over all chain elements and the mean-square end-to-end distance $\overline{r^2}$, it will be required to evaluate the sum $\sum s_i^2$ in terms of displacement lengths r. To this end, we proceed with the aid of Eq. (C-1) as follows:

$$\sum s_i^2 = \sum s_i \cdot s_i$$
$$= \sum (r_i - Z) \cdot (r_i - Z)$$
$$= \sum r_i^2 + nZ^2 - 2 \sum Z \cdot r_i \qquad \text{(C-5)}$$

The last term in Eq. (C-5) may be written $2Z \cdot \sum r_i$, which reduces according to Eq. (C-3) to $2nZ \cdot Z = 2nZ^2$. Hence

$$\sum s_i^2 = \sum r_i^2 - nZ^2$$

Substituting from Eq. (C-4), we obtain

$$\sum s_i^2 = \sum r_i^2 - (1/n) \sum_i \sum_j r_i \cdot r \qquad \text{(C-6)}$$

Now $r_i \cdot r_j$, representing the projection of one vector on the other multiplied by the scalar length of the latter, must equal

$$(r_i^2 + r^2 - r_{ij}^2)/2$$

where r_{ij} is the length of the vector connecting elements i and j. Making this substitution in Eq. (C-6), we obtain

$$\sum s_i^2 = (1/2n) \sum_i \sum_j r_{ij}$$
$$= (1/n) \sum_i \sum_{j<i} r_{ij}^2 \qquad (\text{C-7})$$

The mean-square distance of an element from the center of gravity of the given configuration is therefore

$$(1/n) \sum s_i^2 = (1/n^2) \sum_i \sum_{j<i} r_{ij}^2 \qquad (\text{C-8})$$

Let us now take the average over all chain configurations.* Indicating this average by the bar superscript, we have

$$\overline{s^2} = (1/n) \sum \overline{s_i^2} = (1/n^2) \sum_j \sum_{i<j} \overline{r_{ij}^2} \qquad (\text{C-9})$$

For the sake of definiteness, let us assume a freely jointed chain consisting of n bonds each of length l. Then the mean-square displacement length $\overline{r_{ij}^2}$ between elements i and j, by analogy with Eq. (13), will be given by $(j-i)l^2$, which yields

$$\overline{s^2} = (l^2/n^2) \sum_j \sum_{i<j} (j - i)$$
$$= (l^2/n^2) \sum_i \sum_{i=1}^{j-1} i$$

The summation $\sum_{i=1}^{j-1} i = j(j-1)/2 \cong j^2/2$. The succeeding summation, $\sum_{j=1}^{n} j^2 = n(n+1)(2n+1)/6 \cong n^3/3$ for large n. Hence,

$$\overline{s^2} = nl^2/6$$

giving, according to Eq. (13),

$$\sqrt{\overline{s^2}} = \sqrt{\overline{r^2}/6} \qquad (14)$$

* It is to be observed that two averaging processes are involved: one over the various elements of the chain as expressed in Eq. (C-8), and the other over all configurations of the chain.

Inasmuch as any real chain may be replaced by an equivalent freely jointed chain, provided the chain is sufficiently long, it is obvious that Eq. (14) is general.

REFERENCES

1. Lord Rayleigh, *Phil. Mag.*, [6], **37**, 321 (1919).
2. S. Chandrasekhar, *Rev. Mod. Phys.*, **15**, 3 (1943).
3. P. Debye, *J. Chem. Phys.*, **14**, 636 (1946).
4. W. Kuhn and F. Grün, *Kolloid Z.*, **101**, 248 (1942).
5. H. M. James and E. Guth, *J. Chem. Phys.*, **11**, 470 (1943).
6. W. Kuhn, *Kolloid Z.*, **76**, 258 (1936); **87**, 3 (1939).
7. H. Eyring, *Phys. Rev.*, **39**, 746 (1932); F. T. Wall, *J. Chem. Phys.*, **11**, 67 (1943).
8. J. D. Kemp and K. S. Pitzer, *J. Am. Chem. Soc.*, **59**, 276 (1937); **60**, 1515 (1938); G. B. Kistiakowski, J. R. Lacher, and F. Stitt, *J. Chem. Phys.*, **7**, 289 (1939).
9. S. Mizushima and T. Simanouti, *J. Am. Chem. Soc.*, **71**, 1320 (1949).
10. W. J. Taylor, *J. Chem. Phys.*, **16**, 257 (1948).
11. H. Benoit, *J. Chim. phys.*, **44**, 18 (1947); H. Kuhn, *J. Chem. Phys.*, **15**, 843 (1947).
12. P. Debye and F. Bueche, *J. Chem. Phys.*, **19**, 589 (1951).
13. P. J. Flory, L. Mandelkern, J. B. Kinsinger, and W. B. Shultz, *J. Am. Chem. Soc.*, **74**, 3364 (1952).
14. W. L. Roth and D. Harker, *Acta Krist.*, **1**, 34 (1948).
15. E. H. Aggarwal and S. H. Bauer, *J. Chem. Phys.*, **18**, 42 (1950).
16. R. O. Sauer and D. J. Mead, *J. Am. Chem. Soc.*, **68**, 1794 (1946).
17. H. Benoit, *J. Polymer Sci.*, **3**, 376 (1948).
18. F. T. Wall, *J. Chem. Phys.*, **11**, 67 (1943).
19. B. H. Zimm and W. H. Stockmayer, *J. Chem. Phys.*, **17**, 1301 (1949).
20. P. J. Flory, *J. Chem. Phys.*, **17**, 303 (1949); P. J. Flory and T. G. Fox, Jr., *J. Am. Chem. Soc.*, **73**, 1904 (1951)

CHAPTER XI

Rubber Elasticity

RUBBER and rubberlike materials are distinguished from other substances by a remarkable combination of two characteristics. In the first place, they are capable of sustaining large deformations without rupture; a maximum elongation of five to ten times the unstretched length is commonplace among typical rubbers. Secondly, the deformed rubber possesses the capacity to recover spontaneously very nearly to its initial dimensions, no appreciable fraction of the deformation remaining permanently after removal of the stress. Rubberlike bodies resemble liquids in respect to their deformability without rupture; they resemble solids in their capacity to recover. This combination of properties, aptly described by the term *long-range elasticity*, is by no means singularly characteristic of a restricted group of hydrocarbon polymers. Through proper choice of temperature and/or plasticization, virtually any long chain polymer may be induced to exhibit typical rubberlike behavior. Polyesters, polyamides, elastic sulfur (supercooled from the liquid), cellulose derivatives, and even certain inorganic gels as well, display the above-mentioned characteristics under appropriate physical conditions.

The unique structural feature common to all rubberlike substances is the presence of long polymer chains. These chains ordinarily are connected to one another by cross-linkages, but the preponderance of the structure consists of the intervening polymer chains each comprising a hundred or more single bonds between points of cross-linkage. That long polymer chains should be required for the fulfillment of high extensibility is immediately apparent. In the unstrained state the chains normally occur in randomly coiled arrangements, but they are able to rearrange to other configurations, and in particular to more highly extended ones. Thus, when the rubber is subjected to an externally applied stress, a large deformation can be accommodated merely through rearrangement of the configurations of the chains.

In the process of elongation, for example, uncoiling of the polymer chains takes place and they tend to become aligned parallel to the axis of elongation. This orientation generally is far from complete, however. No other type of chemical, or physical, structure is known which is capable of undergoing such large deformations without suffering permanent internal rearrangements (as in ordinary liquids).

The development of a restoring force in rubber as it is deformed is of course directly related to the second property which characterizes rubberlike elasticity, namely, the ability to recover its original dimensions. In the light of the structural interpretation of the deformation process set forth above, this restoring force may be presumed to originate in a tendency for the polymer chains, rearranged during deformation, to return to their initial configurations. Thermodynamic analysis of elastic deformation is particularly illuminating on this point, and the subject is therefore considered at length in the present chapter.

The presence of long chains, although a necessary condition for rubberlike behavior, is by no means a *sufficient* one. The system also must possess sufficient internal mobility to allow the required rearrangements of chain configuration during deformation and during recovery as well. More specifically, the polymer must be neither crystalline (to an appreciable degree) nor in the glassy state (see Chap. II); either impedes internal motions, or suppresses them altogether. Both usually may be avoided by maintaining the temperature sufficiently high, or by adding a suitable diluent, or plasticizer, in sufficient quantity.

Internal mobility permitting segmental motions of the chain elements, though essential, is not sufficient to assure genuine rubberlike behavior; a permanence of structure also is required. Otherwise permanent plastic flow rather than elastic recovery would be observed. This permanence of structure usually is achieved through insertion of occasional cross-linkages which join the chains into a space network of "infinite" extent (see Chap. IX). Whereas separate linear molecules could, through spontaneous rearrangements of their configurations, dissipate any orientation induced by deformation, chains bound at either end to the network cannot do so except through restoration of the initial macroscopic dimensions of the sample. Hence the retention of the ability to recover, and indeed the concept of an equilibrium retractive force which is a function of the strain, depend on the existence of a permanent network structure.* The relationship

---

* Linear polymers may exhibit elastic behavior with good recovery if the molecular weight is very high (10^6 or more), owing to the slow rate of relaxation of very long chains.

of stress to strain depends directly on the structure of the network, as will be shown later in this chapter.

A typical stress-strain curve for a "pure gum" natural rubber vulcanizate (i.e., without carbon black or other "fillers") is shown in Fig. 83. The stress rises slowly up to an elongation of about 500 percent (length six times initial length), then rises rapidly to a value at break in the neighborhood of 3000 pounds per square inch based on the initial cross section, or to nearly ten times this figure if the strength is calculated on the final cross section. When compared on an equivalent basis of weight per unit length, the ultimate strength is about half that of steel. Thus natural rubber combines easy extensibility (i.e., a low modulus of elasticity) over a wide range of extension with a high breaking strength. The marked rise in slope, which begins between 300 and 500 percent elongation, is associated with crystallization induced by stretching. In effect, the stretching process induces formation of a fiberlike oriented crystalline structure, which causes the modulus of elasticity to increase accordingly. On release of the stress, the crystalline portions melt and the original state is restored.[1]

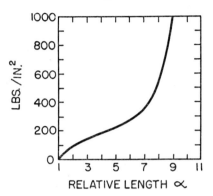

FIG. 83.—Stress-strain curve for gum-vulcanized natural rubber. The tensile force given on the ordinate axis is referred to the initial cross section.

1. THERMODYNAMICS OF RUBBER ELASTICITY

1a. Historical Background.—In 1805 John Gough,[2] a contemporary of John Dalton and a sharp critic of the latter's views on the nature of gases, reported a series of experiments on rubber which were offered in substantiation of the then prevailing calorific fluid theory of heat. These simple experiments, though qualitative only, embrace all of the important thermodynamic features of the elastic behavior of natural rubber.

Gough described his first experiment as follows:

Hold one end of the slip [of rubber] . . . between the thumb and forefinger of each hand; bring the middle of the piece into slight contact with the lips; . . . extend the slip suddenly; and you will immediately perceive a sensation of warmth in that part of the mouth which touches it. . . . For this resin evidently grows warmer the further it is extended; and the edges of the lips possess a high degree of sensibility, which enables them to discover these changes with greater facility than other parts of the body. The increase in temperature, which is perceived upon extending a piece of Caoutchouc, may be destroyed in an instant, by permitting the slip to contract again; which it will do quickly by virtue of its own spring, as oft as the stretching force ceases to act as soon as it has been fully exerted.

In the second experiment he observed:

If one end of a slip of Caoutchouc be fastened to a rod of metal or wood, and a weight be fixed at the other extremity . . . ; the thong will be found to become shorter with heat and longer with cold.

Finally reporting on the results of his third experiment, he concluded as follows:

If a thong of Caoutchouc be stretched, in water warmer than itself, it retains its elasticity unimpaired; on the contrary, if the experiment be made in water colder than itself, it loses part of its retractile power, being unable to recover its former figure; but let the thong be placed in hot water, while it remains extended for want of spring, and the heat will immediately make it contract briskly. . . . The object of the present letter is to demonstrate, that the faculty of this body to absorb the calorific principle, may be lessened, by forcibly diminishing the magnitude of its pores; and this essential point of the theory may be confirmed by experiment: for the specific gravity of a slip of Caoutchouc is increased, by keeping it extended, while it is weighed in water.

The full significance of these observations could not be appreciated in advance of the formulation of the second law of thermodynamics by Lord Kelvin and Clausius in the early 1850's. In a paper[3] published in 1857 that was probably the first to treat the thermodynamics of elastic deformation, Kelvin showed that the quantity of heat Q absorbed during the (reversible) elastic deformation of any body is related in the following manner to the change with temperature in the work $-W_{el}$ required to produce the deformation:

$$Q/T = (dW_{el}/dT) \tag{1}$$

(In the post-Gibbsian era this may be recognized as the equivalent of $(\partial \Delta F/\partial T) = -\Delta S = -Q/T$, where ΔF is the change in Gibbs free energy equal to $-W_{el}$ for a deformation at constant temperature and pressure.) If stretching is conducted adiabatically, Q should be replaced by $-C_p(\Delta T)_{ad}$, where C_p is the heat capacity of the elastic material and $(\Delta T)_{ad}$ is the adiabatic temperature change. If this substitution is made in Eq. (1) and both sides are differentiated with respect to the length L at constant pressure P

$$(C_p/T)(\partial T/\partial L)_{P,ad} = -(\partial^2 W_{el}/\partial T \partial L)_P$$

The force of retraction f equals $(\partial \Delta F/\partial L)_{T,P} = -(\partial W_{el}/\partial L)_{T,P}$; hence

$$C_p(\partial T/\partial L)_{P,ad} = T(\partial f/\partial T)_{P,L} \qquad (2)$$

The adiabatic temperature change with length is thus related to the temperature coefficient of the retractive force. Transforming Eq. (2) to give the change in temperature with applied force (instead of length) as a function of the change of length (instead of force) with temperature, we obtain

$$(\partial T/\partial f)_{P,ad} = -(T/C_p)(\partial L/\partial T)_{P,f} \qquad (3)$$

Equations (2) and (3) were derived by Lord Kelvin,[3] who pointed out that since a metal spring becomes more extensible with increase in temperature, i.e., since $(\partial f/\partial T)_{P,L}$ is negative, the temperature must fall in adiabatic extension; heat will be absorbed therefore during extension at constant temperature. Gough's experiments on the length of stretched rubber as a function of the temperature demonstrated the now-well-established fact that $(\partial L/\partial T)_{P,f}$ is negative and that $(\partial f/\partial T)_{P,L}$ is therefore positive for stretched rubber, in contrast to the values of opposite signs normally found for nonrubberlike bodies. Apparently unaware at this time of Gough's experiments and of the anomalous thermoelastic behavior of rubber, Kelvin drew the erroneous conclusion "that an india rubber band suddenly drawn out (within its limit of perfect elasticity) produces cold, and that, on the contrary, when allowed to contract, heat will be evolved from it. (For it is certain that an india-rubber band with a weight suspended by it will expand in length if the temperature is raised.)"[3] The logic was correct, but the premise false!*

Lord Kelvin's close associate, the expert experimentalist J. P. Joule,[4] set about to test the former's theoretical relationship and in 1859 published an extensive paper on the thermoelastic properties of various solids—metals, woods of different kinds, and, most prominent of all, natural rubber. In the half century between Gough and Joule not only was a suitable theoretical formula made available through establishment of the second law of thermodynamics, but as a result of the discovery of vulcanization (Goodyear, 1839) Joule had at his disposal a more perfectly elastic substance, vulcanized rubber, and most of his experiments were carried out on samples which had been vulcanized. He confirmed Gough's first two observations but contested the third. On stretching vulcanized rubber to twice its initial length, Joule ob-

* It appears from the writings of Joule[4] that Kelvin became aware of Gough's first two observations shortly thereafter and, of course, recognized at once the thermodynamic connection between them.

served, in conformity with the behavior of most solids, a very small decrease in density, from which he concluded that Gough's experiments on this point were excessively inaccurate. The increase in density reported by Gough probably was real nevertheless; the discrepancy may be attributed to differences in experimental conditions: Joule used vulcanized rubber at a rather low extension; Gough had at his disposal only unvulcanized rubber, which he doubtless stretched sufficiently to produce a considerable degree of crystallinity. Whereas rubber dilates slightly up to moderate elongations,[5] as should be expected from

TABLE XXXIII.—THERMOELASTIC MEASUREMENTS OF JOULE[4] ON VULCANIZED RUBBER

f lbs.	L/L_0 at 281°K	$(\partial L/\partial T)_{P,f}/L$ $\times 10^4/°C$	L/L^0 at 281°K mean for interval	$(\partial L/\partial T)_{P,f}/L$ $\times 10^4$ mean for interval	ΔT °C calcd.	ΔT °C obs.
0	1.00	(+2.2)[a]				
			1.07	(+ 0.6)	(−0.001)	−0.004
7	1.13	−1.1				
			1.18	−2.3	+0.004	+0.003
14	1.22	−3.5				
			1.30	−4.1	.009	.015
21	1.38	−4.7				
			1.49	−7.3	.018	.039
28	1.60	−10.0				
			1.73	−12.3	.035	.042
35	1.86	−14.6				
			2.00	−15.2	.050	.042
42	2.14	−15.9				

[a] Joule appears to have assumed $(\partial L/\partial T)_{P,f}/L$ to be zero for $f=0$. Given above in parentheses in column three is the value of the linear thermal expansion coefficient on the basis of which initial values in parentheses in other columns replace those given by Joule (see table on p. 106 of Ref. 4).

the classical theory of elasticity, this normal effect is overshadowed by the higher density of the crystalline fraction produced at higher elongations.[6,*] Moreover, unvulcanized rubber crystallizes more readily than the vulcanized material used by Joule. Both Gough and Joule commented on the already familiar (and technically important)

* This increase in density with elongation[6] signifies a value for Poisson's ratio (the ratio of the change in breadth to the change in length) in excess of one-half, which is usually given as the maximum value possible for any substance. The violation of this rule in the case of rubber at higher elongations can be excused as the result of a phase change (i.e., of crystallization).

loss of elasticity of raw rubber when cooled, and especially when cooled while stretched.   This is now recognized as a further manifestation of crystallization in natural rubber.

Joule's experiments on thermoelastic effects in vulcanized rubber are summarized in Table XXXIII, taken with minor modifications from his 1859 paper.[4] In one set of experiments the length of the sample stretched by a fixed load was measured at various temperatures between 0 and 50°C. The average linear expansion coefficients $(1/L)(\partial L/\partial T)_{P,f}$ obtained from repeated measurements over this temperature range are recorded in the third column for the loads and relative lengths L/L_0 given in the first two columns, respectively.   In another set of experiments a small thermocouple was inserted in the sample and the temperature changes (last column) were observed as the load was increased in increments of seven pounds.   The temperature changes given in the next-to-last column were calculated from the observed linear expansion coefficients using the awkward variant of Eq. (3)

$$\Delta T_{ad} = - (T/jc_p w)[(1/L)(\partial L/\partial T)_{P,f}]\Delta f \tag{4}$$

which Joule acknowledged as Kelvin's formula; c_p is the specific heat expressed in calories per pound and w the weight per unit cross section of the sample of length L, so that $wLc_p = C_p$, the heat capacity of the entire sample between marks; j is the factor for converting calories to food pounds.*   The approximate agreement between ΔT's calculated using Eq. (4) and those observed (Table XXXIII) was hailed by Joule as a confirmation of Kelvin's theoretical formula.   It will be observed that the reversal in sign of $(\partial L/\partial T)_{P,f}$ occurring at a low elongation is accompanied by a corresponding inversion in ΔT, as is required by theory.   This *thermoelastic inversion* in the sign of ΔT may be regarded as a necessary consequence of the fact that the linear expansion coefficient of sufficiently stretched rubber is negative, while that of unstretched rubber is positive (i.e., normal).   The inversion itself is of no particular significance beyond the implications of the sign of ΔT [or of $(\partial L/\partial T)$] at higher elongations.

Addition of the ΔT's in the last column of the table indicates an aggregate temperature rise of about 0.14° for a twofold stretch.   Recalling Gough's observations on the evolution of "calorific principle" during the extension of rubber, his rudimentary experimental method of detection would appear to have been possessed of an amazingly "high degree of sensibility."   Actually, the Gough effect is readily

---

* For Joule's samples $w = 0.2075 (L_0/L)$ lbs./ft.; c_p was taken as 0.415; and $j = 1390$ ft. lbs./cal.

detected in this manner only at elongations considerably exceeding twofold. The temperature rise continues to increase with further elongation, as Joule in fact showed, and it may exceed several degrees.[7] The heat evolved at higher elongations is greatly enhanced by the process of crystallization; the major portion of the temperature rise for higher elongations arises from the latent heat associated with crystallization.

According to the first law of thermodynamics the heat Q absorbed by the system may be equated to the change in internal energy ΔE plus the work W done by the system

$$Q = \Delta E + W$$

The heat $(-Q)$ expelled in the stretching of rubber consists of the difference between the work $(-W)$ performed in deforming the sample and the increase in its internal energy. Joule's experiments show, therefore, that the increase in internal energy during elongation must be less than the work expended. He did not, however, establish a quantitative relationship between the work spent and the heat recovered. From the fact that heat is evolved on stretching, it follows according to the second law of thermodynamics that a part, at least, of the elastic force derives from a decrease in entropy, rather than from an increase in internal energy (see below). Joule seems to have been aware of this far-reaching implication of the thermoelastic behavior of rubber, for he wrote as follows regarding "the evolution of heat by elastic fluids" such as gases: "If the heat given out in this case proved to be the equivalent of the work spent, then the natural inference was that the elastic force of a gas and its temperature are owing to the motion of its constituent particles." More recent work has shown that the heat $(-Q)$ evolved on stretching rubber is very nearly equal to the work $(-W)$ required for extension up to 150 to 250 percent over the unstrained length; hence the internal energy is substantially independent of the length and the elastic force over this range originates almost entirely from a decrease in entropy with length. This will be shown in greater detail below.

These deductions from basic facts of observation interpreted according to the rigorous laws of thermodynamics do not alone offer an insight into the structural mechanism of rubber elasticity. Supplemented by cautious exercise of intuition in regard to the molecular nature of rubberlike materials, however, they provide a sound basis from which to proceed toward the elucidation of the elasticity mechanism. The gap between the cold logic of thermodynamics applied to the thermoelastic behavior of rubber and the implications of its structure was bridged by Meyer, von Susich, and Valkó[8] in 1932, only a few years after the polymeric nature of rubber had been established by the work of Staudinger and others. They advanced the important suggestion that the decrease in entropy with elongation is a consequence of orientation of the molecular chains of the rubber.

1b. Thermodynamic Relationships.—More recent investigators of the influence of temperature on the elasticity of rubber have carried out measurements of the stress as a function of temperature at fixed length, in preference to measuring the length at fixed stress as Joule chose to do. The following thermodynamic analysis is therefore directed toward applications to the former type of experiment. The change in internal energy E accompanying the stretching of an elastic body may be written with complete generality as follows:

$$dE = dQ - dW \tag{5}$$

where dQ is the element of heat *absorbed by* and dW is the element of work *done by* the system on the surroundings. If P represents the external pressure and f the external force of extension

$$dW = PdV - fdL \tag{6}$$

If the process is conducted reversibly, $dQ = TdS$ where S is the entropy of the elastic body. Substitution of this expression for dQ in Eq. (5) will require dW to represent the element of reversible work. In order to comply with this requirement, the coefficients P and f in Eq. (6) must be assigned their equilibrium values. In particular, f will henceforth represent the equilibrium tension for a given state of the system, which may be specified variously by S, V, and L, by T, V, and L, or by T, P, and L. Then

$$dE = TdS - PdV + fdL \tag{7}$$

Introducing the Gibbs free energy defined by

$$F = H - TS = E + PV - TS$$

where $H = E + PV$ is the heat content

$$dF = dE + PdV + VdP - TdS - SdT$$

and substituting for dE from Eq. (7), we obtain

$$dF = VdP - SdT + fdL \tag{8}$$

This equation expresses the differential of the free energy in terms of the differentials of the experimentally most convenient independent variables, P, T, and L. It follows from Eq. (8) that

$$(\partial F/\partial L)_{T,P} = f \tag{9}$$

or

$$f = (\partial H/\partial L)_{T,P} - T(\partial S/\partial L)_{T,P} \tag{10}$$

Similarly

$$(\partial F/\partial T)_{P,L} = -S \tag{11}$$

Since the second derivative obtained by differentiating $(\partial F/\partial L)_{T,P}$ with respect to T at constant P and L is identical with that obtained by differentiating $(\partial F/\partial T)_{P,L}$ with respect to L at constant T and P, we obtain from Eqs. (9) and (11)

$$-(\partial S/\partial L)_{T,P} = (\partial f/\partial T)_{P,L} \tag{12}$$

which replaces Kelvin's equation (2). Upon substitution in Eq. (10) there is obtained (Wiegand and Snyder[9])

$$f = (\partial H/\partial L)_{T,P} + T(\partial f/\partial T)_{P,L} \tag{13}$$

which may be regarded as a thermodynamic equation of state for elasticity, in analogy with the ordinary thermodynamic equation of state

$$P = -(\partial E/\partial V)_T + T(\partial P/\partial T)_V \tag{14}$$

Other forms of the elastic equation of state will appear below.

Before proceeding further it is desirable to point out that $(\partial H/\partial L)_{T,P}$ will differ indiscernibly from $(\partial E/\partial L)_{T,P}$ in any likely application to rubberlike elastic phenomena. This may be seen by observing that the second term on the right-hand side of the relation

$$(\partial H/\partial L)_{T,P} = (\partial E/\partial L)_{T,P} + P(\partial V/\partial L)_{T,P}$$

(obtained from the definition of H) will be negligible compared with the first under all ordinary circumstances, owing to the smallness of $(\partial V/\partial L)_{T,P}$.* With the trivial reservation that the external pressures should not exceed a hundred atmospheres, therefore, in place of Eq. (13) we may use the relation

$$f = (\partial E/\partial L)_{T,P} + T(\partial f/\partial T)_{P,L} \tag{13'}$$

which is a somewhat closer analog of Eq. (14).

* The validity of this approximation is comprehensible from an experimental point of view if it is granted that the elastic behavior should be independent of the external pressure P for moderate pressures. Experiments performed at zero pressure, where the term $P(\partial V/\partial L)_{T, P}$ is literally equal to zero, should on this basis yield results equivalent to those actually obtained at one atmosphere; hence omission of the term in question appears justified.

For the purpose of illustrating the application of the thermodynamic equation of state to experimental data, consider the plot given in Fig. 84 for the retractive force, measured at fixed length, against the absolute temperature for a hypothetical elastic substance. The slope at any temperature T' gives the important quantity $-(\partial S/\partial L)_{T,P}$ according to Eq. (12); an increase in f with T at constant L shows immediately, therefore, that the entropy decreases with increase in length at constant T and P, as was indicated in the preceding section of this chapter. The second term in Eq. (13'), being the product of the absolute temperature and the slope, is represented by the length of the line AB in Fig. 84. Subtracting AB from the value of f at T', we obtain the increase in internal energy with length at constant T and P, i.e., $(\partial E/\partial L)_{T,P}$ (or $(\partial H/\partial L)_{T,P}$) as represented by the length OC. In other words, the slope of the tangent to the f vs. T curve representing measurements

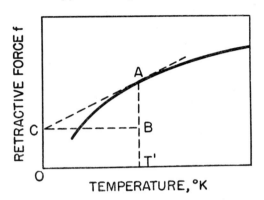

FIG. 84.—The retractive force f of a hypothetical elastic body plotted against the absolute temperature at constant pressure.

made at constant length and pressure gives the entropy decrease with length, and the intercept of the tangent on the ordinate axis gives the change in internal energy with length, both quantities referring to constant temperature and pressure. Curvature in a plot of this type would indicate, of course, changes in the values of $(\partial S/\partial L)_{T,P}$ and $(\partial E/\partial L)_{T,P}$ with temperature, and such changes would imply changes in the heat capacity with length.

Having shown how $(\partial E/\partial L)_{T,P}$ and $(\partial S/\partial L)_{T,P}$ may be deduced directly from force-temperature measurements conveniently carried out at fixed length and pressure, we are obliged to point out that these are *not*, unfortunately, the quantities desired. The ultimate object of the thermodynamic analysis is the deduction of quantities which may be interpreted physically, especially in terms of the structure. The primary structural effect of deformation being an orientation of the polymer chains, one may be inclined to regard $(\partial E/\partial L)_{T,P}$ and $(\partial S/\partial L)_{T,P}$, respectively, as the energy and entropy changes associated with molecular orientation. While such identifications may under some circumstances be acceptable (e.g., at higher elongations), from a quantitative point of view they are generally untenable[10,11] and their adoption may lead to serious misinterpretations. The error originates in the fact that orientation is not the *only* consequence of elongation *at constant pressure;* owing to the decrease in *internal pressure* (not to be confused with the external pressure P) which accompanies the application of a tensile force, the volume of the rubber increases also.[5] Both the energy and entropy change as consequences of this dilation, or, rather, of the increased intermolecular separation accompanying the dilation. In mathematical language

$$(\partial E/\partial L)_{T,P} = (\partial E/\partial L)_{T,V} + (\partial E/\partial V)_{T,L}(\partial V/\partial L)_{T,P} \qquad (15)$$

The derivative $(\partial E/\partial L)_{T,V}$ is appropriately identified with the change in energy arising solely from orientation of the chain structure.[10,11] Inasmuch as the condition of constant volume precludes alteration of the average intermolecular distance, the van der Waals intermolecular (or interunit) energy should remain approximately unchanged for a change in length at constant temperature *and volume*. The derivative which may be evaluated in the manner outlined above, namely $(\partial E/\partial L)_{T,P}$, and the one desired, $(\partial E/\partial L)_{T,V}$, differ by the value of the last term in Eq. (15). Having already pointed out in another connection that $(\partial V/\partial L)_{T,P}$ is very small for rubber, one might be inclined to conclude that this term also is negligible. However, $(\partial E/\partial V)_{T,L}$ is very large; hence the last term in Eq. (15) in general is appreciable;[5,11] up to elongations of 150 to 300 percent it may exceed $(\partial E/\partial L)_{T,V}$.

An expression corresponding to Eq. (15) may be set down for $(\partial S/\partial L)_{T,P}$. Similar considerations apply to this derivative also.

In quest of relationships which will allow evaluation of the desired derivatives of the entropy and energy at constant volume, it is convenient to introduce the work function $A = E - TS$. Proceeding as above

$$dA = dE - TdS - SdT$$
$$= -PdV - SdT + fdL \qquad (16)$$

Then

$$f = (\partial A/\partial L)_{T,V}$$
$$f = (\partial E/\partial L)_{T,V} - T(\partial S/\partial L)_{T,V} \qquad (17)$$

From (16) there may be obtained, in analogy with Eq. (12)

$$(\partial S/\partial L)_{T,V} = -(\partial f/\partial T)_{V,L} \qquad (18)$$

Hence

$$f = (\partial E/\partial L)_{T,V} + T(\partial f/\partial T)_{V,L} \qquad (19)$$

which is another form of the thermodynamic equation of state for elasticity. If an experiment were arranged to measure $(\partial f/\partial T)_{V,L}$, the desired derivatives could be had from Eqs. (18) and (19). This would require measurement of the force of retraction at constant length and *at constant volume*. To fulfill the latter condition, a hydrostatic pressure would have to be applied to nullify the increase in volume due to thermal expansion, and the pressures required would be excessive for any considerable range in temperature—a most unattractive experiment.

It is possible to show (see Appendix A) that

$$- (\partial S/\partial L)_{T,V} \cong (\partial f/\partial T)_{P,\alpha} \tag{20}$$

where α is the *elongation* defined by

$$\alpha = L/L_0 \tag{21}$$

L_0 being the length for zero stress *at the temperature* T and the pressure P. The coefficient $(\partial f/\partial T)_{P,\alpha}$ could be obtained experimentally, therefore, as the change in force with temperature measured at *constant pressure* while the length L is varied in the manner which will maintain a constant ratio between L and the unstressed length L_0 at each temperature. In other words, the length L should be varied slightly with temperature as dictated by the ordinary linear thermal expansion coefficient of the unstretched rubber. The actual measurements may, of course, be carried out at fixed length and the required adjustment introduced as a calculated correction[12] deduced from the change in force with length (i.e., from the stress-strain curve). The approximation involved in Eq. (20) should be extremely small (see Appendix A) under all ordinary circumstances; hence no appreciable error is involved in disregarding it.

Substituting Eq. (20) in (17)

$$f = (\partial E/\partial L)_{T,V} + T(\partial f/\partial T)_{P,\alpha} \tag{22}$$

This version of the thermodynamic equation of state for elasticity is most useful for interpretation of the experimental data discussed below. By measuring the force as a function of temperature at constant pressure and elongation α, one may readily derive $(\partial E/\partial L)_{T,V}$ from Eq. (22) and $(\partial S/\partial L)_{T,V}$ from Eq. (20).

1c. The Results of Stress-Temperature Measurements.—Hysteresis in the stress-strain behavior of rubber and rubberlike materials has presented the most serious problem encountered in the execution of otherwise simple experiments on the change of stress in stretched rubber with temperature at constant length (L) or at constant elongation (α). The stress exerted by the sample held at fixed length decays with time toward a limiting value, but the rate of approach to this limit decreases with time in such a way as virtually to preclude attainment of the actual limiting stress within any reasonable period of time. The rate of equilibration of the stress increases rapidly with temperature, but at higher temperatures chain scission of one sort or another imposes the hazard of irreversible changes in structure.

For the purpose of establishing by experiment the dependence of the stress on temperature, these difficulties may be satisfactorily circumvented by first permitting stress relaxation to proceed at fixed length for a reasonable time at the highest temperature to be employed (which should be safely below that at which chemical changes occur rapidly), then proceeding immediately with measurements at successively lower temperatures at the same length without waiting for further relaxation at these temperatures. Since the rate of relaxation is much slower at the lower temperatures, the measurements may be completed without appreciable further relaxation, and, on returning to the maximum temperature, the initial value of the stress is very nearly reproduced. (Joule used an equivalent procedure in his experiments on the length as a function of temperature at constant load, and subsequent investigators have adopted similar schemes in their more precise measurements.) The sample is then allowed to recover virtually to its initial length before performing the next stress-temperature series at a different length; or, preferably, a new sample is chosen for each series.

Although the stresses observed within each set of measurements carried out at a given length in the manner described above are reversibly related to one another, internal adjustment of the structure to a state of true elastic equilibrium usually will not have been attained. The stresses observed at the various temperatures will be consistently higher, on this account, than their equilibrium values for the given length. The degrees of relaxation attained at *different lengths* generally will not be equivalent; hence comparison of stresses observed for various lengths at a given temperature entails an error to the extent that the various states of deformation are not reversibly related.

The first comprehensive investigation of the stress-temperature behavior of vulcanized rubber, carried out in the manner described above and extending over wide ranges of temperature and elongation, was reported by Meyer and Ferri[12] in 1935. Samples of rubber vulcanized with sulfur alone were used, the stress being observed at various temperatures while the length was held constant. Similar, though somewhat more detailed, measurements were subsequently published by Anthony, Caston, and Guth,[13] who also used rubber vulcanized with

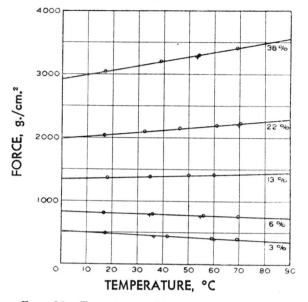

FIG. 85.—Force-temperature curves at constant length obtained by Anthony, Caston, and Guth[13] for natural rubber vulcanized with sulfur for elongations from 3 percent to 38 percent (at 20°C), as indicated.

sulfur. Some of the results obtained by the latter investigators at comparatively low elongations are shown in Fig. 85. In agreement with Meyer and Ferri's results, the force of retraction increases linearly with the temperature, but the force-temperature slope decreases with decreasing elongation, becoming negative below an elongation of about 10 percent. This thermoelastic inversion corresponds to the one observed by Joule in his measurements under constant load. Its explanation is the same, namely, that the normal (*positive*) thermal expansion of the unstressed rubber assumes dominance at low elongations.

The intercepts obtained by linearly extrapolating data such as are shown in Fig. 85 to 0°K are given by the lowest curve in Fig. 86. These represent, of course, $(\partial E/\partial L)_{T,P}$. The force f is given by the uppermost curve in this figure, and the quantity $-T(\partial S/\partial L)_{T,P}$ $= T(\partial f/\partial T)_{P,L}$ obtained by difference according to Eq. (13') is given by the dashed curve.

No thermoelastic inversion should appear in the force-temperature coefficient at constant elongation α, inasmuch as the effect of ordinary thermal expansion is eliminated by fixing α instead of the length L as the temperature is varied. As the elongation approaches unity, both the force and its temperature coefficient $(\partial f/\partial T)_{P,\alpha}$ must van-

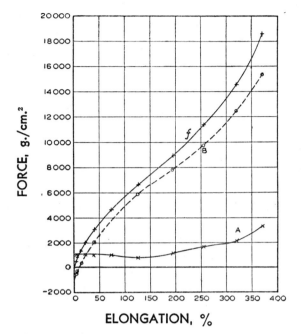

FORCE, g./cm.²

ELONGATION, %

Fig. 86.—The force of retraction f and its components $(\partial E/\partial L)_{T,P}$ (curve A) and $-T(\partial S/\partial L)_{T,P}$ (curve B), as obtained from force-temperature curves at fixed length such as are shown in Fig. 85, plotted against the percent elongation at 20°C. (Anthony, Caston, and Guth.[13])

ish according to the definition of α as the ratio of L to the length L_0 at which $f=0$ at the temperature T. It follows from Eq. (22) that $(\partial E/\partial L)_{T,V}$ also must inevitably vanish as $\alpha\to1$. Meyer and Ferri[12] observed that conversion of their results to fixed elongations not only eliminated the thermoelastic inversion but rendered the force directly proportional to the absolute temperature for elongations up to about 350 percent. In the light of Eq. (22), which was established theoretically in a somewhat different form by Elliott and Lippmann[10] ten years later (1945), this observation indicates that $(\partial E/\partial L)_{T,V}=0$ up to the elongation specified.

Anthony, Caston, and Guth[13] applied similar corrections to their measurements in order to express the force as a function of the temperature at various constant elongations. The results are shown in Figs. 87 and 88. The force-temperature plots are linear within experimental error. Their intercepts at 0°K are near zero, except at the higher elongations. These intercepts, which according to Eq. (22) must represent $(\partial E/\partial L)_{T,V}$, are plotted in the lowest curve of Fig. 88. It is doubtful

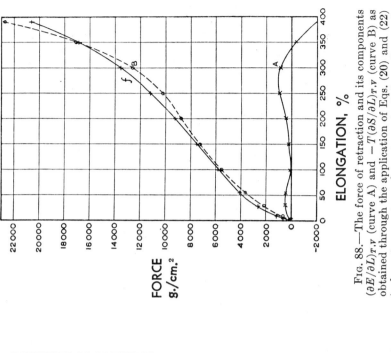

Fig. 88.—The force of retraction and its components $(\partial E/\partial L)_{T,V}$ (curve A) and $-T(\partial S/\partial L)_{T,V}$ (curve B) as obtained through the application of Eqs. (20) and (22) to force-temperature plots at constant elongation, such as are shown in Fig. 87. (Anthony, Caston, and Guth.[19]

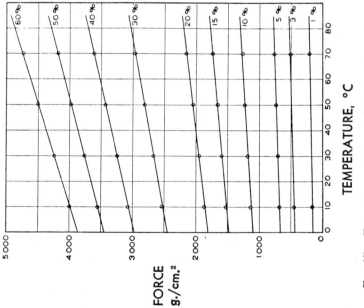

Fig. 87.—Force-temperature results of Anthony, Caston, and Guth[19] corrected to the constant elongations shown in percent.

that the differences from zero for the values obtained at elongations below 200 percent exceed the error of extrapolation. Thus, the persistent positive value of $(\partial E/\partial L)_{T,P}$ is due principally to the contribution from the volume expansion, as expressed by the last term in Eq. (15). Up to an elongation of 250 to 350 percent (for the particular unaccelerated rubber-sulfur vulcanizates used by Anthony, Caston, and Guth) $-T(\partial S/\partial L)_{T,V}$ is therefore very nearly equal to the total force. Thus the change in entropy is almost wholly responsible for the retractive force.

These conclusions have been confirmed by Wood and Roth,[14] who carried out measurements at both constant lengths and at constant elongations using natural rubber vulcanized with sulfur and an accelerator. Their results at constant elongation, to be considered later in connection with the thermodynamics of rubber elasticity at higher elongations, are summarized in Fig. 89.

Meyer and van der Wyk[15] observed that the shearing stress exerted by vulcanized rubber held at fixed shear strain is directly proportional to the absolute temperature within an experimental error not exceeding a few percent.

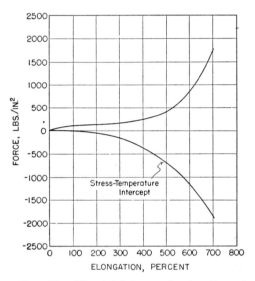

FIG. 89.—The total force of retraction at 25°C and $(\partial E/\partial L)_{T,V}$ obtained from the force-temperature intercepts at constant *elongation* for natural rubber gum-vulcanized using an accelerator. (Wood and Roth.[14])

In pure shear no decrease in internal pressure is involved; hence the complications caused by volume changes at constant pressure in linear extension did not enter. Meyer and van der Wyk vulcanized their sample of rubber between concentric cylinders open at either end. By rotating one cylinder with respect to the other, the stresses for strains less than 10 percent were easily measured. Their results offer the best available basis for concluding that the elastic retractive force in vulcanized natural rubber at low deformations originates in a decrease in entropy with elongation, and not at all from changes in energy, within the limit of accuracy of experiment.

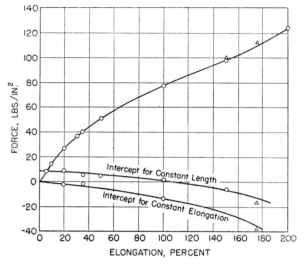

FIG. 90.—The force of retraction at 25°C and its internal energy component for gum-vulcanized GR-S synthetic rubber. Upper curve, total force f; middle curve, $(\partial E/\partial L)_{T,P}$ from the intercepts of force-temperature plots at constant length; lower curve, $(\partial E/\partial L)_{T,V}$ from the intercepts of stress-temperature plots at constant elongation. (Roth and Wood.[16])

Turning to synthetic rubbers, Roth and Wood[16] obtained the results shown in Fig. 90 for vulcanized butadiene-styrene copolymer ("GR-S" synthetic rubber; proportions of monomers by weight, 3:1). The internal energy coefficient $(\partial E/\partial L)_{T,V}$, given by the lowest curve, becomes increasingly negative with elongation. However, it amounts to only a small fraction of the total force; hence the decrease in entropy with elongation again is primarily responsible for the retractive force. Peterson, Anthony, and Guth[17] have reported similar results for several other synthetic rubbers.

1d. The Ideal Rubber.—The data available at present as summarized above show convincingly that for natural rubber $(\partial E/\partial L)_{T,V}$ is equal to zero within experimental error up to extensions where crystallization sets in (see Sec. 1e). The experiments of Meyer and van der Wyk[15] on rubber in shear indicate that this coefficient does not exceed a few percent of the stress even at very small deformations. This implies not only that the energy of intermolecular interaction (van der Waals interaction) is affected negligibly by deformation at constant volume—which is hardly surprising inasmuch as the average intermolecular distance must remain unchanged—but also that contributions from deformation of valence bonds, or from hindered rotations about bonds, are likewise very small.

If the rubber were composed of ideal polymer chains constructed out of bonds having their lengths and valence angles fixed and with perfectly free rotation prevailing about all single chain bonds, then we should predict that $(\partial E/\partial L)_{T,V}$ would be exactly equal to zero. Accordingly, an *ideal rubber* may be defined as one for which $(\partial E/\partial L)_{T,V} = 0$. The correspondence to an ideal gas for which $(\partial E/\partial V)_T = 0$ is immediately apparent. If the condition for elastic ideality applies over a range of temperatures, then according to the equation of state for elasticity, Eq. (22), the force of retraction should be proportional to the absolute temperature, i.e.

$$f = T(\partial f/\partial T)_{P,\alpha} = - T(\partial S/\partial L)_{V,T} \tag{23}$$

Similarly, the condition $(\partial E/\partial V)_T = 0$ for an ideal gas demands direct proportionality between P and T, for according to Eq. (14) we then have

$$P = T(\partial P/\partial T)_V \tag{24}$$

Since $(\partial P/\partial T)_V = (\partial S/\partial V)_T$

$$P = T(\partial S/\partial V)_T$$

for an ideal gas. Thus, just as the pressure of an ideal gas may be attributed solely to the increase in its entropy with volume, the force of retraction in an *ideal* rubber is due entirely to the decrease in entropy with increase in length. To paraphrase Joule, each is owing to the thermal motions of its "particles."

1e. Natural Rubber at High Elongations.—At the maximum elongations recorded in Fig. 88 the curve representing $(\partial E/\partial L)_{T,V}$ begins to turn rapidly downward. Simultaneously, the curve representing $- T(\partial S/\partial L)_{T,V}$ increases in slope. Both changes set in in the region just preceding the fairly abrupt rise in the stress-strain curve as ordinarily obtained by isothermal stretching (Fig. 83, p. 434). We have already identified this region as that in which strain-induced crystallization commences. Results of stress-temperature measurements[14] extended to higher elongations are shown in Fig. 89. The rubber in this case was vulcanized with the aid of an accelerator; hence less sulfur was combined with the rubber (see Sec. 2a), and this fact accounts for the onset of crystallization at a lower elongation. The pronounced negative values assumed by the change in internal energy with length at the higher elongations and the simultaneous enhancement in the decrease of the entropy with length are clear indications of an ordering process accompanied by a decrease in energy. These are the essential thermodynamic criteria for a crystallization process. X-ray studies show conclusively that crystallization does in fact commence in this range. One may estimate the degree of crystallinity reached at a given elongation[18] from the integral of $(\partial E/\partial L)_{T,V}$, which may be obtained, for example, from the area between the axis of α and the curve of Fig. 89. This integral represents, approximately at least, the internal energy change due to crystallization. Division by the heat of fusion (determined independently) gives the degree of crystallinity.

Rubbers which do not crystallize when stretched give no evidence of pronounced decreases in either $(\partial E/\partial L)_{T,V}$ or $(\partial S/\partial L)_{T,V}$ at higher elongations. (Compare Fig. 90.) Except when mixed with rather large quantities of a suitable "pigment" of fine particle size such as carbon black, noncrystallizing rubbers characteristically exhibit very low strengths (see Sec. 5a). It might be assumed that the intervention of rupture at comparatively low elongations in such rubbers forestalls the development of crystallinity and associated thermodynamic features, which would otherwise be observed at greater elongations. As later discussion will show, however, this interpretation is untenable; the so-called noncrystallizing rubbers truly lack the inherent capacity to crystallize when stretched, and this deficiency is responsible for the failure of the sample to withstand a large tensile stress.

That the net effect of crystallization on the stress-strain relationship is to increase the stress may seem to violate the principle of Le Châtelier. If the application of a stress is responsible for inducing crystallization, one would expect this transformation to occur in a manner which would facilitate compliance with the stress, rather than oppose it. Similarly, from a structural point of view, the development of axially oriented crystallites in which the molecular chains are nearly fully extended along the axis of elongation (fiber axis) should cause the length to be greater than would otherwise be observed at the same stress; hence it would seem that crystallization should lower the stress-strain curve instead of raise it. The clue to this apparent anomaly lies in the fact that the state of partial crystallinity ordinarily generated by isothermal stretching is not one of equilibrium. The first crystallites formed by elongation probably do contribute an incremental increase to the elongation (or a decrease in the stress). They also act as giant cross-linkages each of which binds together many chains.[19] Since these remain more or less permanently bound together by the crystallite as the elongation is increased, the chains extending into the amorphous regions from each crystallite are subject to much greater orientation with further elongation than would otherwise be their lot. Consequently, the decrease in entropy with further elongation is enhanced, as is also the increase in retractive force. Additional crystallization occurs with each increment in the elongation; hence the increase in the retractive force f accelerates with each successive elongation increment. If, on the other hand, the rubber is first stretched at a higher temperature such that crystallization does not occur during the stretching process, and subsequently crystallization is allowed to take place on cooling at fixed length, the stress *diminishes* as the temperature is lowered; it may even become negative, as manifested by an actual spontaneous increase in length at lower temperatures.[20]

In further confirmation of the extreme nonequilibrium disposition of the crystalline arrangement obtained during stretching, Wood and Roth[14] observed marked hysteresis in their stress-temperature studies on natural rubber stretched to higher elongations. Successive heating and cooling cycles at fixed length caused the stress to diminish with each cycle without approaching an apparent limit. Quantitative significance ordinarily cannot be attached to results obtained in this region, therefore. However, the coefficient $(\partial E/\partial L)_{T,V}$ remains strongly negative throughout successive temperature cycles; hence the qualitative conclusions drawn above from stress-temperature measurements at higher elongations appear to be valid. Magnitudes of stresses observed in this region, on the other hand, are of limited significance inasmuch as they may be expected to depend markedly on the process by which the given state of strain is reached.

If the temperature is reduced below the melting point for the crystalline polymer, (oriented) crystallites become stable even in the absence of stress. This results, qualitatively, in a shift of the origin of coordinates for the stress-strain curve (Fig. 89) to a point well to the right and in the region of steep ascent. For like reasons, a high melting crystalline polymer with fiberlike orientation (e.g., polyethylene and the polyamides) may be expected to exhibit at ordinary temperatures a stress-strain curve resembling the final phase of that for a rubber which crystallizes on stretching. Similar thermoelastic effects also may be expected, and under certain circumstances a resemblance actually is observed.[21] If additional crystallization is induced by stretching, both $(\partial E/\partial L)_{T,V}$ and $(\partial S/\partial L)_{T,V}$ should be negative and large.

Muscle proteins in the resting state exhibit the thermoelastic properties to be expected for an oriented semicrystalline polymer.[22] The thermoelastic coefficient $(\partial f/\partial T)_L$ is positive and large up to moderate extensions, indicating that both $(\partial S/\partial L)_T$ and $(\partial E/\partial L)_T$ are large in magnitude. The latter increases (algebraically) with extension, beginning with a large negative value at $\alpha = 1$, reaching zero at a higher elongation (80 percent extension in the case of *ligamentum nuchae*[22]), and becoming positive with further elongation. This indicates that crystallization may become nearly complete at high extensions, beyond which the internal energy increases owing presumably to deformations of bonds and of the crystal lattice.

2. THE STRUCTURE OF VULCANIZED RUBBER

2a. Vulcanization Processes.—The vulcanization of rubber, already cited in Chapter IX as an example of the generation of a space network by cross-linking, is an extremely complex process the principal features of which are accounted for by the sequence of reactions shown on p. 455 and based largely on the investigations of Farmer and his co-workers.[23]   The first step (i) consists in removal of an allylic hydrogen from the polymer chain by a free radical supplied by an accelerator; or in the absence of an accelerator the required radical may be furnished by sulfur in the chain form $\cdot S_y \cdot$.   In the second step a sulfur molecule (presumably cyclic S_8) combines with the radical with release of free sulfur.   The product (III) may then react with the isoprene unit of another chain according to step (iii) or (iv).   In the former an allylic hydrogen atom is removed; in the latter, addition occurs at the double bond with the formation of a cross-linkage subsequently stabilized by hydrogen transfer according to (vii).   In either case, a radical II is regenerated; hence the over-all process assumes the character of a chain reaction.   Reaction (iv) also may occur intramolecularly with formation of a cyclic structure.   The alternative intermediate (IV) may yield a cross-linkage according to step (vi), or, after loss of sulfur, cyclization may occur via step (v).   The polysulfide cross-linkages formed at low temperatures according to steps (iv) and (vi) may contain up to six sulfur atoms (probably in chain form), but on further heating at vulcanization temperatures (ca. 140°C) x decreases and the surplus sulfur combines in some way with other isoprene units without forming additional cross-linkages.

The reaction scheme presented above is supported by investigations of the products obtained in the reaction of olefins with sulfur[23,24] at temperatures of 140° or so, and especially by the reactions with the diisoprenoid hydrocarbons, dihydromyrcene and geraniolene, and with the hexamer, squalene.[23]

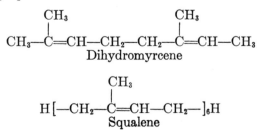

$$CH_3-\overset{\overset{\displaystyle CH_3}{|}}{C}=CH-CH_2-CH_2-\overset{\overset{\displaystyle CH_3}{|}}{C}=CH-CH_3$$
Dihydromyrcene

$$H[-CH_2-\overset{\overset{\displaystyle CH_3}{|}}{C}=CH-CH_2-]_6H$$
Squalene

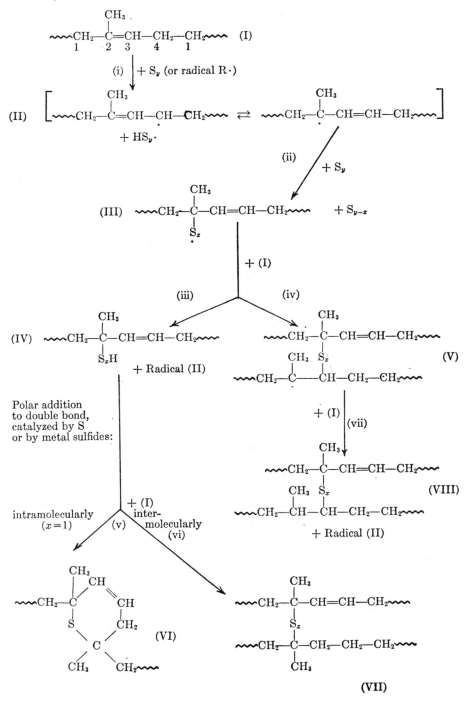

While free radical attack in step (i) is by no means confined to carbon atom 4, the products obtained in the reactions involving the lower polyisoprenes indicate that this process is the dominant one.   Likewise in step (ii) sulfur may frequently add at carbon atom 4 rather than at atom 2.   Addition in the manner shown is indicated, however, by infrared spectra,[25] which reveal the formation of —CH=CH— groups during vulcanization.   The scheme accounts also for the observed constancy of the C/H ratio during vulcanization and for the relatively low efficiency of utilization of sulfur in the formation of cross-linkages in the absence of accelerators.[26]   A preponderance of the sulfur is involved in addition without formation of cross-linkages; a considerable fraction of the thus-combined sulfur may occur in five- and six-membered heterocyclic rings formed by the mechanisms indicated.

While the action of various accelerators is not altogether clear, many of them (tetramethylthiuram disulfide, for example) may decompose with the release of radicals which accelerate step (i).   Others (e.g., the diphenylguanidine type) appear to promote the normal reactions of sulfur, perhaps by catalyzing the conversion of cyclic sulfur (S_8) to a more soluble and reactive form.[26,27]   The efficiency of sulfur utilization in forming cross-linkages generally is greater in accelerated recipes,[28] and this must be due, in part at least, to the zinc oxide and other ingredients such as zinc stearate and stearic acid normally present in them.   Zinc compounds probably form mercaptides $Zn(SR)_2$ either by reaction with the radicals (III)[28] or with the mercaptans (IV); the mercaptides decompose to cross-linked mono- or disulfides (VII) and ZnS.   The intramolecular cyclization reaction (v) may thus be circumvented with a resultant increase in the efficiency of sulfur utilization.   The amount of zinc sulfide formed may be a fair index of the degree of cross-linking under some conditions.[24]

In addition to the two major processes, cross-linking and chain modification (or cyclization), chain scission doubtless occurs also to varying degrees during conventional vulcanizations.   Processes of this nature are not difficult to envisage in the presence of free radicals. The radical intermediate (II) may, for example, undergo β-fission as follows:

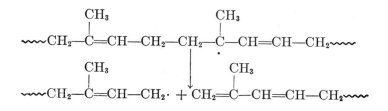

Chain scission will have the effect of decreasing the primary chain length, which, of course, is a highly important factor in respect to the vulcanizate network structure.

Various other chemical agents which by their nature are capable of producing cross-linkages between polymer chains effect the same changes in physical properties that are observed in sulfur vulcanization. One of the best known of these agents is sulfur monochloride, which readily combines with two molecules of an olefin (the mustard gas reaction). Applied to rubber, it induces vulcanization even at moderate temperatures, the probable structure of the cross-linkage being

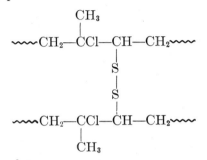

Ethyl azodicarboxylate

$$C_2H_5O—CO—N=N—CO—OC_2H_5$$

combines readily with a number of unsaturated compounds including rubber. Consequently bis-azodicarboxylates[29] containing two such groupings introduce cross-linkages which are presumed to be of the following nature:

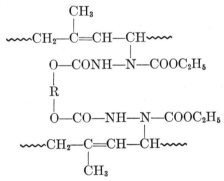

A small amount of the bis-azo compound—sufficient to react with only 1 or 2 percent of the isoprene units—converts the specimen to a material having all of the physical properties characteristic of vulcanized rubber.

Polychloroprene may be vulcanized by the action of metal oxides such as those of magnesium or zinc. These are presumed to act by the removal of halogen atoms from two chains, which are then joined through an ether linkage. Rubberlike polyesters may be vulcanized through the use of organic peroxides, the action of which is believed to depend on release of free radicals.[30] These may remove hydrogen atoms from α-methylene groups of the acid radical of the ester; cross-linkages are then formed by combination of the α-methylenic radicals thus produced. Other more conventional rubbers, notably GR-S and even natural rubber,[31] are susceptible to carbon-carbon cross-linking induced in this way by radicals released from peroxides.

The examples cited serve to emphasize that the process known as vulcanization should be regarded generally as a structural change and not merely as the result of chemical action induced by a restricted group of chemical substances.

2b. Quantitative Characterization of Network Structures.[32,33]—The changes in polymer structure occurring during cross-linking of long polymer chains are indicated schematically in Fig. 91. Regardless of the chemical method employed, the formation of cross-linkages consists in the joining together of pairs (or possibly larger groups) of structural units belonging to different primary molecules.

Since no Maxwell Demon is at hand to officiate in these unions, each pair of units participating in the formation of a cross-linkage will be selected at random; it is required merely that the partners be in suitable proximity at the instant of formation of the linkage. From a chemical point of view, cross-linkages are scattered throughout the bulk of the

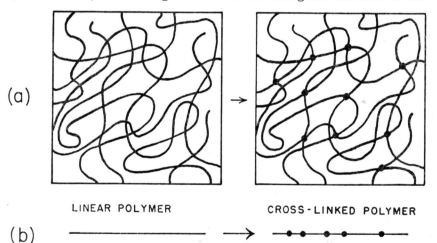

(a)

LINEAR POLYMER   CROSS-LINKED POLYMER

(b)

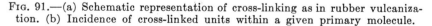

FIG. 91.—(a) Schematic representation of cross-linking as in rubber vulcanization. (b) Incidence of cross-linked units within a given primary molecule.

polymer at random points where pairs of polymer units are properly situated with respect to each other. Chain scission, to the extent that it is involved, will likewise occur at random.

Conventional vulcanization introduces cross-linkages in considerable excess of the number of primary (linear) molecules. Ordinarily, about one repeating unit in 50 to 100 is cross-linked. Since the primary molecules may consist of 1000 to 2000 units, an average of 10 to 40 cross-linked units per primary molecule is indicated. This is far in excess of the one per molecule required for incipient gelation in a homogeneous primary polymer (see Chap. IX); hence the network, or gel, should comprise almost all of the polymer, there being virtually no sol fraction. This is indeed observed. Only in certain synthetic rubbers having primary molecular weight distributions which include appreciable amounts of very low polymers is the fraction of the sol significantly large.

The structure of a random network may be characterized satisfactorily by two quantities: the density of cross-linking, previously designated (see Chap. IX) by the fraction ρ of the structural units engaged in cross-linkages, and the proportion of free chain ends. The latter quantity may be expressed on a corresponding basis as ρ_t, representing that fraction of the total units which occur as terminal units (i.e., which are connected to the structure by only one bond). Of course, both ρ and ρ_t may be converted to equivalents per unit weight, 10^6 grams being a convenient unit for this purpose. In general, the functionality of the interchain linkage should be designated also. In ordinary cross-linking, with which we shall be mainly concerned, this functionality is four. Having expressed the cross-linking density in terms of the number of *units* cross-linked, rather than in the number of cross-linkages, the terms chosen are immediately applicable to networks of other functionalities.

The two quantities ρ and ρ_t are satisfactory for the quantitative description of any random network structure. Alternative quantities sometimes are used to advantage, however. Instead of ρ_t, one may prefer to specify the number N of primary molecules

$$N = N_0\rho_t/2 \qquad (25)$$

and the number ν of cross-linked units

$$\nu = N_0\rho \qquad (26)$$

where N_0 is the total number of units. Since it has become customary to convert measurements which actually yield N or ρ to number average molecular weights, owing perhaps to a predilection for large figures, we note also that

$$N = N_0M_0/M = V/\bar{v}M \tag{27}$$

and

$$\nu = N_0M_0/M_c = V/\bar{v}M_c \tag{28}$$

where M_0 is the (mean) molecular weight per structural unit, V is the total volume, and $\bar{v}$ is the specific volume of the polymer. The primary molecular weight (number average) M is related uniquely to each of the quantities N/N_0 and ρ_t in accordance with the relations given. The molecular weight per cross-linked unit, M_c, is an analogous substitute for either ν or ρ as the variable defining the number of cross-linked units.

A hypothetical *perfect* network may be defined as one having no free chain ends; that is, the primary molecular weight M for a perfect network would be infinite. Any actual network could be formed (hypothetically) from the perfect one having the same degree of cross-linking ρ (or ν/N_0) by severing chains at random, or otherwise as may be required by the primary molecular weight distribution. These free chain ends may be regarded as flaws in the structure, about which more will be said later.

The perfect network will contain ν *chains*, a chain being defined here as in Chapter IX as a portion of the structure extending from a cross-linkage to the next one occurring along the given primary molecule. The cross-linkages represent fixed points of the *structure* in the sense that at each of them four (or f) chain ends are required to meet regardless of whatever displacements in space the cross-linkage may sustain. The network junctions, or cross-linkages, are not fixed in space, however, except below the glass transition temperature, or when restricted by copious crystallization, of course. A junction may diffuse over a limited region of space about its mean position through rearrangements in the configurations of structurally associated chains. A change in the configuration of one of the chains of the structure can be accomplished only by altering the configurations of several, at least, of the near-neighbor chains of the network (i.e., neighbors in the structural sense, not merely in space). When the specimen is deformed, the network junctions assume a new array of average spatial positions relative to one another. The coordinates expressing their relative average positions must change in proportion to the changes in macroscopic dimensions of the specimen (an affine transformation). This assertion follows from the isotropy of the network structure. It is fundamental to the theory of rubber elasticity to be developed in the following section.   The extension of these considerations to real networks should be considered before proceeding with the development, however.

Any real network must contain *terminal* chains bound at one end to a cross-linkage and terminated at the other by the end ("free end") of a primary molecule.   One of these is indicated by chain AB in Fig. 92, a.   Terminal chains, unlike the internal chains discussed above, are subject to no permanent restraint by deformation; their configurations may be temporarily altered during the deformation process, but rearrangements proceeding from the unattached chain end will in time re-

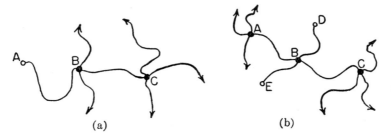

(a)   (b)

Fig. 92.—Effects of flaws consisting of ends of primary molecules on the network structure. ● indicates a cross-linkage, ○ the terminus of a primary molecule, and → signifies continuation of the network structure.[32]

store them to the random state.   Obviously these terminal chains, once they have relaxed, will contribute nothing toward elastic recovery nor to the change in any other property which depends on orientation induced by deformation.

The total number of chains, both internal and terminal, is $\nu + N$, as is evident from Fig. 91,b.   For every primary molecule there will be two terminal chains, i.e., a total of $2N$.   The number of internal chains must therefore be $\nu - N$.   Hence a fraction of the chains given by

$$s_a = (\nu - N)/(\nu + N) \tag{29}$$

are subject to permanent alteration by deformation.   By substitution from Eqs. (27) and (28), this may be converted to the more convenient expression

$$s_a = 1 - 2M_c/(M + M_c) \tag{29'}$$

Under all ordinary circumstances the average lengths of terminal chains and of internal chains will be the same, or nearly so; hence s_a represents the weight fraction of the structure which is "active" in deformation. For certain applications, as for example to crystallization induced by stretching, s_a is an appropriate measure of the effective portion of the network structure.   It is not, however, the proper correction factor to use in the theoretical account of rubber elasticity, as will now be shown.

For the treatment of rubber elasticity we require to know the total *number* of separate chain elements which are subject to orientation by deformation of the sample; their lengths are of secondary importance only.   The cross-linkage at B in the portion of a network structure shown in Fig. 92,b divides the short primary molecule DE into two inactive chains, and it also divides internal chain AC into two internal chains according to previous definition.   The former two (DB and BE) are properly discounted by the preceding analysis.   Whereas AB and BC would be counted as two chains, actually B does not represent a point of permanent restraint at all, and AC should be regarded as a single elastic element.   One may question also whether B in Fig. 92,a should be counted as a full-fledged cross-linkage, inasmuch as it has only three connections to the main body of the network.   The chain end imperfections lead to still other structural patterns which obscure the clear-cut distinction we have sought to draw between elastically active and inactive chains.   A different approach turns out to be better adapted to the quantitative derivation of the effects of network imperfections owing to free chain ends.

Let us for this purpose return to a consideration of the formation of the network by the incorporation of cross-linkages among the N primary molecules.[32,33]   We may imagine the process to be conducted in such a manner that the cross-linkages are introduced intermolecularly until this is no longer possible.   After $N-1$ cross-linkages have been introduced in this manner, the system will consist of one giant molecule.   It will not, however, be a network and it may conceivably dissipate all orientations temporarily imposed by a deformation if allowed enough time for rearrangement of the configuration of the giant structure.   Additional cross-linkages must necessarily be intramolecular, and as they are added the structure immediately acquires the characteristics of a network.   For each such cross-linkage added, exactly one additional closed circuit is imparted to the structure. In general, these circuitous paths will be very large in extent, and each of them becomes the equivalent of two elastic elements (the two sides of the circuit).

It is now an easy matter to express the total number $\nu_e/2$ of effective, or intrastructural, cross-linkages: it will equal the actual number $\nu/2$ of cross-linkages less the threshold number, which may be taken as N. Hence

$$\nu_e/2 = \nu/2 - N$$

For the elastically effective number of chain elements, taken as twice the number of closed circuits, we have

$$\nu_e = \nu - 2N = \nu(1 - 2N/\nu) \tag{30}$$

or

$$\nu_e = \nu(1 - 2M_c/M) \tag{30'}$$

of which use will be made later. The correction factor occurring in parentheses in Eqs. (30) and (30') resembles s_a and differs little from it for $M \gg M_c$.

For anything as complex as a random network structure, this account unquestionably is an oversimplification. Experiments which will be

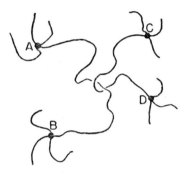

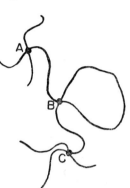

Fig. 93.—Entanglement involv-   Fig. 94.—Intramolecular
ing a pair of internal chains.[32]   cross-linking.[32]

discussed presently suggest that the internal chains may be subject to restraints other than those imparted at the network junctions. Entanglements of the sort indicated in Fig. 93 impose additional configurational restraints on the chains AB and CD. One such entanglement cannot be expected to increase the internal orientation during deformation as much as would a cross-linkage, but an average of several per chain may be presumed to increase appreciably the effective number of elastic elements in the structure.

It is important to emphasize that these entanglements become permanent only in the presence of definite cross-linkages and the network structure to which they give rise. Linear polymer chains of high molecular weight normally are *almost* hopelessly entangled with their neighbors. Given enough time, however, such a molecule may extricate itself from entanglements with a given group of neighbors and thereby dissipate such restraints as may have been imposed by them.

A terminal chain may rearrange similarly, but each of a pair of entangled internal chains is prevented from doing so by the permanent
chemical structure, as is clear from Fig. 93.   Inasmuch as the terminal
chains are not thus involved, we should expect the correction factors
for terminal chains occurring in Eqs. (30) and (30′) to be valid.   The
correction for entanglements should be expected to enter as a factor
modifying the entire expression for ν_e.   Unfortunately, there is at
present no theoretical basis from which to estimate this factor, except
to note that it presumably will be a function of ρ.

Another type of network imperfection, resulting from cross-linking
of two units not distantly related structurally, is indicated in Fig. 94.
Cross-linkages such as B are wasted (except insofar as the loop may be
involved in entanglements not otherwise operative).   The proportion
of these "short path" cross-linkages should be small ordinarily but
could become very large if the cross-linking process were carried out
in a dilute solution of the polymer.

3. THE STATISTICAL THEORY OF RUBBER ELASTICITY[32,34,35,36]

The formal thermodynamic analogy existing between an ideal rubber
and an ideal gas carries over to the statistical derivation of the force
of retraction of stretched rubber, which we undertake in this section.
This derivation parallels so closely the statistical-thermodynamic deduction of the pressure of a perfect gas that it seems worth while to
set forth the latter briefly here for the purpose of illustrating clearly
the subsequent derivation of the basic relations of rubber elasticity
theory.

The quantity to be calculated for the gas in a container of volume V_0
is the probability Ω that all of the gas molecules will spontaneously
move to a portion of the container having a volume V.   The probability
that any given molecule occurs in this volume is V/V_0, and the probability that all of them, ν in number, are there simultaneously is

$$\Omega = (V/V_0)^\nu$$

According to the Boltzmann relation, the entropy change ΔS for the
process of compression is given by

$$\Delta S = k \ln \Omega \tag{31}$$

where k is the Boltzmann constant.   Hence, for the compression of
the gas from V_0 to V.

$$\Delta S = k\nu \ln(V/V_0) \tag{32}$$

Introducing the thermodynamic relation

$$P = - (\partial A/\partial V)_T = - (\partial E/\partial V)_T + T(\partial S/\partial V)_T$$

and recalling that for a perfect gas $(\partial E/\partial V)_T = 0$, we have

$$P = T(\partial S/\partial V)_T$$
$$= kT\nu/V \tag{33}$$

which is the ideal gas law.

Turning now to a network composed of ν chains (ν_e for an imperfect network), it will be assumed that the actual cross-linking process is conducted in the isotropic (unoriented) polymer unswollen by diluents. Since cross-linking is a random process, the chains created by cross-linking will occur in random configurations. Their end-to-end vectors will be distributed according to the probability density function $W(x, y, z)$ of the preceding chapter, which for present purposes is given with adequate accuracy by the Gaussian function, Eq. (X-7). Such fluctuations in configuration as are possible (see p. 460) following cross-linking may alter the vectors of *individual chains*, but these alterations will occur in a random manner; hence the over-all distribution for the entire system of chain vectors will not be affected by them.

After the formation of the network structure has been completed, let the sample be subjected to any type of homogeneous strain (including swelling, to be treated in Chap. XIII) which may be described as an alteration of its dimensions X, Y, and Z by factors α_x, α_y, and α_z, respectively. As was pointed out on p. 460, the coordinates of the mean position of any junction of the network relative to any other must change by the same factors. Hence a chain i characterized by an end-to-end vector $\mathbf{r}_i$ having components x_i, y_i, z_i *after* deformation must have had components x_i/α_x, y_i/α_y, z_i/α_z *before* deformation. The number of chains having specified coordinates before deformation may be calculated from the probability density function $W(x, y, z)$ referred to above. Thus the number $\nu_i(x_i, y_i, z_i)$ of chains having end-to-end vector components from x_i to $x_i+\Delta x$, y_i to $y_i+\Delta y$, and z_i to $z_i+\Delta z$ *after deformation* must be given by

$$\nu_i(x_i, y_i, z_i) = \nu W(x_i/\alpha_x, y_i/\alpha_y, z_i/\alpha_z)\Delta x \Delta y \Delta z/\alpha_x\alpha_y\alpha_z$$

Substituting from Eq. (X-7),

$$\nu_i(x_i, y_i, z_i) = \nu(\beta/\pi^{1/2})^3$$
$$\exp\{- \beta^2[(x_i/\alpha_x)^2 + (y_i/\alpha_y)^2 + (z_i/\alpha_z)^2]\}\Delta x \Delta y \Delta z/\alpha_x\alpha_y\alpha_z \tag{34}$$

If, for example, the sample is stretched in the x-direction, its volume remaining constant, the x-components of all vectors will be increased by the factor α_x, while the y- and z-components will be decreased as is

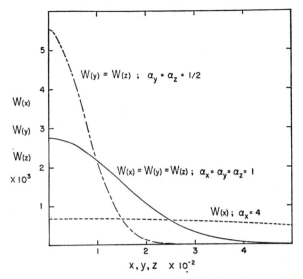

FIG. 95.—Transformation of the distributions of the x, y, and z components of chain displacement vectors by fourfold stretch along the x-axis at constant volume ($\alpha_x = 4$; $\alpha_y = \alpha_z = 1/2$) for chain molecules consisting of 10^4 freely jointed segments each of length $l = 2.5$ Å (compare Fig. 77). Initial distribution ($\alpha_x = \alpha_y = \alpha_z = 1$) shown by solid curve. Other curves represent x, y, and z components after deformation as indicated.

required by the constancy of the volume. Thus, if transverse contraction is allowed to occur equally in each of these directions ($\alpha_y = \alpha_z$), the y- and z-components must decrease by factors $1/\alpha_x^{1/2}$. The resulting transformation of the component distributions is illustrated in Fig. 95 for $\alpha_x = 4$. While the x-component distribution is broadened, that for the y- and for the z-components is narrowed.

In order to arrive ultimately at the entropy change accompanying deformation, we now proceed to calculate the configurational entropy change involved in the formation of a network structure in its deformed state as defined by α_x, α_y, and α_z. (We shall avoid for the present the stipulation that the volume be constant, i.e., that $\alpha_x\alpha_y\alpha_z = 1$.) Then by subtracting the entropy of network formation when the sample is undeformed ($\alpha_x = \alpha_y = \alpha_z = 1$), we shall have the desired entropy of deformation. As is obvious, explicit expressions will be required only for those terms in the entropy of network formation which are altered by deformation.

In analogy with the consideration of the probability of a spontaneous compression of the ideal gas to a volume V, we must find the probability Ω that the un-cross-linked polymer will occur spontaneously in a configuration consistent with the formation of a deformed network having numbers of chains ν_i in "states" x_i, y_i, z_i, as required by the deformation α_x, α_y, α_z. In order to gain the advantage of dealing with a set of ν chains which are uniquely defined, let it be supposed that ν of the units of the polymer molecules have been designated in advance for participation in the formation of cross-linkages. Introduction of this artifice is permissible for present purposes inasmuch as it will modify the entropy of formation of the undeformed network to the same extent. To be acceptable for the formation of a network characterized by the required deformation, an arrangement of the ν chains must fulfill two conditions:[36] (1) the proper chain vector distribution, specified by the ν_i, must occur, and (2) the units designated for cross-linking must occur in appropriate juxtaposition. The probabilities that these conditions are fulfilled will be written Ω_1 and Ω_2, respectively, with $\Omega = \Omega_1 \Omega_2$.

The probability that any given chain has the components x_i, y_i, z_i within the ranges Δx, Δy, Δz, may for convenience be written

$$\omega_i = W(x_i, y_i, z_i) \Delta x \Delta y \Delta z \tag{35}$$

The probability that *each chain* in the entire system complies with certain specified coordinates will be given by the product of such factors. By grouping together those with the same coordinates, this product may be written

$$\prod_i \omega_i^{\nu_i}$$

Since the particular selection of chains to comply with given end-to-end coordinates is of no importance, this expression must be multiplied by the number of permutations of the chains over the specified distribution. Multiplying the expression given above by this factor, $\nu! / \prod \nu_i!$, we obtain

$$\Omega_1 = \nu! \prod (\omega_i^{\nu_i}/\nu_i!)$$

On taking logarithms and introducing Stirling's approximation for the factorials,

$$\ln \Omega_1 = \sum \nu_i \ln(\omega_i \nu/\nu_i) \tag{36}$$

According to Eqs. (34) and (35) with Eq. (X-7) substituted for $W(x_i, y_i, z_i)$ in the latter

$$\ln(\omega_i \nu/\nu_i) = \beta^2 [x_i^2(1/\alpha_x^2 - 1) + y_i^2(1/\alpha_y^2 - 1) + z_i^2(1/\alpha_z^2 - 1)]$$
$$+ \ln(\alpha_x \alpha_y \alpha_z)$$

Introducing this expression into Eq. (36) together with Eq. (34) for ν_i, and replacing the summation by appropriate integrations, we obtain

$$\ln \Omega_1 = \nu\left[(\beta^3/\pi^{3/2})/(\alpha_x\alpha_y\alpha_z)\right] \int\int\int_{-\infty}^{\infty} \exp\left\{-\beta^2\left[(x/\alpha_x)^2\right.\right.$$
$$+ (y/\alpha_y)^2 + (z/\alpha_z)^2\right]\}$$
$$\times \left\{\beta^2\left[x^2(1/\alpha_x^2 - 1) + y^2(1/\alpha_y^2 - 1) + z^2(1/\alpha_z^2 - 1)\right]\right.$$
$$\left. + \ln(\alpha_x\alpha_y\alpha_z)\right\}dxdydz \qquad (37)$$

the subscript i having been omitted. After carrying out the integrations,*

$$\ln \Omega_1 = -\nu\left[(\alpha_x^2 + \alpha_y^2 + \alpha_z^2 - 3)/2 - \ln(\alpha_x\alpha_y\alpha_z)\right] \qquad (38)$$

The probability that any given one of the ν units designated for cross-linking has another such unit situated next to it within the required volume element δV is $(\nu-1)\delta V/V$, where V is the total volume. The probability that another unit selected from the remaining $\nu-2$ designated units is similarly paired is $(\nu-3)\delta V/V$, and so forth. The probability that all of them are paired is[36]

$$\Omega_2 = (\nu - 1)(\nu - 3) \cdots (1)(\delta V/V)^{\nu/2}$$
$$\cong (\nu/2)!(2\delta V/V)^{\nu/2}$$

Replacing V with $\alpha_x\alpha_y\alpha_z V_0$, where V_0 is the volume of the undeformed sample, we arrive at the result

$$\ln \Omega_2 = -(\nu/2)\ln(\alpha_x\alpha_y\alpha_z) + \text{const.} \qquad (39)$$

The constant consists of $\ln(\nu/2)! + (\nu/2)\ln\delta V - (\nu/2)\ln(V_0/2)$. All of these terms are independent of the deformation; hence their composition is of no concern.

If Eqs. (38) and (39) are substituted in the Boltzmann expression for the entropy of formation of the deformed network[36]

$$S = k \ln \Omega = k \ln \Omega_1 + k \ln \Omega_2$$

then

$$S = \text{Const.} - (k\nu_e/2)\left[\alpha_x^2 + \alpha_y^2 + \alpha_z^2 - 3 - \ln(\alpha_x\alpha_y\alpha_z)\right] \qquad (40)$$

where ν has been replaced by the effective number ν_e of chains. The entropy change involved in deformation, obtained by subtracting from Eq. (40) the value of S for $\alpha_x = \alpha_y = \alpha_z = 1$, is

$$\Delta S = -(k\nu_e/2)\left[\alpha_x^2 + \alpha_y^2 + \alpha_z^2 - 3 - \ln(\alpha_x\alpha_y\alpha_z)\right] \qquad (41)$$

It is worth noting that from Eq. (38) onward the parameter β,

* The integrals occurring in Eq. (37) are of the types $\int_{-\infty}^{\infty}\exp(-\beta^2x^2/\alpha_x^2)dx = (\alpha_x/\beta)\sqrt{\pi}$ and $\int_{-\infty}^{\infty}\exp(-\beta^2x^2/\alpha_x^2)x^2dx = (\alpha_x/\beta)^3\sqrt{\pi}/2$.

which alone characterizes the given Gaussian probability distribution and which depends on the (average) contour length of the chains (see Eq. X-6), has vanished from the equations. The only quantity pertaining to the network structure which is retained is ν_e. Only the total number of (effective) chains matters; their contour lengths and statistical qualities, beyond adherence to the Gaussian form for the distribution of their displacement lengths, are of no importance. It follows that variations in the contour lengths of the chains of the network may be disregarded, within the scope of the present treatment at any rate.

For ordinary deformations of rubberlike substances (excluding swelling phenomena to be discussed in the following chapter) it is permissible to assume constant volume, i.e., $\alpha_x \alpha_y \alpha_z = 1$. The logarithmic term in Eq. (41) then disappears. In the particular case of *elongation* at constant volume, $\alpha_y = \alpha_z = \alpha_x^{-1/2}$, giving for the entropy of deformation

$$\Delta S = - (k\nu_e/2)(\alpha^2 + 2/\alpha - 3) \tag{42}$$

where α is written for α_x.

The elastic retractive force for an ideal rubber $((\partial E/\partial L)_{T,V} = 0)$ according to Eq. (17) is

$$f = - T(\partial S/\partial L)_{T,V}$$

which may be written*

$$f = - (T/L_0)(\partial S/\partial \alpha)_{T,V} \tag{43}$$

The constant volume condition having already been introduced, Eq. (42) may be used to obtain

$$(\partial S/\partial \alpha)_{T,V} \equiv (\partial \Delta S/\partial \alpha)_{T,V} = - k\nu_e(\alpha - 1/\alpha^2)$$

On substituting in Eq. (43) and dividing by the initial cross-sectional area V_0/L_0, or V/L_0, we obtain for the retractive force τ per unit initial cross-sectional area[34]

$$\tau = (RT\nu_e/V)(\alpha - 1/\alpha^2) \tag{44}$$

where ν_e is expressed in moles. By the use of Eq. (30) or (30'),

$$\tau = RT(\nu/V)(1 - 2N/\nu)(\alpha - 1/\alpha^2) \tag{45}$$

---

* An approximation, which actually is quite negligible, is involved in letting $(\partial S/\partial L)_{T,V} = (\partial S/\partial \alpha)_{T,V}/L_0$, since α is defined as L/L_0 where both lengths are measured at the same temperature and *pressure*. See Eq. (21).

or[32]

$$\tau = RT(1/\bar{v}M_c)(1 - 2M_c/M)(\alpha - 1/\alpha^2) \tag{45'}$$

Equations (44), (45), and (45') are alternative expressions for the theoretical equation of state for an ideal rubber. Equations of state for swollen networks, derived in Appendix B, have the same form. Only the magnitude predicted for τ at a given elongation is affected by swelling.

Other types of deformation may be handled similarly. Shear, for example,[1,34] may be treated as a homogeneous strain involving an increase in one coordinate (x) while another (z) remains constant, the volume being constant also. Thus, we may let $\alpha_x = \alpha$, $\alpha_y = 1/\alpha$, and $\alpha_z = 1$. On substitution of these conditions in Eq. (41), the deformation entropy *per unit volume* becomes

$$\Delta S^v = - (R\nu_e/2V)(\alpha^2 + 1/\alpha^2 - 2) \tag{46}$$

The shear strain γ is $\alpha - 1/\alpha$. Hence

$$\Delta S^v = - (R\nu_e/2V)\gamma^2 \tag{47}$$

The shearing stress (treated as simple shear[1]) is given by

$$\tau_s = - T(\partial \Delta S^v/\partial \gamma)$$

hence

$$\tau_s = (RT\nu_e/V)\gamma \tag{48}$$

The shearing stress should therefore be proportional to the shear strain, i.e., the rubber should obey Hooke's law in shear whereas this is not true of elongation.[1,34] The quantity $RT\nu_e/V$ is the modulus of rigidity.

4. EXPERIMENTAL STRESS-STRAIN BEHAVIOR OF VULCANIZED RUBBERS AT MODERATE ELONGATIONS

The theoretical equation of state for an ideal rubber in tension, Eq. (44) or (45), equates the tension τ to the product of three factors: RT, a structure factor ν_e/V, (or ν_e/V_0, the volume of the rubber being assumed constant), and a deformation factor $(\alpha - 1/\alpha^2)$ analogous to the bulk compression factor V_0/V for the gas. The equation of state for an ideal gas, which for the purpose of emphasizing the analogy may be written $P = RT(\nu/V_0)(V_0/V)$, consists of three corresponding factors. Proportionality between τ and T follows necessarily from the condition $(\partial E/\partial L)_{T,V} = 0$ for an ideal rubber. Results already cited for real rubbers indicate this condition usually is fulfilled almost within experimental error. Hence the propriety of the temperature factor need not be discussed further. The structure factor as expressed by Eqs. (45) and (45′) consists of two component factors, one (ν/V) representing the concentration of cross-linked units and the other ($1-2N/\nu$) depending on the relative proportion of flaws occurring as ends of primary molecules. The deformation factor prescribes a particular form for the stress-strain curve. The extent to which these predictions of rubber elasticity theory are confirmed by experimental results will now be considered.

4a. The Stress-Strain Curve.—Stress-strain results obtained by Treloar[37] on natural rubber vulcan-

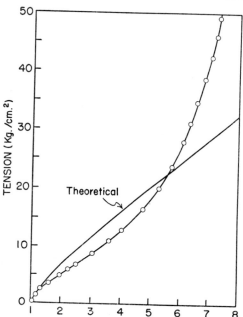

Fig. 96.—Theoretical and experimental stress-strain curves for simple elongation of gum-vulcanized rubber. (Treloar.[37])

ized with sulfur alone are shown in Fig. 96 in comparison with a theoretical curve calculated using the function $\alpha - 1/\alpha^2$ with an arbitrarily chosen constant of proportionality (representing $RT\nu_e/V$ of Eq. 44). The influence of crystallization previously discussed is in evidence beyond $\alpha = 4$ or 5. Appreciable deviations from the theoretical curve occur at lower elongations. It would appear that a better compromise could have been achieved in fitting the data to Eq. (44) from $\alpha = 1$ to 4. However, the particular constant of proportionality used was chosen in consideration of data obtained at still lower "elongations," i.e., at $\alpha < 1$.

In turning to the last-mentioned results, it is to be noted that the theory given above should, in principle, be equally applicable to a deformation for which $\alpha < 1$, provided that expansions are allowed to occur in each of the transverse directions to the extent $\alpha^{-1/2}$. The deformation thus described consists of a compression of the sample in the x-direction with simultaneous *uniform* expansions in both the y- and the z-directions as required by constancy of the volume. If the deformation were to be carried out by pressing the sample in the x-

direction between flat surfaces, uniform lateral adjustment would be prevented by friction at the surfaces. To circumvent complications of this nature inherent in any compression method, Treloar[37] resorted to the pneumatic inflation of a rubber sheet. In this way the rubber was stretched equally in the two directions parallel to the plane of the sheet, without inducing extraneous tangential forces on the surface. If each of these elongations is represented by $\alpha^{-1/2}$, the com-

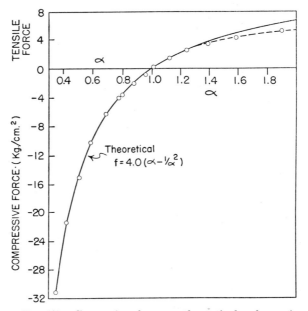

Fig. 97.—Comparison between theoretical and experimental stress-strain curves for vulcanized rubber for elongations $\alpha = 0.40$ to 2.0. Points are experimental data; those for $\alpha < 1$ were obtained by inflating a rubber sheet. (Treloar.[37])

pression normal to the surface of the sheet is $\alpha(<1)$ and the equation of state previously derived for elongation should apply. Treloar's results treated in this manner are shown in Fig. 97; those included for $\alpha > 1$ are repeated from Fig. 96, and the theoretical curves are the same. The fit of the calculated curve to the experimental points is good from $\alpha = 0.4$ to 1.2. At higher elongations the deviation becomes severe.

Anthony, Caston, and Guth[13] obtained considerably better agreement between the experimental stress-strain curve for natural rubber similarly vulcanized and the theoretical equation over the range $\alpha = 1$ to 4. Kinell[38] found that the retractive force for vulcanized polychloroprene increased linearly with $\alpha - 1/\alpha^2$ up to $\alpha = 3.5$.

Throughout all of the experiments cited above on isothermal stretching, time and patience alone were invested in attempting to arrive at an equilibrium stress (or strain, for measurements made at constant stress). These are not enough, and the results quoted undoubtedly are in error to some extent on this account. In more recent work close approach to equilibrium has been achieved by taking advantage of the pronounced increase in internal mobility brought about by absorption of a diluent. In the procedure used by Gee[11] the sample is allowed to absorb the vapor of a volatile liquid like petroleum ether while subjected to the assigned load. The diluent is then removed in vacuo and the length observed at once. Alternatively,[29] the sample may be held at fixed length while it is subjected to swelling by a suitable volatile liquid with which it is surrounded for a brief period. The absorbed liquid may then be removed using a stream of dry nitrogen after which the stress is observed. That the results obtained by each of these procedures represent states of equilibrium is confirmed by the close agreement of the lengths observed when the given stress is approached in either direction, i.e., from a higher or from a lower stress.

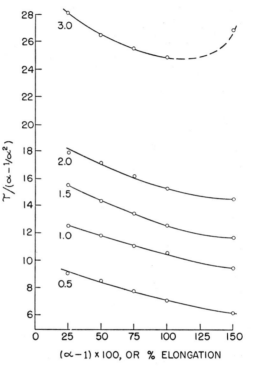

FIG. 98.— $\tau/(\alpha-1/\alpha^2)$ for natural rubber, cross-linked to the densities ($\rho \times 100$) indicated with each curve using a bis-azo cross-linking agent. (Flory, Rabjohn, and Shaffer.[29])

For the purpose of testing the validity of what we have called the deformation factor, it is convenient to plot $\tau/(\alpha-1/\alpha^2)$ vs. α; if rubbers swollen permanently (not merely for purposes of equilibration) are included, the quantity $\tau_0 v_2^{1/3}/(\alpha-1/\alpha^2)$ is to be preferred, where τ_0 is the tension referred to the cross section when unswollen and unstretched, and v_2 is the volume fraction of polymer in the swollen mixture (see Appendix B, Eq. B-5). According to theory, both of these quantities should be independent of the deformation, and the latter should be independent of the degree of swelling as well. On the basis of careful stress measurements on natural rubber equilibrated in the manner previously described, Gee[11] found that $\tau_0 v_2^{1/3}/(\alpha - 1/\alpha^2)$ decreased with extension, especially in the range from $\alpha = 1$ to 2. The decrease with elongation was diminished by swelling (permanently) with toluene. Results[29] obtained similarly on natural rubber vulcanized to varying extents with decamethylene-bis-methyl azodicarboxylate shown in Fig. 98 lead to equivalent conclusions. The approach to theory was again somewhat better for the swollen systems (results not shown here).

These experimental results show conclusively that the deformation factor occurring in the theoretical equation of state offers only a crude approximation to the form of the actual equilibrium stress-strain curve. The reasons behind the observed deviation are not known. It does appear, however, from observations on other rubberlike systems that the type of deviation observed is general. Similar deviations are indicated in "butyl" rubber[33] (essentially a cross-linked polyisobutylene) and even in polyamides having network structures[39] and exhibiting rubberlike behavior at high temperatures (see Sec. 4b).

Considerably better agreement with the observed stress-strain relationships has been obtained through the use of empirical equations first proposed by Mooney[40] and subsequently generalized by Rivlin.[41] The latter showed, solely on the basis of required symmetry conditions and independently of any hypothesis as to the nature of the elastic body, that the stored energy associated with a deformation described by α_x, α_y, α_z at constant volume (i.e., with $\alpha_x \alpha_y \alpha_z = 1$) must be a function of two quantities: $(\alpha_x^2 + \alpha_y^2 + \alpha_z^2)$ and $(1/\alpha_x^2 + 1/\alpha_y^2 + 1/\alpha_z^2)$. The simplest acceptable function of these two quantities can be written

$$W = C_1(\alpha_x^2 + \alpha_y^2 + \alpha_z^2 - 3) + C_2(1/\alpha_x^2 + 1/\alpha_y^2 + 1/\alpha_z^2 - 3) \qquad (49)$$

This is Mooney's equation[40] for the stored elastic energy per unit volume. The constant C_1 corresponds to the $kT\nu_e/2V$ of the statistical theory; i.e., the first term in Eq. (49) is of the same form as the theoretical elastic free energy per unit volume $\Delta F^v = -T\Delta S/V$ where ΔS is given by Eq. (41) with $\alpha_x \alpha_y \alpha_z = 1$. The second term in Eq. (49) contains the parameter C_2 whose significance from the point of view of the structure of the elastic body remains unknown at present. For simple extension, $\alpha_x = \alpha$, $\alpha_y = \alpha_z = 1/\alpha^{1/2}$, and the retractive force per unit initial cross section, given by $dW/d\alpha$, is

$$\tau = 2C_1(\alpha - 1/\alpha^2) + 2C_2(1 - 1/\alpha^3) \qquad (50)$$

the first term of which corresponds to the theoretical equation (44). By arbitrarily choosing an appropriate value for C_2, a much better fit to experimental data is possible than through the use of the first

term alone.* Stress-strain relations may be derived for other types
of deformation as well, and the agreement with experimental data is
likewise improved.†

4b. The Force of Retraction in Relation to Network Structure.—If
the structure of the network as specified by N and ν is known, the
theoretical equation of state, Eqs. (44) or (45), permits explicit calculation of the stress for a given elongation and temperature; there are no
arbitrary parameters. A
direct comparison between
observed and calculated
stresses is indicated, therefore, as a crucial test of the
statistical theory. Experiments of this nature have
received less attention than
is deserved owing to the almost total lack of inde-
pendent methods for ascertaining the structure of a
network. It has been nec-
essary to choose cross-linking procedures which permit
quantitative manipulation
of the number of cross-
linkages formed. Through
the use of the bis-azodi-
carboxylates referred to
above, for example, the
production of a number of
cross-linkages equal to the

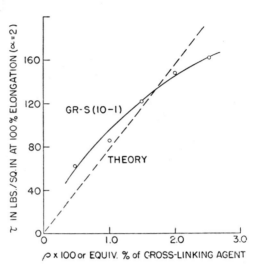

FIG. 99.—Dependence of the force of retraction on the degree of cross-linking ρ (expressed as equivalents per butadiene unit in
copolymer containing 72 percent of butadiene
by weight) in GR-S cross-linked with bis-azo
compound. (Flory, Rabjohn, and Shaffer.[29])

number of molecules of cross-linking agent used seems reasonably well
assured. The reaction is a rapid one which goes smoothly to completion. Limited compatibility with the rubber presents a more serious
problem, for the agent may not be properly distributed through the
rubber on this account.[29]

The equilibrium tension τ at $\alpha = 2$ for GR-S synthetic rubber vulcanized in this manner is plotted in Fig. 99 against the mole fraction
($\rho \times 100$) of units cross-linked. The straight line has been calculated

* R. F. Blackwell, *Trans. Inst. Rubber Ind.*, **28**, 75 (1952), has shown that very
good agreement with Eq. (50) is obtained for a wide range of vulcanized natural
rubbers using the same value for C_2. The constant C_1 varies, of course, with
the degree of vulcanization.

† The reader is referred to Treloar's splendid discussion of this subject in
Chapter VII of Ref. 1.

according to Eq. (45) assuming $N/\nu = 0$; actually N/ν was about 0.2×10^{-2}, which would result in a slight shift of the theoretical line to the right. The experimentally observed stresses increase with the degree of cross-linking about as predicted, except in the higher range where the quantitative reliability of the cross-linking method is dubious. Most important of all, the magnitudes of the stresses agree approximately with those calculated directly from theory. This is a remarkable achievement, particularly in view of the apparent complexity of network structures.

Similar results have been obtained[29] for natural rubber vulcanized in like manner. Here also the observed equilibrium stress tends to be a little higher than that calculated. Studies of the sulfur-accelerator vulcanization reaction, though less definite from a quantitative point of view, again show about the expected increase in "modulus" with the estimated degree of cross-linking.

Bardwell and Winkler[42] adopted a different procedure for investigating the relationship between force of retraction and structure in GR-S vulcanizates. They introduced cross-linkages into the dispersed particles of the GR-S latex by treating the latex with potassium persulfate. The "vulcanized" latex particles were subsequently coagulated, then bonded together in a coherent network by heating at 80°C. The degree of cross-linking ρ was calculated from the percent of sol, as determined by extraction, using Eqs. (IX-53) and (IX-51) (see p. 379). The degree of cross-linking ρ'' in the gel was then deduced using Eq. (IX-55). Since the product was obtained as a heterogeneous agglomerated mass entirely unsuitable for quantitative elasticity measurements, the equilibrium degree of swelling (see Chap. XIII) in toluene was measured instead, and the force of retraction for $\alpha = 4$ was esti-

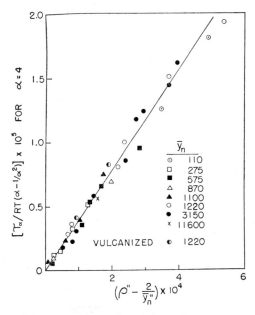

FIG. 100.—Network activity, $\tau/RT(\alpha - 1/\alpha^2)$ $= \nu_e/V$, of GR-S cross-linked with persulfate plotted against the effective degree of cross-linking. The values of $\bar{y}_n$ indicated represent primary degrees of polymerization. (Results of Bardwell and Winkler.[42])

mated from the swelling measurement, using for this purpose a correlation between swelling and τ independently established on samples vulcanized in uniform sheets. Bardwell and Winkler applied this rather round-about method to a series of GR-S polymers covering a wide range in primary degrees of polymerization $\bar{y}_n$. Their results are shown in Fig. 100, treated according to an alternate form of Eq. (45)

$$\tau/RT(\alpha - 1/\alpha^2) = (1/\bar{v}M_0)(\rho'' - 2/\bar{y}_n'') \qquad (45'')$$

where ρ'' is the cross-linking density of the gel (i.e., $\rho''/\bar{v}M_0 = \nu/V$) and $\bar{y}_n''$ is the primary number average degree of polymerization in the network. Despite wide variations both in the degree of cross-linking (ρ'') and in the proportion of primary molecule ends $(2/\bar{y}_n'')$ for the various samples investigated, the points describe a single straight line. Its slope, however, is two and one-half times that predicted by theory as expressed by Eq. (45'').

Perhaps the most convincing test of the theoretical equation of state is provided by the work of Schaefgen[39] on so-called multilinked polyamides, which would not ordinarily be classified as rubberlike. These were prepared by efficiently interlinking pairs of carboxyl end groups in fibers extruded from tetrachain and from octachain poly-ϵ-caproamides (see p. 331), using a diamine for this purpose. The products retain typical fiber properties below their melting points, ca. 225°C, but when heated above this temperature they melt to highly elastic rubbers. The strain $(\alpha - 1)$ was measured at several temperatures in this range as a function of the stress. A typical set of results is compared in Fig. 101 with the theoreti-

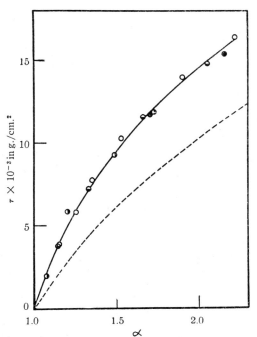

Fig. 101.—Experimental and theoretical (dashed line) stress-strain curves for tetralinked polyamide with $(\nu/V) \times 10^4 = 1.34$ at low elongations. Temperatures (in °C) were 229° ⬤; 241° ◯; 253° ◖; 281° ◑. The range is too small to show a definite temperature coefficient beyond the experimental error. (Schaefgen and Flory.[39])

cal curve (dashed line) calculated according to theory. The stresses observed exceed those calculated by about 50 percent, and the shape of the curve deviates from the $\alpha - 1/\alpha^2$ function much as do the corresponding curves for natural rubber. Similar sets of data were obtained from each of a series of the multilinked polyamides prepared using varying proportions of the multifunctional units, hence with varying ν/V. Stresses τ interpolated from the experimental stress-strain curve for each of several elongations α are plotted in Fig. 102 against ν/V. A sustained, approximately linear increase in τ with the degree of cross-linking is indicated, but the slope exceeds somewhat that predicted by theory. At higher elongations these polyamides crystallize with stretching even at temperatures up to 280°C. The close parallel to the behavior of natural rubber extends also to the tensile strength at elevated temperatures,[39] which is roughly the same as that of rubber at ordinary temperatures.

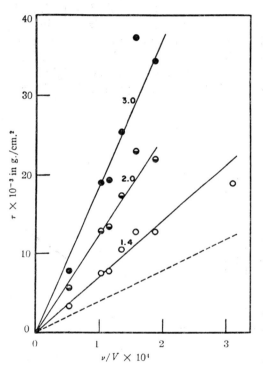

Fig. 102.—Tension at the elongations α indicated vs. degree of cross-linking for multilinked polyamides. The theoretical line (dashed) is shown for $\alpha = 1.4$ where the discrepancy is greatest. (Schaefgen and Flory.[39])

The results cited above show with a convincing measure of consistency that (1) the stress for a given strain increases about linearly with the degree of cross-linking, (2) the magnitudes of the stresses observed stand in remarkably close agreement with the predictions of the statistical theory of rubber elasticity, but (3) they invariably tend to exceed theoretical predictions somewhat, a mean discrepancy of about 50 percent being indicated by an assessment of various results. As one is usually obliged unhappily to allot more space to concern over discrepancies than to the fruition of concurrences, we may point out that network entanglements provide a convenient scapegoat for the discrepancy observed, small though it be in view of the imponderables involved. As must be fully obvious, the notion of entanglements was introduced earlier in this chapter with this object in view. Perhaps it is nonetheless plausible.

A further point which remains to be examined more carefully is the dependence of the tension on the proportion of flaws, or ends of primary molecules, in the network. Bardwell and Winkler's[42] results cover a wide range in primary molecular weight as well as in degree of cross-

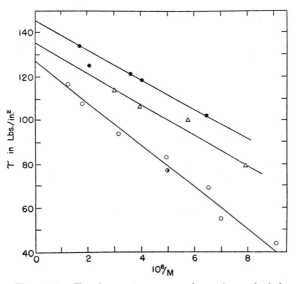

FIG. 103.—Tension τ at $\alpha = 4$ vs. the reciprocal of the primary molecular weight for three butyl rubber series differing in degree of cross-linking (or in M_c). (Flory.[33])

linking; hence the coincidence of their results, shown in Fig. 100, substantiates the procedure adopted in deriving Eqs. (30) and (30′) for the effective number of chains. A more direct test is offered by results obtained by the author[33] using butyl rubbers, which are copolymers of isobutylene with from 1 to 2 percent of isoprene. During vulcanization only the isoprene units are involved in cross-linking, and these units are distributed equitably over all molecular species. Hence by fractionating the unvulcanized polymer a series of samples susceptible to identical degrees of cross-linking ρ but varying in M may be obtained. Elasticity measurements on vulcanizates so prepared should yield tensions τ varying linearly as $1/M$ according to Eq. (45′) since M_c is constant throughout the series. Results confirming this prediction are shown in Fig. 103 for samples from three series differing in cross-linking density, and hence in M_c. Not only are the relationships linear as predicted, but the slopes and intercepts (see Eq. 45') lead to M_c values in excellent agreement with those independently determined from the percentage of sol (see Chap. IX) remaining after vulcanization of the samples of lower molecular weights. The value of M_c calculated from the slope and intercept of the line drawn through the lowest set of data shown in Fig. 103, for example, is 37,000, as compared with 35,000 obtained from the sol-gel relationship.[33]

4c. Effects of Fillers in Rubber Vulcanizates.—Fillers consisting of finely divided solids, usually carbon blacks, alter the physical properties of a noncrystallizing vulcanized rubber in two principal respects: they increase the modulus of elasticity, i.e., raise the stress-strain curve, and they also increase the ultimate tensile strength. A partially satisfactory account of the former effect has been secured through application of the theory of elasticity to a system consisting of rigid inclusions in an otherwise homogeneous elastic medium.[43,44] The treatment corresponds identically with that applied to the viscosity of a suspension of rigid particles in liquids, as developed originally by Einstein (see Chap. XIV) and extended by Guth and Gold;[44] all that is required is the replacement of the viscosity by the modulus of elasticity. Thus Young's modulus μ (which is equal to $d\tau/d\alpha$) for a system consisting of rigid spherical particles wetted by the surrounding elastic medium should depend on the volume fraction v_f of filler as follows:[43,44]

$$\mu = \mu_0(1 + 2.5v_f + 14.1v_f^2) \tag{51}$$

where μ_0 is the modulus for $v_f = 0$. Equation (51), strictly speaking, should be applied only at low elongations within which Hooke's law holds. However, the shape of the stress-strain curve up to elongations of about twofold resembles very closely that for the rubber without filler; the ratio of the stresses with and without filler is nearly constant over this range.[44] Hence it is permissible to write

$$\tau = \tau_0(1 + 2.5v_f + 14.1v_f^2) \tag{52}$$

τ and τ_0 being stresses for the same extension α with and without filler, respectively. Although these equations have been derived for spherical particles, the correction required for asymmetric (e.g., elongated) particles becomes important only for large asymmetries.

The results of Cohan[45] on the force of retraction τ at $\alpha = 1.5$ for GR-S synthetic rubbers vulcanized with various proportions of a calcium carbonate filler are shown in Fig. 104. The agreement with the theoretical curve drawn according to Eq. (52) is good. In further confirmation of the theory, variations in average particle diameter

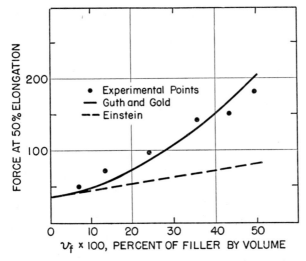

FIG. 104.—Tension τ at $\alpha = 1.5$ for GR-S synthetic rubber containing various proportions of calcium carbonate (particle diameter 3900 mμ), but vulcanized under otherwise identical conditions. The solid curve has been calculated according to Eq. (52); the broken curve by neglecting the third term in this equation. (Cohan.[45])

over an 80-fold range had no effect on the modulus (τ) at a given filler concentration by volume. Calcium carbonates were chosen by Cohan because of their approximate spherical shape and especially because, unlike carbon blacks (see below), they presumably do not enter into permanent bonds with the rubber chains at their surfaces, nor do they otherwise affect the normal vulcanization reaction. Hence the degree of cross-linking of the rubber matrix should be unaltered by the presence of the filler. These same conditions should be approximately fulfilled for butyl rubber mixed with carbon black and vulcanized, inasmuch as the available unsaturated units are few and a large fraction of them are involved in the normal vulcanization reactions. Zapp and Guth[46] observed stress increases owing to the filler particles in this system which were in fair agreement with Eq. (52).

When rubber containing carbon black is vulcanized using sulfur and accelerators, extensive bonding (i.e., cross-linking of a sort) of polymer units to the filler surface occurs.[47] The carbon black surface, being unsaturated, appears to participate in vulcanization reactions. The bonds so formed constitute additional points of constraint on the rubber matrix. Consequently, the rubber layer next to the surface of the carbon black is more highly cross-linked, i.e. ν/V is greater, than elsewhere in the medium. This layer will possess a correspondingly higher modulus of elasticity. The effective thickness of the thus-modified layer should be of the order of the displacement length of a chain, or about 50 Å. At a volume fraction $v_f = 0.20$ of filler of particle diameter 300 Å, these layers constitute a considerable fraction of the total rubber; hence it may be legitimate to consider that the matrix as a whole is stiffened by the formation of bonds between the rubber chains and the filler surface. Under these conditions Eq. (52) cannot be applied unambiguously. Specifically, the value of τ_0 for the rubber matrix will exceed the retractive force for the rubber vulcanized under similar conditions without filler.[47,48]

5. THE STRESS-STRAIN CURVE AT HIGH ELONGATIONS

5a. Crystallinity and Tensile Strength.—Most of the foregoing discussion has been concerned with that portion of the stress-strain curve (Fig. 83, p. 434) over which the stress remains low. It has been pointed out that the subsequent steep rise is associated with oriented crystallization induced by stretching. Indeed, the region in which the stress-strain curve assumes positive curvature coincides with the onset of crystallization as revealed by X-ray diffraction, by density change, or by the abrupt decrease in $(\partial E/\partial L)_{T,V}$ (see above). If the tensile strength is to reach a high value, the stress-strain curve must necessarily assume a high slope and this steep rise must be maintained up to the range of the (large) breaking stress. If manifestation of a high slope in the stress-strain curve requires crystallization in a given polymer system, then one may justifiably look for a correlation between crystallinity, or the ability to crystallize on stretching, and tensile strength. This approach to the problem of tensile strength in rubberlike polymers invites inquiry into the tensile strength of rubbers which do not crystallize on stretching, owing, for example, to lack of symmetry in their structures. Significant in this connection are the very low tensile strengths observed for (gum-vulcanized) synthetic polymers and copolymers of butadiene or of isoprene consisting of a plurality of structural units (including the isomeric diene units; see Chap. VI) no one of which is present in sufficient preponderance to crystallize on stretching. These and other noncrystallizing rubbery copolymers usually exhibit tensile strengths not exceeding a few hundred pounds per square inch. Furthermore, the tensile strength of vulcanized natural rubber falls sharply to a similar low value when the temperature of measurement is elevated above that (ca. 120°C) at which crystallization fails to take place on stretching. Butyl rubber exhibits a similar drop in tensile strength, but at a considerably lower temperature (ca. 50°C)

These observations show that the development of high tensile strength in rubberlike polymers ordinarily is intimately related to crystallization. As a matter of fact, this generalization applies in large measure to nonrubberlike polymers where crystallization is known to enhance markedly the strength (especially the impact strength, or the work of rupture equal to the total area under the stress-strain curve). As it has already been pointed out that natural rubber stretched to the region of high crystallinity resembles a highly oriented fiber, both in its thermoelastic characteristics and in its morphology, this extension of the generalization requires little further comment. Highly stretched rubber acquires the physical properties of a typical fiber when the temperature is lowered; properly cross-linked fibers may exhibit rubberlike behavior at elevated temperatures. These features are mentioned here for the purpose of calling attention to the plausibility of considering the basic characteristics of the tensile behavior of rubberlike polymers as being broadly applicable to a wide variety of polymers many of which show no superficial resemblance to rubbers. Vulcanized rubbers probably are the materials best suited for the investigation of the fundamental factors underlying tensile strength in polymers.

5b. Experimental Results on the Relationship between Tensile Strength and Network Structure.—The tensile strengths of a series of vulcanized butyl rubber polymers differing solely in primary molecular

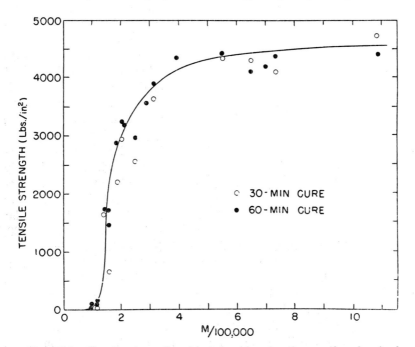

FIG. 105.—Tensile strengths of butyl rubber fractions each vulcanized to the same cross-linking density ρ, corresponding to $M_c = 37,000$, plotted against the primary molecular weight M.[33]

weight are shown in Fig. 105. In obtaining these results,[33] a series of fractions of restricted molecular weight range, each having the same proportion of isoprene units (about 1.5 mole percent), was prepared by fractional precipitation of suitable isobutylene-isoprene copolymers. These were then vulcanized according to a standard procedure to give vulcanizates each of which possessed a degree of cross-linking ρ of 0.16×10^{-2} (per structural unit), which corresponds to $M_c = 37,000$ (see Eqs. 26 and 28). The condition for incipient formation of a network (see Chap. IX) as applied to these fractions, each being considered molecularly homogeneous, is $\gamma = 1$ where γ is the number of cross-linked units per primary molecule, i.e., $\gamma = \rho M / M_0 = M / M_c$. Since ρ, and likewise M_c, are the same for all fractions, infinite networks should occur in those for which M exceeds 37,000, the fraction of the vulcanized polymer occurring as part of the network increasing rapidly as the molecular weight of the fraction exceeds this figure. According to the results shown in Fig. 105, appreciable tensile strength is developed only at molecular weights considerably beyond that required for incipient gelation. The curve rises sharply with molecular weight above $M = 100,000$, then approaches an asymptotic limit with further increase in M.

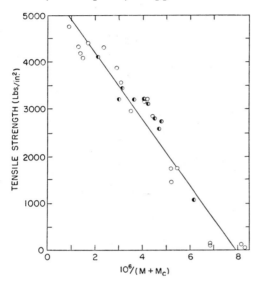

Fig. 106.—Tensile strengths of the vulcanized butyl rubber fractions (Fig. 105) plotted against $(M+M_c)^{-1}$. This quantity is proportional to the fraction $1-s_a$ of the network which is "inactive."[33] Individual fractions, $\bigcirc$; mixture of fractions taking $M = \overline{M}_n$ for the mixture, $\bullet$.

In Fig. 106 these same tensile strengths are plotted against the quantity $(M+M_c)^{-1}$, which is proportional to the inactive fraction $1-s_a$ of the network, where s_a is the fraction of the network which is subject to permanent orientation by stretching, as given by Eq. (29'). The points follow a straight line, within the rather large experimental error characteristic of tensile strength measurements. The equation of the straight line corresponds to

$$\text{T.S.} = 9500(s_a - 0.41) \tag{53}$$

in pounds per square inch referred to the initial cross-sectional area. Thus, if the network is so imperfectly constructed that less than 41 percent of the structure is subject to permanent orientation by deformation, the tensile strength is negligible, but it increases approximately linearly with s_a for increasing degrees of perfection. Experiments on vulcanizates prepared from mixtures of fractions differing in molecular weight give results which correlate satisfactorily (Fig. 106) with those for the individual fractions, provided M is taken as the number average for the mixture.

When the butyl rubber was compounded with up to 30 percent of polyisobutylene, which, lacking the unsaturated isoprene units, did not enter into the cross-linking reaction, the tensile strengths were, of course, considerably reduced.[33] They were found nevertheless to be accurately represented by the same equation, (53), provided merely that s_a is taken as the fraction of the composite specimen consisting of network chains subject to orientation. Thus, in this case

$$s_a = v_2[1 - 2M_c/(M + M_c)]$$

where v_2 is the volume fraction consisting of the butyl rubber and $1 - v_2$ is the fraction of the polyisobutylene "diluent." Use of a heavy mineral oil in place of the polyisobutylene produced an equal effect within experimental error. Thus it becomes apparent that the tensile strength in a typical rubber which crystallizes on stretching is a function of the fraction of the composition subject to sustained orientation by deformation and, further, that the functional relationship is approximately linear.

The influence of the other structural variable, namely, the degree of cross-linking ρ, on the tensile strength is illustrated most clearly by the results[49] shown in Fig. 107. These were obtained by cross-linking natural rubber having a number average molecular weight of about 500,000 with varying proportions of decamethylene-bis-methyl azodicarboxylate as indicated by the ρ values expressed in percent along the abscissa scale. As was mentioned earlier, vulcanization by this method may be presumed to occur with a minimum of chain scission, hence ρ is virtually the sole variable for the series shown (open circles). At the lowest degree of cross-linking, $\rho = 0.10 \times 10^{-2}$, for which measurements were made, s_a is already about 0.78 (assuming no chain scission whatever). The initial steep rise in tensile strength with ρ may be attributed to the increase in s_a, but the passage of the curve through a rather well-defined maximum at higher degrees of cross-linking requires an explanation of a different sort.

The decrease in tensile strength with "over-vulcanization" by con-

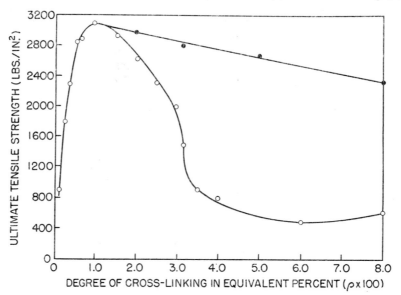

FIG. 107.—Tensile strengths of natural rubber plotted against the degree of cross-linking with bis-azo vulcanizing agent (O), expressed as equivalent percent ($\rho \times 100$). Upper curve (●): sample prepared using one equivalent percent of bis-azo compound plus monoreactive ethyl azodicarboxylate for the total degrees of modification of the units indicated on the abscissa scale. (Flory, Rabjohn, and Shaffer.[49])

ventional sulfur-accelerator combinations has been attributed frequently to conversion of a fraction of the structural units, either by cross-linking or by auxiliary reactions with sulfur or other ingredients in the vulcanization recipe, to derivatives incapable of entering the crystal lattice.   Thus one may consider natural rubber to be converted by vulcanization to a copolymer of *cis* units of isoprene with a minority of modified isoprene units.   However, only a very small fraction of the units are so modified, and, in opposition to the above-mentioned hypothesis, studies on other polymer systems (e.g., linear polyesters and polyamides) show that the suppression of crystallization owing to a *small* proportion of a copolymerized ingredient does not reduce the tensile strength markedly.   Furthermore, experiments in which up to 7 percent of the non-cross-linked isoprene units in the azo-vulcanized rubbers were modified (without cross-linking them) failed to bring about an appreciable decrease in tensile strength.   The results of these experiments are included in Fig. 107 (points ●).   These vulcanizates were prepared by incorporating, in addition to the optimum quantity (1 equivalent percent) of decamethylene-bis-methyl azodicarboxylate, varying proportions as indicated of monoreactive ethyl azodicarboxylate, each molecule of which combines with only one isoprene unit. Such drastically modified isoprene units certainly cannot enter the crystal lattice characteristic of the *cis* unit of isoprene; yet the resulting suppression of crystallization decreases the tensile strength no more than may be attributed to the effect of dilution by the adduct. One is forced to conclude, therefore, that cross-linking beyond about 1 equivalent percent is inherently detrimental to tensile strength for other reasons.

A more plausible explanation[49] appears on careful examination of the mechanics of induction of crystallization by stretching. The network produced by higher degrees of vulcanization consists of shorter chains which are more susceptible to orientation by elongation than the longer ones occurring in networks of low degrees of cross-linking. This follows from the fact that the root-mean-square distance from beginning to end of a chain varies as the square root of the number x of units in the chain. Hence the ratio of the displacement length of a chain in the unoriented state to its fully extended length varies as $x^{-1/2}$; the greater x, the greater the elongation required to bring the undisturbed chain to full extension. From this it follows that crystallization should begin at lower elongations for higher degrees of cross-linking. Experimental evidence bears out this deduction.

We have called attention previously to the fact that by no means all of the chains of a network are lengthened, in the sense that their ends are moved farther apart, by elongation of the sample; those with displacement vectors situated transverse to the axis of elongation are shortened. The proportion of the chains which are lengthened by a macroscopic elongation α of the sample can be shown[49] to be given by

$$1 - [(\alpha - 1)/(\alpha^3 - 1)]^{1/2}$$

which increases with the macroscopic deformation but depends not at all on the degree of cross-linking.* Those chains which happen to be oriented with vectors nearly parallel to the axis of elongation may be expected to join in the formation of crystallites beginning at an elongation α which, in the light of the conclusion reached in the preceding paragraph, will be lower the greater the degree of cross-linking. However, if crystallization begins at a comparatively low elongation, (e.g., $\alpha = 2$ to 3), as in the case of a relatively high degree of cross-linking, the fraction of the structure oriented favorably for crystallization will

---

* Whether or not a given chain is lengthened by the deformation depends only on the orientation of its end-to-end vector and not on its initial length. Hence the fraction of chains whose lengths are increased is likewise independent of the lengths of the chains.

be small, and a large fraction of the chains will remain ineligible for crystallization. The development of a high over-all degree of crystallization is favored by postponement of crystallization until a high elongation has been reached. In this way it is possible to arrive at an explanation for the rather rapid decline in tensile strength with crosslinking beyond about $\rho = 1 \times 10^{-2}$ in natural rubber. This explanation represents an extension of the previous generalization that the tensile strength is related to the degree of participation of the structure in (oriented) crystallization. The reported increase in tensile strength observed in gum-vulcanized rubber with elevation of the temperature up to about 100°C,[50] in which manner the occurrence of crystallization is shifted to higher elongations, substantiates this viewpoint. It should not, however, be concluded that a decline in tensile strength with overvulcanization by conventional methods occurs solely as a result of excessive cross-linking; chain scission accompanying the vulcanization process is a contributing factor, which often may be of greater importance.

5c. Tensile Strength in Noncrystallizing Rubbers.—The characteristically low tensile strengths of rubbers which do not crystallize on stretching has already been mentioned. Through the incorporation of finely divided particles, usually carbon blacks, it is possible however to raise the tensile strengths of noncrystallizing rubbers from a few hundred pounds per square inch to a figure of 3000 or even 4000. In contrast to the effect of a filler on the elastic modulus (or the force of retraction at low elongations) which depends only on the total volume fraction of filler (see Eqs. 51 and 52), the tensile strength depends markedly on the particle size. Fillers of very small particle size, i.e., less than 500 Å in diameter, are required for a pronounced increase in tensile strength. The tensile strength reaches a maximum for a volume fraction of such a filler of about 0.20 to 0.25. The mean distance between nearest-neighbor particles at this optimum concentration is somewhat less than their radii, provided the particles are completely dispersed. The mechanism by which fine filler particles effect such marked increases in tensile strength is obscure. It is apparent, however, that in this respect they perform the function of crystallites, and it is not improbable that the mechanism of their reinforcing action is similar also.

Below their glass transition temperatures T_g, noncrystalline polymers (without fillers) usually exhibit high tensile strengths, although lower than those of typical crystalline polymers if the strength is referred to the cross section at rupture in both cases. A brittle type of fracture occurs, which probably is initiated by a different mechanism than that for kinetically mobile polymer systems above their glass transition temperatures. The deformation at break is low, and the *energy* of rupture consequently is never large. The tensile strength may be sustained at a moderately high value over a range of 20° or somewhat more above the transition temperature T_g even in the total absence of fillers, if a network structure is introduced, by vulcanization for example, in order to suppress plastic flow at such temperatures. Borders and Juve[51] have shown that the tensile strengths of various synthetic rubbers, mixed with carbon black and vulcanized, can be correlated with the difference between the tensile test temperature T and T'_g. A single correlation between the tensile strength and $T - T_g$ holds for changes both in the nature of the rubber (i.e., in T_g) and in the test temperature. These observations point to the importance of a rate mechanism governing the observed tensile strength in noncrystallizing rubbers.

APPENDIX A

The Force-Temperature Coefficient at Constant Elongation.—The purpose here is to show that

$$- (\partial S/\partial L)_{T,V} \cong (\partial f/\partial T)_{P,\alpha} \qquad (20)$$

and to indicate the nature of the approximations involved in the derivation of this relation which has been used in the text to obtain the important equation (22). The following treatment resembles in some respects those given by Elliott and Lippmann[10] and by Gee[11] in their discussions of the thermodynamics of high elasticity.

The derivative occurring on the left of Eq. (20) may be expanded in terms of the independent variables T, P, and L as follows:

$$- (\partial S/\partial L)_{T,V} = - (\partial S/\partial L)_{T,P} - (\partial S/\partial P)_{T,L}(\partial P/\partial L)_{T,V}$$

Substituting from Eq. (12) for $(\partial S/\partial L)_{T,P}$ and making use of the additional relationship $(\partial S/\partial P)_{T,L} = - (\partial V/\partial T)_{P,L}$ (which may be derived from Eq. 8)

$$- (\partial S/\partial L)_{T,V} = (\partial f/\partial T)_{P,L} + (\partial V/\partial T)_{P,L}(\partial P/\partial L)_{T,V}$$

$$= (\partial f/\partial T)_{P,\alpha} + \Delta \qquad (A\text{-}1)$$

where

$$\Delta = (\partial f/\partial \alpha)_{T,P}(\partial \alpha/\partial T)_{P,L} + (\partial V/\partial T)_{P,L}(\partial P/\partial L)_{T,V} \qquad (A\text{-}2)$$

In order to arrive at Eq. (20), it is required to show that this quantity Δ is negligible.

Since $\alpha = L/L_0$ where L_0 is a function of pressure and temperature only,

$$(\partial f/\partial \alpha)_{T,P} = L_0(\partial f/\partial L)_{T,P}$$
$$= -L_0(\partial f/\partial P)_{T,L}(\partial P/\partial L)_{T,f}$$

It follows from Eq. (8) that $(\partial f/\partial P)_{T,L} = (\partial V/\partial L)_{T,P}$; hence

$$(\partial f/\partial \alpha)_{T,P} = -L_0(\partial V/\partial L)_{T,P}(\partial P/\partial L)_{T,f} \qquad (A\text{-}3)$$

Furthermore,

$$(\partial \alpha/\partial T)_{P,L} = -\alpha(\partial \ln L_0/\partial T)_P$$
$$= -\alpha \alpha_T/3 \qquad (A\text{-}4)$$

where α_T is the bulk thermal expansion coefficient of the unstretched rubber, which is three times the linear expansion coefficient $(\partial \ln L_0/\partial T)_P$, assuming, of course, that the unstretched rubber is isotropic. The last term in Eq. (A-2) may be converted to

$$-(\partial V/\partial T)_{P,L}(\partial P/\partial V)_{T,L}(\partial V/\partial L)_{T,P}$$

Introducing this expression along with Eqs. (A-3) and (A-4) into Eq. (A-2), we have

$$\Delta = (\partial V/\partial L)_{T,P}\alpha_T[(L/3)(\partial P/\partial L)_{T,f} - (1/\alpha_T)(\partial V/\partial T)_{P,L}(\partial P/\partial V)_{T,L}]$$

For reasons which appear below, it is preferable to express this relation as follows

$$\Delta = (\partial V/\partial L)_{T,P}(\alpha_T/\kappa)[(\kappa L/3)(\partial P/\partial L)_{T,f}$$
$$- (\kappa/\alpha_T)(\partial V/\partial T)_{P,L}(\partial P/\partial V)_{T,L}] \qquad (A\text{-}5)$$

where κ is the bulk compressibility, i.e.

$$\kappa = -(\partial \ln V/\partial P)_{T,\alpha=1} = -3(\partial \ln L_0/\partial P)_T$$

The first term in brackets in Eq. (A-5) may be elucidated by considering its reciprocal

$$(3/\kappa L)(\partial L/\partial P)_{T,f} = (3/\kappa \alpha)(\partial \alpha/\partial P)_{T,f} + (3/L_0\kappa)(\partial L_0/\partial P)_T$$
$$= (3/\kappa \alpha)(\partial \alpha/\partial P)_{T,f} - 1 \qquad (A\text{-}6)$$

The derivative $(\partial \alpha/\partial P)_{T,f}$ must be very nearly zero, for it is inconceivable that the extensibility may depend appreciably on the pressure; if this were the case, it would be necessary to postulate a large change in volume with elongation, and this is contradicted by experiment.[5] Hence we may assume with confidence that

$$(\kappa L/3)(\partial P/\partial L)_{T,f} \cong -1 \qquad (A\text{-}7)$$

With reference to the second term in the brackets in Eq. (A-5), the product of the derivatives occurring therein is the ratio of the expansion coefficient to the negative of the compressibility *both at constant length*. This term might be written, therefore

$$(\kappa/\alpha_T)(\alpha_T/\kappa)_L$$

the subscript L referring to the deformed rubber. For a noncrystalline rubber $(\alpha_T/\kappa)_L$ must be very nearly equal to α_T/κ; such small effects as may be introduced into either coefficient as a result of anisotropy are likely to affect both coefficients similarly. Hence the bracketed expression in Eq. (A-5) must be very nearly zero.

Before concluding that Δ also is negligible, the magnitude of the factor $(\partial V/\partial L)_{T,P}(\alpha_T/\kappa)$ should be compared with $(\partial f/\partial T)_{P,\alpha}$ of Eq. (A-1). The latter quantity, as has been shown in Section 1c, is approximately equal to f/T for a noncrystalline rubber. The volume change accompanying the stretching of natural rubber may be attributed to the dilation resulting from the decrease in internal pressure. This decrease in internal pressure, being the hydrostatic component of the applied tension, equals one-third of this tension (in force/actual cross section) according to the theory of elasticity. Hence

$$\Delta P_{\text{int}} = - fL/3V$$

and the resulting dilation should be

$$\Delta V = - \kappa V \Delta P_{\text{int}} = \kappa fL/3$$

giving

$$(\partial V/\partial L)_{T,P} = \kappa f/3 \tag{A-8}$$

Gee, Stern, and Treloar[5] have shown that the volume changes in rubber stretched up to elongations of 100 percent are accounted for quantitatively in this manner, thus affording strong evidence for the absence of anisotropy to an extent which might seriously vitiate the approximations introduced in the preceding paragraphs.

Introducing Eq. (A-8) into (A-5), we obtain

$$\Delta \cong (f\alpha_T/3)\left[(\kappa/\alpha_T)(\alpha_T/\kappa)_L - 1\right] \tag{A-9}$$

For natural rubber at ordinary temperatures $\alpha_T \cong 6.6 \times 10^{-4}$ per °K. The quantity $f\alpha_T/3 \cong 2.2 \times 10^{-4} f$ is therefore much less than $(\partial f/\partial T)_{P,\alpha}$ of Eq. (A-1), the latter being of the order of f/T, or $3 \times 10^{-3} f$ at ordinary temperature. The value of Δ is further diminished by the factor in brackets in Eq. (A-9); this factor cannot differ much from zero. We conclude, therefore, that in any case $\Delta \ll f/T \cong (\partial f/\partial T)_{P,\alpha}$.

APPENDIX B

Deformation of a Swollen Network.[52]—Let us assume that the network formed at volume V_0 is subsequently swollen isotropically by a diluent to a volume V such that the volume fraction of polymer is $v_2 = V_0/V$. In the subsequent deformation the volume is assumed constant. Letting α_x, α_y, and α_z represent the changes in dimensions resulting from the combination of *swelling and deformation*, we have therefore that $\alpha_x \alpha_y \alpha_z = 1/v_2$ is constant during the deformation. Confining the development given here to a simple elongation in the direction of the x-axis, we let α represent the length in this direction relative to the *swollen, unstretched* length $L_{0,s} = (V/V_0)^{1/3} L_0$, since the swelling is isotropic. Then

$$\alpha_x = \alpha(V/V_0)^{1/3} = \alpha/v_2^{1/3} \qquad \text{(B-1)}$$

and since $\alpha_x \alpha_y \alpha_z = v_2^{-1}$

$$\alpha_y = \alpha_z = (v_2 \alpha_x)^{-1/2} = 1/\alpha^{1/2} v_2^{1/3} \qquad \text{(B-2)}$$

With the introduction of these relations into Eq. (41), the configurational entropy change (the entropy of mixing solvent and polymer excluded; see Chap. XII) relative to the initial state $\alpha_x = \alpha_y = \alpha_z = 1$ becomes

$$\Delta S = - (kv_e/2v_2^{2/3})[\alpha^2 + 2/\alpha - 3 + \ln v_2] \qquad \text{(B-3)}$$

Application of Eq. (43) with L_0 replaced by $L_{0,s}$ yields the following expression for the tension τ expressed as the force per unit area of the *swollen, unstretched* sample:

$$\tau = (RTv_e/Vv_2^{2/3})(\alpha - 1/\alpha^2) \qquad \text{(B-4)}$$

$$= (RTv_e/V_0)v_2^{1/3}(\alpha - 1/\alpha^2) \qquad \text{(B-4')}$$

where v_e is given in moles. The tension τ_0 referred to the *unswollen, unstretched* cross section is

$$\tau_0 = \tau(V/L_{0s})/(V_0/L_0) = \tau/v_2^{2/3}$$

$$= (kTv_e/V_0)v_2^{-1/3}(\alpha - 1/\alpha^2) \qquad \text{(B-5)}$$

Comparison of Eq. (B-5) with (44) shows, since $v_2^{-1/3} > 1$, that a greater force is required for a given elongation of the swollen rubber than would have been required to elongate the same sample, unswollen, by the same factor α. If, however, samples having the same cross-sectional area are compared, the one swollen and the other unswollen, both having been selected from the same vulcanizate, a lesser force should be required according to Eq. (B-4') to elongate the former by the given factor α than is required to elongate the latter to the same degree α. While these predictions are qualitatively confirmed by experiment, Gee's[11] results on rubber show that the simple dependence on $v_2^{-1/3}$ according to Eq. (B-5) is not quantitatively valid. The reasons for this discrepancy probably are related to those responsible for the failure of τ_0 to vary as $\alpha - 1/\alpha^2$. On the other hand, the stresses observed by Wiederhorn and Reardon[53] on stretching swollen collagen fibers are in good accord with these equations, over a wide range in swelling and at elongations up to $\alpha = 2$.

REFERENCES

1. For a more extensive discussion of the physical properties of rubber, the reader is referred to *The Physics of Rubber Elasticity*, by L. R. G. Treloar (Oxford, Clarendon Press, 1949).
2. J. Gough, *Proc. Lit. and Phil. Soc.*, Manchester, 2d ser., **1**, 288 (1805).
3. Lord Kelvin, *Quarterly J. Math.*, **1**, 57 (1857).
4. J. P. Joule, *Trans. Roy. Soc.* (London), **A149**, 91 (1859). See also, *Phil. Mag.*, **14**, 227 (1857).
5. G. Gee, J. Stern, and L. R. G. Treloar, *Trans. Faraday Soc.*, **46**, 1101 (1950).
6. W. L. Holt and A. T. McPherson, *J. Research Nat. Bur. Stand.*, **17**, 657 (1936); T. G. Fox, P. J. Flory, and R. E. Marshall, *J. Chem. Phys.*, **17**, 704 (1949).
7. S. L. Dart, R. L. Anthony, and E. Guth, *Ind. Eng. Chem.*, **34**, 1340 (1942).
8. K. H. Meyer, G. von Susich, and E. Valkó, *Kolloid Z.*, **59**, 208 (1932).
9. W. B. Wiegand and J. W. Snyder, *Trans. Inst. Rubber Ind.*, **10** (No. 3), 234 (1934).
10. D. R. Elliott and S. A. Lippmann, *J. Applied Phys.*, **16**, 50 (1945).
11. G. Gee, *Trans. Faraday Soc.*, **42**, 585 (1946).
12. K. H. Meyer and C. Ferri, *Helv. Chim. Acta*, **18**, 570 (1935).
13. R. L. Anthony, R. H. Caston, and E. Guth, *J. Phys. Chem.*, **46**, 826 (1942).
14. L. A. Wood and F. L. Roth, *J. Applied Phys.*, **15**, 781 (1944).
15. K. H. Meyer and A. J. A. van der Wyk, *Helv. Chim. Acta*, **29**, 1842 (1946).
16. F. L. Roth and L. A. Wood, *J. Applied Phys.*, **15**, 749 (1944).
17. L. E. Peterson, R. L. Anthony, and E. Guth, *Ind. Eng. Chem.*, **34**, 1349 (1942).
18. B. B. S. T. Boonstra, *Ind. Eng. Chem.*, **43**, 362 (1951).
19. P. J. Flory, *J. Chem. Phys.*, **15**, 397 (1947).
20. W. H. Smith and C. P. Saylor, *J. Research Nat. Bur. Stand.*, **21**, 257 (1938).
21. H. J. Woods, *J. Colloid Sci.*, **1**, 407 (1946).

22. E. Wöhlisch and co-workers, *Kolloid Z.*, **104**, 14 (1943); E. Guth, *Annals N. Y. Acad. Science*, **47**, 715 (1947).

23. E. H. Farmer and F. W. Shipley, *J. Polymer Sci.*, **1**, 293 (1946); R. F. Naylor, *ibid.*, **1**, 305 (1946); G. F. Bloomfield, *ibid.*, **1**, 312 (1946). See also *J. Chem. Soc.*, **1947**, 1519, 1532, 1546, 1547. G. F. Bloomfield and R. F. Naylor, *Proceedings of the XIth International Congress of Pure and Applied Chemistry*, Vol. II, "Organic Chemistry, Biochemistry," p. 7 (1951).

24. R. T. Armstrong, J. R. Little, and K. W. Doak, *Ind. Eng. Chem.*, **36**, 628 (1944); C. M. Hull, S. R. Olsen, and W. G. France, *ibid.*, **38**, 1282 (1946); B. C. Barton and E. J. Hart, *Ind. Eng. Chem.*, **44**, 2444 (1952).

25. N. Sheppard and G. B. B. M. Sutherland, *J. Chem. Soc.*, **1947**, 1699.

26. G. F. Bloomfield, *J. Soc. Chem. Ind. (London)*, **68**, 66 (1949); *ibid.*, **67**, 14 (1948).

27. M. Gordon, *J. Polymer Sci.*, **7**, 485 (1951).

28. L. Bateman, private communication.

29. N. Rabjohn, *J. Am. Chem. Soc.*, **70**, 1181 (1948); P. J. Flory, N. Rabjohn, and M. C. Schaffer, *J. Polymer Sci.*, **4**, 225 (1949).

30. W. O. Baker, *J. Am. Chem. Soc.*, **69**, 1125 (1947).

31. E. H. Farmer and C. G. Moore, *J. Chem. Soc.*, **1951**, 142.

32. P. J. Flory, *Chem. Revs.*, **35**, 51 (1944).

33. P. J. Flory, *Ind. Eng. Chem.*, **38**, 417 (1946).

34. F. T. Wall, *J. Chem. Phys.*, **10**, 485 (1942); **11**, 527 (1943).

35. L. R. G. Treloar, *Trans. Faraday Soc.*, **39**, 36, 241 (1943). See also Ref. 1.

36. P. J. Flory, *J. Chem. Phys.*, **18**, 108 (1950).

37. L. R. G. Treloar, *Trans. Faraday Soc.*, **40**, 59 (1944).

38. P-O. Kinell, *J. Phys. Colloid Chem.*, **51**, 70 (1947).

39. J. R. Schaefgen and P. J. Flory, *J. Am. Chem. Soc.*, **72**, 689 (1950).

40. M. Mooney, *J. Applied Phys.*, **11**, 582 (1940); **19**, 434 (1948).

41. R. S. Rivlin, *Trans. Royal Soc. (London)*, **A240**, 459, 491, 509 (1948); **241**, 379 (1948).

42. J. Bardwell and C. A. Winkler, *Can. J. Research*, **B27**, 116, 128, 139 (1949).

43. H. M. Smallwood, *J. Applied Phys.*, **15**, 758 (1944).

44. E. Guth, *J. Applied Phys.*, **16**, 20 (1945); E. Guth and O. Gold, *Phys. Rev.*, **53**, 322 (1938).

45. L. H. Cohan, *India Rubber World*, **117**, 343 (1947).

46. R. L. Zapp and E. Guth, *Ind. Eng. Chem.*, **43**, 430 (1951).

47. R. S. Stearns and B. L. Johnson, *Ind. Eng. Chem.*, **43**, 146 (1951).

48. A. M. Bueche, *J. Applied Phys.*, **23**, 154 (1952).

49. P. J. Flory, N. Rabjohn, and M. C. Shaffer, *J. Polymer Sci.*, **4**, 435 (1949).

50. G. Gee, *J. Polymer Sci.*, **2**, 451 (1947).

51. A. M. Borders and R. D. Juve, *Ind. Eng. Chem.*, **38**, 1066 (1946).

52. H. M. James and E. Guth, *J. Chem. Phys.*, **11**, 455 (1943); P. J. Flory and J. Rehner, Jr., *ibid.*, **11**, 521 (1943).

53. N. M. Wiederhorn and G. V. Reardon, *J. Polymer Sci.*, **9**, 315 (1952).

CHAPTER XII

Statistical Thermodynamics

of Polymer Solutions

THE ideal solution law, as embodied in Raoult's law, provides the basis for the treatment of simple molecule solutions. Although such solutions seldom behave ideally over wide ranges in concentration, the correlation usually is adequate to justify adoption of the classically defined ideal solution as a standard for comparison. Solutions in which the solute is a polymer of high molecular weight exhibit very large deviations from ideality. Only at extreme dilutions where, as we have shown in Chapter VII, the ideal law must be approached as an asymptotic limit, does the polymer solution conform approximately with ideality. At concentrations exceeding a few percent, deviations from ideality usually become so great that the ideal law is of little value as a basis for rationally correlating the thermodynamic properties of polymer solutions. Some other relationship is therefore needed.

According to Raoult's law, the activity a_1 of the solvent in the solution should equal its mole fraction N_1. In a binary solution consisting of solvent and a polymer having a molecular weight a thousand times or more that of the solvent, only a very small percentage by *weight* of the solvent is sufficient to bring its *mole* fraction N_1 very close to unity. Hence, according to Raoult's law (with $a_1 = P_1/P_1^0$) the partial pressure P_1 of the solvent in the solution should be very nearly equal to that of the pure solvent P_1^0 over the greater portion of the composition range. Experiments do not confirm this prediction. The smoothed curve representing observed activities of benzene mixtures with rubber[1] as a function of the volume fraction v_1 of solvent is shown in Fig. 108. The broken curve appearing in the upper right-hand portion of the diagram represents Raoult's law; the lower broken straight line represents the replacement of Raoult's law with the arbitrary relationship $a_1 = v_1$. The experimental curve lies between the two. The

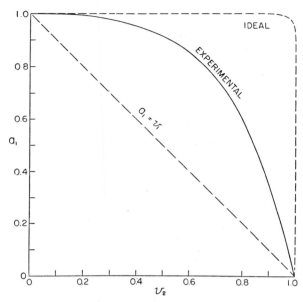

FIG. 108.—The activity of benzene in solution with rubber plotted against the volume fraction of rubber. The solid curve represents smoothed experimental data of Gee and Treloar.[1] The upper dashed curve represents the calculated ideal curve for an ideal solution of a solute with M =280,000 dissolved in benzene. The diagonal dashed curve corresponds to $a_1 = v_1$. (From the data of Gee and Treloar.[1])

latter is asymptotic with the Raoult's law curve in the limit of infinite dilution, as already mentioned. However, the activity a_1 according to either curve lies so close to unity at high dilutions that the difference $1 - a_1$ is imperceptible in Fig. 108. At higher concentrations the activity correlates better with the volume fraction than with the mole fraction, although it is not numerically equal to it. The underlying basis for the failure of the ideal solution law lies in its use of the mole fraction as the composition variable. It consequently presupposes that the effect of a large polymeric solute molecule on the activity of the solvent should be equal to that of an ordinary solute molecule, which may be no larger than one unit in the polymer chain.

Ideal solution behavior over extended ranges in both composition and temperature requires that the following conditions be fulfilled: (i) the entropy of mixing must be given by*

$$\Delta S_M = - k(n_1 \ln \mathrm{N}_1 + n_2 \ln \mathrm{N}_2) \tag{1}$$

where n_1 and n_2 are the numbers of molecules of solvent and solute, respectively, and N_1 and N_2 their mole fractions; and (ii) the heat of mixing ΔH_M must equal zero. Deviations from ideality may arise from failure of either of these conditions. Early work[2] on polymer solutions revealed that the deviations from ideality were not strongly temperature-dependent; hence it was concluded that condition (i),

* See an appropriate text.

at least, is not fulfilled, a conclusion abundantly confirmed by more recent work. We shall therefore first derive an expression for the entropy of mixing polymer and solvent with which to replace Eq. (1).

1. GENERAL THERMODYNAMIC RELATIONS FOR POLYMER SOLUTIONS

1a. The Entropy of Mixing according to Liquid Lattice Theory.[3,4,5]— The ideal entropy of mixing expression, Eq. (1), may be derived by considering a binary solution consisting of two types of molecules virtually identical in size, spatial configuration, and external force field. In such a mixture a molecule of one type may be replaced by one of the other without affecting the circumstances of immediate neighbors in the solution. The greater entropy of the solution as compared with the pure components arises entirely from the greater number of arrangements possible in the solution. The number of permissible arrangements in so simple a binary system is easily calculated.

The molecules in the pure liquids and in their solution are considered to be arranged with enough regularity to justify approximate representation by a lattice, as is indicated schematically in Fig. 109. In a simple liquid consisting of nearly spherical molecules, the first neighbors of a given molecule in the liquid will occur at a distance from its center which is fairly well defined, although not as precisely as in a crystal. The tier of second neighbors will occur at less accurately specified distances, and so forth. Since we shall be concerned only with the first coordination sphere about a given molecule (or polymer segment, see below), the rapid deterioration of the lattice as a valid representation with increasing distance from the reference point is unimportant. The adoption of a

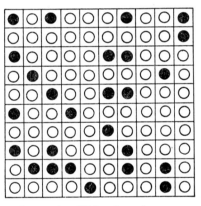

FIG. 109.—Molecules of a monomeric solute distributed over the lattice used to describe a binary solution.

lattice scheme is not in itself, therefore, a necessarily hazardous idealization. The succeeding postulate, namely, that the *same* lattice may be used to describe the configurations of *both* pure components and of the solution, is a much more serious one from the standpoint of application to real solutions. It requires (among other things) that the geometry of the two molecular species be virtually identical.

These assumptions having been accepted as required, we consider the total number of ways of arranging the n_1 identical molecules of the solvent and n_2 identical molecules of the solute on the lattice comprising $n_0 = n_1 + n_2$ cells. This is just the number of combinations of n_0 things taken n_1 at a time, or

$$\Omega = n_0!/n_1!n_2! \tag{2}$$

Whereas the pure components may be arranged in their respective lattices in only one way, the number of arrangements possible in the solution is given by the very large number Ω. It follows according to the Boltzmann relation that the entropy of mixing should be given by $\Delta S_M = k \ln \Omega$. With the introduction of Stirling's approximation, $\ln n! = n \ln n - n$, for the factorials

$$\Delta S_M = k\big[(n_1 + n_2) \ln(n_1 + n_2) - n_1 \ln n_1 - n_2 \ln n_2\big]$$

which reduces by suitable rearrangement to Eq. (1). In spite of the sweeping approximations involved, the ideal entropy expression has proved to be a most useful generalization for solutions of simple molecules, even in instances where the components differ considerably in molecular shape and as much as twofold in size.

This treatment, resting essentially on the assumed approximate interchangeability of molecules of solvent and solute in the solution, cannot possibly hold for polymer solutions in which the solute molecule may be a thousand or more times the size of the solvent. The long chain polymer may be considered to consist of x chain *segments*,* each of which is equal in size to a solvent molecule; x is, of course, the ratio of the molar volumes of the solute and solvent. A segment and a solvent molecule may replace one another in the liquid lattice. In other respects the assumptions required are equivalent to those used above. The polymer solution differs from that containing an equal proportion of monomeric solute in the one important respect that sets of x contiguous cells in the lattice are required for accommodation of polymer molecules, whereas no such restriction applies to the solution of the monomeric solute. The situation is illustrated in Fig. 110.

We wish to calculate first the total *configurational entropy* of the poly-

---

* The symbol x is used here and in following pages in a somewhat different sense than in earlier portions of the book, where it represents the number of structural units. The segment employed in mixing problems often is conveniently defined as that portion of a polymer molecule requiring the same space as a molecule of solvent; it is unrelated to the size of the structural unit, which is of no interest here. The present x, like the previous one, defines the size of a polymer species, however.

mer solution arising from the variety of ways of arranging the polymer and solvent molecules. Hence the initial, or reference, states will be taken as the pure solvent and the pure, perfectly ordered polymer; i.e., the polymer chains will be taken to be initially in a perfect crystallike arrangement. The polymer molecules are then added successively to the lattice consisting of $n_0 = n_1 + xn_2$ cells which are to be occupied by the solution; n_1 and n_2 are the numbers of solvent and solute molecules, respectively, as before. In order to arrive at a proper value for the total number Ω of arrangements, the number of ways in which each polymer chain may be inserted in the lattice will be estimated. Suppose that i polymer molecules have been inserted previously at random. There remains a total of $n_0 - xi$ vacant cells in which to place the first segment of molecule $i+1$. Let z be the lattice *coordination number* or number of cells which are first neighbors to a given cell; we may expect z to be in the range of six to twelve, but its actual value is of little importance. The second segment could be assigned to any of the z

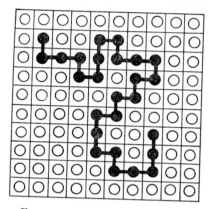

Fig. 110.—Segments of a chain polymer molecule located in the liquid lattice.

neighbors of the cell occupied by the first segment, except, of course, those which happen to be occupied by segments of some of the preceding i polymer molecules. Let f_i represent the expectancy that a given cell adjacent to a previously vacant one (such as that assigned to the first segment of molecule $i+1$) is occupied. Postponing temporarily the evaluation of f_i, we observe that the expected number of cells available for the second segment will be $z(1 - f_i)$. The expected number of cells available to the third segment will be $(z-1)(1-f_i)$, since one of the cells adjacent to the second segment is occupied by the first. For each succeeding segment the expected number of permissible alternative assignments can be taken also as $(z-1)(1-f_i)$, disregarding those comparatively infrequent instances in which a segment other than the immediately preceding one of the same chain occupies one of the cells in question. Hence the expected number of sets of x contiguous sites available to the molecule is

$$\nu_{i+1} = (n_0 - xi)z(z - 1)^{x-2}(1 - f_i)^{x-1} \tag{3}$$

If each of the n_2 polymer molecules to be added to the lattice were differentiable from every other, the number of ways in which all of them could be arranged in the lattice would be given by the product of the ν_i for each molecule added consecutively to the lattice, i.e., by

$$\prod_{i=1}^{n_2} \nu_i$$

Arrangements in which the sets of x contiguous lattice cells chosen for occupation by polymer molecules are identical but which differ only in the permutation of the polymer molecules over these sets would be counted as different in this enumeration scheme. Since the polymer molecules actually are identical, it is appropriate to eliminate this redundancy. We require merely the number Ω of ways in which n_2 sets of x consecutively adjacent cells may be chosen from the lattice; the order in which the sets are filled by polymer molecules is immaterial. Since n_2 differentiable polymer molecules (labeled, for example, in the order of their insertion into the lattice) could be assigned to a given arrangement of n_2 sets of sites in $n_2!$ different ways, it follows that

$$\prod_{i=1}^{n_2} \nu_i = n_2!\Omega$$

or

$$\Omega = (1/n_2!)\prod_{i=1}^{n_2} \nu_i \tag{4}$$

The expectancy f_i that a cell (ξ) is occupied by a segment of a preceding molecule when vacancy of an adjoining cell $(\xi-1)$ has been established beforehand in the course of seeking out the sets of x contiguous lattice cells available to molecule $i+1$, is not exactly equal to the average expectancy $\bar{f}_i$ of occupation of a cell selected at random. This latter expectancy is, of course, equal to the fraction of all cells in the lattice which are occupied by segments of the i molecules; or

$$1 - \bar{f}_i = (n_0 - xi)/n_0 \tag{5}$$

The expectancy f_i will be somewhat less than $\bar{f}_i$ because the manner in which a molecule can be assigned to the lattice so that one of its segments occupies cell ξ is restricted by the fact that an adjoining segment of the same molecule cannot have been assigned to the neighboring cell $\xi-1$ already ascertained to be vacant.* However, differentiation be-

---

* The expectancy f_i may be evaluated in the following manner. The probability that cell ξ is vacant irrespective of the status of neighboring cell $\xi-1$ is, of course, equal to $1-\bar{f}_i$. It may also be equated to the product of the probability that both of these cells are *not* occupied by consecutive segments of the same

tween f_i and $\bar{f}_i$ represents a refinement not warranted by approximations inherent in the lattice model.[5]

Substituting Eq. (5) for $1-f_i$ in Eq. (3) and replacing the lone factor z with $z-1$, we obtain

$$\nu_{i+1} = (n_0 - xi)^x [(z-1)/n_0]^{x-1}$$

which may be further approximated for convenience, and with an error which will be imperceptible, by

$$\nu_{i+1} = \{(n_0 - xi)!/[n_0 - x(i+1)]!\} [(z-1)/n_0]^{x-1} \qquad (6)$$

Substitution of Eq. (6) in (4) gives for the total number of ways of arranging n_2 identical polymer molecules on the lattice consisting of n_0 sites

$$\Omega = [n_0!/(n_0 - xn_2)!n_2!][(z-1)/n_0]^{n_2(x-1)} \qquad (7)$$

If each solvent molecule may occupy one of the remaining lattice sites, and in only one way, Ω represents also the total number of configurations for the solution, from which it follows that the configurational entropy of mixing the perfectly ordered pure polymer and the pure solvent is given by $S_c = k \ln \Omega$. Introduction of Stirling's approximations for the factorials occurring in Eq. (7) for Ω, replacement of n_0 with $n_1 + xn_2$, and subsequent simplification yields

$$S_c = - k\{n_1 \ln[n_1/(n_1 + xn_2)] + n_2 \ln[n_2/(n_1 + xn_2)]$$
$$- n_2(x-1) \ln[(z-1)/e]\} \qquad (8)$$

The formation of the solution may be conceived to occur in two steps: disorientation of the polymer molecules and mixing of the disoriented polymer with solvent. The separate entropy changes are readily obtained as follows: The first is given by Eq. (8) with $n_1 = 0$, i.e.

---

chain and the probability $1 - f_i$ that cell ξ is vacant when $\xi - 1$ also is vacant. The probability that both *are* occupied by consecutive segments of the same chain is $\bar{f}_i 2(x-1)/xz$; hence the equality annunciated above can be written

$$1 - \bar{f}_i = [1 - 2\bar{f}_i(x-1)/xz](1 - f_i)$$

or, by substitution from Eq. (5)

$$1 - f_i = (n_0 - xi)/[n_0 - 2(x-1)i/z] \qquad (5')$$

The value of $1 - f_i$ approaches $1 - \bar{f}_i$, as given by Eq. (5), for large values of the lattice coordination number z. The results of the more accurate treatments of the lattice model given by Huggins,[3] Miller,[6,7] Orr,[8] and Guggenheim[9] are obtained by using Eq. (5') instead of (5) in the development presented above. The resulting expressions, though much more complicated, offer no better agreement with experiment, however.[5]

$$\Delta S_{\text{disorientation}} = kn_2\{\ln x + (x - 1)\ln[(z - 1)/e]\} \qquad (9)$$

If x is large, the first term is negligble compared with the second, and the entropy of disorientation *per segment* reduces to

$$(1/x\, n_2)\Delta S_{\text{disorientation}} \cong k\ln[(z - 1)/e] \qquad (9')$$

As a result of having assumed that the pure polymer and the pure solvent, and all intermediate compositions as well, can be accommodated in the same lattice, the number x of segments in a polymer molecule is taken implicitly as the ratio of the molecular volumes of the polymer and the solvent. Consequently, the entropy of disorientation expressions (9) and (9') contain the parameter x which depends jointly on the solute and the solvent, although the latter has nothing to do with an actual disorientation (e.g., melting) process. As a more reasonable procedure for the estimation of the entropy of disorientation, the actual chain might be replaced by its hypothetical, equivalent freely jointed chain, as discussed in Chapter X, the size and number of units being so chosen as to match the contour length and the root-mean-square displacement length ($\sqrt{\overline{r^2}}$) of the actual chain in its unperturbed state. The number x' of such units (n' in Chap. X) would then replace x in Eqs. (9) and (9'). Since, however, the lattice coordination number z which should be used in conjunction with this equivalent chain cannot be evaluated in any independent manner, such relationships would be of little quantitative significance. It is possible to say merely that the entropy of disorientation should be greater the more flexible the chain and the more disordered the packing of the segments in the liquid (i.e., the larger z). According to Eq. (9'), it should be of the order of R calories per mole of equivalent elements. This entropy of disorientation of the chain skeleton may be expected to constitute a major contribution to the entropy of fusion, a subject to be discussed in the following chapter.

The entropy of mixing disoriented polymer and solvent may be obtained, according to the original assumptions pertaining to the lattice model, by subtracting Eq. (9) from (8). The result reduces to[4]

$$\Delta S_M{}^* = - k(n_1 \ln v_1 + n_2 \ln v_2) \qquad (10)$$

where v_1 and v_2 are the volume fractions of solvent and solute, i.e.

$$\left.\begin{array}{l} v_1 = n_1/(n_1 + xn_2) \\ v_2 = xn_2/(n_1 + xn_2) \end{array}\right\} \qquad (11)$$

An asterisk is appended to the symbol $\Delta S_M{}^*$ as a reminder that it represents only the *configurational entropy* computed by considering the external arrangement of the molecules and their segments without regard for the internal situations of the latter.   Possible contributions to the entropy resulting from specific interactions between near neighbors will be considered later.

Comparison of Eq. (10) with (1) reveals an interesting analogy to the ideal entropy of mixing; mole fractions occurring in the ideal expression are replaced with volume fractions in the formula for mixing of molecules dissimilar in size.   The ideal mixing "law" can be derived only for the case in which solvent and solute molecular volumes are equal.   But the volume fractions and mole fractions are identical when this is true, and Eqs. (10) and (1) are then equivalent.   Thus the ideal mixing relation appears as a special case of the more general mixing expression, Eq. (10).

Extension of this treatment to heterogeneous polymers[10,11] comprising a range of homologous molecular species leads to the following generalization of Eq. (10):

$$\Delta S_M{}^* = - k \left(n_1 \ln v_1 + \sum_i{}' n_i \ln v_i \right) \tag{12}$$

where the prime attached to the summation is intended to indicate that polymer species only are included therein.   The term $n_1 \ln v_1$ representing the solvent could, of course, be included in this summation as well, to yield a more compact expression.   However, the form adopted above, in which the solvent and polymer contributions are given separately, will ordinarily be preferred inasmuch as the heat of mixing term, to be introduced later, may depend on the total polymer concentration $\sum' v_i = v_2 = 1 - v_1$.   Since the solvent term is identical in form with that for any other species, it is evident that the solvent need not necessarily consist of a single segment.   The solvent may, in fact, be composed of any number of segments (i.e., its molecular volume is unrestricted) without altering the form of either Eq. (10) or (12).

1b. Approximations and Limitations of the Lattice Treatment.—The simplicity of the preceding expressions for the entropy of mixing pure disoriented polymer(s) with solvent is most attractive, but before we proceed with their applications it is desirable to examine carefully the limitations to be expected from the assumptions and approximations involved, explicitly or implicitly, in the derivation.   Foremost of these is the acceptance of a single lattice to characterize both solvent and polymer and all intermediate compositions.   As we have already mentioned, the use of a lattice merely as a scheme for treating the relationships between a given molecule and its immediate neighbors in the pure liquid is not particularly objectionable, but use of the *same* lattice for different components is seldom justified by the spatial requirements of the species involved.   There seems to be no way to circumvent this assumption in treating the lattice model; it constitutes the most serious compromise with reality for the sake of expediency in the lattice approach to polymer solution properties.   Mathematical approximations introduced for the purpose of simplifying the treatment appear to be of trivial significance compared with the physical unrealities of the model itself.

Most efforts toward improvement of the theory, however, have been directed at refinement of the mathematical treatment of the model, and the imperfections inherent in the model itself too often have been ignored.   These mathematical refinements are of two sorts.   The one consists in the use of the more accurate expression for f_i (see p. 500, footnote) in the derivation of an expression for ν_i, or in otherwise avoiding the equating of f_i to $\bar{f}_i$.[3,6,8,9]   The equations which result are considerably more complicated than those given above, without any real evidence of improved agreement with experiment.[5]   The other refinement in the treatment of the lattice model is concerned with the assumption, which we have adopted implicitly, that the solution configuration is random.   If preferential attractions occur between like or unlike species, the chance of finding a solvent molecule, for example, in the cell adjacent to a polymer segment will be modified by these attractions.   Expressions for the entropy of mixing have been derived by Orr[8] and by Guggenheim[12] which take into consideration the influence of preferential interactions.   Their results show that, for such interactions as may ordinarily occur in polymer solutions, the modification in the calculated entropy of mixing caused by nonrandomness is scarcely significant in comparison with the effects of other approximations.

One further assumption[13] common to all treatments of the lattice model, but not mentioned in the course of the preceding development, remains to be considered.   This one, of a rather more subtle nature, was ignored by most investigators for a number of years.   In seeking the number of sequences of x-sites in the lattice which are available to molecule $i+1$, it has been assumed that segments of the i molecules previously added are distributed essentially at random throughout the lattice.   Actually, of course, they occur in sequences of x consecutively adjacent cells.   If the concentration of polymer molecules already added to the lattice is large, they will be intertwined with one another to such an extent that this assumption probably is legitimate.   However, if the solution is so dilute that the polymer molecules are well separated from one another, cells occupied by segments must occur in loose clusters of low population density, and intervening regions will be entirely empty.

This state of affairs in dilute solutions is illustrated in Fig. 34 (p. 268). The solution consists of two roughly distinguishable regions at high dilutions: one completely unoccupied by polymer and the other containing a low (average) concentration of segments. This concentration within the domain of a molecule depends on the molecular configuration of the polymer chain and not on the composition of the solution as a whole. If molecule $i+1$ is added to the former region, it encounters no segments of the i previously added molecules; the effective f_i in this region is zero. If it is so located as to overlap one (or more) of the domains occupied by other polymer molecules, the effective f_i will exceed its *average* throughout the very dilute solution as a whole. Except in a poor solvent (see Sec. 2b) the molecule $i+1$ is much more likely to locate in the former region where it is subject to no competition for lattice locations for its segments. This is true, in spite of the low segment density in a molecular coil, because of the very large number (x) of segments in the molecule $i+1$ for which locations must be found (see Sec. 2). Consequently, in a good solvent, where the polymer segments find the solvent environment energetically at least as favorable as that provided by other polymer segments, conflicts arising from tendencies of segments belonging to different molecules to occupy the same site will be far less frequent than we have calculated above by assuming a random distribution of segments over the whole volume. The polymer molecules may even be considered in rough approximation as noninterpenetrating spheres.[13] The configurational statistics of a dilute solution of such spheres are incompatible with the theory developed above, as will be shown more fully later in the present chapter. Of importance here is the realization that the theory as developed so far is inappropriate, generally speaking, for dilute polymer solutions.

1c. An Alternate Derivation.—It will be observed that Eqs. (10) and (12) contain no lattice-dependent parameters, which suggests that they may enjoy greater validity than the artificialities of the lattice model would warrant.* As a matter of fact, Hildebrand[14] has shown that the entropy of mixing expression (10) may be derived without resort to a liquid lattice model, and in a manner which is more enlightening as to the significance of the terms appearing in this relationship. Instead of assuming identical lattices for polymer and solvent, he as-

---

* If the more accurate expression (5′) for f_i given in the footnote on p. 500 is used, additional terms containing the lattice coordination number z appear in the entropy of mixing expression.

sumes that the *free volume* available to the molecules per unit volume of liquid is the same for the polymer as for the solvent.* Furthermore, the free volume in any solution of the polymer and solvent is assumed to be the same fraction v_f of the total volume. The free volumes available to the pure components according to the above assumptions will then be

$$V_{f,1} = n_1 V_1 v_f \tag{13}$$

and

$$V_{f,2} = n_2 V_2 v_f \tag{14}$$

where V_1 and V_2 are the respective molecular volumes. That for the solution can be written

$$V_{f,12} = (n_1 V_1 + n_2 V_2) v_f \tag{15}$$

According to the concept of free volume as the effective volume over which the centers of gravity of the molecules are distributed, the entropy may be taken as that of a perfect gas composed of the same number of molecules confined to a volume equal to the free volume. Since the entropy of a perfect gas consisting of n molecules depends on its volume as $nk \ln V$, the increase in entropy owing to the greater free volume available to the solvent molecules in the solution will be

$$n_1 k \ln(V_{f,12}/V_{f,1})$$

and the corresponding term for the solute will be

$$n_2 k \ln(V_{f,12}/V_{f,2})$$

The sum of these expressions represents the entropy of mixing. The result obtained by substituting for the free volumes from Eqs. (13), (14), and (15), is identical with Eq. (10). According to this simple derivation, the terms appearing in Eq. (10) represent contributions to the entropy which originate in the greater spatial freedom of the molecules in the solution. By a simple extension of the derivation to a mixture of polymer species, Eq. (12) may be obtained.

An obvious refinement suggested by this second derivation would consist in ascribing different free volume fractions v_f to the two pure

---

* The free volume is considered to represent the difference between the actual volume of the liquid (or the amorphous polymer) and the minimum volume which it would occupy if its molecules were packed firmly in contact with each other. Incompressible molecules with rigid dimensions are implied in this definition of a free volume. The unrealistic nature of this implication undermines precise determination, or even an exact definition, of the free volume. The concept has proved useful nevertheless.

components, with a further stipulation as to the change in v_f with composition of the solution; it might be assumed to vary linearly with the volume fraction of polymer, for example. The free volume fraction of the polymer ordinarily must be considerably smaller than that of the monomeric solvent; hence, as may be easily verified by re-examining the derivation just given, the configurational entropy of mixing will on this account be less than is given by Eq. (10). Incorporation of this refinement into the theory complicates subsequent equations considerably, and additional arbitrary parameters appear which cannot be evaluated intelligently from existing data. For these reasons we shall adhere to the simpler mixing expression (10), realizing that the resulting entropies (including the partial molar entropies; see Sec. 1e) are likely to be somewhat too large.

1d. The Heat and Free Energy of Mixing.—Intermolecular interactions are large in the liquid state owing to the close proximity of the molecules. Since the pure solvent and pure liquid polymer are taken as reference states for the treatment of the solution, we are interested only in the *difference* between the total interaction energy in the solution as compared with that for the pure liquid components; the absolute magnitudes of the interaction energies, either in the solution or in the pure components, are of no direct concern. In particular, we need to express the dependence of this difference, or heat of mixing ΔH_M, on the concentration. Inasmuch as the forces between uncharged molecules decrease rapidly with the distance of separation, it suffices to restrict our considerations to the energies developed by first neighbor molecules, or chain segments, in the solution; interactions between elements which are not immediate neighbors contribute trivially to the total interaction energy. The heat of mixing, therefore, can be considered to originate in the replacement of some of the contacts between like species in the pure liquids with contacts between unlike species in the solution. According to the lattice model with each cell able to accommodate either a solvent molecule or a polymer segment, three types of first neighbor contacts, conveniently represented by the self-explanatory symbols, [1,1], [2,2], and [1,2], will occur. The formation of the solution may be likened to a chemical reaction in which bonds of the latter type are formed at the expense of an equal number of the former two according to the stoichiometric equation

$$1/2[1,1] + 1/2[2,2] = [1,2] \tag{16}*$$

If w_{11}, w_{22}, and w_{12} are the energies associated with these respective

---

* This representation is a restricted variation of the *quasi-chemical method* used by Guggenheim[12] and by Orr[8] in the derivation of the complete free energy.

pair contacts, or "bonds," the change in energy for the formation of an unlike contact pair according to Eq. (16) is

$$\Delta w_{12} = w_{12} - (1/2)(w_{11} + w_{22}) \tag{17}$$

If in a uniquely specified arrangement of the molecules in the solution there are p_{12} pairs of unlike neighbors, i.e., 1,2 contacts, the heat of formation of that particular configuration from the pure components is

$$\Delta H_M = \Delta w_{12} p_{12} \tag{18}$$

In order to determine the average value of p_{12} in a solution of given composition, we observe that the probability that a particular site adjacent to a polymer segment is occupied by a solvent molecule is approximately equal to the volume fraction v_1 of solvent in the solution.* The total number of contacts between a polymer molecule and all of its neighbors will be $z-2$ per chain unit plus two additional ones for the terminal units, making a total of $(z-2)x+2$. To about the same approximation as that involved in the use of the volume fraction v_1 above,* this number of contacts per polymer molecule may be replaced by zx. Hence the total number of 1,2 contacts in the solution becomes $zxn_2v_1 \equiv zn_1v_2$, and the heat of mixing the two components may accordingly be expressed as

$$\Delta H_M = z\Delta w_{12} n_1 v_2 \tag{19}$$

This is just the well-known van Laar expression for the heat of mixing in any two-component system. Thus, within the limits of the approximations used, the polymeric character of the solute does not alter the form of the heat of mixing expression. To generalize somewhat, let the solvent molecule contain x_1 segments instead of only one. Eq. (19) is then replaced by

$$\Delta H_M = z\Delta w_{12} x_1 n_1 v_2 \tag{19'}$$

which we shall find it advantageous to recast in the form

$$\Delta H_M = kT\chi_1 n_1 v_2 \tag{20}$$

---

* Rigorous treatment of the model assumed would dictate the use of a "surface fraction," or "site fraction," in place of the volume fraction v_1. The site fraction of the solvent may be defined as the number of intermolecular site locations adjacent to solvent molecules divided by the total number of such sites for both solvent and polymer molecules. Thus, if the solvent consists of a single segment its site fraction is

$$zn_1/\{zn_1 + [(z - 2)x + 2]n_2\}$$

which approaches v_1 for large z.

where

$$\chi_1 = z\Delta w_{12}x_1/kT \tag{21}$$

is a dimensionless quantity which characterizes the interaction energy per solvent molecule divided by kT. The quantity $kT\chi_1$ represents merely the difference in energy of a solvent molecule immersed in the pure polymer $(v_2 \cong 1)$ compared with one surrounded by molecules of its own kind, i.e., in the pure solvent. The heat of mixing expression, Eq. (20), like the entropy of mixing expression (10), retains no parameters of the hypothetical lattice. Eq. (20) is equally suitable for heterogeneous polymers for which $v_2 = \sum' v_i$.

(The following definition of χ_1 is sometimes preferred

$$\chi_1 = Bv_1/RT \tag{21'}$$

where $B = z\Delta w_{12}/V_s$, v_1 is the *molar* volume of the solvent, and V_s denotes the *molecular* volume of a segment. Then B represents the interaction energy density characteristic of the solvent-solute pair.)

If the configurational entropy ΔS_M^* is assumed to represent the total entropy change ΔS_M on mixing, the free energy of mixing is simply obtained by combining Eqs. (10) and (20). That is,

$$\Delta F_M = \Delta H_M - T\Delta S_M = \Delta H_M - T\Delta S_M^*$$
$$= kT[n_1 \ln v_1 + n_2 \ln v_2 + \chi_1 n_1 v_2] \tag{22}$$

In the case of a heterogeneous polymer comprising molecular species differing in size but otherwise equivalent chemically, substitution of Eq. (12) for ΔS_M^* gives

$$\Delta F_M = kT[n_1 \ln v_1 + \sum' n_i \ln v_i \chi_1 + n_1 v_2] \tag{23}$$

where $v_2 = \sum' v_i$, the prime designating summation over the solute species only. These equations express the total free energy change for the formation of the solution from pure, disoriented polymer (i.e., amorphous or liquid polymer) and pure solvent.

The effect of accepting the configurational entropy of mixing ΔS_M^* as a proper expression for the *total* entropy of mixing ΔS_M is to neglect possible contributions which may arise from specific interactions between neighboring components (solvent molecules and polymer segments) of the solution. We have thus far considered these interactions to lead solely to a heat of mixing. There is no a priori justification, however, for dismissing the possibility that, owing to orienting influences on the components in the solution which differ from those existing in the pure component, or for other reasons, there may not also be a contribution to the entropy from first neighbor interactions.

To put the matter another way, the mixing process as represented by Eq. (16) should, in general, be characterized[15] by a standard state entropy change* as well as by a change in energy or heat content.   The entropy change associated with first neighbor interactions must be proportional to the number of pair contacts developed in the solution, just as for the heat change.   Hence the omission of this entropy contribution is conveniently rectified by considering the Δw of Eqs. (17) through (21) to consist of two parts, one representing the change in heat content for the process (16) and the other the product of the absolute temperature T and the entropy change for conversion of "reactants" to "products" in the standard state.   One might then write

$$\Delta w_{12} = \Delta w_h - T\Delta w_s$$

whereupon Δw_{12} acquires the character of a standard state free energy change for unit displacement of the process.[15]   Consequently, the parameter χ_1 defined by Eq. (21) will contain an entropy contribution (divided by k) in addition to a heat of mixing term (divided by kT), and the quantity $kT\chi_1 n_1 v_2$ should be regarded as the standard state free energy change, rather than as the heat of mixing only. Since the total free energy change ΔF_M must consist of this standard state free energy change plus the configurational free energy change, $-T\Delta S_M{}^*$, *the form of the free energy function*, as expressed by either equation (22) or (23), *is unaltered by these considerations*.   It is necessary merely to reappraise the significance of the parameter χ_1 in the manner indicated.

Expressions for the entropy and heat of mixing which are appropriate in the light of the above revision may be obtained from ΔF_M through use of the standard thermodynamic relations

$$\Delta S_M = - (\partial \Delta F_M/\partial T)_P$$

$$\Delta H_M = - T^2(\partial(\Delta F_M/T)/\partial T)_P$$

Thus, from Eq. (22)

$$\Delta S_M = - k\{n_1 \ln v_1 + n_2 \ln v_2 + [\partial(\chi_1 T)/\partial T]n_1 v_2\} \qquad (24)$$

$$\Delta H_M = - kT^2(\partial\chi_1/\partial T)n_1 v_2 \qquad (25)$$

If Δw_{12} is independent of T, that is, if Δw_{12} contains no entropy contribution, $\chi_1 = z\Delta w_{12}x_1/kT$ will be inversely proportional to T.   Then the third term in Eq. (24) is zero and $\Delta S_M = \Delta S_M{}^*$; also, $-T(\partial\chi_1/\partial T)$

---

* The standard state entropy change refers to the hypothetical process so conducted as to convert pure 1 and 2 to a uniquely specified array of 1, 2 pairs; e.g., to the pure 1, 2 "compound."   In short, the standard state entropy change is just that portion of ΔS_M other than $\Delta S_M{}^*$.

$= \chi_1$ and ΔH_M reduces to the previous expression, Eq. (20). If Δw_{12} is not independent of T, i.e., if interaction between near neighbors contributes to the entropy, the more complicated equations (24) and (25) must be used.

In conclusion, we may expect Eqs. (22) and (23) to express the free energy of mixing satisfactorily as a function of the concentration, within the limits imposed by the assumptions introduced in the derivation of the configurational entropy of mixing as discussed in the preceding section. It will be recalled, in the first place, that the most serious of these is the assumption of a single lattice to describe the configurational character of both pure components and their solutions as well; in the Hildebrand derivation the corresponding assumption was that of constancy of the free volume fraction. Current solution theories in general are predicated on the hope that no grave error is committed in making this assumption. Secondly, owing to the discontinuous nature of very dilute polymer solutions, the expression derived above for ΔH_M, like that for $\Delta S_M{}^*$, is inapplicable in principle at high dilutions quite apart from the question of the validity of the lattice model. Hence all of the thermodynamic relations derived thus far are appropriate in general only at concentrations such that the randomly coiled molecules overlap one another extensively.* Finally, the *total* entropy of mixing found experimentally may contain a contribution from $[\partial(\chi_1 T)/\partial T]$, which it would be difficult to evaluate separately. Hence the theoretical *configurational* entropy of mixing $\Delta S_M{}^*$ cannot be compared in an unambiguous manner with the experimentally accessible quantity ΔS_M. It should be noted that the various difficulties encountered, aside from those precipitated by the character of dilute polymer solutions, are not peculiar to polymer solutions but are about equally significant in the theory of solutions of simple molecules as well.

1e. Partial Molar Quantities.—The chemical potential μ_1 of the solvent in the solution relative to its chemical potential μ_1^0 in the pure liquid is obtained by differentiating the free energy of mixing, ΔF_M, with respect to the number n_1 of solvent molecules. Differentiation of Eq. (22) for ΔF_M with respect to n_1 (bearing in mind that v_1 and v_2 are functions of n_1) and multiplication of the result by Avogadro's number N in order to obtain the chemical potential per mole (or relative partial molar free energy, $\Delta \overline{F}_1$) gives

$$\mu_1 - \mu_1^0 = RT \left[\ln(1 - v_2) + (1 - 1/x)v_2 + \chi_1 v_2^2\right] \qquad (26)$$

* This condition should be fulfilled at concentrations exceeding several percent unless the molecular weight is less than 10^5, in which case somewhat higher concentrations may be required.

From Eq. (23) for heterogeneous polymers we find

$$\mu_1 - \mu_1^0 = RT\left[\ln(1 - v_2) + (1 - 1/\bar{x}_n)v_2 + \chi_1 v_2^2\right] \tag{27}$$

which differs from Eq. (26) only in the replacement of x by the number average degree of polymerization $\bar{x}_n$ of the heterogeneous polymer. These equations separate easily into contributions from the configurational entropy and from the first neighbor interactions. Thus (26) may be written

$$\mu_1 - \mu_1^0 = -T\Delta\bar{S}_1{}^* + RT\chi_1 v_2^2 \tag{26'}$$

where

$$\Delta\bar{S}_1{}^* = -R\left[\ln(1 - v_2) + (1 - 1/x)v_2\right] \tag{28}$$

is the relative partial molar *configurational* entropy of the solvent in the solution. It may be obtained directly by differentiation of Eq. (10). If χ_1 varies inversely with T (i.e., if the interaction contributes an energy only), the first two terms occurring in Eq. (26) represent the relative partial molar entropy, or the entropy of dilution, and the last term the relative partial molar heat content, or heat of dilution

$$\Delta\bar{H}_1 = RT\chi_1 v_2^2 \tag{29}$$

If χ_1 contains an entropy contribution, the form of the chemical potential is unaltered but its resolution into entropy and heat contributions must be carried out according to operations like those applied above to the free energy of mixing.

From the chemical potential we may at once set down expressions for the activity a_1 of the solvent and for the osmotic pressure π of the solution, using standard relations of thermodynamics. For the activity

$$\ln a_1 = (\mu_1 - \mu_1^0)/RT = \ln(1 - v_2) + (1 - 1/x)v_2 + \chi_1 v_2^2 \tag{30}$$

Since the pure solvent has been chosen as the standard state, $a_1 = P_1/P_1^0$, to the approximation that the vapor may be regarded as an ideal gas. For the osmotic pressure $\pi v_1 = -(\mu_1 - \mu_1^0)$ where v_1 is the molar volume of the solvent. Thus, according to Eq. (26)

$$\pi = -(RT/v_1)\left[\ln(1 - v_2) + (1 - 1/x)v_2 + \chi_1 v_2^2\right] \tag{31}$$

The osmotic method is most useful, of course, in dilute solutions where the theory which we have developed above is invalid. Disregarding this deficiency for the moment, we may expand the logarithmic term in series with the retention only of terms in low powers of v_2. Then

$$\pi = (RT/v_1)\left[v_2/x + (1/2 - \chi_1)v_2^2 + v_2^3/3 + \cdots\right] \tag{31'}$$

It is more convenient to use the concentration c in grams per ml.   Now $v_2 = c\bar{v}$, where $\bar{v}$ is the (partial) specific volume of the polymer, and since x is the ratio of the molar volumes of polymer and solvent, we have $v_2/xv_1 = c\bar{v}/xv_1 = c/M$.   Hence

$$\pi/c = RT/M + RT(\bar{v}^2/v_1)(1/2 - \chi_1)c + RT(\bar{v}^3/3v_1)c^2 + \cdots \quad (31'')$$

The first term on the right is the ideal, or van't Hoff, term.   At infinite dilution, π/c must necessarily approach this limit, as may be shown (see Chap. VII) by compelling thermodynamic arguments.   The higher terms represent the deviations from ideality predicted by the foregoing theory.   It is these deviation terms which are subject to error as a result of the aforementioned limitations of this theory in dilute solutions.

Differentiation of Eq. (22) with respect to n_2 yields for the chemical potential of the polymeric solute relative to the pure liquid polymer as standard state

$$\mu_2 - \mu_2^0 = RT\big[\ln v_2 - (x - 1)(1 - v_2) + \chi_1 x(1 - v_2)^2\big] \quad (32)$$

The chemical potential of the species of size x in a heterogeneous polymer, obtained by differentiating Eq. (23) with respect to n_x, is

$$\mu_x - \mu_x^0 = RT\big[\ln v_x - (x - 1) + v_2 x(1 - 1/\bar{x}_n) + \chi_1 x(1 - v_2)^2\big] \quad (33)$$

which reduces to Eq. (32) if the polymer is homogeneous; in this event $v_x = v_2$ and $\bar{x}_n = x$.   Eq. (33) is useful in treating problems involving the partitioning of individual polymer species between two phases in equilibrium, such as occurs in fractionation.   In other circumstances, where the polymer is confined as a whole to one phase, as in osmotic equilibrium or in solution-vapor equilibrium, the chemical potential averaged over all polymer species is required.   Then Eq. (32) should be used with x replaced by its *number average* $\bar{x}_n$.

For some purposes, as for example in the treatment of crystallization, it is more convenient to deal with the chemical potential per mole of structural units instead of per mole of polymer, as expressed in the formulas given above.   Dividing Eq. (32) by the number of units per polymer molecule, which is xv_1/v_u where v_1 and v_u are the molar volumes of the solvent and of the structural unit, respectively, we obtain for the chemical potential difference per unit

$$\mu_u - \mu_u^0 = RT(v_u/v_1)\big[(\ln v_2)/x - (1 - 1/x)(1 - v_2) + \chi_1(1 - v_2)^2\big] \quad (34)$$

It will be observed that the chemical potentials depend appreciably on the chain length, represented by x, only at low concentrations.   The influence of the chain length vanishes with increase in concentration of polymer, and the more rapidly so the higher its molecular weight. This is evident in Eq. (26) for the chemical potential of the solvent, and also in the series expansion forms (31') and (31'') derived from it. For large values of x and at higher concentrations it is permissible to use in place of Eq. (26)

$$\mu_1 - \mu_1^0 \cong RT[\ln(1 - v_2) + v_2 + \chi_1 v_2^2] \qquad (26'')$$

which yields for the activity

$$a_1 \cong v_1 e^{v_2 + \chi_1 v_2^2} \qquad (29')$$

If the amount of solvent is very small, $v_2 \cong 1$ and

$$a_1 \cong v_1 e^{1 + \chi_1} \qquad (29'')$$

or

$$P_1 \cong P_1^0 v_1 e^{1 + \chi_1}$$

which establishes $e^{1 + \chi_1}$ as the limiting Henry's law constant when the concentration of the diluent is expressed by its volume fraction.

In a similar way the chemical potential of the polymer unit readily reduces for large x to

$$\mu_u - \mu_u^0 \cong - RT(\mathrm{v}_u/\mathrm{v}_1)[(1 - v_2) - \chi_1(1 - v_2)^2] \qquad (34')^*$$

1f. Experimental Results.—Inasmuch as the thermodynamic equations given above for binary systems contain a single arbitrary parameter χ_1, a value for it may be obtained from a single measurement of the activity or of the osmotic pressure of a solution containing polymer of known molecular weight (or x) at the specified concentration v_2. The validity of the expressions derived for the free energy of mixing and for the chemical potential may be judged, in part at least by the constancy of this parameter over extended ranges in composition. Values of χ_1 calculated from the solvent activities according to

---

* It will be observed from Eqs. (34) and (34') that $\mu_u - \mu_u^0$ reaches a finite limit at a polymer concentration of zero provided $x = \infty$. Thus the activity of a polymer unit remains finite (i.e., greater than zero) even when its concentration is zero, provided the chain length is infinite. One of the consequences of this peculiar circumstance is virtual complete disappearance of polymer from the equilibrium

$$\text{monomer} \rightleftarrows \text{polymer}$$

beyond a critical temperature. Ring-chain equilibria (Chaps. III and VIII) represent examples of this nature. Another is provided by the interconversion of cyclic sulfur, S_8, to linear polymeric sulfur of high chain length (see G. Gee, *Trans. Faraday Soc.*, **48**, 515 [1952]); all of the latter disappears below a temperature of about 159°C at equilibrium.

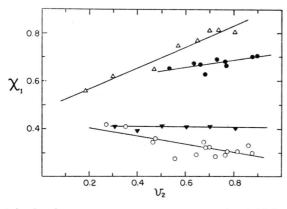

Fɪɢ. 111.—Experimental values of the interaction parameter χ_1 plotted against the volume fraction v_2 of polymer. Data for polydimethylsiloxane ($M = 3850$) in benzene, △ (Newing[16]); polystyrene in methyl ethyl ketone, ● (Bawn $et\ al.$[17]); and polystyrene in toluene, ○ (Bawn $et\ al.$[17]) are based on vapor pressure measurements. Those for rubber in benzene, ▼ (Gee and Orr[18,19]) were obtained using vapor pressure measurements at higher concentrations and isothermal distillation equilibration with solutions of known activities in the dilute range.

Eq. (30) are plotted in Fig. 111 against the volume fraction of polymer for several systems. Results at very low concentrations, i.e., in the range of osmotic measurements, have been omitted from Fig. 111 because of the inadequacy of the theory in this region. Over the concentration range covered, the molecular weight of the polymer is of little significance provided it is large. Only in the case of the silicone of molecular weight 3850 has it been necessary to include the $1/x$ term occurring in Eq. (30), and in related expressions, in making the calculations.

The rubber-benzene system, for which χ_1 is remarkably constant[18] over a very wide concentration range, was the first system studied accurately.[1] The excellent agreement found in this case engendered optimism regarding the quantitative reliability of the theory, which subsequent studies on other systems have failed to verify. In no other system so far investigated is the agreement so good as for rubber-benzene. The change in χ_1 with concentration is particularly marked for the silicone in benzene[16] and for polystyrene in methyl ethyl ketone.[17] Data for the latter system are quite scattered, but we are justified in concluding that the indicated decrease in χ_1 with decrease in concentration is real on the grounds that polystyrene, regardless of its molecular weight, and methyl ethyl ketone are miscible in all proportions at ordinary temperatures. From this fact it must follow (see Chap. XIII) that χ_1 is no greater than 0.50 at lower polymer concentrations. Prager and Long[20] weighed the amounts of hydrocarbon vapors absorbed by polyisobutylene ($M \sim 10^6$) as functions of the vapor activity, or pressure. Values calculated for χ_1 were found to be quite constant. With n-pentane at 35°, for example, $\chi_1 = 0.63 \pm 0.01$ for concentrations over the range $v_1 = 0.01$ to 0.60. However, polyisobutylene is miscible with n-pentane in all proportions; hence χ_1 must decrease at solvent concentrations beyond those investigated.

Available data are too few to justify generalizations, but the results cited above suggest that the free energy expression derived for polymer solutions gives rather good agreement for nonpolar systems, except of course, at low concentrations. In those cases where either the polymer unit or the solvent possesses a dipole, as in the silicone-benzene and in the polystyrene–methyl ethyl ketone systems, χ_1 appears to vary throughout the concentration range. For such systems the free energy of mixing expression (22) and the chemical potentials derived from it, i.e., Eqs. (26) and (34), may nevertheless be used as semi-quantitative approximations.

The chemical potential difference $\mu_1 - \mu_1^0$ may be resolved into its heat and entropy components in either of two ways: the partial molar heat of dilution may be measured directly by calorimetric methods and the entropy of dilution calculated from the relationship $\Delta \overline{S}_1 = (\Delta \overline{H}_1 - \Delta \overline{F}_1)/T$ where $\Delta \overline{F}_1 \equiv \mu_1 - \mu_1^0$; or the temperature coefficient of the activity (hence the temperature coefficient of the chemical potential) may be determined, and from it the heat and entropy of dilution can be calculated using the standard relationships

$$\Delta \overline{H}_1 = - RT^2 (\partial \ln a_1 / \partial T)_{P,v_2} \tag{35}$$

$$\Delta \overline{S}_1 = - R [\partial (T \ln a_1)/\partial T]_{P,v_2} = (\Delta \overline{H}_1 - \Delta \overline{F}_1)/T \tag{36}$$

The extremely low rates of solution of polymers and the high viscosities of their solutions present serious problems in the application of the delicate calorimetric methods required to measure the small heats of mixing or dilution.* This method has been applied successfully only to polymers of lower molecular weight where the rate of solution is rapid and the viscosity of the concentrated solution not intolerably great.[22] The second method requires very high precision in the measurement of the activity in order that the usually small temperature coefficient can be determined with sufficient accuracy.

The results of determinations of heats of dilution for several polymer-solvent systems over wide ranges in concentration are shown in Fig.

---

* The integral heat of mixing is, of course, the quantity directly measured in the calorimetric method. However, the heat change on diluting a solution of the polymer with an additional amount of solvent may sometimes be measured in preference to the mixing of pure polymer with solvent.[21] In either case, the desired partial molar quantity $\Delta \overline{H}_1$ must be derived by a process of differentiation, either graphical or analytical.

112. In order to minimize confusion, only the curves representing the smoothed results are shown for squalene-benzene, polyisoprene-benzene, and rubber-benzene. Calorimetric methods were applied to those polymers of comparatively low molecular weight; temperature coefficients of the activity were used for the rubber-benzene mixtures. The ratio of the heat of ᵢdilution $\Delta \overline{H}_1$ to the square of the volume fraction v_2, which is plotted against v_2 in Fig. 112, should be independent of the concentration according to the treatment of interactions

Fig. 112.—Heat of dilution ($\Delta \overline{H}_1$) results for various polymers in benzene. Polydimethylsiloxane $M = 3850$, $\triangle$, and $M = 15,700$, $\blacktriangle$ (Newing[16]); squalene, curve a (Gee and Orr[19]); polyisoprene $M = 4000$, curve b (Gee and Orr[19]). Curve c is for rubber in benzene (Gee and Orr[19]) based on temperature coefficients of activities determined from vapor pressures or vapor equilibration; osmotic pressure-temperature coefficients establish the lower intercept for rubber.

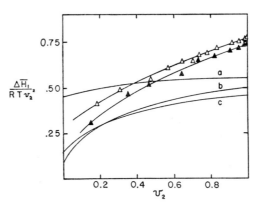

given above. Only in the case of squalene, a comparatively low molecular weight analog of rubber consisting of six isoprene units, is this quantity reasonably constant. For all of the bona fide polymers for which data are available, $\Delta \overline{H}_1/v_2^2$ increases with v_2. The use of a site fraction, or surface fraction, instead of the volume fraction, would effect an improvement, but such revision would not introduce a correction of the size required to bring about constancy in the heat of mixing parameter. Furthermore, the conversion to site fractions would introduce an almost equal modification of the results for squalene, where little or no correction is required. The fall in $\Delta \overline{H}_1/v_2^2$ at low concentrations doubtless is in part due to the necessarily nonuniform distribution of segments over the dilute solution, but there are no grounds at present for considering that this feature persists at the higher concentrations where $\Delta \overline{H}_1/v_2^2$ continues to rise.

Entropies of dilution for several systems are compared in Fig. 113 with the theoretical line (broken curve) for $\Delta \overline{S}_1{}^*$ according to Eq. (28). An infinite molecular weight has been assumed in making this calculation; over the range covered by the curve, the molecular weight (i.e., x) is of no significance provided it is large. The fit for rubber-benzene is reasonably good; the pronounced deviation at low concen-

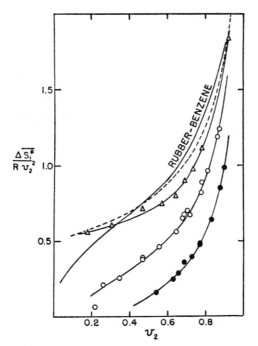

Fig. 113.—Comparison of observed entropies of dilution (points and solid lines) with results calculated for $\Delta \bar{S}_1^*$ according to Eq. (28) (broken line). Data for polydimethylsiloxane, $M = 3850$, in benzene, $\triangle$ (Newing[16]), obtained from measured activities and calorimetric heats of dilution. Entropies for polystyrene (Bawn et al[17]). in methyl ethyl ketone, $\bullet$, and in toluene, $\bigcirc$, were calculated from the temperature coefficient of the activity. The smoothed results for benzene solutions of rubber, represented by the solid curve without points, were obtained similarly.[1,18,19]

trations appears to be due to the nonuniformity of the dilute solution to which repeated reference has been made. Other systems—polydimethylsiloxane-benzene, polystyrene-toluene, and polystyrene–methyl ethyl ketone—show larger deviations, which for the last two mentioned are very large indeed. Hope for the theory as a quantitative interpretation which was inspired by the accord demonstrated by Gee and Treloar[1] in their early comprehensive investigation of rubber-benzene unfortunately has not received support from work on other systems. Either the theoretical treatment of the configurational entropy of dilution $\Delta \bar{S}_1^*$ suffers from gross inaccuracies, or the first neighbor interactions make contributions to the entropy of dilution which are comparable to $\Delta \bar{S}_1^*$ (calcd.), or both of these sources of error may contribute to the deviations observed.

Gee and Orr[19] have pointed out that the deviations from theory of the heat of dilution and of the entropy of dilution are to some extent mutually compensating. Hence the theoretical expression for the free energy affords a considerably better working approximation than either Eq. (29) for the heat of dilution or Eq. (28) for the configurational entropy of dilution. One must not overlook the fact that, in spite of its shortcomings, the theory as given here is a vast improvement over classical ideal solution theory in applications to polymer solutions.

2. DILUTE POLYMER SOLUTIONS

An acceptable theory of dilute polymer solutions must take account of the inherent nonuniformity of the polymer segment concentration at high dilutions, to which attention has already been directed in pointing out the deficiencies of the theory developed above. To this end, the very dilute polymer solution may be regarded as a dispersion of clouds, or dilute clusters, of segments (see Fig. 34, p. 268), the region between swarms consisting of pure solvent. Each such cloud will be approximately spherical, with an average density which is a maximum at the center and decreases in a continuous manner with the distance from the center. Since the volume pervaded by a polymer molecule comprising x segments varies roughly as $x^{3/2}$, the mean density of segments in the cloud must decrease with increase in chain length. If we define $\bar{\rho}$ as the average *number* of segments per unit volume and let V_s be the volume of a segment, then the mean density $V_s\bar{\rho}$ will be of the order of $x^{-1/2}$.* The probability that *all* of a given sequence of x lattice sites located in the domain of a polymer molecule are vacant and hence available for occupancy by another polymer molecule, will be

$$(1 - V_s\bar{\rho})^x \cong (1 - x^{-1/2})^x \cong e^{-x^{1/2}}$$

which is very small for large values of x. Thus, in spite of the low density of a molecular cluster of segments, acceptable locations for a polymer are far less abundant in the domains already occupied by other molecules than in the intervening regions consisting of pure solvent (where every set of x consecutively contiguous sites is vacant, and the aforementioned probability is therefore unity). On this basis we may conclude that extensive overlapping of the domain of one molecule by another will be avoided insofar as the space available permits. This conclusion requires modification, however, if the forces of attraction between polymer and solvent are much less than those between like species, i.e., if χ_1 is sufficiently positive. Then the preference for polymer-polymer contacts will counteract the virtual repulsion between polymer molecules arising from their spatial requirements, to an extent depending on the magnitude of the interaction.

In general, each molecule in a very dilute solution in a good solvent (low χ_1) will tend to exclude all others from the volume which it occupies. This leads to the concept of an *excluded volume* from which a given

---

* To a better approximation (see Chap. XIV), the "volume" of the cloud goes as x^{1+a}, a being the exponent in the intrinsic viscosity relationship Eq. (VII-52); a usually is in the vicinity of 0.6 to 0.75. Hence the density $V_s\bar{\rho}$ is of the order of x^{-a}, or $x^{-0.6}$ to $x^{-0.75}$.

polymer molecule effectively excludes all others. In order to calculate the excluded volume, which will be called u, it will be necessary
to evaluate the interaction between a pair of molecules the centers of
gravity of which are separated by a distance a, as is illustrated in Fig.
114,a. The distance a in situations of interest will be no greater
than the order of magnitude of the ill-defined diameter of either polymer molecule. Hence any volume element δV in the vicinity of the
pair of molecules may be expected to contain segments belonging to
both molecules k and l. We shall attempt first to evaluate the resulting interactions in such a volume element   The sum of these for all
volume elements will equal the total intermolecular interaction due to
the proximity of one molecule to the other. Once this interaction is
known as a function of the intermolecular distance a, the excluded
volume u can be calculated by standard methods, and we may proceed
to the derivation of thermodynamic relations suitable for the dilute
polymer solution.

It is worth while to present in some detail the derivation of appropriate expressions for the thermodynamic interactions between segments occurring in the same volume element. Not only are such
expressions essential for the present objective, they will be used also in
the treatment of *intra*molecular interactions given in Chapter XIV.

**2a. Thermodynamic Interaction within a Region of Uniform Average
Segment Concentration.**—Let us undertake the derivation of a general
expression for the total free energy of interaction between *all* of the
segments in the volume element δV of Fig. 114, including those between segments belonging to the same molecule. In the enlargement
of the volume element shown in Fig. 114,b portions of two polymer
chains are indicated. No attempt has been made to differentiate
chain sections belonging to the respective molecules. Since the volume
element is small, the expected segment density may be considered to
be the same for all portions of δV. Hence the objection raised to the
previous treatment on account of the nonuniformity of the dilute solution as a whole should not apply to the limited region of a volume element, wherein the expectation of encountering a polymer segment can
be taken to be a constant throughout. The previous theory may consequently be adapted to the present purpose.

An important departure from the previous theory arises from the
fact that portions only of the polymer molecules, and never an entire
molecule, occur within a volume element. It will be recalled that, in
the estimation of the total number of configurations in the entire
solution, one segment (the first one in the chain) of each molecule was
allowed the opportunity to locate in any vacant cell of the lattice.

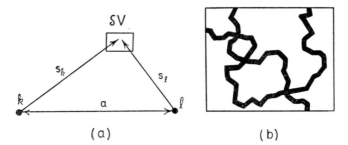

FIG. 114.—(a) Interaction between polymer molecules k and l.
(b) Enlargement of the volume element δV.

Each succeeding segment was restricted in its choice of locations to the lattice sites immediately neighboring its predecessor in the chain. Of these, the expected number vacant was given as $(z-1)(1-f_i)$. Inasmuch as the volume element considered here contains brief sections only of the polymer molecules, *every* segment concerned is restricted to locations adjacent to its predecessor. Never is there the opportunity for a segment to enjoy a wider range of choice. To be sure, we have under consideration two independent molecules; hence at some stage in the development they must be assigned locations in the total volume. We have, however, already specified their distance apart (a), and the location of the pair (with a fixed) in space is of no concern in the calculation of the excluded volume.

Let there be δx segments from both molecules in the volume element δV. The number of permissible arrangements of them, subject to the restriction discussed above, is simply

$$\Omega_{\delta V} = \prod_{=0}^{j=\delta x} (z-1)(1-f_j) \tag{37}$$

where f_j is the expectancy that a cell of the coordination sphere is occupied when j segments have been added to δV. Letting $1-f_j = (\delta n_0 - j)/\delta n_0$, where δn_0 is the number of sites in δV, we obtain in place of Eq. (7)

$$\Omega_{\delta V} = [\delta n_0!/(\delta n_0 - \delta x)!][(z-1)/\delta n_0]^{\delta x} \tag{38}$$

and finally by a series of steps corresponding to those leading from Eq. (7) to (10)

$$\delta(\Delta S_M{}^*) = -k\delta n_1 \ln v_1 \tag{39}$$

where δn_1 and v_1 are the number and volume fraction of solvent molecules in the volume element. This equation may be derived from the previous Eq. (10) by observing that the number of polymer molecules in the volume element is zero. It is as if the molecular weight were infinite, in which case n_2 of Eq. (10) would be zero.*

The free energy of mixing polymer segments with solvent in the volume element δV, obtained from $\delta(\Delta S_M^*)$ in conjunction with a term $kT\chi_1\delta n_1 v_2$ representing the standard state free energy of mixing (see Eq. 20), is

$$\delta(\Delta F_M) = kT\big[\delta n_1 \ln(1 - v_2) + \chi_1\delta n_1 v_2\big] \tag{40}$$

The volume fraction v_2 appearing here and in the following equations refers to the volume element and not to the solution as a whole.

The chemical potential of the solvent in the volume element, obtained by differentiating Eq. (40), is

$$(\mu_1 - \mu_1^0)_E = RT\big[\ln(1 - v_2) + v_2 + \chi_1 v_2^2\big] \tag{41}$$

which also follows directly from the previous Eq. (26) when x is taken to be infinite. Expanding in series.

$$(\mu_1 - \mu_1^0)_E = -RT\big[(1/2 - \chi_1)v_2^2 + v_2^3/3 + \cdots\big] \tag{42}$$

These expressions comprise the nonideal terms in the previous equations for the chemical potential, Eqs. (30) and (31'). They may therefore be regarded as the *excess* relative partial molar free energy, or chemical potential, frequently used in the treatment of solutions of nonelectrolytes;[23] i.e., the chemical potential in excess (algebraically) of the ideal contribution, which is $-RTv_2/M$ in dilute solutions.

It is possible to show from the most general thermodynamic arguments, which are independent of any assumed model, that the excess chemical potential of the solvent must vary with the square and higher powers of the solute concentration. If the concentration of polymer in the volume element is very small, and this condition will prevail under all situations of present interest, higher terms in such a series expansion may be neglected, the term in the square of the solute concentration alone being retained. With *complete generality*, therefore, and without apologies for any particular model, we may write with neglect of higher terms

$$(\mu_1 - \mu_1^0)_E = RT(\kappa_1 - \psi_1)v_2^2 \tag{43}$$

where κ_1 and ψ_1 are heat and entropy parameters such that

---

* Eq. (39) may be derived very simply by the "free volume" procedure discussed on p. 505. Since there are no polymer molecules with centers of gravity to be distributed over the volume element, the second term in Eq. (10) obviously will not appear.

$$\Delta \overline{H}_1 = RT\kappa_1 v_2^2$$

$$\Delta \overline{S}_1 = R\psi_1 v_2^2 \tag{44}$$

The parent free energy of mixing expression, Eq. (40), is equally general.

Within the limits of validity of the theories developed above on the basis of idealized models and of the simplifying assumptions, χ_1 of the preceding treatment may be related to these parameters by comparing Eq. (42) with (43). Then

$$\kappa_1 - \psi_1 = \chi_1 - 1/2 \tag{45}$$

Mention of this connection is incidental only. It will not be required to identify the quantity $\kappa_1 - \psi_1 + 1/2$ with the parameter χ_1 of the theory based on an idealized model.

Frequently it is preferred to use as a parameter the "ideal" temperature Θ defined by

$$\Theta = \kappa_1 T/\psi_1 \tag{46}$$

such that

$$\psi_1 - \kappa_1 = \psi_1(1 - \Theta/T)$$

Hence the excess chemical potential may be written

$$(\mu_1 - \mu_1^0)_E = -RT\psi_1(1 - \Theta/T)v_2^2 \tag{43'}$$

In a poor solvent, where both κ_1 and κ_1/ψ_1 generally are positive, Θ also will be positive. At the temperature $T = \Theta$, the chemical potential due to segment-solvent interactions is zero according to Eq. (43'). Hence the temperature Θ is that at which the excess chemical potential is zero and deviations from ideality vanish. The free energy of interaction of the segments within a volume element is therefore zero also.

2b. The Excluded Volume.[24]—In order to calculate the excluded volume, we must deduce the change in free energy when a polymer molecule such as l in Fig. 114,a is brought from a location remote from molecule k to a distance a from k. As already indicated, we first consider the free energy change which takes place within a restricted volume element δV. In a more exact sense, we consider two volume elements δV_l and δV_k so situated in the respective molecules when they are far apart as to cause them to coincide with the element δV when the distance of separation is reduced to a. If ρ_k and ρ_l are the densities of polymer segments (expressed as numbers of segments per unit volume) in the respective elements when the molecules are well separated from one another and V_s is the volume of a segment, then the volume fractions of polymer in the separate elements are, respectively

$$v_{2k} = \rho_k V_s \quad \text{and} \quad v_{2l} = \rho_l V_s$$

When the molecules are brought to the distance of separation a, the combined concentration of polymer segments belonging to both molecules in the corresponding volume element δV is

$$v_{2kl} = (\rho_k + \rho_l) V_s$$

Letting V_1 represent the volume of a solvent molecule, the corresponding numbers of solvent molecules are

$$\delta n_{1k} = \delta V (1 - \rho_k V_s)/V_1; \qquad \delta n_{1l} = \delta V (1 - \rho_l V_s)/V_1$$

and

$$\delta n_{1kl} = \delta V (1 - \rho_k V_s - \rho_l V_s)/V_1$$

Then from Eq. (40) the free energy of mixing segments and solvent in the initial state in which the two volume elements are well separated is

$$
\begin{aligned}
\delta(\Delta F_M)_k + \delta(\Delta F_M)_l = kT(\delta V/V_1) \big[& (1 - \rho_k V_s) \ln(1 - \rho_k V_s) \\
& + (1 - \rho_l V_s) \ln(1 - \rho_l V_s) \\
& + \chi_1 \rho_k V_s (1 - \rho_k V_s) + \chi_1 \rho_l V_s (1 - \rho_l V_s) \big]
\end{aligned}
\tag{47}
$$

Similarly for the volume element when the molecules overlap

$$
\begin{aligned}
\delta(\Delta F_M)_{k,l} = kT(\delta V/V_1) \big[& (1 - \rho_k V_s - \rho_l V_s) \ln(1 - \rho_k V_s - \rho_l V_s) \\
& + \chi_1(\rho_k V_s + \rho_l V_s)(1 - \rho_k V_s - \rho_l V_s) \big]
\end{aligned}
\tag{48}
$$

The quantity of interest is the difference between these free energies in the initial and final states, for it represents the change in free energy within the volume element δV when the distance between the polymer molecules is reduced from infinity to a. Since the concentrations (ρV_s) of polymer segments will invariably be small, we may expand the logarithms in Eqs. (47) and (48), retaining no terms beyond those in the squares of the concentrations. Subtraction of Eq. (47) from (48) then gives for this change in free energy

$$\delta(\Delta F_a) = 2kT(1/2 - \chi_1)\rho_k \rho_l (V_s^2/V_1)\delta V \tag{49}$$

or

$$\delta(\Delta F_a) = 2kT(\psi_1 - \kappa_1)\rho_k \rho_l (V_s^2/V_1)\delta V \tag{50}$$

$$= 2kT\psi_1(1 - \Theta/T)\rho_k \rho_l (V_s^2/V_1)\delta V \tag{50'}$$

It will be observed that the sign of the free energy change within any given volume element when the molecules are brought closer together depends on the sign of $\psi_1 - \kappa_1$. Taking the entropy of dilution parameter ψ_1 to be positive, as ordinarily will be the case, the sign of $\delta(\Delta F_a)$ depends on whether or not the heat of dilution parameter κ_1 is less or greater (algebraically) than ψ_1; otherwise stated, it depends on whether or not Θ/T is less than or greater than unity. Thus, in an athermal solvent ($\kappa_1 = 0$; or $\Theta = 0$), the decrease in entropy resulting from the increase in segment concentration in δV opposes the bringing of the molecules together, and $\delta(\Delta F_a)$ is positive on this account. The increase in entropy with dilution originates, according to the lattice model, from the increase in the number of permissible arrangements subject only to the rule that two segments may not occupy the same site. In a poor solvent ($\kappa_1 > 0$), the preference of the polymer segments for contacts with other segments, rather than with solvent molecules, counteracts this tendency toward dilution; if $\kappa_1 = \psi_1$, i.e., at $T = \Theta$, these two terms are in exact balance and $\delta(\Delta F_a) = 0$ for every volume element.

In order to render the expression for $\delta(\Delta F_a)$ in a usable form, it remains to evaluate ρ_k and ρ_l. We have already pointed out that the average segment density of a molecule will be greatest at the center of gravity and that it will decrease smoothly as the distance s (Fig. 114,a) from the center is increased. While the distribution will not be exactly[25] a Gaussian function of s, it may be so represented without introducing an appreciable error in our final result, which can be shown to be insensitive to the exact form assumed for the radial dependence of the segment density. Hence we may let

$$\rho = x(\beta'/\pi^{1/2})^3 e^{-\beta'^2 s^2} \tag{51}$$

where x is the total number of segments in a molecule; the subscripts k and l are to be applied to ρ and s as required. The value of the parameter β' may be established by stipulating that the mean-square distance $\overline{s^2}$ from the center of gravity, averaged over all segments of the molecule, must coincide with its value as calculated from the theory of chain configuration. Assuming the molecule to be a linear chain, then according to Eq. (14) of Chapter X (see also Appendix C, Chap. X)

$$\overline{s^2} = \overline{r^2}/6 = 1/4\beta^2$$

where β is the parameter in the Gaussian probability distribution for the end-to-end distance r. But from Eq. (51) we obtain

$$\overline{s^2} = \int_0^\infty (\rho/x)4\pi s^4 ds$$

$$= 3/2\beta'^2$$

which by comparison with the preceding expression yields

$$\beta' = \sqrt{6}\,\beta = 3/\sqrt{\overline{r^2}} \tag{52}$$

Thus we may retain the root-mean-square end-to-end distance $(\overline{r^2})^{1/2}$ as a measure of the size of the random-coiling polymer chain, and the parameter β' required to characterize the spatial distribution of polymer segments (not to be confused with the end-to-end distribution) can be calculated from $(\overline{r^2})^{1/2}$. It should be noted that the r used here refers to the actual r and not to the unperturbed r_0, which would be found in the absence of intramolecular interactions (see p. 425, Chap. X; also Chap. XIV); i.e., $(\overline{r^2})^{1/2} = \alpha(\overline{r_0^2})^{1/2}$ where α represents the expansion due to these interactions. Similarly, the parameter β refers to the actual distribution; i.e., $\beta = \beta_0/\alpha$, where β_0 characterizes the unperturbed end-to-end distribution.

The total free energy change accompanying the process of bringing the molecules to the distance of separation a may be obtained as the sum of the changes for each of the volume elements, or

$$\Delta F_a = \sum_{all\,\delta V} \delta(\Delta F_a)$$

$$= 2kT\psi_1(1 - \Theta/T)(V_s^2/V_1) \int \rho_k\rho_l\delta V \tag{53}$$

Upon introduction of the Gaussian distributions, Eq. (51), for the ρ's and performance of the required integration[24] (see Appendix) over the total volume, there is obtained for the required free energy of the pair relative to their free energy when infinitely far apart

$$\Delta F_a = kTJ\xi^3 e^{-y} \tag{54}$$

where

$$J = (\psi_1 - \kappa_1)\bar{v}^2/V_1 \equiv \psi_1(1 - \Theta/T)\bar{v}^2/V_1 \tag{55}$$

$$\xi = \beta'm^{2/3}/2^{1/6}\pi^{1/2} \tag{56}$$

$$y = \beta'a/2^{1/2} = (\pi^{1/2}/2^{1/3})\xi a/m^{2/3} \tag{57}$$

$\bar{v}$ being the (partial) specific volume of the polymer and $m = xV/\bar{v}$ the mass of a polymer molecule. The sign and magnitude of the total free energy change ΔF_a for a given distance a depends, through J, on $\psi_1 - \kappa_1$, or on $\psi_1(1 - \Theta/T)$, exactly as we have pointed out previously for $\delta(\Delta F_a)$. Whatever the value of J may be, the *magnitude* of the free energy change increases monotonically as a decreases, reaching the maximum value $kTJ\xi^3$ at $a=0$.

The probability of finding the center of gravity of polymer molecule l in a volume element δV far removed from any other polymer molecule naturally will be proportional to the size of the element. (This volume element should not be confused with the δV represented in Fig. 114 and employed above in the investigation of segment interactions in the vicinity of the pair k,l.) Presuming $\psi_1 - \kappa_1$ to be positive, the probability that molecule l is found near another molecule, such as k, will be diminished to an extent depending on ΔF_a for the pair. If we consider volume elements of the same size, one at a finite distance a from molecule k and the other far away from any polymer molecule, the relative probability that the center of molecule l will occur in the former compared with the latter is

$$f_a = e^{-\Delta F_a/kT} \tag{58}$$

In other words, f_a is the relative probability per unit volume that molecule l is located a distance a from molecule k (no other polymer molecules being in the vicinity), the relative probability f_∞ per unit volume at an infinite distance being taken equal to unity. In a more elegant discussion, f_a might be referred to as the partition function for the pair, and ΔF_a as the potential of average force. Every spherical shell, or volume element, $4\pi a^2 da$ about the center of molecule k should be weighted by the factor f_a representing its accessibility to molecule l. The product $f_a \cdot (4\pi a^2 da)$ represents the volume effectively offered by this element to molecule l. Conversely, the product $(1-f_a) \cdot 4\pi a^2 da$ represents the volume within the element which is, in effect, excluded to molecule l. The sum of the latter products for all volume elements represents the total *excluded volume u*. In mathematical form

$$u = \int_0^\infty (1 - f_a)4\pi a^2 da \tag{59}$$

The result obtained by substituting Eq. (54) for ΔF_a in Eq. (58), and this in turn in Eq. (59), can be put in the form[24]

$$u = 2Jm^2\mathcal{J}(J\xi^3) \tag{60}$$

where,

$$\mathcal{J}(X) = (4\pi^{-1/2})X^{-1}\int_0^\infty (1 - e^{-Xe^{-y^2}})y^2 dy \tag{61}$$

(with X replacing $J\xi^3$). This may be transformed through integration by parts to the more convenient expression

$$\mathcal{J}(X) = (8/3\pi^{1/2}) \int_0^\infty e^{-(y^2 + Xe^{-y^2})} y^4 dy \qquad (61')$$

or by series expansion and term-by-term integration to

$$\mathcal{J}(X) = 1 - X/2!2^{3/2} + X^2/3!3^{3/2} - \cdots \qquad (61'')$$

The excluded volume unfortunately does not reduce to a simple expression. According to Eq. (60), it depends first of all on the thermodynamic parameters as expressed through J; it depends also on

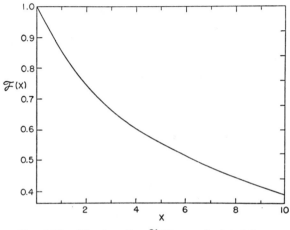

Fig. 115.—The function $\mathcal{J}(X)$ as calculated from (Eq. 61') plotted against X.[24]

the square of the molecular weight and on a function of $J\xi^3$. The latter, calculated from Eq. (61'), is shown in Fig. 115. The quantity $J\xi^3$ can be transcribed to a somewhat more intelligible form by substituting $\beta' = 3/(\overline{r^2})^{1/2}$ from Eq. (52) into Eq. (56). Then

$$J\xi^3 = (3^3/2^{1/2}\pi^{3/2})\psi_1(1 - \Theta/T)(\bar{v}^2/v_1 N)M^2/(\overline{r^2})^{3/2} \qquad (62)$$

where v_1 is the molar volume of the solvent and N is Avogadro's number. If we replace $\overline{r^2}$ by $\alpha^2 \overline{r_0^2}$, where α is the factor by which the linear dimensions of the chain molecule are increased owing to intramolecular interactions (see Chap. XIV), then

$$J\xi^3 = 4C_M\psi_1(1 - \Theta/T)M^{1/2}/\alpha^3 \qquad (63)$$

where

$$C_M = (3^3/2^{5/2}\pi^{3/2})(\bar{v}^2/v_1 N)(M/\overline{r_0^2})^{3/2} \qquad (64)$$

Since $\bar{r^2}/M$ is a constant characteristic of the polymer and independent of the solvent, $v_1 C_M$ likewise should depend only on the polymer type and not on its molecular weight, nor on the solvent in which it is dissolved. Thus from Eq. (62) we may conclude that $J\xi^3$ depends primarily on the thermodynamic parameters; however, it should increase gradually with molecular weight as $M^{1/2}/\alpha^3$ increases. As the "ideal" temperature Θ is approached in a poor solvent, J vanishes (Eq. 55), and $\mathcal{F}(J\xi^3)$ reaches unity. The better the solvent, the larger $J\xi^3$ and the smaller the value of $\mathcal{F}(J\xi^3)$.

(In Chapter XIV, we shall derive an expression for α from consideration of the *intra*molecular interactions between segments of a given polymer molecule. The same thermodynamic parameters, ψ_1 and κ_1 (or Θ) characterizing the underlying segment interactions, serve also to describe the intramolecular interaction. Combining the result as given by Eq. (XIV-10) with Eq. (63) above,

$$J\xi^3 = 2(\alpha^2 - 1) \qquad (65)$$

This simple relation rests on literal validity of *both* the intermolecular and the intramolecular interaction theories; hence its unqualified acceptance is open to question).

Although we have dwelt at some length on the function $\mathcal{F}(J\xi^3)$, it is to be observed that the excluded volume u is dominantly dependent, according to Eq. (60), on J. Thus, as the solvent is made poorer, which is manifested in the equations by a decrease in $\psi_1(1-\Theta/T)$, the excluded volume decreases, and at $T=\Theta$ it vanishes altogether. In other words, the poorer the solvent the less the polymer molecules repel one another, and at the Θ-point they telescope one another freely without any net interaction. The polymer molecules then distribute themselves over the volume like hypothetical point molecules which exert no forces on one another. At temperatures below Θ, they attract one another and the excluded volume is negative. If the temperature is much below Θ, precipitation occurs (see Chap. XIII).

Before concluding this discussion of the excluded volume, it is desirable to introduce the concept of an equivalent impenetrable sphere having a size chosen to give an excluded volume equal to that of the actual polymer molecule. Two such hard spheres can be brought no closer together than the distance at which their centers are separated by the sphere diameter d_e. At all greater distances the interaction is considered to be zero. Hence $f_a=1$ for $a \geq d_e$, and $f_a=0$ for $a < d_e$. Then

$$u = (4\pi/3)d_e^3$$

which is eight times the volume of either sphere. In order to maintain the equivalence to the polymer molecule, the diameter of the equivalent sphere must diminish as the solvent is made poorer, and at $T = \Theta$, $d_e = 0$. At temperatures $T < \Theta$, u becomes negative, and the required equivalent sphere is an imaginary one of negative volume.

2c. Thermodynamic Relations for Dilute Polymer Solutions.—Let us consider the total number of ways in which n_2 identical polymer molecules of excluded volume u may be distributed over a volume V of solution. The solution will be assumed to be so dilute that $n_2 u$ is much less than V, thus allowing ample room for all molecules without forced overlapping of their molecular domains. The number of locations available for the center of gravity of the first polymer molecule introduced into the solution will be proportional to V. The first molecule effectively excludes a volume u; consequently the number of locations available to the second molecule will be proportional to $V - u$. The volume available to the third molecule will be $V - 2u$, and so forth. Provided the final solution is sufficiently dilute to justify assumption of independent volume exclusion on the part of the individual molecules, the total excluded volume will remain additive in the number of polymer molecules, and the total number of arrangements for n_2 polymer molecules can be written

$$Q = \text{Const.} \times \prod_{i=0}^{n_2-1} (V - iu) \tag{66}$$

On the other hand, if the solution becomes too concentrated the same volume element may often be subject to exclusion by two (or more) different molecules; hence the total volume excluded is somewhat less than $V - iu$.* Eq. (66) takes account only of binary encounters. Ternary and higher encounters, which become increasingly prominent with increase in concentration, are neglected. Eq. (66) is limited accordingly to low concentrations.

The total partition function of the system is represented by Q of Eq. (66); hence

$$\Delta F_M = - kT \ln Q$$

$$= - kT \sum_{i=0}^{n_2-1} \ln(V - iu) + \text{Const.} \tag{67}$$

---

* This is especially clear in the case of hard spheres each of which excludes a volume $(4\pi/3)d_e^3$ which is eight times its net volume. Thus, the portions of the spherical domains at distances between $d_e/2$ and d_e of the centers of two nearby molecules may overlap, with the result that the volume actually excluded by these two molecules is less than $2(4\pi/3)d_e^3$.

which may be written

$$\Delta F_M = - kT \left[n_2 \ln V + \sum_{i=0}^{n_2-1} \ln(1 - iu/V) \right] + \text{Const.}$$

Since iu/V is always much less than unity in the sufficiently dilute solution, the logarithmic terms of the summation may be expanded in series with neglect of higher terms to give

$$\Delta F_M \cong - kT \left[n_2 \ln V - (u/V) \sum_{i=0}^{n_2-1} i \right] + \text{Const.}$$

$$\cong - n_2 kT [\ln V - (u/2)(n_2/V)] + \text{Const.} \tag{68}$$

The first term in Eq. (68) represents the ideal free energy of mixing term in dilute solutions, as will be apparent below. If the entire volume were available to *all* of the molecules, which would be an acceptable assumption if either V were very large or u were very small, it would be the only term, for then $Q = \text{Const.}\ (V)^{n_2}$. The second term in Eq. (68) represents the first-order deviation from ideality.

The osmotic pressure π may be deduced from ΔF_M by standard thermodynamic operations as follows:

$$\pi = - (\mu_1 - \mu_1^0)/v_1 \equiv - N(\partial \Delta F_M/\partial n_1)_{T,P,n_2}/v_1$$

where $(\mu_1 - \mu_1^0)$ [not to be confused with $(\mu_1 - \mu_1^0)_E$] represents the chemical potential difference for the solution as a whole. Then

$$\pi = - N(\partial \Delta F_M/\partial V)_{T,P,n_2}(\partial V/\partial n_1)_{T,P,n_2}/v_1$$

Since $(\partial V/\partial n_1)_{T,P,n_2} = v_1/N$

$$\pi = - (\partial \Delta F_M/\partial V)_{T,P,n_2} \tag{69}$$

which is the solution analog of the relation $P = -(\partial F/\partial V)_T$ for a closed system. It follows from Eq. (68), therefore, that

$$\pi \cong kT [n_2/V + (u/2)(n_2/V)^2] \tag{70}$$

or, substituting $n_2/V = cN/M$, where c is the concentration expressed in grams per unit volume

$$\pi/c \cong RT [1/M + (Nu/2M^2)c] \tag{71}$$

A more accurate treatment which takes account of multiple interactions between molecules would yield additional terms indicative of the general form

$$\pi/c = RT [A_1 + A_2 c + A_3 c^2 + \cdots] \tag{72}$$

where A_1, A_2 and so forth, are coefficients the first two of which we have evaluated above, i.e.

$$A_1 = 1/M \tag{73}$$

$$A_2 = Nu/2M^2 \tag{74}$$

From Eq. (60) for u

$$A_2 = (J/N)\mathcal{J}(J\xi^3) \tag{75}$$

$$A_2 = (\bar{v}^2/v_1)\psi_1(1 - \Theta/T)\mathcal{J}(J\xi^3) \tag{75'}$$

The analogy with the virial expansion of PV for a real gas in powers of $1/V$, where the excluded volume occupies an equivalent role, is obvious.   If the gas molecules can be regarded as point particles which exert no forces on one another, $u=0$, the second and higher virial coefficients (A_2, A_3, etc.) vanish, and the gas behaves ideally.   Similarly in the dilute polymer solution when $u=0$ (i.e., at $T=\Theta$), Eqs. (70), (71), and (72) reduce to van't Hoff's law

$$\pi/c = RT/M$$

which, of course, rearranges to $\pi V = n_2 RT$, in analogy to the perfect gas law.   The temperature $T=\Theta$ for a polymer solution is thus seen to be the analog of the Boyle point of a real gas (i.e., the temperature at which a real gas obeys the relation $PV=nRT$, except for terms in the square and higher powers of $1/V$).

The general theory of polymer solutions, in which the nonuniformity at high dilutions was disregarded, yielded Eq. (31''), which is of the same form as Eq. (72).   The coefficients of the first terms in the two expansions are identical, of course.   In view of Eqs. (45) and (46), the second coefficient as given by Eq. (75') differs from that in Eq. (31'') by the factor $\mathcal{J}(J\xi^3)$.   Thus the dilute solution theory predicts a slope for the π/c vs. c plot which is smaller by the factor $\mathcal{J}(J\xi^3)$.   In a good solvent this may be considerably less than unity, but it approaches unity as poorer solvents are chosen, and in a Θ-solvent (i.e., at $T=\Theta$ in a poor solvent) $\mathcal{J}(J\xi^3)=1$.   This leads to the important conclusion that the general theory as previously developed (Sec. 1) should hold at the Θ point.   The same conclusion may be reached on physical grounds by noting that at $T=\Theta$ the molecules overlap one another without prejudice; hence the frequency of segment-segment contacts in the total solution is unperturbed by the discontinuous character of the solution.

In applications to osmotic data the following series form for π/c usually is preferred:

$$(\pi/c) = (\pi/c)_0[1 + \Gamma_2 c + \Gamma_3 c^2 + \cdots] \qquad (76)$$

where $(\pi/c)_0 = RT/M$, and

$$\Gamma_2 = A_2/A_1 = (JM/N)\mathcal{F}(J\xi^3) \qquad (77)$$

In the approximation that the molecules can be replaced by equivalent noninterpenetrating spheres, it may be shown that $\Gamma_3 = (5/8)\Gamma_2^2$.[24] This result is indicative of a strong dependence of the third coefficient on the second. A more detailed analysis[26] shows that the numerical coefficient 5/8 should be replaced for polymer molecules by a slowly increasing function of Γ_2—one that is less than 5/8 and that vanishes as Γ_2 goes to zero. Calling this slowly varying function g

$$\Gamma_3 = g\Gamma_2^2 \qquad (78)$$

and

$$\pi/c = (\pi/c)_0[1 + \Gamma_2 c + g\Gamma_2^2 c^2 + \cdots] \qquad (79)$$

which was given previously in Chapter VII, Eq. (13). The important conclusion remains that Γ_3 should rapidly vanish as Γ_2 decreases toward zero. Thus, when osmotic pressure-concentration ratios π/c are plotted against c, the decrease in slope for poorer solvents should be paralleled by a more rapid disappearance of curvature. Both slope and curvature reach zero at $T = \Theta$. The curvature predicted by the present theory is much greater for good solvents than that given in Eq. (31′) or (31″) of the general theory.

Osmotic pressure measurements ordinarily are carried out for the purpose of determining either the molecular weight (number average) or the solvent-polymer interaction as embodied in Γ_2. The former requires evaluation of the first term of the series in Eq. (79); the latter depends on evaluation of the second term also. The third term of the series is important for this purpose only as it aids in the accurate evaluation of the coefficients of the preceding terms. Hence an approximate value for the third coefficient will suffice. Inasmuch as it makes a negligible contribution in poor solvents throughout the concentration range covered osmotically, it suffices to take for g its appropriate value in a good solvent, that is, 0.25[26] and to treat it as a constant.[27] Neglecting higher terms, Eq. (79) reduces to

$$\pi/c \cong (\pi/c)_0[1 + (\Gamma_2/2)c]^2 \qquad (79')$$

Having taken g to be constant, there remain two adjustable parameters, $(\pi/c)_0$ and Γ_2, which are the ones it is desired to evaluate from experimental data.[27]

Other thermodynamic functions may be derived from the partition function Q, or from the expression for the osmotic pressure. The chemical potential of the solvent in the solution (not to be confused with the excess chemical potential $(\mu_1 - \mu_1^0)_E$ within a region of uniform segment expectancy, or density) is given, of course, by $\mu_1 - \mu_1^0 = -\pi v_1$. The heat and entropy of dilution may be derived by differentiation, but the resulting expressions are unwieldy.[24]   It is preferable to undertake the evaluation of Γ_2, or of A_2, at different temperatures and then to deduce the primary entropy and heat of dilution parameters ψ_1 and κ_1 by means of the equations given above (see below).

The expressions for the turbidity τ which may be derived from the osmotic expansion, Eq. (79), have been given previously; see Eqs. (VII-37') and (VII-37''), pp. 299 and 301.

The dilute solution treatment set forth here may be extended to polymers comprising a series of homologs differing in molecular weight.[24] The resulting equations for heterogeneous polymers assume the same general form, but numerical evaluation of the second coefficient, A_2 or Γ_2, involves formidable summations over the entire distribution. Molecular weights M occurring in the first term of the osmotic expressions must, of course, be replaced by number averages, $\overline{M}_n$. Dilute solutions of two chemically different polymer species also have been treated.[28]

2d. Application to Experimental Data.—Plots of π/c against c (where c is expressed in grams/100 ml.) were presented in Chapter VII, Figs. 38 and 39 (pp. 278, 279). Similar results obtained by Krigbaum[29] for polystyrene fractions in toluene are shown in Fig. 116. The curves drawn in each of these plots have been calculated from Eq. (79) by choosing values of the parameters $(\pi/c)_0$ and Γ_2 (taking $g = 0.25$ in all cases) giving the best fit. The methods described on

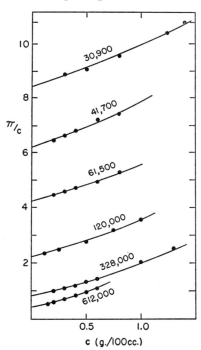

FIG. 116.—π/c plotted against c for several fractions of polystyrene in toluene at 30°C. Molecular weights of the fractions are indicated by the numbers appearing with each curve. The osmotic pressure π is expressed in g./cm.[2] (Results of Krigbaum.[29])

p. 280 of Chapter VII have been used for the purpose of evaluating these parameters. The experimental points confirm the predicted positive curvature in good solvents (e.g., polyisobutylene in cyclohexane in Fig. 38, or polystyrene in toluene, Fig. 116). There is no sign of curvature in poor solvents (e.g., for polyisobutylene in benzene at 30°, Fig. 38), which is in accord with the prediction that the third coefficient vanishes as the second approaches zero. We may conclude that the dependence of the osmotic pressure of dilute polymer solutions on concentration is in satisfactory agreement with Eq. (79). This expression is restricted to low concentrations, however, for reasons already given. It should not be expected to apply beyond concentrations at which π/c exceeds three or four times the value of $(\pi/c)_0$.[27,29] Hence the upper concentration limit is smaller the higher the molecular weight (i.e., the lower the value of $(\pi/c)_0$). At concentrations above this limit, π/c values fall below those calculated, as some of the data in the figures mentioned show.

That the turbidities of dilute polymer solutions agree with the corresponding theoretical relationship, given by Eq. (VII-37), was shown in Figs. 47 and 48.

The initial slope of the π/c vs. c plot may be calculated from the parameters Γ_2 and $(\pi/c)_0$ evaluated in the manner referred to above. This slope is given by RTA_2, which is equal to $\Gamma_2(\pi/c)_0$. The parameter A_2 is given according to theory by Eq. (75) or (75'). The quantity $(\bar{v}^2/v_1)\psi_1(1-\Theta/T) = J/N$ occurring in these equations should be independent of the molecular weight, but the factor $\mathcal{J}(J\xi^3)$ decreases slowly with M, inasmuch as $J\xi^3$ is proportional to $M^{1/2}/\alpha^3$ according to Eq. (63), and the function $\mathcal{J}(J\xi^3)$ decreases with increase in the value of its argument in the manner shown in Fig. 115. It follows that the slope RTA_2 of the π/c plot against c should decrease gradually with increase in M, i.e., with decrease in $(\pi/c)_0$, for a series of polymer homologs in the same good solvent at the same temperature. This prediction is borne out by the results shown in Fig. 38 for polyisobutylene in cyclohexane and in Fig. 116 for polystyrene in toluene. The initial slopes calculated from the parameters used in obtaining the curves in Fig. 38 for the former pair are shown in Fig. 117 where $\log RTA_2$ is plotted against $\log M$. A similar plot is shown in Fig. 118 for the system polystyrene-toluene, as derived from the results of Fig. 116. Log-log plots are used merely for convenience.

The thermodynamic behavior of the dilute polymer solution depends on three factors: (1) the molecular weight, (2) the thermodynamic interaction parameters ψ_1 and κ_1, or ψ_1 and Θ, which characterize the segment-solvent interaction, and (3) the configuration, or "size," of the molecules in solution. The coefficient A_1 depends only on the first of these, which is customarily determined from this coefficient. The second coefficient A_2 (and likewise Γ_2) depends on all three factors. Consequently, in seeking to evaluate the primary thermodynamic parameters contained in this *second virial coefficient* A_2, one is obliged to resort to an independent determination of one sort or another of the size of the molecule in solution. Osmotic measurements alone will not suffice for the resolution of A_2 into its components. This may be made clearer by re-examination of the relationships involved.

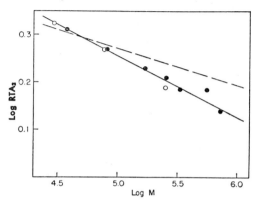

FIG. 117.—Log (RTA_2) for polyisobutylene fractions in cyclohexane at 30°C plotted against log M. The filled circles represent the initial slopes (RTA_2) of the curves shown in Fig. 38. The open circles are from earlier results[30] on the same system. Dashed line calculated as described in text. (Krigbaum[29]).

The combination of the thermodynamic parameters $\psi_1(1-\Theta/T)$ occurs in the defined quantity J (see Eq. 55), which occupies the major role in the expression, Eq. (75), for A_2. A logical first objective is the evaluation of J, therefore. But according to Eq. (75), A_2 depends also on the function $\mathcal{J}(J\xi^3)$; hence it is necessary to evaluate ξ^3. This latter quantity, defined by Eq. (56), depends on the parameter β' characterizing the spatial distribution of segments. Introducing Eq. (52) for β' into Eq. (56)

$$\xi^3 = (3^3/2^{1/2}\pi^{3/2})M^2/\overline{(r^2)}^{3/2}N^2 \tag{80}$$

which may also be written

$$\xi^3 = (3^3/2^{1/2}\pi^{3/2})N^{-2}(\overline{r_0^2}/M)^{-3/2}M^{1/2}/\alpha^3 \tag{80'}$$

The quantity $\overline{r_0^2}$ is directly proportional to M (see Chap. X); hence for any linear polymer homologous series their ratio, $\overline{r_0^2}/M$, is a constant characterizing the unperturbed configuration of the chain molecules of the given type. Both this ratio and the value of the expansion factor α^3 may be determined from suitable measurements of intrinsic viscosities (see Chap. XIV), or they may be determined more directly, though with less precision, from light-scattering dissymmetry measurements. Thus, ξ^3 may be calculated independently of the osmotic measurements. Solving Eq. (75) by trial, the value of J may then be deduced from A_2 (osmotic) and ξ^3 (configuration). Then $\psi_1(1-\Theta/T)$ follows from the equation (55) defining J. If determinations are made at different temperatures, the former quantity may be resolved into its separate components ψ_1 and Θ, and the analysis is complete. A corresponding analysis may of course be applied to values of A_2 ascertained from light-scattering intensity measurements (see Eq. VII-37').

When this procedure is applied to the data shown for polystyrene in Fig. 116 and to those for polyisobutylene shown previously in Fig. 38 of Chapter VII, the values obtained for $\psi_1(1-\Theta/T)$ decrease as the molecular weight increases. The data for the latter system, for example, yield values for this quantity changing from 0.087 at $M=38,000$ to 0.064 at $M=720,000$. This is contrary to the initial definition of the thermodynamic parameters, according to which they should characterize the inherent segment-solvent interaction independent of the molecular structure as a whole.

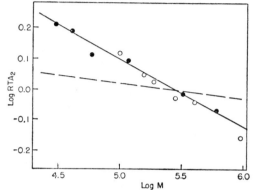

Fig. 118.—Plot of log (RTA_2) against log M for polystyrene fractions in toluene. Filled circles represent slopes of the curves shown in Fig. 116. Open circles are from the results of Frank and Mark.[31] Dashed line calculated as described in text. (Krigbaum.[29])

The values obtained for $\psi_1(1-\Theta/T)$ actually are fairly close to those calculated from intrinsic viscosities by methods to be described in Chapter XIV, and this suggests an alternative method of treating the dependence of the coefficient A_2 on molecular weight. Through the use of the expansion factor α established from the intrinsic viscosity studies, $J\xi^3$ may be calculated from Eq. (65.) Relying also on the intrinsic viscosity results for the value of $\psi_1(1-\Theta/T)$, an A_2 may be computed for each polymer fraction through the use of Eq. (75'). The dashed lines shown in Figs. 117 and 118 were calculated by Krigbaum[29] in this way. Lack of numerical agreement may be excused on the grounds that neither the intramolecular theory of the configuration of a single molecule nor the theory of intermolecular interactions given above is exact. The significantly larger decreases in A_2 with M than would be predicted by theory are not so easily dismissed, however; they represent what appears to be a regrettable deficiency of present theory. Estimation of the segment interactions by considering the polymer molecule as a cloud of chain segments without specifically taking into account the continuity of the chain from one segment to the next may be responsible for the inaccuracy of the theory. No better approximation amenable to theoretical treatment has been suggested, however.

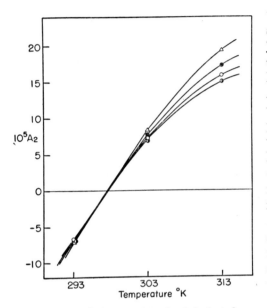

FIG. 119.—A_2 for a series of polyisobutylene fractions in benzene plotted against the absolute temperature. The molecular weights of the fractions are as follows: △, 102,000; ●, 193,000; ○, 210,000; ◐, 723,000. (Results of Krigbaum.[32])

At temperatures near Θ, the quantity $J\xi^3$ is very small (see Eq. 63); hence $\mathcal{F}(J\xi^3)$ is close to unity, and $\psi_1(1-\Theta/T)$ may be calculated unambiguously from A_2. By plotting A_2 against the reciprocal absolute temperature, or simply against the temperature over a limited range in the vicinity of Θ, this quantity may readily be resolved into its components Θ and ψ_1. Krigbaum's[32] results for a series of polyisobutylene fractions in benzene are shown in Fig. 119. The temperature at which A_2 is zero corresponds to Θ, and it is gratifying to observe that this occurs at 24.5 (± 0.5)°C for each of the fractions, which is in accordance with theory. From the slope at $T=\Theta$, ψ_1 is found to be about 0.30 for this system.[32] The spread of the curves at higher temperatures is due to the decrease in $\mathcal{F}(J\xi^3)$, to an extent depending on the molecular weight, as J increases.

In conclusion, the intermolecular interaction theory offers a reliable expression for the dependence of the osmotic pressure on the concentration in very dilute solutions, and it succeeds in predicting a decrease in the value of the initial slope of π/c plotted against c as the molecular weight is increased. The decreases observed are significantly larger, however, than the theory would allow. The prediction of a slope of zero for all polymers of a given series, irrespective of the molecular weight of the polymer, at the same temperature $T=\Theta$ in a given poor solvent, is well borne out by experimental results. Unequivocal determination of the fundamental thermodynamic parameters from osmotic or light-scattering measurements is an unfortunately difficult matter owing to the secondary dependence of the observed thermodynamic behavior on the molecular configuration.

APPENDIX

Integration of the Interaction Free Energy for a Pair of Molecules.[24] —We require, according to Eq. (53), the integral $\int \rho_k \rho_l \delta V$ over the total volume, when the centers of the polymer molecules k and l are separated by the distance a; ρ_k and ρ_l, the segment density distributions for the respective molecules about their centers, are assumed to be given with sufficient accuracy by Gaussian distributions (Eq. 51). Since the chain lengths of the molecules are taken to be identical, the Gaussian parameters β' characterizing their distributions will be identical. It is convenient to choose cylindrical coordinates, r^* and θ with origin midway between the molecules. Then

$$s_k^2 = a^2/4 + ar^* \cos \theta + r^{*2}$$
$$s_l^2 = a^2/4 - ar^* \cos \theta + r^{*2} \tag{A-1}$$

Substituting these relations in the Gaussian expression, Eq. (51), for ρ_k and ρ_l, and inserting them in the required integral

$$\int \rho_k \rho_l \delta V = (x^2 \beta'^6 / \pi^3) \int_0^\infty \int_0^\pi \exp[- \beta'^2 (a^2/2 + 2r^{*2})] 2\pi r^{*2} \sin \theta dr^* d\theta$$

$$= (4x^2 \beta'^6 / \pi^2) \exp(- \beta'^2 a^2/2) \int_0^\infty \exp(- 2\beta'^2 r^{*2}) r^{*2} dr^*$$

$$= [x^2 \beta'^3 / (2\pi)^{3/2}] \exp(- \beta'^2 a^2/2) \tag{A-2}$$

Substituting this result in Eq. (53), we have for the free energy change associated with the transfer of molecule l from infinity to a distance a from molecule k

$$\Delta F_a = kT\psi_1 (1 - \Theta/T)(\beta'^3/2^{1/2}\pi^{3/2})(x^2 V_s^2/V_1) \exp(- \beta'^2 a^2/2) \tag{A-3}$$

Replacement of the polymer molecular volume xV_s by $m\bar{v}$, where m is the mass of the molecule and $\bar{v}$ its specific volume, leads at once to Eq. (54) in which the quantities J, ξ, and y are defined by Eqs. (55), (56), and (57).

REFERENCES

1. G. Gee, and L. R. G. Treloar, *Trans. Faraday Soc.*, **38**, 147 (1942).
2. K. H. Meyer and R. Lühdemann, *Helv. Chim. Acta*, **18**, 307 (1935);

K. H. Meyer, *Z. physik. Chem.*, **B44**, 383 (1939); *Helv. Chim. Acta*, **23**, 1063 (1940).

3. M. L. Huggins, *J. Phys. Chem.*, **46**, 151 (1942); *Ann. N.Y. Acad. Sci.*, **41**, 1 (1942); *J. Am. Chem. Soc.*, **64**, 1712 (1942).

4. P. J. Flory, *J. Chem. Phys.*, **10**, 51 (1942).

5. P. J. Flory and W. R. Krigbaum, *Annual Review of Physical Chemistry*, **2**, 383 (1951).

6. A. R. Miller, *Proc. Cambridge Phil. Soc.*, **39**, 54 (1943)

7. A. R. Miller, *The Theory of Solutions of High Polymers* (Clarendon Press, Oxford, 1948).

8. W. J. C. Orr, *Trans. Faraday Soc.*, **40**, 320 (1944).

9. E. A. Guggenheim, *Proc. Roy. Soc.* (London), **A 183**, 203 (1944).

10. P. J. Flory, *J. Chem. Phys.*, **12**, 425 (1944).

11. R. L. Scott and M. Magat, *J. Chem. Phys.*, **13**, 172 (1945).

12. E. A. Guggenheim, *Proc. Roy. Soc.* (London), **A 183**, 213 (1944).

13. P. J. Flory, *J. Chem. Phys.*, **13**, 453 (1945).

14. J. H. Hildebrand, *J. Chem. Phys.*, **15**, 225 (1947).

15. E. A. Guggenheim, *Trans. Faraday Soc.*, **44**, 1007 (1948).

16. M. J. Newing, *Trans. Faraday Soc.*, **46**, 613 (1950).

17. C. E. H. Bawn, R. F. J. Freeman, and A. R. Kamaliddin, *Trans. Faraday Soc.*, **46**, 677 (1950).

18. G. Gee, *J. Chem. Soc.*, **1947**, 280.

19. G. Gee and W. J. C. Orr, *Trans. Faraday Soc.*, **42**, 507 (1946).

20. S. Prager, E. Bagley, and F. A. Long, *J. Am. Chem. Soc.*, **75**, 2742 (1953).

21. H. Tompa, *J. Polymer Sci.*, **8**, 51 (1952).

22. J. Ferry, G. Gee, and L. R. G. Treloar, *Trans. Faraday Soc.*, **41**, 340 (1945).

23. J. H. Hildebrand and R. L. Scott, *The Solubility of Non-Electrolytes*, 3d ed. (Reinhold Publishing Corp., New York, 1950).

24. P. J. Flory and W. R. Krigbaum, *J. Chem. Phys.*, **18**, 1086 (1950); P. J. Flory, *ibid.*, **17**, 1347 (1949).

25. P. Debye and F. Bueche, *J. Chem. Phys.*, **20**, 1337 (1952).

26. W. H. Stockmayer and E. F. Casassa, *J. Chem. Phys.*, **20**, 1560 (1952).

27. W. R. Krigbaum and P. J. Flory, *J. Polymer Sci.*, **9**, 503 (1952).

28. W. R. Krigbaum and P. J. Flory, *J. Chem. Phys.*, **20**, 873 (1952).

29. W. R. Krigbaum and P. J. Flory, *J. Am. Chem. Soc.*, **75**, 1775 (1953).

30. P. J. Flory, *J. Am. Chem. Soc.*, **65**, 372 (1943).

31. H. P. Frank and H. Mark, *J. Polymer Sci.*, **6**, 243 (1951).

32. W. R. Krigbaum and P. J. Flory to be published.

CHAPTER XIII

Phase Equilibria in Polymer Systems

1. PHASE EQUILIBRIA IN LIQUID SYSTEMS

IF THE solvent chosen for a given polymer becomes progressively poorer as the temperature is lowered, eventually a temperature may be reached below which the solvent and the polymer no longer are miscible in all proportions. At each lower temperature, mixtures of the polymer and solvent over a certain composition range will separate into two phases. If the polymer consists of a sufficiently narrow range of molecular species, e.g., if it has been carefully fractionated, then it may be regarded as a single component and the mixtures of solvent and polymer may be treated according to the well-established procedures applicable to binary small molecule systems. The temperature-composition phase diagram consists of a boundary curve distinguishing the homogeneous from the heterogeneous region. Mixtures falling inside this so-called binodial curve separate into two liquid phases having compositions on the binodial and connected by the familiar tie line. The maximum in the binodial curve is a critical point for the binary system; the compositions of the coexisting phases merge at this point, and at temperatures above the critical temperature T_c the system is homogeneous for all compositions.

To the extent that the polymer consists of a range of homologous species, it must be regarded as a mixture of components, and its mixtures with a solvent should be looked upon as a polycomponent system. The maximum temperature for coexistence of two phases in the system formed from the given polymer (mixture) and solvent is not in general a true critical point. Its location will depend on the distribution of species in the polymer, and the compositions of the two coexisting phases immediately below this temperature may differ considerably owing to partitioning of the polymer components between the phases.

Separation of a polymer solution into two phases may also be brought

541

about by the addition of a precipitant to the mixture. In this case no less than three components are involved, and the use of a conventional triangular ternary diagram is indicated.

The characteristic features of phase equilibria in polymer-solvent mixtures will be examined in the present section, the discussion being confined to systems having both phases completely liquid. Equilibria involving a polymer-rich phase in which the polymer is semicrystalline will be the subject of the following section.

We shall find, provided the molecular weight is not too low (i.e., $M > 10,000$), that incipient liquid-liquid phase separation in a dilute polymer solution invariably occurs near the temperature Θ (usually a little below Θ) at which π/c is independent of concentration over a considerable range (see Chap. XII). Over the molecular weight range of interest the critical temperature T_c will not differ very much from Θ. The net segment interaction is then very small, and, as has been stressed on p. 532, the dilute solution theory converges for this reason to the simpler general theory presented in the first section of Chapter XII. Consequently, it is legitimate to use the latter for the treatment of liquid-liquid phase equilibria. In particular, we shall employ Eqs. (XII-26) and (XII-32) for the chemical potentials of the solvent and polymer, respectively.

1a. Theory of Binary Systems.[1]—The conditions for equilibrium between two phases in a binary system are expressed by stipulating equality of the chemical potentials in the two phases; that is

$$\left.\begin{aligned} \mu_1 &= \mu_1' \\ \mu_2 &= \mu_2' \end{aligned}\right\} \tag{1}$$

where the prime is adopted as the designation for the more concentrated phase. Before proceeding from these conditions to the deduction of the concentrations of the two phases in equilibrium, it is instructive to examine the requirements for the occurrence of incomplete miscibility. Eqs. (XII-26) and (XII-32) express the chemical potentials μ_1 and μ_2 as functions of the volume fraction v_2 of polymer. The single parameter χ_1 occurs in these functions, and we wish to ascertain the range of values of this parameter permitting coexistence of two phases. Fulfillment of the former of the conditions (1) requires that there be two concentrations at which the chemical potential μ_1 has the same value, and this requires that μ_1 pass through a minimum and then a maximum as v_2 is increased from zero to unity. Similarly, in order to comply with the second condition (1), μ_2 must exhibit a maximum and a minimum. Since μ_1 and μ_2 are derived by differentiating the same free energy function, Eq. (XII-22), it is easy to show that the one must pass through a maximum where the other is at its minimum,* hence it will suffice to consider either chemical potential alone. Confining attention to μ_1, therefore, we note further that a point of inflection, characterized by the condition of zero curvature $(\partial^2\mu_1/\partial v_2^2)_{T,P} = 0$, must necessarily occur between the minimum and the maximum in the curve. At both the maximum and the minimum $(\partial\mu_1/\partial v_2)_{T,P} = 0$. Manifestation of these characteristics in the function representing μ_1 constitutes the necessary and sufficient condition for incomplete miscibility. If the curve representing μ_1 as a function of composition decreases monotonically with v_2, total miscibility is assured. In passing to progressively poorer solvents, eventually a minimum and a maximum will appear in the curve; their appearance signifies incomplete miscibility. Ordinarily a poor solvent for a given polymer becomes poorer with lowering of the temperature; hence an initially totally miscible system at higher temperatures may be transformed to one of limited miscibility by lowering the temperature. Thus, at some critical temperature T_c, incipient phase separation will be encountered, and at this point the previously monotonic curve must begin to exhibit a minimum, a maximum, and an inflection. At T_c, these features will occur simultaneously at the same concentration. Hence the conditions for incipient phase separation are

$$(\partial\mu_1/\partial v_2)_{T,P} = 0 \Big\rbrace$$
$$(\partial^2\mu_1/\partial v_2^2)_{T,P} = 0 \Big\rbrace \qquad (2)\dagger$$

The magnitude of the *decrease* in the chemical potential of the solvent below that of the pure liquid, calculated according to Eq. (XII-26), is plotted in Fig. 120 against v_2 for a polymer of size $x = 1000$ and for the several values of χ_1 indicated. These curves illustrate the features discussed in the preceding paragraphs. For small values of χ_1, the chemical potential decreases (i.e., $-(\mu_1 - \mu_1^0)/RT$ increases) with v_2 throughout the concentration range. As χ_1 is increased (corresponding ordinarily to a decrease in temperature), a minimum, an inflection, and a maximum eventually appear. At critical interaction, represented by the curve for $\chi_1 = 0.532$ in this case, these characteristics are coincident at the critical point.

Application of the critical conditions (2) to the chemical potential

---

* The argument required here is, of course, the equivalent of the Gibbs-Duhem relation.

† These conditions are equivalent to $(\partial\mu_2/\partial\mu_2)_{T,P} = 0$ and $(\partial^2\mu_2/\partial v_2^2)_{T,P} = 0$, according to the Gibbs-Duhem equation. Other thermodynamic functions such as the activity or the osmotic pressure could have been used in place of the chemical potential.

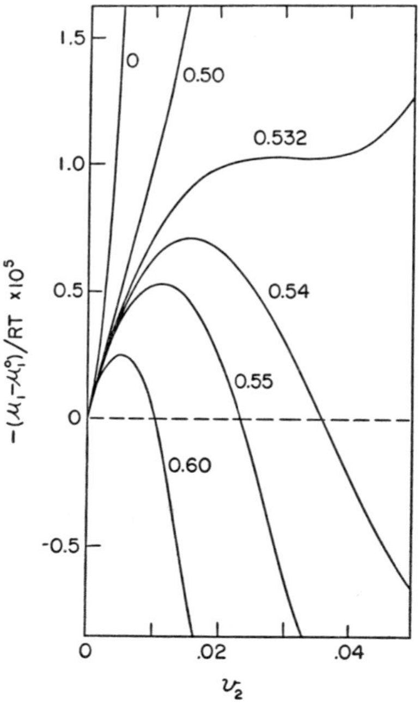

$-(\mu_1-\mu_1^0)/RT \times 10^5$

v_2

FIG. 120.—The chemical potential of the solvent in a binary solution containing polymer at low concentrations (v_2). Curves have been calculated according to Eq. (XII-26) for $x = 1000$ and the values of χ_1 indicated with each curve.[1]

as given by Eq. (XII-26) yields

$$1/(1 - v_2) - (1 - 1/x) - 2\chi_1 v_2 = 0$$

and

$$(3)$$

$$1/(1 - v_2)^2 - 2\chi_1 = 0$$

Eliminating χ_1, there is obtained for the critical composition[1]

$$v_{2c} = 1/(1 + x^{1/2}) \qquad (4)$$

which for large x reduces to

$$v_{2c} = 1/x^{1/2} \qquad (4')$$

The critical concentration at which phase separation first appears on passage to the two-phase region is thus predicted to occur at a very small volume fraction of polymer; for a polymer having a molecular weight of a million ($x \cong 10^4$), for example, $v_{2c} \cong 0.01$. Substituting Eq. (4) in either of the equations (3), we obtain

$$\chi_{1c} = (1 + x^{1/2})^2/2x \qquad (5)$$

$$\cong 1/2 + 1/x^{1/2} \qquad (5')$$

Thus the critical value of χ_1 will exceed 1/2 by a small increment depending on the chain length, and at infinite molecular weight it must equal 1/2.

The first definition of χ_1 given by Eq. (XII-21) indicates that it is inversely proportional to the temperature. According to the more general interpretation, in which Δw_{12} is regarded as a free energy, χ_1 should be (approximately) a linear function of $1/T$, although generally not proportional to $1/T$. This is apparent in the subsequent formulation expressed by Eqs. (XII-45) and (XII-46). Adopting this dilute solution notation, we should have

$$- \psi_1(1 - \Theta/T_c) = 1/x^{1/2} + 1/2x \tag{6}$$

$$1/T_c = (1/\Theta)[1 + (1/\psi_1)(1/x^{1/2} + 1/2x)] \tag{7}$$

which for large x can be written

$$1/T_c = (1/\Theta)(1 + b/M^{1/2}) \tag{7'}$$

The constant b is given according to theory by

$$b = (v_1/\bar{v})^{1/2}/\psi_1 \tag{8}$$

inasmuch as $x = M\bar{v}/v_1$, where $\bar{v}$ is the specific volume of the polymer and v_1 is the molar volume of the solvent. Theory thus predicts that the reciprocal of the critical temperatue (in °K) for incipient miscibility should vary linearly with the reciprocal of the square root of the molecular weight in a given polymer-solvent system. Furthermore, Θ may now be identified as the *critical miscibility temperature in the limit of infinite molecular weight*.

Before comparing these predictions regarding the critical point with experimental results, we may profitably examine the binodial curve of the two-component phase diagram required by theory. The following useful approximate relationship between the composition v_2 of the more dilute phase and the ratio $\gamma = v_2'/v_2$ of the compositions of the two phases may be derived[2] (see Appendix A) by substituting Eq. (XII-26) on either side of the first of the equilibrium conditions (1), using the notation v_2 for the volume fraction in the more dilute phase and v_2' for that in the more concentrated phase, and similarly substituting Eq. (XII-32) for μ_2 and μ_2' in the second of these conditions:

$$v_2 \cong \{- (\gamma + 1)h + [(\gamma + 1)^2h^2 + 4(\gamma - 1)^3h]^{1/2}\}/2(\gamma - 1)^3 \tag{9}$$

where

$$h = (12/x)[(\gamma + 1)(\ln \gamma)/2 - (\gamma - 1)]$$

The approximation involved is inconsequential for concentrations up to ten times the critical concentration v_{2c} (see Appendix A). The corresponding value of χ_1 is given by

$$\chi_1 = \frac{(\gamma - 1)(1 - 1/x) + (\ln \gamma)/v_2 x}{2(\gamma - 1) - v_2(\gamma^2 - 1)} \tag{10}$$

By substituting chosen values of the ratio γ in Eq. (9), together with the value of h corresponding to γ and the given value of x, one may calculate the composition of the more dilute phase. The composition of the more concentrated phase is obtained immediately as $v_2' = \gamma v_2$. Finally, the corresponding value of χ_1 may be calculated from γ, x, and v_2 according to Eq. (10). It is possible thus to calculate the binodial as a function of χ_1. The dashed curves shown in Fig. 121 have been calculated in this manner. The conversion of χ_1 to the temperatures of the ordinate scale has been accomplished by an empirical "calibration" of the relationship of χ_1 to T based on the change in the observed critical temperature T_c with molecular weight (see below).[3]

1b. Experimental Results on Binary Systems.—Temperatures at which precipitation occurs on cooling diisobutyl ketone solutions of three polyisobutylene[3] fractions are shown by the points plotted against the composition in Fig. 121. Liquid-liquid precipitation temperatures such as these are easily reproducible; they are not dependent on the rate of cooling, and the temperatures for incipient phase separation obtained on warming are the same as those obtained on cooling, within the experimental error (ca. $\pm 0.1°$). The general features predicted by theory for the binodial are confirmed. The maximum in the curve, representing the critical point, occurs at a decidedly low polymer concentration, although not quite so low as predicted by Eq. (4). The critical concentrations are roughly twice those predicted, and the experimental curves are broader than those calculated as shown by the dashed curves. At temperatures not much below the critical temperature T_c for incipient restricted miscibility, even

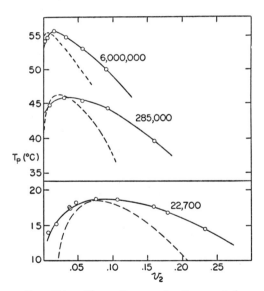

FIG. 121.—Phase diagram for three polyisobutylene fractions (molecular weights indicated) in diisobutyl ketone. Solid curves are drawn through the experimental points. The dashed curves have been calculated from theory. (Shultz and Flory.[3])

the more concentrated phase contains much more solvent than solute (polymer). At temperatures appreciably below T_c, the more dilute phase may retain an inappreciable amount of solute. These characteristics, which are typical of systems consisting of a polymer fraction in a single solvent component, are more marked the higher the molecular weight. This is evident in Fig. 121.

It is worth remarking at this point on the physical reasons for the striking dissymmetry of the phase diagram as manifested, for example by the occurrence of the critical composition at a point far from the middle region of the diagram, where it normally is found in binary systems of simple liquids. The great disparity in the sizes of the molecules of the two components is, of course, responsible for the dissymmetry. The first term (RTA_1) in the series expansion of the osmotic pressure is very small owing to the small number of solute particles per unit volume, i.e., owing to the large molecular weight. Higher terms in the expansion (e.g., RTA_2; see Chap. XII) assume values which resemble those for simple liquids. At a temperature below Θ, the polymer segments find their own environment more satisfying than that provided by the solvent. Only the greater volume of the more dilute phase offers an inducement for them to disperse in it. Inasmuch as the number of particles (molecules) to be distributed over the volume is very small, this inducement is likewise small unless the volume is made very great. The solvent molecules, being much greater in number, are more effective in equalizing their concentration (number per unit volume) in both phases; hence both phases are dilute.

Reciprocals of the critical temperatures, i.e., the maxima in curves such as those in Fig. 121, are plotted in Fig. 122 against the function $1/x^{1/2}+1/2x$, which is very nearly $1/x^{1/2}$ when x is large. The upper line represents polystyrene in cyclohexane and the lower one polyisobutylene in diisobutyl ketone.[3] Both are accurately linear within experimental error. This is typical of polymer-solvent systems exhibiting limited miscibility. The intercepts represent Θ. Values obtained in this manner agree within experimental error ($<1°$) with those derived from osmotic measurements,[4] taking Θ to be the temperature at which A_2 is zero (see Chap. XII). Precipitation measurements carried out on a series of fractions offer a relatively simple method for accurate determination of this critical temperature, which occupies an important role in the treatment of various polymer solution properties.

Entropy of dilution parameters ψ_1 are calculable, according to Eq. (7), from the slopes of the lines in Fig. 122. Values obtained in this manner are 0.65 and 1.055 for the polyisobutylene and the polystyrene systems, respectively. These are considerably higher than the values

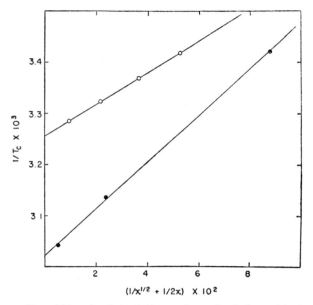

Fig. 122.—A plot of the reciprocal of the critical temperature against the molecular size function occurring in Eq. (7) for polystyrene fractions in cyclohexane (○) and for polyisobutylene fractions in diisobutyl ketone (●). (Shultz and Flory.[3])

indicated by osmotic measurements on dilute solutions. On the other hand, Shultz[5] has shown that interaction parameters deduced from phase equilibrium studies agree remarkably well with those for simple molecule analogs in the same solvents (e.g., for cyclohexane-toluene compared with cyclohexane-polystyrene). This would suggest that the ψ_1 values deduced from dilute solution thermodynamic measurements in accordance with the intermolecular theory (or from intrinsic viscosity studies; see Chap. XIV) are appreciably in error. Whatever its origin may be, the discrepancy serves as a further reminder that existing thermodynamic theory, though generally reliable in predicting the form of observed results, often is inaccurate numerically.

1c. Ternary Systems Consisting of a Single Polymer Component in a Binary Solvent Mixture.—Three conditions must be satisfied for equilibrium between two liquid phases in a system of three components. In place of the conditions (1) we have

$$\left.\begin{array}{l} \mu_1 = \mu_1' \\ \mu_2 = \mu_2' \\ \mu_3 = \mu_3' \end{array}\right\} \tag{11}$$

The chemical potentials derived by an extension of the procedure used to obtain Eqs. (XII-26) and (XII-32) may be put in the forms*

$$\mu_1 - \mu_1^0 = RT\big[\ln v_1 + (1 - v_1) - v_2(x_1/x_2) - v_3(x_1/x_3)$$
$$+ (\chi_{12}v_2 + \chi_{13}v_3)(v_2 + v_3) - \chi_{23}(x_1/x_2)v_2v_3\big] \tag{12}$$

$$\mu_2 - \mu_2^0 = RT\big[\ln v_2 + (1 - v_2) - v_1(x_2/x_1) - v_3(x_2/x_3)$$
$$+ (\chi_{21}v_1 + \chi_{23}v_3)(v_1 + v_3) - \chi_{13}(x_2/x_1)v_1v_3\big] \tag{13}$$

$$\mu_3 - \mu_3^0 = RT\big[\ln v_3 + (1 - v_3) - v_1(x_3/x_1) - v_2(x_3/x_2)$$
$$+ (\chi_{31}v_1 + \chi_{32}v_2)(v_1 + v_2) - \chi_{12}(x_3/x_1)v_1v_2\big] \tag{14}$$

where x_1, x_2, and x_3 represent the numbers of segments per molecule in the respective species, and χ_{ij} are pair interaction parameters corresponding to χ_1 in the two-component system. Specifically

$$\chi_{ij} = z\Delta w_{ij}x_i/kT \tag{15}$$

where $\Delta w_{ij} \equiv \Delta w_{ji}$ is the analog of Δw_{12} defined by Eq. (XII-17). It will be observed that the parameter χ_{ij} defined in this way is proportional to the molecular size of species i. Thus, if this species is a polymer, χ_{ij} may be very large indeed. As a more appropriate measure of the intensity of the interaction, we may use χ_{ij}/x_i, which, according to literal interpretation of the theory, represents the interaction intensity per segment. It is to be noted further that the six χ_{ij}'s for a system of three components reduce to three independent parameters by the relations

$$\chi_{ji} = \chi_{ij}(x_j/x_i) = \chi_{ij}(v_j/v_i) \tag{16}$$

where v_i and v_j are the molar volumes. Moreover, the x's enter equations (12) to (14) as ratios, representing in each case a ratio of molar volumes; hence one of them may be assigned the arbitrary value, unity, without loss of generality. Since the component 1 ordinarily will be a solvent of low molecular weight, it is appropriate to take $x_1 = 1$.

* Eqs. (12) to (14) are readily obtained by taking as the generalized expressions for the entropy and heat of mixing in a polycomponent system

$$\Delta S_M^* = -k\sum n_i \ln v_i$$

(see Eq. XII-12) and

$$\Delta H_M = z\sum_{i<j} x_i n_i v_j \Delta w_{ij} = kT\sum_{i<j} n_i v_j \chi_{ij}$$

where the summations include all pairs of unlike species (see Eqs. XII-19′ and XII-20). For a three-component system this gives

$$\Delta F_M = kT\big[n_1 \ln v_1 + n_2 \ln v_2 + n_3 \ln v_3 + \chi_{12}n_1v_2 + \chi_{13}n_1v_3 + \chi_{23}n_2v_3\big]$$

Differentiation yields Eqs. (12) to (14).

If the chemical potentials in each of the two phases are expressed according to Eqs. (12), (13), and (14) and if these equations are substituted in the equilibrium conditions (11) (in analogy with the treatment of the two-component system; see Appendix A), three simultaneous equations are obtained involving four independent concentration variables, two for each phase. An acceptable set of variables would be, for example, v_1, v_2, v_1' and v_2' (v_3 is defined by v_1 and v_2, since $v_1+v_2+v_3=1$; similarly, v_1' and v_2' define v_3'). If one of the four independent composition variables is specified, the other three are fixed by the three equations. It should be possible, therefore, to compute the binodial curve representing isothermal two-phase equilibrium compositions for a three-component system, given the values of the parameters (i.e., the χ's and the x's). The equations cannot be solved explicitly, however, and one is obliged to resort to numerical methods. Even in special cases leading to simplification of the equations, the calculation of the isothermal three-component phase diagram is a tedious undertaking.

Binodials calculated by Tompa[6] are shown in Fig. 123,a for the special case of a nonsolvent [1], a solvent [2], and a polymer [3] with $v_1=v_2$, $\chi_{23}=0$, and $\chi_{12}=\chi_{13}=1.5$. Otherwise stated, the nonsolvent-solvent and the nonsolvent-polymer segment free energies of interaction are taken to be equal, while that for the solvent and polymer is assumed to be zero. It is permissible, then, to take $x_1=x_2=1$ and $x_3=v_3/v_1$. The number of parameters is thus reduced for this special case from five to two. Binodial curves are shown in Fig. 123,a for $x_3=10$, 100, and ∞; tie lines are shown for the intermediate curve only. The critical points for each curve, shown by circles, represent the points at which the tie lines vanish, i.e., where the compositions of the two phases in equilibrium become identical.

The critical points occur at low polymer concentrations—the lower the higher the molecular weight. As the molecular weight goes to infinity (see dotted curve in Fig. 123,a), the critical point moves into the nonsolvent-solvent axis; i.e., the critical concentration of polymer is zero in the limit of infinite molecular weight. Near the critical point when x is large but finite, both phases are dilute in polymer.

Mergence of the binodial with the nonsolvent-solvent axis shows that the polymer concentration in the more dilute phase becomes vanishingly small when the proportion of nonsolvent exceeds appreciably that at the critical point. These features clearly parallel those observed in two-component systems, with the nonsolvent-solvent ratio assuming the role of temperature in the latter. It may be shown that they are not critically dependent on the particular values assigned to the parameters. Quite different characteristics appear, however, when major changes are made in the interaction parameters, as for example when the solvent and nonsolvent are incompletely miscible with one another (which occurs when $\chi_{12} > 2$), or when both components 1 and 2 are nonsolvents for the polymer ($\chi_{13} > 1/2$ and $\chi_{23} > 1/2$) while over a certain range their mixtures are solvents for the polymer.[6,7]

Such experiments as are available seem to confirm qualitatively the predictions of theory.[5,6] However, in analogy with the two-component systems, the observed binodials are considerably broader than the calculated ones such as are shown in Fig. 123.

The composition of the nonsolvent-solvent mixture representing the critical composition of the ternary system with polymer of infinite molecular weight (see Fig. 123,a) possesses unique significance. Scott[7] and Tompa[8] have shown that the composition at the critical point in the limit of infinite molecular weight is specified by

$$v_3 = 0$$

and

$$1 - 2\chi_{13}(1 - v_2) - 2\chi_{23}v_2 + [2(\chi_{21}\chi_{13} + \chi_{12}\chi_{23} + \chi_{13}\chi_{23})$$
$$- (x_2/x_1)\,\chi_{12}^2 - (x_2/x_1)\,\chi_{13}^2 - (x_1/x_2)\,\chi_{23}^2]v_2(1 - v_2) = 0 \quad (17)^*$$

They showed further that the limiting slope (RTA_2) of the plot of the osmotic pressure–concentration ratio π/c against the polymer concentration in a binary solvent mixture should be proportional to the value of the quantity on the left side of Eq. (17),† with v_2 representing the volume fraction of solvent in the nonsolvent-solvent mixture which is in osmotic equilibrium with the solution. The composition of the liquid medium outside the polymer molecules in a dilute solution must likewise be given by v_2. The composition of the solvent mixture within the domains of the polymer molecules may differ slightly from that outside owing to selective "absorption" of solvent in preference to the nonsolvent. This internal composition is not directly of concern here. If the solution is made sufficiently dilute, the external nonsolvent-solvent composition ($v_2 = 1 - v_1$) will be practically equal to the over-all solvent composition for the solution as a whole. Hence

---

* Scott[7] derived corresponding results for the special case $x_1 = x_2 = 1$; i.e., $v_1 = v_2$. Tompa[8] has carried out a more general treatment for unrestricted values of the size parameters, with results which coincide with Eq. (17), when expressed in the present notation.

† Modifications required by the discontinuous nature of the polymer solution were disregarded in the derivation of this result. They are unimportant in the poor solvent mixtures which are under consideration here.

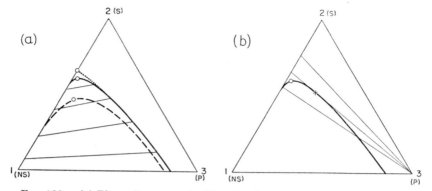

Fig. 123.—(a) Phase diagram calculated for three-component systems consisting of nonsolvent [1], solvent [2], and polymer [3] taking $x_1 = x_2 = 1$ and x_3 equal to 10 (dashed curve), 100 (solid curve), and ∞ (dotted curve); $\chi_{12} = \chi_{13} = 1.5$ and $\chi_{23} = 0$. All critical points ($\bigcirc$) are shown and tie lines are included for the $x_3 = 100$ curve. (Curves calculated by Tompa.[5]) (b) The binodial curve for $x_3 = 100$ and three solvent ratio lines. The precipitation threshold is indicated by the point of tangency $\times$ for the threshold solvent mixture.

we are justified in asserting that the *initial* slope of the π/c vs. c plot will be proportional to the quantity appearing on the left in Eq. (17), with v_2 representing the volume fraction of solvent in the nonsolvent-solvent mixture as a whole. Since Eq. (17) requires this quantity to equal zero at the critical point for the ternary system in the limit of $M = \infty$, the solvent composition for which this is true must also be that for which the second coefficient (A_2) in the virial expansion of the osmotic pressure is zero (irrespective of M but at the same temperature, of course). Hence, this *limiting critical point is the analog of* Θ *in the system of two components.*

It must not be inferred from these analogies to two-component systems that a polymer solution in a mixed solvent may be treated in the same way as a solution in a single solvent. Except in rare instances it is not legitimate, for example, to regard the mixed solvent as if it were a single component and then to treat precipitation measurements according to the procedures outlined for solutions of a polymer in a single solvent.[7] As a forceful reminder of the fallacy of such a procedure, we may note from Fig. 123,a that the nonsolvent-solvent ratios differ markedly for the two phases in equilibrium whose compositions are indicated by the ends of the tie line. All mixtures of a given solvent ratio occur on the same straight line running from the polymer apex [3] to the opposite side [1,2]. In Fig. 123,b three constant "solvent ratio" lines are shown together with the binodial for $x_3 = 100$ from Fig. 123,a. The difference between the solvent ratios in the two phases depends of course on the interaction parameters, and especially on the difference between χ_{13} and χ_{23}. If χ_{13} and χ_{23} are nearly equal, the tie lines will tend to run parallel to solvent ratio lines, and the solvent composition will be approximately the same in the two phases in equilibrium. It will be exactly so for all tie lines only if $\chi_{13} = \chi_{23}$ *and if* $\chi_{12} = 0$.[7] The mixed solvent in this special case behaves as if it were a single solvent; i.e., the so-called *single liquid* approximation applies and such a system could be treated as one of two components. A large difference between χ_{13} and χ_{23} favors selective absorption of the better solvent in the polymer-rich phase.

To emphasize further the severity of the misinterpretations which may arise from treatment of the three-component system as though it were a binary solution of polymer in a single solvent, consider a series of solutions prepared by choosing a given mixture of nonsolvent (1) and solvent (2) and adding varying proportions of the polymer to this mixture. All of the resulting solutions will be homogeneous, i.e., the polymer and the given solvent mixture will be miscible in all proportions, *only if the solvent ratio line does not intersect the binodial.* (See Fig. 123,b.) If it intersects the binodial, as does the lowermost solvent ratio line shown in Fig. 123,b, phase separation must occur for compositions along that portion of the solvent ratio line which lies below the binodial curve. The solvent ratio line which is tangential to the binodial represents the minimum nonsolvent-solvent ratio for which phase separation may occur at the given temperature.[9] At the polymer concentration designated by the point of tangency, (X in Fig. 123,b), the system is on the verge of inhomogeneity; an increase in the proportion of nonsolvent at this point, which will be referred to as the precipitation threshold,[9] would bring about separation into two phases. This composition does not in general coincide with the critical point, where the compositions of the phases in equilibrium become identical. It occurs at a higher polymer concentration and at a lower nonsolvent-solvent ratio than the true critical point. The displacement of the precipitation threshold from the critical point depends, of course, on the values of the χ parameters. The precipitation threshold, in contrast to the critical point, may actually occur at polymer concentrations which are comparatively high.

For the purpose of investigating solubility relations at a given temperature in three-component systems of the type discussed above, the proportion of nonsolvent which must be added to produce incipient precipitation may be observed as a function of the polymer concentration. Isothermal phase diagrams like that shown in Fig. 123 may be constructed in this manner;[3] location of the tie lines requires the additional information afforded by analysis of the composition of one of the phases.   Often it is desired to establish solubility relations for the polymer in nonsolvent-solvent mixtures of fixed ratio.   To this end, three-component mixtures are prepared by adding varying proportions of the polymer to the nonsolvent-solvent mixture in the specified ratio.   Each of the resulting three-component mixtures is warmed to homogeneity, then cooled gradually until precipitation (i.e., phase separation) sets in, and the precipitation temperature T_p is noted. The routine follows that applied to a two-component (single solvent) system for the purpose of establishing its phase diagram.   In like manner, one may plot T_p for the system of three components with fixed nonsolvent-solvent ratio against the polymer concentration as in Fig. 121 (p. 546).   The curve thus obtained does not, however, have the same significance as for the two-component system for the reasons already given; the maximum represents the precipitation threshold and not the critical point.   The solution giving this maximum may contain a large proportion of polymer, in sharp contrast to the single solvent system for which the precipitation threshold and the critical point are identical.   Hence it is important to avoid confusion of the precipitation threshold with the true critical point at which the compositions of the phases in equilibrium become identical.[9]

1d. Ternary Systems Consisting of Two Polymeric Components in a Single Solvent.[6,10]—Before proceeding with the discussion of ternary systems in which two chemically different polymers (components 2 and 3) are mixed with a single solvent (component 1), we may profitably examine the related binary system consisting of two dissimilar polymers in the absence of the solvent.   For simplicity let the molecular weights of the polymers be identical.   Then $x_2 = x_3 = x$ and $\chi_{23} = \chi_{32}$.   If we substitute these relations in Eqs. (13) and (14) and let $v_1 = 0$,

$$\mu_2 - \mu_2^0 = RT[\ln v_2 + \chi_{23} v_3^2] \tag{18}$$

$$\mu_3 - \mu_3^0 = RT[\ln v_3 + \chi_{23} v_2^2] \tag{19}$$

These simple expressions may also be obtained from the chemical potentials according to Eqs. (XII-26) and (XII-32) by appropriately changing subscripts and recalling that x in these equations represents the *ratio* of the molar volumes, which in the present case is unity. Owing to the identity of volume fractions with mole fractions in this case, Eqs. (18) and (19) are none other than the chemical potentials for a *regular* binary solution in which the heat of dilution can be expressed in the van Laar form.   The critical conditions (see Eqs. 2) are satisfied in this case when $v_2 = v_3 = 1/2$ and χ_{23} (crit.) $= 2$ (see Eqs. 4 and 5 with $x = 1$). The binary mixture of two polymer components having the same molecular weight would thus appear to be surprisingly normal. The distinctive feature of such systems becomes apparent on scrutinizing the constitution of the parameter χ_{23} as defined by Eq. (15). This parameter is proportional to the size of the molecule, represented by the number x of segments per molecule. The proper measure of the *intensity* of the interaction is given by the factor of proportionality $z\Delta w_{23}/kT$, representing the interaction free energy per segment divided by kT. (This factor is directly comparable to χ_1 used in the preceding treatment of binary systems, provided $x_1 = 1$.) The value of $z\Delta w_{23}/kT = 2/x$ at the critical point is therefore very small when x is very large. Thus, only a minute, positive, first neighbor interaction free energy is required to produce limited miscibility. The critical value of the interaction free energy is so small for any pair of polymers of high molecular weight that it is permissible to state as a principle of broad generality that *two high polymers are mutually compatible with one another only if their free energy of interaction is favorable, i.e., negative.* Since the mixing of a pair of polymers, like the mixing of simple liquids, in the great majority of cases is endothermic,* incompatibility of chemically dissimilar polymers is observed to be the rule and compatibility is the exception. The principal exceptions occur among pairs possessing polar substituents which interact favorably with one another. The practical operation of this principle will be dealt with in the course of the discussion of the ternary system containing a solvent.

The physical reason for the inherent lack of incentive for mixing in a polymer-polymer system is related to that already cited in explanation of the dissymmetry of the phase diagram for a polymer-solvent binary system. The entropy to be gained by intermixing of the polymer molecules is very small owing to the small numbers of molecules involved. Hence an almost trivial positive free energy of interaction suffices to counteract this small entropy of mixing.

Suppose now that there is chosen as the third component (component 1) a monomeric substance in which each of the polymer components (2 and 3) is separately miscible in all proportions in the absence of the other. In order that this condition may be fulfilled, both χ_{12} and χ_{13} are required to be less than one-half. Aside from this stipulation the actual values of these parameters are of minor importance only; hence we may let $\chi_{12} = \chi_{13}$. As before we take $x_2 = x_3 = x$,

---

* This results from the fact that w_{ij} of Eq. (XII-17) usually exceeds (algebraically) the mean of w_{ii} and w_{jj}.

and we let $x_1 = 1$ without loss of generality as already noted. Three parameters χ_{12}, χ_{23}, and x characterize the three-component system thus defined since $\chi_{21} = \chi_{31} = x\chi_{12}$ and $\chi_{32} = \chi_{23}$ (see Eq. 16). Inasmuch as the relationship of each polymer component to the solvent is the same, symmetry requires that the concentration of solvent in each phase at equilibrium must be the same and, moreover, that the concentration of component 2 in one phase must equal that of component 3 in the other, and vice versa. That is

$$\left.\begin{array}{l} v_1 = v_1' \\[4pt] v_2 = v_3' \\[4pt] v_3 = v_2' \end{array}\right\} \tag{20}$$

If the chemical potentials μ_2 and μ_2' in the two phases are expressed by Eq. (13) with parameters as specified above, then the result obtained on equating μ_2 to μ_2' according to the equilibrium conditions (11) may be rearranged to the following form,[10] after taking advantage of the relations (20) applicable to this case:

$$\ln \phi_2 + \chi_{23}(1 - v_1)\phi_3^2 = \ln \phi_2' + \chi_{23}(1 - v_1)\phi_3'^2 \tag{21}$$

where

$$\phi_2 = v_2/(v_2 + v_3)$$

$$\phi_2' = v_2'/(v_2' + v_3') = v_3/(v_2 + v_3) = 1 - \phi_2$$

and of course $\phi_3 = 1 - \phi_2$ and $\phi_3' = 1 - \phi_2'$. Thus ϕ_2 and ϕ_2' are the volume fraction compositions of the polymer constituents apart from the solvent. Eq. (21) corresponds to the equilibrium condition in the binary system consisting of the two polymers, as may be seen by comparing the result obtained on equating μ_2 to μ_2' as given by Eq. (18). The modified volume fractions ϕ_2 and ϕ_2' enter in place of v_2 and v_2', and the interaction parameter is reduced by the factor $(1 - v_1)$. The proportions of the polymer constituents in the two phases are readily calculable from Eq. (21) for any given solvent concentration, which is the same in both phases under the conditions assumed.

Ternary equilibrium curves calculated by Scott,[10] who developed the theory given here, are shown in Fig. 124 for $x = 1000$ and several values of χ_{23}. Tie lines are parallel to the 2,3-axis. The solute in each phase consists of a preponderance of one polymer component and a small proportion of the other. Critical points, which are easily derived from the analogy to a binary system, occur at

$$v_{2c} = v_{3c} = (1 - v_{1c})/2$$

$$\chi_{23} = 2/(1 - v_{1c}) \tag{22}$$

It will be observed that the value
of $\chi_{12} = \chi_{13}$ is not involved. Thus
the magnitude of the solvent-
polymer interaction is of no importance according to theory
(provided the solvent is completely miscible with each polymer alone; see above). Phase
separation results solely from
the polymer-polymer interaction. The value of χ_{23} which
may be tolerated without incidence of phase separation is
greater than that in the absence
of the solvent by the factor
$1/(1-v_{1c})$. However, the intensity of the interaction at the critical point, as represented by
$z\Delta w_{23}/kT$, remains very small provided $(1-v_{1c}) \ll 1/x$.

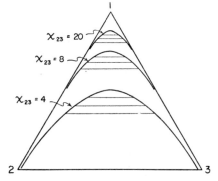

Fig. 124.—Three-component system
consisting of solvent [1], polymer [2],
and polymer [3], in which the chemically different polymers 2 and 3 interact
equally with the solvent and the polymer-polymer interaction parameter χ_{23}
assumes the values indicated. $x_2 = x_3$
$= 1000$ (Scott.[10])

Dobry and Boyer-Kawenoki[11] have investigated a number of solvent-
polymer-polymer systems, with results which confirm all of the qualitative predictions of the theory. Fig. 125 shows their experimental
results for the benzene-rubber-polystyrene system with coordinates
expressed in weight percent. The symmetry resulting from the
stipulations $x_2 = x_3$ and $\chi_{12} = \chi_{13}$ in the case treated theoretically
above of course is not dis-
played. However, each of the
two coexisting phases contains
a preponderance of one solute in
large excess over the other, except in the vicinity of the critical
point. Even for the low intensity of interaction which must be
involved between these two hydrocarbon polymers, the repulsion is sufficient to prevent the
existence of a phase containing
both solutes in concentrations
exceeding about 1 percent. Of
the 35 pairs of polymers examined by Dobry and Boyer-
Kawenoki, only three [nitro-

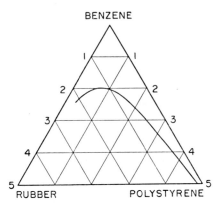

Fig. 125.—The ternary system benzene-rubber-polystyrene according to
Dobry and Boyer-Kawenoki[11] in the
dilute region. Scales are in weight percentages of the polymeric components.

cellulose–poly-(vinyl acetate), nitrocellulose–poly-(methyl methacrylate), and benzyl cellulose–polystyrene] were definitely compatible up to moderate concentrations.   Two polymers incompatible in one solvent were found to be incompatible also in other solvents, in agreement with the predicted secondary importance of the solvent-polymer interactions (χ_{12} and χ_{13}).   The mutual precipitation of one polymer by another was more nearly complete the higher the molecular weight of the former, as should be expected.

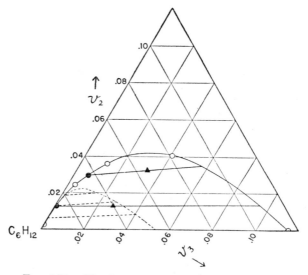

Fig. 126.—The ternary system consisting of cyclohexane and two polystyrene fractions ($x_2 = 770$ and $x_3 = 11,000$) at 28.2°C. Solid lines are drawn through the experimental points; results of theoretical calculation are shown by the dotted lines. (Shultz and Flory.[3])

Preluding the discussion of fractionation of heterogeneous polymers which follows, we may now consider briefly ternary systems comprising a solvent and two polymer homologs.[3,6]   The system cyclohexane-polystyrene($x_2 = 770$)-polystyrene($x_3 = 11,000$),   results[3]   for   which are presented in Fig. 126, is of this type.   The solid curve and tie lines represent the experimental results at 28.2°C.   Triangular points represent the average compositions of the mixtures separating along the tie lines shown.   Each tie line was located from this composition and that of the more dilute phase as determined by analysis and represented in Fig. 126 by a filled circle.   The dotted curve and tie lines have been calculated theoretically according to the following conditions appropriate in this case: $\chi_{23} = 0$ and $\chi_{12} = \chi_{13} = \chi_1$, where χ_1 is the interaction parameter for the binary polymer-solvent system at the same temperature, 28.2°C. The value of χ_1 obtained from the dependence of the critical temperature in the binary cyclohexane-polystyrene system on x (see Fig. 122, p. 548) was used. The experimental curve is much broader than the one calculated. This divergence from theory corresponds to that found in binary systems (see Fig. 121). Qualitative features are consistent with the theory, however. It may be shown, for example, that the critical ratios of the two polymer components are in approximate agreement with theoretical prediction.[3]

1e. Theory of Polymer Fractionation.*—A solution consisting of the broad array of polymer homologs present in any ordinary polymer dissolved in a single solvent is adequately characterized by a single interaction parameter $\chi_1 = \chi_{12}$ (all $x \neq 1$); only the size parameter x varies from one species to another. The chemical potential expressions (XII-27) and (XII-33) were designed for just such systems. Substituting these in the conditions for two-phase equilibrium, namely

$$\mu_1 = \mu_1' \quad \text{(for the solvent)}$$

$$\mu_x = \mu_x' \quad \text{(for each polymer species)}$$

it is possible, in principle, to deduce complete details of the equilibrium for a given value of χ_1 and a specified initial size distribution. The molecular weight distribution of the solute in each phase could be calculated in this way, and we should then have answered the question of the extent to which higher components are selectively transferred, under given conditions, to the "precipitated" phase, i.e., to the more concentrated phase which we designate with a prime. A complete calculation of the phase equilibrium from the equations would be a staggering task. The principal issue involved in fractionation, namely, the efficiency of the sorting of species between the two phases may be settled by the much less ambitious procedure set forth below.

The result obtained by equating μ_x to μ_x' using Eq. (XII-33) can be expressed very simply as

$$\ln(v_x'/v_x) = \sigma x \tag{23}$$

where

$$\sigma = v_2(1 - 1/\bar{x}_n) - v_2'(1 - 1/\bar{x}_n')$$
$$+ \chi_1[(1 - v_2)^2 - (1 - v_2')^2] \tag{24}$$

* Techniques for fractionating polymers have been discussed in Chapter VIII, pp. 339 ff.

It is sufficient for our purpose to accept the existence of a quantity σ such that the partitioning of every component between the two phases is governed by $v'_x/v_x = e^{\sigma x}$ according to Eq. (23); we need not undertake the incomparably more involved calculation of σ. Even if the latter step were carried out, the numerical value obtained for σ probably would be subject to a considerable error owing to imperfections of the theory. The simple relationship (23) in which σ enters as an undetermined parameter may, on the other hand, be derived from the most general consideration of the free energy change involved in transferring a molecule of size x from one phase to the other. Hence we may confidently base our treatment on this equation.

Let V and V' represent the volumes of the phases in equilibrium. The fraction f_x of constituent x remaining in the more dilute phase will be given by[2,12]

$$f_x = Vv_x/(Vv_x + V'v'_x) = 1/(1 + \mathcal{R}v'_x/v_x)$$

where $\mathcal{R} = V'/V$. By substitution from Eq. (23)

$$f_x = 1/(1 + \mathcal{R}e^{\sigma x}) \tag{25}$$

The fraction $f'_x = 1 - f_x$ of x-mer occurring in the more concentrated phase is given by

$$f'_x = \mathcal{R}e^{\sigma x}/(1 + \mathcal{R}e^{\sigma x}) \tag{26}$$

These simple relations suffice for demonstration of the essentials of a polymer fractionation.[2]

Suppose the heterogeneous polymer of given molecular size distribution is dissolved in a large volume of a poor solvent at a suitable temperature. On cooling the solution, χ_1 increases and eventually a temperature is reached at which phase separation occurs. Let the initial solution be so dilute that the newly formed phase is much more concentrated than the initial solution. The temperature is so adjusted as to form a small, though adequate, amount of the more concentrated phase. In developing this qualitative picture, we rely on the form of the binary phase diagram as shown in Fig. 121, bearing in mind, of course, the hazards involved in drawing too close an analogy to a binary system. Thus it is assumed that the concentration of the initial solution is well below that of the peak in the plot of the precipitation temperature T_p against concentration (as in Fig. 121). Then the more dilute (unprimed) phase will be the larger of the two. The parameters $\mathcal{R}$ and σ are determined by χ_1, and also by the molecular weight distribution and the proportion of diluent. Now it is particularly significant that according to Eq. (23) *every polymer species is more soluble in the precipitated phase*, i.e., $v'_x > v_x$ for all x. However, the magnitude of v'_x/v_x increases exponentially with x, and herein lies the basis for separation by fractional precipitation, inefficient though it may be. If the volume of the dilute phase is made much larger than that of the more concentrated phase ($\mathcal{R} \ll 1$), most of the smaller, less discriminating species will be retained in the more dilute phase merely as a consequence of its much greater volume. The

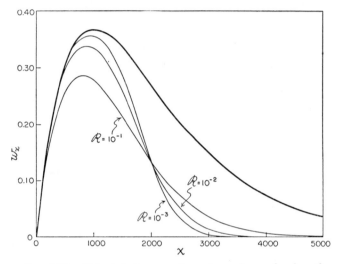

FIG. 127.—Calculated separation of a polymer having the weight fraction (w_x) distribution given by the uppermost curve by precipitation under conditions giving $\mathcal{R} = 10^{-3}$, 10^{-2}, and 10^{-1}. The corresponding values of σ, arbitrarily chosen to give $f_x = 1/2$ at $x = 2000$ in each case, are 3.456×10^{-3}, 2.303×10^{-3}, and 1.151×10^{-3}, respectively.

much larger partition factor v'_x/v_x for the higher species will cause them to be selectively transferred to the precipitated phase, in spite of its relatively small volume.

The character of the calculated partitioning of species between two phases is shown in Fig. 127 for a polymer with $\bar{x}_n = 1000$ having the "most probable" initial molecular weight distribution (Eq. VIII-3) shown by the uppermost curve. The three lower curves represent the composition of the polymer remaining in the dilute phase, as given by $w_x f_x$ where w_x is the weight fraction of x-mer in the initial distribution, calculated according to Eq. (25) for the three values of $\mathcal{R}$ indicated. The value of σ has been adjusted to give $f_x = 1/2$ at $x = 2000$ in each case. The quantity $w_x f'_x$ of each species precipitated may be obtained as the difference between the initial distribution curve (w_x) and that for the dilute phase $(w_x f_x)$. The separation, although never sharp, becomes more efficient as the volume ratio $1/\mathcal{R}$ is increased.

The required small value of $\mathcal{R}$ can be achieved only if the system is very dilute, as has been pointed out in Chapter VIII. The initial concentration should be much less than that of the peak in the plot of T_p against v_2. Otherwise, the phases which separate necessarily will be similar in concentration and hence in amount; or, if the initial concentration exceeds that at the peak, the volume of the precipitated phase actually will exceed that of the more dilute phase (i.e., $\mathcal{R} > 1$). The position of the peak depends predominantly on the species of higher molecular weight;[3] hence as a rough rule one may expect it to occur in the vicinity of $v_2 = 1/x^{1/2}$ (see Eq. 4'), where x lies somewhat above the weight average for the polymer as a whole. Thus, for a polymer in the molecular weight range of 10^6, x will be of the order of 10^4 and the initial concentration of the solution from which fractionation is conducted should be much less than 1 percent in polymer, and preferably not over about 0.1 percent.

In the case of simple fractionation in which each fraction is obtained by a single precipitation, the distribution in the first fraction is given by $w_x \cdot f_x'(1)$, that in the second by $w_x \cdot f_x(1) \cdot f_x'(2)$, that in the third by $w_x \cdot f_x(1) \cdot f_x(2) \cdot f_x'(3)$, etc., where $f_x(1)$, $f_x(2)$, etc., and $f_x'(1)$, $f_x'(2)$, etc., are the separation factors previously defined for the first, second, etc., precipitations. Thus, given the values of $\mathcal{R}$ and σ for each step,

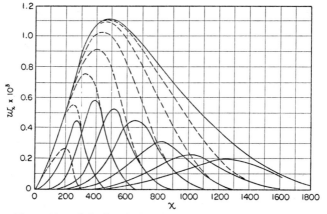

Fig. 128.—Calculated distribution for each of eight fractions separated from the initial distribution shown by the uppermost curve. Dotted lines represent distribution of polymer remaining in the dilute phase after each successive precipitation with $\mathcal{R} = 10^{-3}$ in each case. The distribution for each fraction, obtained as the differences between successive dotted curves, is shown by a solid curve. (Schulz.[12])

the distribution and amount of each fraction can be calculated. The results of such a calculation carried out by Schulz[12] are shown in Fig. 128 for the unusually favorable volume ratio $\mathcal{R} = 10^{-3}$. Even under these conditions the fractions are by no means sharp and they overlap one another extensively. However, it may be shown that the weight and number average molecular weights within any given fraction depicted in Fig. 128 do not differ more than a few percent. For nearly all purposes it is near coincidence of the different averages which is required, rather than literal confinement of the fraction to a narrow range.

The foregoing discussion refers specifically to fractionation from a single solvent component, the temperature being lowered in suitable increments to obtain successive fractions. The practice is far more common to add a precipitant in successive increments at constant temperature to a dilute solution of the polymer in a solvent. Similar formal relations should apply here also, but it is necessary to look for analogy in the three-component system: nonsolvent-solvent-polymer (see Fig. 123, p. 552). The nonsolvent-solvent ratio may be appreciably smaller in the precipitated than in the dilute phase. This difference will be reflected in the value of σ. Of more direct bearing on the question of fractionation efficiency, the solute composition relations between the phases in equilibrium may differ markedly from those obtained in the single solvent system considered above. We may expect, as a matter of fact, that the concentration of polymer in the precipitated phase will be greater in the present case; hence that $\mathcal{R}$ will be smaller, and the tendency in this direction will be enhanced the greater the disparity in solvent power between precipitant and solvent. On this basis, at least, more favorable fractionation conditions are indicated for solvent-precipitant mixtures. Definite evidence bearing on this prediction does not appear to have been reported, however.

Finally, we have assumed that both phases which separate are amorphous. Specifically, we have disregarded the possibility that the polymer in the precipitated phase may be semicrystalline. As has been pointed out in Chapter VIII, a rather different behavior is encountered if the polymer is high melting, and, hence, precipitation is occasioned by the inclination of the polymer to crystallize. The equations given have no bearing on this case.

2. PHASE EQUILIBRIA IN SEMICRYSTALLINE SYSTEMS

2a. The Nature of the Crystalline State in Polymers.—Polymers having a sufficiently regular chain structure nearly always are susceptible to the spontaneous ordering of their configurations, commonly referred to as crystallization. In the crystalline condition the polymer chains occur in bundles with chain axes parallel to one another. The individual molecules may be fully extended, or they may occur in a somewhat less extended spiral configuration. In either case, not only are the axes of adjacent molecules parallel to one another, but substituents on adjacent chain units are arranged in transverse layers. Moreover, the occurrence in the X-ray diffraction pattern of two distinct spacings transverse to the *fiber axis* reveals two periodicities perpendicular to the chain axis, from which it may be inferred that the angles of rotation of the chain molecules about their long axes also conform to a regular array. Thus the polymer crystal may possess regularity in three dimensions of a type corresponding to that which occurs in the crystals made up of simpler molecules. With respect to the unit cell, the repeating unit of the polymer occupies the role of the molecule in the crystal of a simple organic compound. Aside from the fact that the polymer molecule continues through many consecutive unit cells of the lattice, the structure of the polymer crystal at the level of the unit cell is otherwise completely analogous to crystals of monomeric compounds.

At a higher level of morphology significant differences appear on the basis of which the ordering that occurs in polymers sometimes has not been accepted as a genuine crystallization. In the first place, macroscopic single crystals seldom, if ever, are formed from high polymers (certain proteins excepted). Furthermore, polymers probably never crystallize completely; the submicroscopic crystallites seem to occur imbedded in a residual amorphous matrix. The proportion of crystalline material is variable depending on conditions of crystallization; it may vary from a few percent only to more than 90 percent in some cases. The degree of perfection of the crystallites also is variable, and may be quite low if proper annealing conditions have not been employed. It is difficult to differentiate, by X-ray techniques based on the breadth of the diffraction line, between small crystallite size and crystallite imperfection. If the line breadth is attributed entirely to the finite size of the crystallites, the dimensions indicated by such measurements are in the neighborhood of 100 to 200 Å. Low angle X-ray diffraction of oriented fibers and films[13,14] reveals structural periodicities of about this magnitude, which probably are directly related to the dimensions of the crystallites.[13]

Whether or not the type of ordered arrangement which occurs in many polymers should be accepted as a genuine crystalline state reduces ultimately to the question of the legitimacy of regarding the crystalline regions as a separate phase. The criterion to be applied is essentially a thermodynamic one, and it requires that any given property (e.g., the chemical potential) shall be uniform throughout the phase, and that it shall depend solely on the temperature and pressure (a *pure* crystalline phase being assumed).   Properties of poorly ordered crystalline regions will depend not only on the temperature and pressure but also on the degree of order.   Moreover, if the crystallite is very small, its properties may be modified appreciably by the excess free energy of the boundary layer in the same way that the vapor pressure of very small liquid drops is increased owing to the surface tension, or surface free energy.

Now it is well known that an array of mesomorphic states, intermediate between the completely amorphous liquid and the perfect crystal, can be obtained, depending on the conditions of formation of the crystalline regions.   If the polymer is rapidly cooled from the melt to a temperature far below the melting point, only a few diffuse X-ray reflections may be observed, and, in particular, the lateral spacings between adjacent chains may be poorly resolved.[15,16]   On the other hand, if the polymer is annealed at a temperature not much below its melting temperature, or if it is initially allowed to crystallize over a long period of time close to the melting temperature, the reflections are sharper and more of them are observed, indicating a higher degree of order and larger average size of the crystallites.   In consideration of the coexistence of amorphous and more or less ordered polymer chains, and of the ease with which intermediate degrees of order may be obtained, the conclusion has frequently been drawn that a complete spectrum of degrees of order is typical of the semi-crystalline state in polymers.   Reference to a crystalline *phase* would be meaningless if this description were universally applicable.

The process of crystallization in polymers is complicated by the requirement that many consecutive units of each participating chain must enter systematically in the same crystallite.   A polymer unit, in contrast to a separate monomeric molecule, is not free to join a given growing crystallite in a manner unrelated to its fellow members in the same chain.   Consequently, the growth (and nucleation) of polymer crystals is a slow process.[17]   It appears that the mesomorphic states are metastable ones, formed out of compromise between the time required for more perfect coordination in the arrangement of many units and the strong driving force toward crystallization brought about by the large supercooling.*   The fact that annealing improves

---

* The view that the degree of imperfection depends on the amount of supercooling is borne out by the observations of Bekkedahl and Wood[18] on rubber. They showed that the melting range (for fast melting) is lower the lower the temperature at which the rubber had been allowed to crystallize.   Other crystalline polymers exhibit parallel behavior.

the perfection of the crystalline order (besides increasing the amount of crystallinity) shows unequivocally that this is true. It suggests also that well-ordered crystalline states may actually be realized in polymers under favorable circumstances, and that the degree to which a state of perfect order is approached in a given instance depends on the thermal treatment. A high degree of order necessarily implies, among other things, that the order extends over a considerable range.

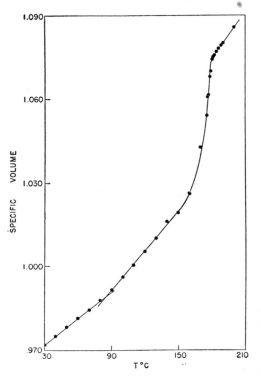

The crystal boundary regions, in which the transition from crystalline order to the amorphous region occurs, must therefore constitute only a negligible fraction of the total crystalline material in the highly ordered state. Boundary regions of intermediate order should melt at lower temperatures than highly ordered internal regions; the melting point should be depressed also if the crystallite is very small, owing to the surface free energy. The extent to which the perfect crystalline state is approached may, for practical purposes, be judged by the sharpness and reproducibility of the melting point.

FIG. 129.—Specific volume of pure poly-(N,N'-sebacoyl piperazine) plotted against the temperature. $T_m = 180–181°C$. A second-order transition appears also at $T_g = 82°C$. (Flory, Mandelkern, and Hall.[19])

The melting transition in polymers is conveniently observed by dilatometric measurements of the volume as a function of the temperature. Results thus obtained for poly-(N,N'-sebacoyl piperazine)[19] are shown in Fig. 129, and for poly-(decamethylene adipate)[20] by the lowest curve in Fig. 130. Most of the melting, as evidenced by the abnormal increase in volume, occurs within a range of 10°, and it terminates abruptly at a temperature which may be defined within ±0.5°. The data shown were obtained by raising the temperature in small increments, e.g., one degree, and holding the temperature unchanged until the volume became constant. This required about 24 hours just below the temperature for completion of melting, which may be called the melting point T_m of the sample. As the less perfect, and therefore less stable, crystallites melted, more stable ones were allowed to form by this procedure; i.e., the conditions used offered abundant opportunity for recrystallization to occur. More rapidly conducted melting, irrespective of the method of observation, usually leads to fictitiously low melting points, which may be in error by as much as 10° or more.

The occurrence of a restricted range within which most of the melting may be confined when conditions conducive to equilibration are adopted lends support to the concept of a crystalline *phase*, a subdivided one notwithstanding, having approximately uniform properties throughout. (It follows also that although the crystallites which melt under the conditions described may be small by ordinary standards, they are not so small as to cause their stabilities to be much diminished by their surface free energies.) To be sure, appreciable pre-melting occurs as much as 10° below T_m, suggesting the persistence

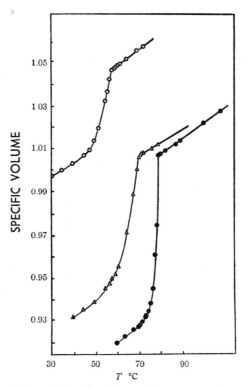

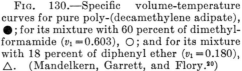

Fig. 130.—Specific volume-temperature curves for pure poly-(decamethylene adipate), ●; for its mixture with 60 percent of dimethylformamide ($v_1 = 0.603$), ○; and for its mixture with 18 percent of diphenyl ether ($v_1 = 0.180$), △. (Mandelkern, Garrett, and Flory.[20])

of crystallites of low stability even with the slowest attainable rates of heating. This is understandable when one considers the structural difficulties of advancing the degree of crystallization in a partially crystalline polymer. Only the sections of polymer chain extending between existing crystallites are available for further crystallization, and these are limited in length and constrained in position.   Hence the formation of additional large, perfect crystallites may be quite impossible. Further growth of existing crystallites may encounter similar difficulties.   As melting progresses, these restraints are relaxed and the last crystallites to melt should be not only the most perfect, but they should also be unrestricted in this respect, apart from such irregularities (e.g., copolymerized units or chain ends) as may be inherent in the polymer structure itself.   The abrupt completion of the melting process is in harmony with this point of view.   We conclude, therefore, that the temperature T_m for completion of melting under the slow heating conditions described ordinarily will lie very nearly at the equilibrium melting temperature for a hypothetical macroscopic, perfect crystal; the difference between the true equilibrium melting temperature and that observed may be no greater than the nominal experimental uncertainty in locating the latter.

2b. Theory of the Melting Point Depression.[21]—At equilibrium between liquid and crystalline polymer, the chemical potentials of the polymer repeating unit in the two phases must be equal, i.e., $\mu_u = \mu_u^c$. The temperature at which this condition is satisfied may be referred to as the melting point T_m, which will depend, of course, on the composition of the liquid phase.   If a diluent is present in the liquid phase, T_m may be regarded alternatively as the temperature at which the specified composition is that of a saturated solution.   If the liquid polymer is pure, $\mu_u \equiv \mu_u^0$ where μ_u^0 represents the chemical potential in the standard state, which, in accordance with custom in the treatment of solutions, we take to be the pure liquid at the same temperature and pressure.   At the melting point T_m^0 of the pure polymer, therefore, $\mu_u^0 = \mu_u^c$.   To the extent that the polymer contains impurities (e.g., solvents, or copolymerized units), μ_u will be less than μ_u^0.   Hence μ_u after the addition of a diluent to the polymer at the temperature T_m^0 will be less than μ_u^c; and in order to re-establish the condition of equilibrium $\mu_u = \mu_u^c$, a lower temperature T_m is required.

The derivation of the quantitative relationship between this equilibrium temperature and the composition of the liquid phase may be carried out according to the well-known thermodynamic procedures for treating the depression of the melting point and for deriving solubility-temperature relations.   The condition of equilibrium between crystalline polymer and the polymer unit in the solution may be restated as follows:

$$\mu_u^c - \mu_u^0 = \mu_u - \mu_u^0 \tag{27}$$

In words, the difference between the chemical potential of the crystalline repeating unit and the unit in the standard state, i.e., the *pure liquid* polymer at the same temperature and pressure, must equal the decrease in chemical potential of the polymer unit in the solution relative to the same standard state. Now the former difference appearing on the left-hand side of Eq. (27) is simply the negative of the free energy of fusion ΔF_u. Hence it may be written

$$\mu_u^c - \mu_u^0 = - \Delta F_u = - (\Delta H_u - T\Delta S_u)$$

where ΔH_u and ΔS_u are the heat and entropy of fusion per repeating unit. This relation is preferably expressed as follows:

$$\mu_u^c - \mu_u^0 = - \Delta H_u(1 - T/T_m^0) \tag{28}$$

where the ratio $\Delta H_u^0/\Delta S_u^0$ is assumed to be constant over the temperature range from T_m^0 to T, and therefore equal to T_m^0, since $\mu_u^c - \mu_u^0 = 0$ at $T = T_m^0$. We require also an expression for $\mu_u - \mu_u^0$, which represents the lowering of the chemical potential of the unit in the liquid phase owing to the presence of diluent therein. For this purpose we may use Eq. (XII-34'), which has been derived by differentiation of the general free energy of mixing relation, Eq. (XII-23), with subsequent simplifications appropriate when the molecular weight is very large. By substituting Eqs. (28) and (XII-34') in Eq. (27) and replacing T by T_m, the equilibrium melting temperature for the mixture, we obtain

$$1/T_m - 1/T_m^0 = (R/\Delta H_u)(v_u/v_1)(v_1 - \chi_1 v_1^2) \tag{29}$$

which relates T_m to the composition as represented by the volume fraction v_1 of the diluent. The quantity $1/T_m - 1/T_m^0 = (T_m^0 - T_m)/T_m T_m^0$ is approximately proportional to the depression of the melting point by the diluent. It depends according to Eq. (29) on the heat of fusion ΔH_u per repeating unit, on the ratio of the molar volumes of this unit and the diluent, and, secondarily, on the interaction parameter χ_1. A relationship expressing the equilibrium solubility v_2 as a function of the temperature may be obtained by substituting $1 - v_2$ for v_1 and solving the quadratic.

Eq. (29) is closely related to the classical melting point depression and solubility expression for solutions of simple molecules. In the case of the ideal solution, for example, $\mu_2 - \mu_2^0 = -RT \ln N_2$, N_2 being the mole fraction of the crystallizing component (which is the solvent if melting point depressions are considered and the solute if the results are tabulated as solubilities). By combining this relation with Eq. (28), we arrive at the solubility relationship for an ideal solution

$$1/T_m - 1/T_m^0 = - (R/\Delta H_2) \ln N_2$$

where ΔH_2 is written for the heat of fusion of the major component, 2.

If the solution contains only a very small fraction of component 1, $\ln \text{N}_2 \cong - \text{N}_1 \cong (v_1/v_1)/[(1-v_1)/v_2] \cong v_1 v_2/v_1$. Hence

$$1/T_m - 1/T_m^0 \cong (R/\Delta H_2)(v_2/v_1)v_1 \tag{30}$$

For an ideal solution $\chi_1 = 0$; hence Eq. (29) for the polymer-diluent system and the melting point relationship for the classical ideal solution reduce to the same form for small proportions of the noncrystallizing component. It is possible, as a matter of fact, to derive this limiting form on the most general grounds for any two-component system, including systems in which one (or both; see below) of the components is a polymer unit.* The other component may be of any nature whatever, provided only that it is distributed throughout the system at random. Hence we may use Eq. (30) as the limiting form of Eq. (29) at high dilution, irrespective of the validity of assumptions employed in the preceding chapter to derive Eq. (XII-34′).

The component which brings about the depression in T_m may be a constituent of the polymer itself. In a copolymer consisting of A units which crystallize and B units which do not, with the two units occurring in random sequence along the chain, it is easy to show[21] that the latter should depress the melting point of the former according to the equation

$$1/T_m - 1/T_m^0 = - (R/\Delta H_u) \ln \text{N}_\text{A} \tag{31}$$

where N_A is the "mole" fraction of A units in the random copolymer. This expression reduces to the limiting form (30) for a small proportion of B units, with $v_1 = v_\text{B}$ in this case. Similarly, the end groups of polymer chains of finite length appear not to occur in the crystallites,[21,22] and hence they may be regarded as a foreign component. In this way it is possible to derive a relationship between the melting temperature and the degree of polymerization. For polymers having the most probable molecular weight distribution, the following relationship is obtained:

$$1/T_m - 1/T_m^0 = (R/\Delta H_u)(2/\bar{x}_n) \tag{32}$$

where $\bar{x}_n$ is the number average degree of polymerization. This equation likewise reduces for large chain lengths to the limiting expression (30).† Thus diluents, copolymerized units, and end groups

* To do so we need only specify a dilution sufficient to assure that the activity of component 1 is proportional to its concentration. Application of the Gibbs-Duhem equation leads to Eq. (30).

† Equations (31) and (32) apply only in case the foreign units (copolymer units B, or end groups) are distributed *at random* along the polymer chains.[21] If

should have an equivalent effect on the melting point when the concentration of each is low.   In the light of prevalent values for heats of fusion of crystalline polymers, molar proportions of an impurity less than 1 percent have no appreciable effect on T_m.   A polymer having a degree of polymerization exceeding about 100 should therefore melt at a temperature not sensibly below that for an infinitely high polymer.

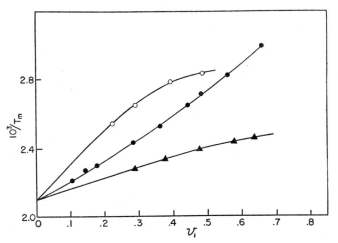

FIG. 131.—The reciprocal melting point plotted against the volume fraction of diluent for cellulose tributyrate in benzophenone (●), in hydroquinone monomethyl ether (○), and in ethyl laurate (▲).   (Mandelkern and Flory.[23])

2c. Experimental Results.—Specific volume measurements on two polyester-diluent mixtures carried out dilatometrically in the manner previously described for the pure polymer are shown in the upper curves of Fig. 130.[20]   The melting range is somewhat broader than that observed for the undiluted polymer, as should indeed be expected,[21] but the termination of the melting process is sufficiently sharp to permit definition of T_m within $\pm 1°$.

According to Eq. (29), the initial slope of a plot of $1/T_m$ against the diluent concentration v_1 should be inversely proportional to the heat of fusion ΔH_u; such curvature as may occur should be indicative of the magnitude of χ_1.   Results for cellulose tributyrate are plotted

the A and the B units of a copolymer tend to occur in separate sequences, the melting point depression will be less than Eq. (31) specifies; if they tend to alternate with the A units along the chain, thereby reducing the abundance of long sequences of A units which are required for crystallization, the depression for a given average composition will exceed that calculated from Eq. (31).

in this fashion in Fig. 131. The melting point T_m^0 cannot in this case be precisely measured directly owing to thermal decomposition; hence the value of $T_m^0 = 206\text{-}7°C$ has been obtained by extrapolation.

Once T_m^0 has been determined, the values of the other two parameters are more readily deduced by plotting the quantity $(1/T_m - 1/T_m^0)/v_1$ against v_1. Or, considering χ_1 to be inversely proportional to the temperature and recalling Eq. (XII-21'), we may plot this quantity against v_1/T_m.*

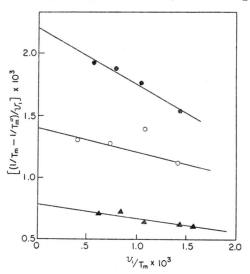

FIG. 132.—$(1/T_m - 1/T_m^0)/v_1$ plotted against v_1/T_m for mixtures of cellulose tributyrate with hydroquinone monomethyl ether (●), dimethyl phthalate (○), and ethyl laurate (▲). (Mandelkern and Flory.[23])

Results for cellulose tributyrate in three diluents[23] are treated in this manner in Fig. 132. According to Eq. (29), the intercepts should equal $(R/\Delta H_u)(v_u/v_1)$, and the energy of interaction $B = \chi_1 RT/v_1$ (see Eq. XII-21') may be computed from the slope. Results obtained in six different solvents are summarized in Table XXXIV. The agreement between values for the heats of fusion obtained using different solvents substantiates the method; analogous results obtained for other systems provide further confirmation. The values of B serve to classify the quality of the solvent for the given polymer. Three of the solvents listed in Table XXXIV are poor ones for cellulose tributyrate as indicated by positive B values; the other three are indifferent within experimental error, and none of the solvents exhibits a favorable interaction with this polymer.

The nature of the heat of fusion ΔH_u deserves particular attention, for it represents the heat required to melt one mole of *crystalline* units; it does *not* refer to the latent heat ΔH_u^* required to melt such crystallinity as may occur in a given semicrystalline polymer. The depression of the melting point T_m, already defined as the maximum temperature at which crystalline regions may coexist with amorphous poly-

* Discrimination between v_1 and v_1/T_m as the abscissa variable is scarcely justified in view of the small absolute temperature range covered and the limited accuracy with which $1/T_m - 1/T_m^0$ can be determined.

TABLE XXXIV.—HEATS OF FUSION AND ENERGY OF INTERACTION CONSTANTS AS
OBTAINED FROM MELTING POINT MEASUREMENTS ON MIXTURES OF
CELLULOSE TRIBUTYRATE WITH DILUENTS[23]

Diluent	ΔH_u (cal./mole)	B (cal./cc.)
Tributyrin	2,800	0.0
Benzophenone	2,900	0.0
Hydroquinone monomethyl ether	2,800	2.8
Dimethyl phthalate	2,800	1.2
Ethyl benzoate	3,200	0.0
Ethyl laurate	3,100	1.0

mer, necessarily depends on ΔH_u rather than on ΔH_u^*; this is implicit
in the thermodynamic derivation.[21] The heat of fusion ΔH_u cannot
be readily obtained by direct calorimetric determination. Measurement of the heat required to melt the sample gives ΔH_u^*, which depends
of course on the degree of crystallinity, a quantity not easy to determine independently with high accuracy. On the other hand, the
ratio of ΔH_u^* to ΔH_u should equal the degree of crystallinity in the given
sample. Hence melting point depressions used in conjunction with
calorimetric measurements provide the necessary data for assigning
a value to the degree of crystallinity.

In Table XXXV are given the heats and entropies of fusion (ΔS_u
$= \Delta H_u/T_m^0$ where T_m^0 is in °K) for various polymers. The melting

TABLE XXXV.—HEATS AND ENTROPIES OF FUSION OF POLYMERS

Polymer	Repeating unit	T_m °C	ΔH_u cal /unit	ΔS_u cal./deg. /unit	ΔS_u per bond cal./deg.
Polyethylene	—CH₂—	ca. 140	785	1.90	1.90
Polyethylene oxide[17]	—CH₂—CH₂—O—	66	1,980	5.85	1.95
Poly-(chlorotrifluoro-ethylene)[24]	—CF₂—CFCl—	210	1,200	2.50	1.25
Poly-(decamethylene adipate)[20]	—O(CH₂)₁₀O—CO(CH₂)₄CO—	79.5	10,200	29	1.60
Poly-(decamethylene sebacate)[22]	—O(CH₂)₁₀O—CO(CH₂)₈CO—	80	12,000	34	1.55
Poly-(N,N′-sebacoyl piperazine)[19]	—N⟨CH₂—CH₂ / CH₂—CH₂⟩N—CO(CH₂)₈CO—	180	6,200	13.7	1.25
Cellulose tributyrate[22]	—(C₆H₇O₂)(OCOC₃H₇)₃—	207	3,000	6.2	3.1
Poly-(chloroprene)[25]	—CH₂—C(Cl)=CH—CH₂— (trans)	80ᵃ	2,000	5.7	1.9

ᵃ Melting point obtained by extrapolating to hypothetical pure trans polymer.[25]

point depression method has been used in each case, except for polyethylene, where extrapolation of the heats of fusion of higher n-paraffin hydrocarbons offers a more accurate value for ΔH_u. In consideration of the variations in the sizes and structures of the repeating units listed, it is not surprising that the values for the heats ΔH_u and the entropies ΔS_u of fusion per unit vary over wide ranges. Inasmuch as the major contribution to the entropy of fusion presumably comes from the increase in the number of configurational arrangements of the chain segments on melting, ΔS_u should be expected to increase with the number of independently orienting members in the unit. With the aim of establishing a basis for comparing results for such diverse units as the methylene unit and the decamethylene sebacate unit containing 1 and 22 chain single bonds, respectively, we may divide ΔS_u by the number of these single bonds about which rotation may occur per repeating unit. The resulting figures, given in the last column of Table XXXV, are remarkably similar. Most of them are in the range from 1.5 to 2.0 cal./deg./bond. The higher value for cellulose tributyrate, having only two chain bonds permitting rotation within each large unit, may be due to a major contribution from disorientation of the butyrate substituents.

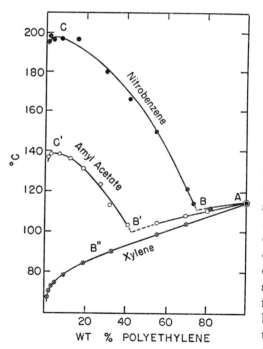

FIG. 133.—Phase diagram for polyethylene with each of the three diluents indicated. (Richards.[26])

According to the lattice theory of polymer solutions developed in the preceding chapter (see Eq. XII-9'), the entropy of disorientation per mole of segments should be $R \ln[(z-1)/e]$. If each group between successive single bonds in the chain is regarded as an independent segment, a reasonable value of about six for the coordination number z is required to match the observed entropy of fusion. This correlation and its implications should not be carried too far, however, for there may be appreciable contributions to the entropy of fusion from sources other than the randomness in the chain configuration. Consider, for example, the change in the relative arrangement of neighboring units which must accompany the increase in volume on melting.

If there are included among the diluents mixed with the crystalline polymer some which are sufficiently poor solvents, the phase diagram may then exhibit liquid-liquid phase separation, in addition to the liquid-crystal boundary curve. Examples are shown in Figs. 133

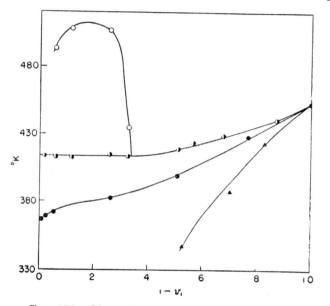

FIG. 134.—Phase diagram for poly-(N,N'-sebacoyl piperazine) with each of the diluents: m-cresol (▲), o-nitrotoluene (●), and diphenyl ether (◖ crystallization, ○ liquid-liquid separation). (Flory, Mandelkern, and Hall.[19])

and 134. With reference to the former, the polyethylene-xylene system investigated by Richards[26] exhibits only the crystal-liquid curve, xylene being a relatively good solvent. The boundary of the region of homogeneity is described by two curves when either of the poor solvents nitrobenzene or amyl acetate is the diluent. One of these, AB or AB', represents crystalline-liquid equilibrium, and the other, BC or B'C', defines the liquid-liquid phase equilibrium. The points of intersection, B and B', are critical points in accordance with the phase rule. Similar results for the polyamide, N,N'-sebacoyl piperazine, in diphenyl ether are shown by the upper curves in Fig. 134.[19] Consider a mixture having a polymer concentration less than that at the critical intersection of the liquidus and solidus curves.* Let the mixture initially be at a temperature high enough to allow complete homogeneity. Then when it is cooled, liquid-liquid phase separation, observed visually and indicated by the open circles, will precede crystallization. On further lowering of the temperature, crystallization (also observed visually in this more dilute range) occurs at the same (critical) temperature independent of the proportion of diluent within this range. This constancy is, of course, required by the phase rule since two liquid phases are present in addition to the crystalline phase being formed.

Recalling the previous assertion that efficient fractionation requires liquid-liquid phase separation, we conclude that nitrobenzene and amyl acetate should be satisfactory solvents from which to fractionate polyethylene by successively lowering the temperature and that the better solvent xylene should be avoided for this purpose. The character of the phase diagram may, in fact, be used as a criterion of the efficacy of a given solvent for fractionation (see Chap. VIII, p. 344). If the curve representing the precipitation temperature plotted against concentration rises monotonically, crystalline separation is clearly indicated; if it passes through a maximum at a low concentration, liquid-liquid separation is virtually assured, and the solvent may be assumed to be a satisfactory one to use for fractionation.

The curves of Figs. 133 and 134 may be regarded as plots of solubilities against temperature. It must be borne in mind however, that the dissolved phase is interspersed with the crystalline phase when polymer is present in excess of its solubility limit. Even in the more dilute solutions from which the crystalline polymer may settle out, the "precipitate" will contain some amorphous polymer and diluent. In short, these curves are useful primarily in defining the maximum amount of polymer which may be totally dissolved as a function of the temperature.

3. SWELLING OF NETWORK STRUCTURES

A three-dimensional network polymer such as vulcanized rubber, although incapable of dispersing completely, may nevertheless absorb a large quantity of a suitable liquid with which it is placed in contact.

* The critical concentration at the maximum of the liquid-liquid curve in Fig. 134 occurs at a higher concentration than in the other systems discussed (Fig. 121, p. 547, and Fig. 133) owing, in part at least, to the comparatively low molecular weight of the polyamide.

Swelling occurs under these conditions for the same reason that the solvent mixes spontaneously with an analogous linear polymer to form an ordinary polymer solution; the swollen gel is in fact a solution, although an elastic rather than a viscous one. Thus an opportunity for an increase in entropy is afforded by the added volume of the polymer throughout which the solvent may spread. This mixing tendency, expressed as the entropy of dilution, may be augmented ($\chi_1 < 0$) or diminished ($\chi_1 > 0$) by the heat (or first neighbor interaction free energy) of dilution. As the network is swollen by absorption of solvent, the chains between network junctions are required to assume elongated configurations, and a force akin to the elastic retractive force in rubber consequently develops in opposition to the swelling process. As swelling proceeds, this force increases and the diluting force decreases. Ultimately, a state of equilibrium swelling is reached in which these two forces are in balance.

A close analogy exists between swelling equilibrium and osmotic equilibrium. The elastic reaction of the network structure may be interpreted as a pressure acting on the solution, or swollen gel. In the equilibrium state this pressure is sufficient to increase the chemical potential of the solvent in the solution so that it equals that of the excess solvent surrounding the swollen gel. Thus the network structure performs the multiple role of solute, osmotic membrane, and pressure-generating device.

3a. Theory of Swelling.[27,28]—The free energy change ΔF involved in the mixing of pure solvent with the initially pure, unstrained (i.e., isotropic) polymeric network is conveniently considered to consist of two parts: the ordinary free energy of mixing ΔF_M, and the elastic free energy ΔF_{el} consequential to the expansion of the network structure. Thus we may write

$$\Delta F = \Delta F_M + \Delta F_{el} \tag{33}$$

A suitable expression for ΔF_M may be obtained from Eq. (XII-22), bearing in mind that the number n_2 of polymer molecules is to be equated to zero owing to the absence of individual polymer *molecules* in the network structure. Thus

$$\Delta F_M = kT(n_1 \ln v_1 + \chi_1 n_1 v_2) \tag{34}$$

By analogy with the deformation of rubber, the deformation process during swelling, considered apart from the actual mixing with solvent, must occur without an appreciable change in internal energy of the network structure. Hence ΔF_{el} may be equated to $-T\Delta S_{el}$ where ΔS_{el}, representing the entropy change associated with the change in configuration of the network, is given by Eq. (XI-41). If we let α_s represent the linear deformation factor (see Chap. XI), then by the condition of isotropy $\alpha_x = \alpha_y = \alpha_z = \alpha_s$, and according to Eq. (XI-41)

$$\Delta F_{el} = (kT\nu_e/2)(3\alpha_s^2 - 3 - \ln \alpha_s^3) \tag{35}$$

where ν_e is the effective *number* of chains in the network.

The chemical potential of the solvent in the swollen gel is given by

$$\mu_1 - \mu_1^0 = N(\partial \Delta F_M/\partial n_1)_{T,P} + N(\partial \Delta F_{el}/\partial \alpha_s)_{T,P}(\partial \alpha_s/\partial n_1)_{T,P} \tag{36}$$

where N is Avogadro's number. In order to evaluate $(\partial \alpha_s/\partial n_1)$, we note that

$$\alpha_s^3 = V/V_0$$

where V_0 is the volume of the relaxed network, i.e., the volume occupied by the polymer when the cross-linkages were introduced into the random system (see Chap. XI), and V is the volume of the swollen gel. Ordinarily the cross-linkages will have been introduced in the unswollen polymer. Assuming this to have been the case, V_0 will represent the volume of the unswollen polymer, and $V_0/V = v_2$. Assuming further that mixing occurs without an appreciable change in the total volume of the system (polymer plus solvent)

$$\alpha_s^3 = 1/v_2 = (V_0 + n_1 v_1/N)/V_0 \tag{37}$$

It follows that

$$(\partial \alpha_s/\partial n_1)_{T,P} = v_1/3\alpha_s^2 V_0 N$$

Evaluating the other two derivatives occurring in Eq. (36) by differentiating Eqs. (34) and (35) and expressing ν_e in moles, we obtain[28]

$$\mu_1 - \mu_1^0 = RT[\ln(1 - v_2) + v_2 + \chi_1 v_2^2 + v_1(\nu_e/V_0)(v_2^{1/3} - v_2/2)] \tag{38}$$

The first three terms occurring in the right-hand member of Eq. (38), represent $\partial \Delta F_M/\partial n_1$; they correspond to $\mu_1 - \mu_1^0$ according to Eq. (XII-26) for a polymer of infinite molecular weight (i.e., $x = \infty$). The last member introduces the modification of the chemical potential due to the elastic reaction of the network structure.* The activity a_1

---

* Until recently[28] the last term in the brackets in Eq. (38) was given erroneously as $(v_1\nu_e/V_0)v_2^{1/3}$. This error resulted from the use of incorrect elastic entropy and free energy expressions in which the $\ln \alpha_s^3$ term of Eq. (35) was omitted. This term takes account of the entropy of distribution of the $\nu_e/2$ effective cross-linkages over the volume $V_0\alpha_s^3 = V$.

The treatment given here, like that of rubber elasticity in Chapter XI, is developed for a network in which the ends of the chains are united tetrafunctionally, of the solvent is specified also by Eq. (38) through the relationship $\ln a_1 = (\mu_1 - \mu_1^0)/RT$.

If the chemical potential difference $\mu_1 - \mu_1^0$ calculated according to Eq. (38) is plotted against v_2, it will be found that, owing to the positive contribution of the elastic term (with $\nu_e > 0$), the chemical potential μ_1 exceeds μ_1^0 for the pure solvent for all concentrations below a certain polymer concentration v_{2m}. In other words, the activity a_1 would exceed unity for compositions with $v_2 < v_{2m}$. This region therefore represents an unstable one, which, if somehow formed, would spontaneously exude pure solvent until the concentration in the gel increased to v_{2m}, at which the activity equals unity. The swollen gel would then be in equilibrium with the surrounding pure solvent. Hence, v_{2m}, defined as the concentration (>0) at which the activity of the solvent is unity, or at which $\mu_1 = \mu_1^0$, represents the composition at *swelling equilibrium*. To locate this composition we equate $\mu_1 - \mu_1^0$ of Eq. (38) to zero, obtaining thereby[28]

$$- \left[\ln(1 - v_{2m}) + v_{2m} + \chi_1 v_{2m}^2\right] = v_1(\nu_e/V_0)(v_{2m}^{1/3} - v_{2m}/2) \qquad (39)$$

or, adopting the terminology used in Chapter XI (see Eqs. XI-28 and XI-30′)

$$- \left[\ln(1 - v_{2m}) + v_{2m} + \chi_1 v_{2m}^2\right]$$
$$= (v_1/\bar{v}M_c)(1 - 2M_c/M)(v_{2m}^{1/3} - v_{2m}/2) \qquad (39')$$

where M_c is the molecular weight per cross-linked unit and M is the primary molecular weight. The factor $(1 - 2M_c/M)$, it will be recalled, expresses the correction for network imperfections resulting from chain ends. For a perfect network ($M = \infty$) it reduces to unity. The left-hand member in these equations represents the lowering of the chemical potential owing to mixing of polymer and solvent; that on the right gives the increase from the elastic reaction of the network. The latter corresponds to the increase πv_1 in the chemical potential resulting from an osmotic pressure π at equilibrium.

It is customary to employ the *swelling ratio* q equal to the ratio V/V_0 of the volumes of the swollen and unswollen structures. Thus, $q = 1/v_2$. At swelling equilibrium, we may replace $1/v_{2m}$ by q_m, the subscript m indicating maximum, or equilibrium, swelling. At low degrees of cross-linking, i.e., at large M_c values of 10,000 or more, q_m in a good solvent will exceed ten. Then $v_{2m}^{1/3}$ is considerably greater than $v_{2m}/2$, and we may as a first approximation neglect the latter com-

---

i.e., by conventional cross-linkages. For networks in which the junctions are f-functional, it is necessary merely to replace $v_2/2$ in Eq. (38) with $2v_2/f$.[28]

pared with the former. To a similar approximation the higher terms in the series expansion of the left-hand member of Eq. (39) may be neglected. The swelling equilibrium equation may then be solved for $v_{2m} = 1/q_m$ with the following result:[27]

$$q_m^{5/3} \cong (V_0/\nu_e)(1/2 - \chi_1)/v_1 \qquad (40)$$

or, from Eq. (39')

$$q_m^{5/3} \cong (\bar{v}M_c)(1 - 2M_c/M)^{-1}(1/2 - \chi_1)/v_1 \qquad (40')$$

These simplified relationships offer a clearer insight into the dependence of the equilibrium swelling ratio q_m on the quality of the solvent as expressed by χ_1, and on the extent of cross-linking. Because of the nature of the approximations introduced to obtain Eqs. (40) and (40'), their use as quantitative expressions must be limited to networks of very low degrees of cross-linking in good solvents.

It has been shown in Chapter XI that the force of retraction in a stretched network structure depends also on the degree of cross-linking. It is possible therefore to eliminate the structure parameter (ν_e/V_0) by combining the elasticity and the swelling equations, and thus to arrive at a relationship between the equilibrium swelling ratio and the force of retraction at an extension α (not to be confused with the swelling factor α_s). In this manner we obtain from Eq. (XI-44)* and Eq. (39)

$$\tau_\alpha = - RT(\alpha - 1/\alpha^2)[\ln(1 - v_{2m}) + v_{2m} + \chi_1 v_{2m}^2]/v_1(v_{2m}^{1/3} - v_{2m}/2) \qquad (41)$$

where T refers to the temperature of the stress measurement. If the equilibrium swelling is very large $(v_{2m} \ll 1)$, we may introduce approximations corresponding to those which yielded Eq. (40). Then

$$\tau_\alpha \cong RT(\alpha - 1/\alpha^2)(1/2 - \chi_1)/v_1 q_m^{5/3} \qquad (42)$$

This equation calls attention to the well-established inverse relationship between the degree of equilibrium swelling of a series of rubber vulcanizates in a given solvent and the forces of retraction, or "moduli," which they exhibit on stretching. The indicated approximate dependence of q_m on the inverse three-fifths power of the "modulus" has been confirmed.[29,30]

In using Eq. (XI-44) to derive Eq. (41), we have, in effect, accepted the former as a valid representation of the dependence of the force of retraction on the extension. Experiments cited in Chapter XI showed

_____

* The total volume V occurring in Eq. (XI-44) is to be identified with the present V_0 inasmuch as the volume was assumed to remain constant during elastic deformation.

this theoretical relationship to be disturbingly inaccurate. The resulting quantitative limitations of Eqs. (41) and (42) must not be overlooked. Better agreement with experiment could be expected through the use of the semiempirical stress-strain relation, Eq. (XI-50), instead of Eq. (XI-44) in the derivation of Eq. (41).

3b. Experimental Results on the Swelling of Nonionic Network Systems.—The degree of swelling observed at equilibrium in a good solvent invariably decreases with increasing degrees of cross-linking.[29,30,31,32] It also decreases with increase in the primary molecular weight M as should be expected according to Eqs. (39') and (40'); as a matter of fact, quantitative proportionality between $q_m^{5/3}$ and the network imperfection factor $(1-2M_c/M)$ has been verified.[29] The dependence of the equilibrium swelling ratio on the network structure need not be pursued further. Instead, we shall focus the discussion on the connection between q_m and the force of retraction τ. The relationship of the latter quantity to the network structure, as embodied in ν_e/V_0 or in M_c and M, was discussed in detail in Chapter XI. Hence the relationship between q_m and the structure is implicit in the discussion of the quantitative connection between q_m and τ, and its separate treatment would represent an unnecessary duplication.

The results shown in Fig. 135 for a series of multilinked polyamides[33] illustrate the relationship between the equilibrium swelling ratio and the equilibrium force of retraction τ_α for the stretched unswollen specimen. Swelling measurements were made in m-cresol at 30°C; and the forces of retraction were measured on the unswollen polymers at 241°C at the several extension ratios $\alpha = 1.4$, 2.0, and 3.0 as indicated. The range in the degree of cross-linking (ν_e) covered by these data is about sixfold. The log-log plot is suggested by the approximate Eq. (42). Although the points describe straight lines within experimental error in accordance with this relation, the negative slopes are somewhat greater than the value 5/3 it prescribes. The lines drawn actually are slightly curved, for they have been calculated from the more accurate relationship given by Eq. (41) rather than from Eq. (42).

Because of the previously mentioned inadequacy of the function $\alpha - 1/\alpha^2$, a different value for the parameter χ_1 is required for the set of points (Fig. 135) at each elongation α. These values are -0.90, -0.73, and -0.56 for $\alpha = 1.4$, 2.0, and 3.0, respectively. If the function $\alpha - 1/\alpha^2$ were replaced by an empirical representation of the shape of the stress-strain curve, a single value of χ_1 would suffice to represent all of the data within experimental error. This limitation of Eq. (41) relates to an unexplained feature of the stress-strain curve and is

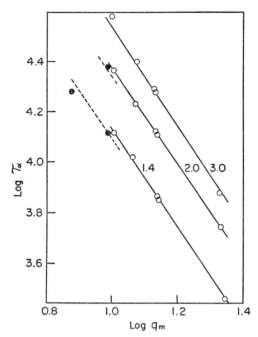

FIG. 135.—The relationship between the equilibrium retractive force τ_α (in lbs./in.²) at 241°C for various multilinked poly-(ϵ-caproamides) at the extensions (α) indicated, and their equilibrium swelling ratios q_m in m-cresol at 30°C. O, tetralinked polymers; ●, octalinked polymers. The lines have been calculated according to Eq. (41), with appropriate revision for the octafunctional case (broken lines), an arbitrary value being assigned to the parameter χ_1 for each elongation. (Schaefgen and Flory.[33])

not peculiar to the swelling behavior, which is of primary concern here. Dashed curves shown in Fig. 135 for the octafunctional network polymers (filled circles) were calculated from the suitably modified form of Eq. (41) which applies in this case,* using the same value of χ_1.

The experimental points in Fig. 135 for each elongation are seen to be in good agreement with the theoretically predicted correlation between q_m and τ_α. These results, which are supported by similar comparisons on vulcanized rubbers,[29,30,34] show that the elastic reaction to the isotropic dilation accompanying swelling and to that induced by stretching without (appreciable) dilation are uniquely related, and hence that they are of common origin. That is to say, the same network is responsible for both. The often-postulated "association bonds" between neighboring chains, which might have been assumed to contribute to the elastic retractive force, must surely be inoperative as permanent junctions in the presence of the swelling agent. The correlation between q_m and τ_α confirms the view that such bonds likewise do not contribute to rubberlike elasticity, at equilibrium at any rate.

According to these results, swelling measurements on a series of

* See p. 578, footnote.

chemically related network structures in the same solvent may be used to ascertain the degrees of cross-linking, at least in a relative way, in the different structures. The same information may, of course, be obtained from elasticity measurements. Equilibrium swelling often is easier to measure, however, and for this reason it may be preferred.

The degree of swelling of the same network structure in different solvents provides a convenient index of the solvent power of each solvent for polymers of the given chemical type. The structure factor ν_e/V_0 (see Eq. 39) is fixed throughout such a series of measurements, and the swelling $q_m = 1/v_{2m}$ depends only on the known molar volume v_1 of the solvent and on the unknown parameter χ_1. If a value can be assigned to χ_1 for one solvent, possibly on the basis of thermodynamic measurements on solutions of the linear polymer analog, then an apparent value of ν_e/V_0 can be calculated and χ_1 may be deduced from q_m for any other solvent. This method has found wide use.[30,32,35] While the absolute values of the parameters thus obtained may be open to question, there is little doubt that the method should classify various solvents in proper order with respect to polymers of the given type.

In the application of the swelling method for determining solvent-polymer interactions it is obviously desirable to eliminate the necessity of the calibration mentioned in the preceding paragraph and involving reliance on an interaction parameter χ_1 deduced from thermodynamic measurements on a linear polymer analog in a selected solvent. Independent determination of ν_e/V_0 is to be preferred. Use of the literal value of ν_e/V_0 established through quantitative synthesis would be likely to lead to error on account of contributions to the elastic response of the network arising from imperfections such as the network entanglements discussed in Chapter XI. There it was shown that the observed forces of retraction tend always to be somewhat greater than those calculated from the number of actual effective chains ν_e in the structure. Hence, for the purpose of evaluating the interaction parameter, a corrected ν_e which takes account of such deviations is needed. Lacking a method for introducing this correction in any other way, one may resort to elasticity measurements[30] of τ_α, from which an apparent value of ν_e/V_0 may be calculated according to Eq. (XI-44). This apparent value of ν_e/V_0 is still subject to ambiguity arising from deviations from the $\alpha - 1/\alpha^2$ function; i.e., the result obtained will depend to some extent on the value of α at which the force of retraction is measured. Gee[36] has shown that the stress-strain behavior of swollen networks is in much closer accord with this function; hence an apparent value of ν_e/V_0 may be determined with less ambiguity from the elastic response of the swollen network. (See Eq. B-4 in Appendix B of Chap. XI.)

Gee[30] has applied this method to the determination of the interaction parameters χ_1 for natural rubber in various solvents. Several rubber vulcanizates were used. The effective value of ν_e/V_0 for each was determined by measuring its extension under a fixed load when swollen in petroleum ether. Samples were then swollen to equilibrium in other solvents, and χ_1 was calculated from the swelling ratio q_m in each. The mean values of χ_1 for the several vulcanizates in each solvent are presented in Table XXXVI,* where they are compared with the χ_1's calculated (Eq. XII-30) from vapor pressure measurements on solutions of unvulcanized rubber in some of the same solvents. The agreement is by no means spectacular, though perhaps no worse than the experimental error in the vapor pressure method.

TABLE XXXVI.—SOLVENT INTERACTION WITH NATURAL RUBBER[30]

Solvent	χ_1 from q_m	χ_1 from vapor pressure
Carbon tetrachloride	0.29	0.28
Chloroform	.34	.37
Carbon disulfide	.425	.49
Benzene	.395	.43
Toluene	.36	.42
Petroleum ether	.54	.43
n-Propyl acetate	.62	
Ethyl acetate	.78	
Methyl ethyl ketone	.94	
Acetone	1.37	

3c. Swelling of Ionic Networks.[37,38]—If the polymer chains making up the network contain ionizable groups, the swelling forces may be greatly increased as a result of the localization of charges on the polymer chains. Ion exchange resins are of this type, but we shall be mainly concerned with ionic networks in which the density of cross-linking is much lower than is common in them. Cross-linked poly-(acrylic acid)[39] and poly-(methacrylic acid)[39,40] afford appropriate examples for the present discussion. When neutralized with sodium hydroxide, either partially or totally, the negatively charged carboxyl groups attached to the polymer chains set up an electrostatic repulsion

---

* Gee[30] used the earlier version of Eq. (39) in which the term $v_{2m}/2$ was omitted from the right-hand member. His values for χ_1, particularly for the poorer solvents, are somewhat too large, therefore.

which tends to expand the network. However, the exceedingly large electrostatic repulsions which would prevail if the fixed carboxylate ions were the only ions present may never be realized because of the inevitable presence of other ions—the sodium ions, for example, and those of other electrolytes which may be present (including ions from the solvent, e.g., H^+ and OH^- for water). These ions, by screening the fixed charges, reduce the electrostatic repulsion tremendously as compared with what it would be in their absence.

The exchange of ions and solvent between a swollen ionic network and the surrounding electrolyte is represented in Fig. 136, where the fixed ion is taken to be a cation. It is apparent that the equilibrium between the swollen ionic gel and its surroundings closely resembles Donnan membrane equilibria.[41] The polymer acts as its own membrane preventing the charged substituents, which are distributed essentially at random through the gel much as they would be in an ordinary solution, from diffusing into the outer solution. The swelling force resulting from the presence of these fixed charges may be identified with the swelling pressure, or net osmotic pressure, across the semipermeable membrane in a typical Donnan equilibrium. The quantitative treatment of this force

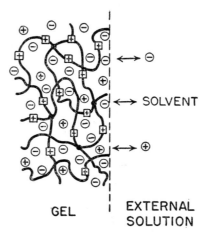

Fig. 136.—Diagram of swollen ionic gel in equilibrium with electrolyte solution. Fixed charges are represented by ⊞ .

may be carried out in either of two ways. We first introduce the conditions for equilibrium between the mobile ion species inside and outside the gel in either case. According to one of the alternative procedures, the electric potential difference between the interior of the gel and its surroundings is calculated through consideration of the ionic equilibria. From the charge density in the gel and its potential relative to the outer solution, the Coulombic energy is easily computed, and from the latter the expansive force may be found. In the second method it is observed that the concentration of *mobile* ions will always be greater in the gel than outside because, in qualitative terms, of the attracting power of the fixed charges. Consequently, the osmotic pressure of the solution inside will exceed that of the external solution. The expansive force may be equated to this difference in osmotic pressures for the two solutions. The two methods yield identical results. The latter one is used here.[38]

Suppose the polymer contains substituents which dissociate completely in the presence of the solvent to yield anions A^{z-} of charge $z_-\epsilon$, where ϵ is the electronic charge. The proportion of these ionizing groups is such as to leave an average residual positive charge $i\epsilon$ per structural unit of the polymer, where i is the degree of ionization multiplied by the valency of the cationic group attached to the polymer chain. Let the concentration of these fixed charges in the swollen gel at equilibrium be ic_2 faradays per unit volume; thus $c_2 = v_2/\mathrm{v}_u$, where v_u is the molar volume of a structual unit. Suppose further that the volume of the external solution with which the cross-linked polyelectrolyte is in equilibrium is very much greater than the volume of the gel and that it contains a strong electrolyte $M_{\nu_+}A_{\nu_-}$ at the concentration c_s^*. This electrolyte is completely dissociated into ν_+ cations M^{z+} and ν_- anions A^{z-}. Necessarily, $\nu_+ z_+ = \nu_- z_-$. Without sacrifice of generality, the anion of the strong electrolyte of the external solution may be taken to be identical with that of the polyelectrolyte gel.

The gel initially contains no M^{z+} cations; hence some of them will tend to diffuse from the external solution into the gel in an effort to establish, as it were, an individual ion equilibrium between the two phases. However, even an immeasurably minute excess of positive ions in the gel would produce a large positive potential relative to the surrounding solution. The potential developed as a result of the migrating tendency of the cation consequently will attract anions A^{z-} from the external electrolyte, and they will be carried into the gel along with the cations. Owing to the large electrostatic potential which would otherwise develop, the anions will migrate in such quantity as to be in exact equivalence with the migrating cations, to an approximation which in the stoichiometric sense is altogether negligible. Having adopted this condition of electroneutrality, we shall have no further need for the potential difference (i.e., membrane potential) which evidently must exist between the gel and the external solution. It suffices to observe that ultimately a compromise will be reached between the unequal diffusion forces acting on the cations and on the anions, and that at this state of equilibrium the gel will have acquired a concentration c_s of the mobile electrolyte. The concentrations of mobile cations and anions in the gel, including among the latter those contributed by dissociation from the polymer, will be

$$\left. \begin{array}{l} c_+ = \nu_+ c_s \\ c_- = \nu_- c_s + ic_2/z_- \end{array} \right\} \qquad (43)$$

The total mobile ion concentration $(c_+ + c_-)$ inside the gel at equilibrium will inevitably exceed that in the external solution, $c_+^* + c_-^* = \nu c_s^*$ where $\nu = \nu_+ + \nu_-$. This must result in an osmotic pressure difference which tends to drive solvent into the gel from the less concentrated external solution. (We neglect for the moment the osmotic effects of polymer itself.) The osmotic pressure π_i arising from the difference in mobile ion concentrations will be given approximately (see Appendix B), assuming the solutions to be dilute, by

$$\pi_i = RT(c_+ + c_- - c_+^* - c_-^*)$$

$$= RT[ic_2/z_- - \nu(c_s^* - c_s)] \tag{44}$$

But for any arbitrary polymer concentration c_2 there will be another osmotic pressure π_0 due to the previously considered polymer-solvent interaction and to the associated elastic reaction of the network. Recalling the general relationship $\pi = -(\mu_1 - \mu_1^0)/v_1$, we may calculate π_0 from Eq. (38). At equilibrium the total osmotic pressure arising from the effects of all solutes must be zero, i.e., $\pi_i = -\pi_0$. Hence from Eqs. (44) and (38) we have at equilibrium

$$ic_{2m}/z_- - \nu(c_s^* - c_s) = (1/v_1)[\ln(1 - v_{2m}) + v_{2m} + \chi_1 v_{2m}^2]$$

$$+ (\nu_e/V_0)(v_{2m}^{1/3} - v_{2m}/2) \tag{45}$$

The use of two different polymer concentrations, c_2 and v_2, in preceding equations results from deference to custom in expressing electrolyte concentrations in molarities (or in molalities). Confusion might have been minimized in Eq. (45) by substituting $c_{2m} = v_{2m}/v_u$.

It remains to evaluate the quantity $c_s^* - c_s$. Since an explicit general solution is not to be had, we resort to the consideration of special cases. First, suppose that the external electrolyte concentration c_s^* is very small compared with the concentration ic_2/z_- of the gegen ions belonging to the polymer and occurring in the gel. Then the second term in the left-hand member of Eq. (45) may be neglected in comparison with the first. Furthermore, the very large ionic osmotic pressures developed in such cases will cause v_{2m} to be very small, thus justifying adoption of the dilute solution approximations (see, for example, Eq. 40) for the right-hand member. The equilibrium relation reduces in this case to

$$iv_{2m}/z_- v_u \cong (\nu_e/V_0)v_{2m}^{1/3} - (1/2 - \chi_1)v_{2m}^2/v_1 \tag{46}$$

If $v_u = 100$, $v_1 = 100$, $z_- = 1$, $i > 0.1$, and $1/2 - \chi_1 \leqq 0.2$, the last term in Eq. (46) may be neglected, giving to a further approximation

$$q_m^{2/3} \cong (i/z_- v_u)/(\nu_e/V_0) \tag{47}$$

Since $v_u \ll (V_0/v_e)$ for a loosely cross-linked gel, this equation predicts a very large swelling ratio at appreciable degrees of ionization i in the absence of a comparable concentration of other electrolyte.

Approximate results obtained by Katchalsky, Lifson, and Eisenberg[40] on the swelling in water of poly-(methacrylic acid) gels cross-linked with divinylbenzene are shown as a function of the degree of neutralization in Fig. 137. The marked effect of even small degrees of neutralization is in accord with theory. The rapid approach of q_m to a maximum with neutralization indicates the not unexpected approach to full extension of the chains of the network. The elastic theory of Chapter XI, from which the swelling relationships given here have been derived, was based on the Gaussian approximation for the representation of the chain length distribution. This approximation is unacceptable for extensions exceeding about half the fully extended length; hence quantitative treatment of swelling of highly charged polyelectrolyte gels in the absence of added electrolytes requires use of the chain length distribution function involving the inverse Langevin function[40,42] (see Chap. X, Appendix B).

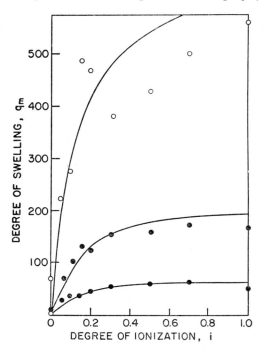

FIG. 137.—Equilibrium swelling ratio q_m of poly-(methacrylic acid) gels prepared by copolymerizing methacrylic acid with 1, 2, and 4 percent (upper, middle, and lower curves, respectively) of divinylbenzene plotted against degree of neutralization i with sodium hydroxide. (Katchalsky, Lifson, and Eisenberg.[40])

It is interesting to note that Breitenbach and Karlinger,[39] who made similar observations on the swelling of poly-(methacrylic acid) gels cross-linked with smaller proportions of divinylbenzene, found that such gels disintegrated at higher degrees of neutralization. Evidently the osmotic (or the electrostatic) forces may be great enough to induce rupture of carbon-carbon bonds. They also observed a marked deswelling action by as little as 0.01 N NaCl.

In the opposite case to that considered above, $c_s^* > ic_2$ and the difference in concentration $c_s^* - c_s$ of the mobile electrolyte inside and outside the gel may be comparable in magnitude to the concentration ic_2/z_- of counter-anions. Hence the ion osmotic pressure π_i is greatly reduced. Calculation of $c_s^* - c_s$ for this case (see Appendix B) gives for the osmotic pressure due to the mobile ions

$$\pi_i \cong RT(ic_2)^2/2wvc_s^* \tag{48}$$

where c_s^* is the electrolyte concentration in the external solution and $w = z_+z_-$ is the valency factor of the electrolyte. Eq. (48) may be rewritten as follows through the use of the ionic strength S^* defined by $S^* = vwc_s^*/2$

$$\pi_i \cong RT(ic_2)^2/4S^* \tag{48'}$$

Introducing the equilibrium condition $\pi_i = -\pi_0$ and using Eq. (38) as before to express $-\pi_0$, we have at swelling equilibrium

$$(iv_{2m}/v_u)^2/4S^* \cong (1/v_1)\left[\ln(1 - v_{2m}) + v_{2m} + \chi_1 v_{2m}^2\right]$$
$$+ (\nu_e/V_0)(v_{2m}^{1/3} - v_{2m}/2) \tag{49}$$

In the dilute polymer solution approximation, which may not necessarily be appropriate if the concentration of electrolyte in the external solution is relatively large

$$q_m^{5/3} \cong \left[(i/2v_u S^{*1/2})^2 + (1/2 - \chi_1)/v_1\right]/(\nu_e/V_0) \tag{50}$$

Observing that i/v_u is the concentration of fixed charge referred to the unswollen network, we see that the variable of importance is the ratio of this concentration to the square root of the ionic strength in the external solution. The five-thirds power of the swelling ratio should therefore increase as the square of the fixed charge and as the reciprocal of the ionic strength S^*, or of the concentration c_s^*, in the external solution. Results which seem to confirm the latter of these deductions have been obtained by Vermaas and Hermans[37] on cellulose xanthate gels. In the limit of a large ratio of the ionic strength compared with the fixed charge concentration (i/v_u) referred to the unswollen gel, the ionic terms in Eqs. (49) and (50) vanish, and the swelling relationship reverts to that for the uncharged network. From a physical point of view, this may be interpreted alternatively as a suppression of Coulombic repulsions by the efficient shielding of the fixed charges provided by the mobile electrolyte ions, or as a suppression of the unbalance in mobile ion concentrations by the large excess of simple electrolyte.

APPENDIX A

Calculation of the Binodial in the Phase Diagram for the Two-Component System Comprising a Solvent and a Single Polymer Species.[2]—Substituting Eq. (XII-26) in the first of the conditions (1) and Eq. (XII-32) in the second, we obtain

$$\ln\left[(1 - v_2')/(1 - v_2)\right] + (v_2' - v_2)(1 - 1/x) + \chi_1(v_2'^2 - v_2^2) = 0 \qquad \text{(A-1)}$$

and

$$(1/x) \ln (v_2'/v_2) + (v_2' - v_2)(1 - 1/x)$$
$$- \chi_1(v_2' - v_2)\left[2 - (v_2' + v_2)\right] = 0 \qquad \text{(A-2)}$$

The latter equation rearranges to Eq. (10), with the insertion of $\gamma = v_2'/v_2$. Elimination of χ_1 from Eqs. (A-1) and (A-2) leads to

$$\left[1 - (v_2' + v_2)/2\right]\left[\ln (1 - v_2') - \ln (1 - v_2)\right] + (v_2' - v_2)$$
$$= - \left[(v_2' + v_2)/2x\right] \ln (v_2'/v_2) + (v_2' - v_2)/x \qquad \text{(A-3)}$$

Series expansion of the logarithms occurring on the left-hand side of Eq. (A-3) and combination of terms gives for the left-hand member of the equation

$$\left[(v_2' - v_2)^3/2\right]\{1/2\cdot3 + (2v_2' + 2v_2)/3\cdot4$$
$$+ (3v_2'^2 + 4v_2'v_2 + 3v_2^2)/4\cdot5 + \cdots \}$$

The quantity occurring in braces is very nearly equal to $(1/6)(1 - v_2' - v_2)^{-1}$ provided v_2' is small. The fractional error in the composition calculated using this approximation is of the order of $v_2'^4 x/50$. Substituting this result in Eq. (A-3)

$$(v_2' - v_2)^3/12(1 - v_2' - v_2) = \left[(v_2' + v_2)/2x\right] \ln (v_2'/v_2) - (v_2' - v_2)/x$$

which simplifies with the introduction of $\gamma = v_2'/v_2$ to

$$(\gamma - 1)^3 v_2^2/12\left[1 - v_2(\gamma + 1)\right] = \left[(\gamma + 1)/2x\right] \ln \gamma - (\gamma - 1)/x \qquad \text{(A-4)}$$

Solution of this equation, a quadratic in v_2, yields Eq. (9).

APPENDIX B

Donnan-Type Equilibria in Polyelectrolyte Gels.—In a somewhat more rigorous fashion we consider the reduction of the chemical potential of the solvent in the swollen gel to be separable into three terms which severally represent the changes due to the mixing of polymer and solvent, to the mixing with the mobile ionic constituents, and to the elastic deformation of the network. Symbolically

$$\mu_1 - \mu_1^0 = (\Delta\mu_1)_p + (\Delta\mu_1)_i + (\Delta\mu_1)_{el} \tag{B-1}$$

In essence, we assume that the gel solution is sufficiently dilute to justify the assumption that the first two contributions enter additively. The first and third are given by Eq. (38). Proceeding at once to the case of swelling equilibrium, we observe that fulfillment of the condition $\mu_1 = \mu_1^*$ is required, where μ_1^* is the chemical potential in the external solution. Inserting this condition in Eq. (B-1) and writing $(\Delta\mu_1^*)_i$ for $\mu_1^* - \mu_1^0$,

$$(\Delta\mu_1^*)_i - (\Delta\mu_1)_i = (\Delta\mu_1)_p + (\Delta\mu_1)_{el} \tag{B-2}$$

where the right-hand member is given by Eq. (38). In general

$$\Delta\mu_1^* \equiv \mu_1^* - \mu_1^0 = gRT \ln N_1^*$$

where g is the osmotic coefficient and N_1^* the mole fraction of solvent in the external solution. In dilute solutions $g \cong 1$ and $\ln N_1^* \cong -v_1 \sum_j c_j^*$ where the summation combines the concentrations of all (ionic) solute species. Hence

$$(\Delta\overset{*}{\mu_1})_i \cong - v_1 RT \sum c_j^*$$

Similarly, within the gel

$$(\Delta\mu_1)_i \cong - v_1 RT \sum_j c_j$$

where the summation includes all *mobile* solute species only. Substitution of these expressions together with Eq. (38) in Eq. (B-2) gives

$$RT \sum_j (c_j - c_j^*) = (RT/v_1)\left[\ln(1 - v_{2m}) + v_{2m} + \chi_1 v_{2m}^2\right]$$
$$+ (RTv_e/V_0)(v_{2m}^{1/3} - v_{2m}/2) \tag{B-3}$$

Turning now to the case of equilibrium with an infinite external solution containing c_s^* moles of electrolyte $M_{\nu_+}A_{\nu_-}$ per liter, we note that Eq. (B-3) is then equivalent to Eq. (45). As a further condition of equilibrium, it is required that the activity of this electrolyte be the same inside and outside the gel. The activity of the electrolyte being equal to the product of the activities of the individual ions into which it dissociates, this condition may be stated as follows

$$a_+^{\nu}a_-^{\nu} = a_+^{*\nu_+}a_-^{*\nu_-}$$

or

$$(a_+/a_+^*)^{\nu_+} = (a_-^*/a_-)^{\nu_-} \tag{B-4}$$

Making the customary approximation that activities are equal to concentrations, we obtain by inserting Eqs. (43)*

$$(c_s/c_s{}^*)^{\nu_+} = [c_s{}^*/(c_s + ic_2/\nu_-z_-)]^{\nu_-} \tag{B-5}$$

Since we are interested in the excess mobile electrolyte concentration, we introduce the variable η defined by

$$\eta = (c_s{}^* - c_s)/c_s{}^*$$

Eq. (B-5) may then be written

$$(1 - \eta)^{1+\nu_+/\nu_-} + Y(1 - \eta)^{\nu_+/\nu_-} - 1 = 0 \tag{B-6}$$

where

$$Y = ic_2/\nu_-z_-c_s{}^* \tag{B-7}$$

The left-hand member of Eq. (B-3), which represents the swelling, or osmotic, pressure π_i arising from the difference in mobile ion concentration inside and outside the gel, may be written in accordance with Eq. (44) as follows:

$$\pi_i = RT(ic_2/z_- - \nu\eta c_s{}^*) \tag{B-8}$$

Hence evaluation of η by means of Eq. (B-6) would allow specification of π_i, which could then be inserted in the swelling equilibrium equation (B-3). Unfortunately, the key equation (B-6) is not amenable to general solution, and we are forced to resort to consideration of special cases.

If $Y \gg 1$, corresponding to a small external electrolyte concentration $c_s{}^*$ compared with the concentration ic_2/z_- of gegen ions belonging to the polymer, and if we further restrict ourselves to the case of a binary electrolyte for which $z_+ = z_- = z$, and consequently $\nu_+ = \nu_- = 1$ and $\nu = 2$, then the appropriate series expansion of Eq. (B-6) is

$$\eta = 1 - 1/Y + 1/Y^3 - 2/Y^5 + \cdots \tag{B-9}$$

By substitution in Eq. (B-8)

$$\pi_i = RT\{(ic_2/z) - 2c_s{}^*[1 - zc_s{}^*/ic_2 + (zc_s{}^*/ic_2)^3$$
$$- 2(zc_s{}^*/ic_2)^5 + \cdots]\} \tag{B-10}$$

Substitution of Eq. (B-10) for the left-hand member of Eq. (B-3) gives the complete equation for this case. When $c_s{}^* \ll ic_2/z$, it may be reduced to Eq. (46).

---

* The effect of the pressure on the activities of the ions is disregarded here. It is readily shown to be of negligible importance, even for swelling pressures up to many atmospheres.

In the other case $Y \leqq 1$, corresponding to a comparable or larger external electrolyte concentration c_s^* compared with $ic_2/z_-\nu_-$, the appropriate series expansion is

$$\eta = (1/\nu z_-)(ic_2/c_s^*) - (1/2w\nu^2)(ic_2/c_s^*)^2$$
$$- (1/3w^2\nu^3)(z_+ - z_-)(ic_2/c_s^*)^3$$
$$- (1/8w^3\nu^4)(2z_+^2 - 5z_-z_+ + 2z_-^2)(ic_2/c_s^*)^4 + \cdots \qquad \text{(B-11)}$$

where $\nu = \nu_+ + \nu_-$, and $w = z_+z_-$ is the valency factor (z_+ is not necessarily equal to z_- here). Substituting in Eq. (B-8) and introducing the ionic strength $S^* = w\nu c_s^*/2$

$$\pi_i = RT(i^2c_2^2/4S^*)[1 + (z_+ - z_-)ic_2/3S^*$$
$$+ (2z_+^2 - 5z_-z_+ + 2z_-^2)(ic_2)^2/16S^{*2} + \cdots] \qquad \text{(B-12)}^{38}$$

A relationship equivalent to this one but including only the first two terms in the series was obtained by Vermaas and Hermans.[37] It will be observed that if $z_+ = z_-$, the second term in the series vanishes, but not the third. In any event, if ic_2/S^* is less than unity, Eq. (48) is a valid approximation.

REFERENCES

1. P. J. Flory, *J. Chem. Phys.*, **10**, 51 (1942).
2. P. J. Flory, *J. Chem. Phys.*, **12**, 425 (1944).
3. A. R. Shultz and P. J. Flory, *J. Am. Chem. Soc.*, **74**, 4760 (1952).
4. W. R. Krigbaum, unpublished.
5. A. R. Shultz, Dissertation, Cornell University, 1953.
6. H. Tompa, *Trans. Faraday Soc.*, **45**, 1142 (1949); C. H. Bamford and H. Tompa, *ibid.*, **46**, 310 (1950).
7. R. L. Scott, *J. Chem. Phys.*, **17**, 268 (1949).
8. H. Tompa, private communication.
9. H. Tompa, *Trans. Faraday Soc.*, **46**, 970 (1950).
10. R. L. Scott, *J. Chem. Phys.*, **17**, 279 (1949).
11. A. Dobry and F. Boyer-Kawenoki, *J. Polymer Sci.*, **2**, 90 (1947).
12. G. V. Schulz, *Z. physik. Chem.*, **B46**, 137 (1940); *ibid.*, **B47**, 155 (1940).
13. E. P. H. Meibohm and A. F. Smith, *J. Polymer Sci.*, **7**, 449 (1951).
14. K. Hess and H. Kiessig, *Z. physik. Chem.*, **193**, 196 (1944). I. Fankuchen and H. Mark, *J. Applied Phys.*, **15**, 364 (1944).
15. C. S. Fuller and W. O. Baker, *J. Chem. Education*, **20**, 3 (1943); W. O. Baker, C. S. Fuller, and N. R. Pape, *J. Am. Chem. Soc.*, **64**, 776 (1942).
16. C. W. Bunn and T. C. Alcock, *Trans. Faraday Soc.*, **41**, 317 (1945).
17. L. Mandelkern and P. J. Flory, to be published.
18. L. A. Wood and N. Bekkedahl, *J. Applied Phys.*, **17**, 362 (1946).
19. P. J. Flory, L. Mandelkern, and H. K. Hall, *J. Am. Chem. Soc.*, **73**, 2532 (1951).

20. L. Mandelkern, R. R. Garrett, and P. J. Flory, *J. Am. Chem. Soc.*, **74**, 3949 (1952).

21. P. J. Flory, *J. Chem. Phys.*, **17**, 223 (1949); *ibid.*, **15**, 684 (1947).

22. R. D. Evans, H. R. Mighton, and P. J. Flory, *J. Am. Chem. Soc.*, **72**, 2018 (1950).

23. L. Mandelkern and P. J. Flory, *J. Am. Chem. Soc.*, **73**, 3206 (1951).

24. A. M. Bueche, *J. Am. Chem. Soc.*, **74**, 65 (1952); J. D. Hoffman, *ibid.*, **74**, 1696 (1952).

25. J. T. Maynard and W. E. Mochel, private communication.

26. R. B. Richards, *Trans. Faraday Soc.*, **42**, 10 (1946); H. C. Raine, R. B. Richards, and H. Ryder, *ibid.*, **41**, 56 (1945).

27. P. J. Flory and J. Rehner, Jr., *J. Chem. Phys.*, **11**, 521 (1943).

28. P. J. Flory, *J. Chem. Phys.*, **18**, 108 (1950).

29. P. J. Flory, *Ind. Eng. Chem.*, **38**, 417 (1946); *Chem. Revs.*, **35**, 51 (1944).

30. G. Gee, *Trans. Faraday Soc.*, **42B**, 33 (1946); *ibid.*, **42**, 585 (1946).

31. H. Staudinger, W. Heuer, and E. Husemann, *Trans. Faraday Soc.*, **32**, 323 (1936).

32. R. F. Boyer and R. S. Spencer, *J. Polymer Sci.*, **3**, 97 (1948).

33. J. R. Schaefgen and P. J. Flory, *J. Am. Chem. Soc.*, **72**, 689 (1950).

34. P. J. Flory, N. Rabjohn, and M. C. Shaffer, *J. Polymer Sci.*, **4**, 225 (1949).

35. G. S. Whitby, A. B. A. Evans, and D. S. Pasternack, *Trans. Faraday Soc.*, **38**, 269 (1942); G. Gee, *ibid.*, **38**, 418 (1942); **40**, 468 (1944).

36. G. Gee, *Trans. Faraday Soc.*, **42**, 585 (1946).

37. D. Vermaas and J. J. Hermans, *Rec. trav. chim.*, **67**, 983 (1948).

38. P. J. Flory, unpublished.

39. J. W. Breitenbach and H. Karlinger, *Monatsh.*, **80**, 312 (1949); W. Kuhn and B. Hargitay, *Z. Elektrochem.*, **55**, 490 (1951).

40. A. Katchalsky, S. Lifson, and H. Eisenberg, *J. Polymer Sci.*, **7**, 571 (1951); **8**, 476 (1952).

41. F. G. Donnan and E. A. Guggenheim, *Z. physik. Chem.*, **A162**, 346 (1932); F. H. MacDougall, *Thermodynamics and Chemistry*, 3d ed. (John Wiley and Sons, New York, 1939), p. 329.

42. A. Katchalsky, O. Künzle, and W. Kuhn, *J. Polymer Sci.*, **5**, 283 (1950).

CHAPTER XIV

Configurational and Frictional

Properties of the Polymer Molecule

in Dilute Solution

THE single polymer molecule in a dilute solution is subject to the osmotic action of the surrounding solvent, which tends to swell it to a larger average size than it would otherwise assume. A close parallel exists between this molecular expansion and the swelling of a macroscopic three-dimensional network, discussed at the close of the preceding chapter. Indeed, the single molecule may quite properly be regarded as a submicroscopic prototype of the latter. The polymer molecule consists of a rather large number of units, or segments, all of which are connected together in a single structure, although not in the form of a cross-linked network. Nevertheless, the single polymer molecule is expanded to a larger configuration in dilute solution much as a chain of the network is extended by swelling. To be sure, the segment density within the domain of the single molecule is lower than that in a typical swollen network, and for this reason the osmotic forces acting on it are smaller. Qualitatively, however, the two situations are strikingly similar.

As the polymer molecule is swollen by the osmotic action of the solvent, the chain molecule is spread to a less probable configuration. An elastic reaction consequently develops, like that induced on stretching rubber or in the isotropic swelling of a gel. At equilibrium, this elastic force is in balance with the osmotic forces which tend to swell the molecule. Recalling an analogy used in the preceding chapter, the single molecule may be looked upon as a tiny thermodynamic system consisting of a cloud of segments confined within an imaginary elastic membrane permeable to solvent. Swelling equilibrium will obtain when the osmotic pressure (which decreases as the system is diluted) is equaled by the pressure generated in the membrane as a result of its deformation.   Incorporation of ionizable groups in the polymer molecule leads to effects paralleling those encountered in polyelectrolyte gels.   In particular, the size of the molecule may be greatly increased by the charge, provided the concentration of other electrolytes is small.

In the present chapter we shall be concerned with quantitative treatment of the swelling action of the solvent on the polymer molecule in infinitely dilute solution, and in particular with the factor α by which the linear dimensions of the molecule are altered as a consequence thereof.   The frictional characteristics of polymer molecules in dilute solution, as manifested in solution viscosities, sedimentation velocities, and diffusion rates, depend directly on the size of the molecular domain. Hence these properties are intimately related to the molecular configuration, including the factor α.   It is for this reason that treatment of intramolecular thermodynamic interaction has been reserved for the present chapter, where it may be presented in conjunction with the discussion of intrinsic viscosity and related subjects.

1. INTRAMOLECULAR INTERACTIONS AND THE AVERAGE MOLECULAR EXTENSION[1,2]*

The perturbation of the configuration of the polymer chain caused by its internal interactions may also be considered from the somewhat different viewpoint set forth qualitatively in Chapter X, Section 3. There it was indicated that, because of the obvious requirement that two segments shall not occupy the same space, the chain will extend over a larger volume than would be calculated on the basis of elementary random flight statistics.   As a matter of fact, the overwhelming majority of the statistical configurations calculated without regard for this requirement are found to be unacceptable, on this account,† to

---

* The problem of the influence of intramolecular interactions on the configurations of polymer molecules has been the subject of much controversy, which need not be reviewed here. For treatment of the problem by methods other than the one presented in the following pages, the reader is referred to papers by F. Bueche, *J. Chem. Phys.*, **21**, 205 (1953), and B. H. Zimm, W. H. Stockmayer, and M. Fixman, *J. Chem. Phys.*, **21**, 1716 (1953).   These papers include references to other literature on the subject.

† This statement rests on an argument like that used (p. 519) to show that two polymer molecules in dilute solution will avoid situations in which their domains overlap extensively.   The segment density within the domain of a polymer chain of more or less random configuration is of the order of $x^{-1/2}$; hence the probability that a given segment does not conflict with any of the others is approximately $1 - x^{-1/2}$.   The probability that no conflicts occur in the entire random flight configuration will be of the order of $(1 - x^{-1/2})^{x/2} \cong e^{-(x^{1/2})/2}$, which is a very small quantity.

the real polymer chain (composed of units which occupy a finite volume). Although the mean density of segments within the polymer domain may be very small, e.g., about 1 percent, the large number of segments reduces the chances for no conflict between any pair of segments to a very small value. A relatively larger fraction of the more expanded configurations will be free of interferences and therefore acceptable for the real polymer molecule; hence the average dimensions will exceed those calculated in the random flight approximation.[1] The resulting expansion is subject to modification depending on the difference in energy of interaction between like and unlike components of the solution, which determines whether or not a polymer segment prefers another segment rather than a solvent molecule as a neighbor.

The exclusion of one segment from the space occupied by another will be recognized as the primary consideration underlying the lattice theory of polymer solutions (see Chap. XII). The statistical treatment of the intramolecular volume exclusion effect just outlined is in reality equivalent to the thermodynamic (or osmotic) treatment of segment-solvent intramolecular interactions described in the opening of the present chapter. Equivalent results may be obtained by either procedure. We adopt the latter, somewhat more general, method[2] here. It has already been applied in Chapter XII, p. 523, to the calculation of the excluded volume for a pair of molecules in a dilute solution. There we were concerned with *intermolecular* interactions between segments belonging to different polymer molecules sharing the same region of space; here we consider *intramolecular* interactions between segments of the same molecule. These interactions should be characterized by identical parameters (i.e., the ψ_1 and κ_1, or the ψ_1 and Θ introduced in Chap. XII). The validity of this assertion, which is of foremost importance with respect to the following interpretation of molecular configuration, can scarcely be questioned, for it reduces to the trivial assumption that the interactions between two chain sections (or segments) are the same irrespective of whether they belong to different molecules or are connected in some circuitous fashion to the same molecule.

Proceeding in a manner paralleling the derivation of the excluded volume for a pair of molecules, we consider the polymer molecule to consist of a swarm of segments distributed on the average about the molecular center of gravity in accordance with the Gaussian formula (see Eq. XII-51).* This spatial distribution in the *unperturbed* molecule, as it would exist on the average in the total absence of inter-

---

* The fact that the actual spatial distribution is not exactly Gaussian[3] has no sensible effect on the results.

actions, may be written

$$x_j = x(\beta_0'/\pi^{1/2})^3 \exp(-\beta_0'^2 s_j^2) 4\pi s_j^2 \delta s_j \qquad (1)$$

where x is the total number of segments and x_j is the average number of them to be found in the spherical shell at a distance s_j from the center of gravity and δs_j in thickness. The subscript zero attached to the parameter β_0' signifies that it applies to the unperturbed distribution, thus distinguishing it from β' of Eq. (XII-52) for the real distribution. We have

$$\beta_0' = (3/2\overline{s_0^2})^{1/2} = 3/(\overline{r_0^2})^{1/2} \qquad (2)$$

where $(\overline{s_0^2})^{1/2}$ is the root-mean-square distance from the center of gravity averaged over all segments in the unperturbed state, and $(\overline{r_0^2})^{1/2}$ is the unperturbed root-mean-square end-to-end distance. In view of Eqs. (X-28), $\beta_0' = \alpha\beta'$, where α represents the intramolecular expansion factor* which it is the aim of the following theory to derive. The spatial distribution in the actual molecule is assumed to be expanded uniformly by the factor α as the result of intramolecular interactions. The x_j segments, which in the unperturbed state occupied the spherical shell between s_j and $s_j + \delta s_j$, after expansion by the factor α occupy the region between αs_j and $\alpha(s_j + \delta s_j)$ of volume $\delta V = 4\pi\alpha^3 s_j^2 \delta s_j$. Let ΔF_{Mj} represent the free energy of mixing the x_j segments with the n_{1j} solvent molecules in this region, where†

$$n_{1j} = 4\pi\alpha^3 s_j^2 \delta s_j (1 - v_{2j}) N/\mathrm{v_1} \qquad (3)$$

and v_{2j} is the volume fraction of polymer, N being Avogadro's number. The total free energy of mixing the polymer segments of one molecule with solvent will consist of the sum of the ΔF_{Mj} for each volume element plus a term ΔF_{el} for the free energy change associated with alteration in the molecular configuration. In analogy with Eq. (XIII-33), therefore

$$\Delta F = \sum_j \Delta F_{Mj} + \Delta F_{el} \qquad (4)$$

At equilibrium $(\partial\Delta F/\partial\alpha)_{T,P} = 0$. This derivative may be written

$$\partial\Delta F/\partial\alpha = \sum_j (\partial\Delta F_{Mj}/\partial\alpha) + \partial\Delta F_{el}/\partial\alpha$$

---

* The α used here corresponds to the isotropic swelling factor α_s for the cross-linked gels treated in Chapter XIII. The subscript is dropped for simplicity.

† It is to be noted that s_j is chosen to refer to the distance of the j-th shell from the center of mass *in the unperturbed state*. The subsequent summations (Eqs. 4 and 5) and the ultimate integration are consequently taken over s_j instead of the actual distance αs_j of the j-th shell from the center.

FRICTIONAL PROPERTIES 599

Inasmuch as ΔF_{Mj} refers to the mixing of segments in a small volume of uniform average segment density and the segments belong to the same molecule of preassigned center of gravity, the solvent chemical potential $\partial \Delta F_{Mj}/\partial n_{1j}$ must correspond to the local "excess" chemical potential $(\mu_{1j} - \mu_1^0)_E$ formulated for just such cases in Chapter XII (see Eqs. XII-41, XII-43, and XII-43'). Hence

$$\partial \Delta F/\partial \alpha = \sum_j (\mu_{1j} - \mu_1^0)_E (\partial n_{1j}/\partial \alpha) + (\partial \Delta F_{el}/\partial \alpha) \tag{5}$$

Introducing Eq. (XII-43') (which merely stipulates the inevitable proportionality between this chemical potential and the square of the concentration in dilute solutions)

$$(\partial \Delta F/\partial \alpha) = - RT\psi_1(1 - \Theta/T) \sum_j v_{2j}^2 (\partial n_{1j}/\partial \alpha) + \partial \Delta F_{el}/\partial \alpha \tag{5'}$$

From Eq. (3), since v_{2j} is small compared with unity

$$\partial n_{1j}/\partial \alpha = 12\pi \alpha^2 s_j^2 \delta s_j N/\mathrm{v}_1$$

Furthermore

$$v_{2j} = x_j V_s/4\pi \alpha^3 s_j^2 \delta s_j$$

$$= x V_s(\beta_0'/\alpha \pi^{1/2})^3 \exp(-\beta_0'^2 s_j^2)$$

where V_s is the volume of a segment and $x V_s$ therefore is the molecular volume of a polymer molecule; $x V_s = M\bar{v}/N$ where M is the molecular weight of the polymer and $\bar{v}$ is its specific volume (or partial specific volume). After substituting these expressions in Eq. (5'), replacing the summation by an integral, integrating over all volume elements from $s_j = 0$ to infinity, and eliminating β_0' through the use of Eq. (2), we obtain[2]

$$\partial \Delta F/\partial \alpha = - 6C_M kT\psi_1(1 - \Theta/T)M^{1/2}/\alpha^4 + \partial \Delta F_{el}/\partial \alpha \tag{6}$$

where the parameter

$$C_M = (27/2^{5/2}\pi^{3/2})(\bar{v}^2/N\mathrm{v}_1)(\overline{r_0^2}/M)^{-3/2} \tag{7}$$

is identical with the one of the same symbol which appeared in the intermolecular theory of the thermodynamic properties of dilute solutions. (Compare Eq. XII-64.)

The evaluation of the elastic free energy ΔF_{el} rests on the assumption that the root-mean-square distance between the ends of the chain is distorted by the same factor α representing the linear expansion of the spatial distribution. As in the treatment of the swelling of network structures (see Chap. XIII), we turn once more to the theory of rubber elasticity (see Chap. XI) for the evaluation of that term. This time the situation differs in that the ends of the polymer chain are unconstrained i.e., they are not required to meet other chain ends at network junction points. It will be recalled that the probability Ω of the deformed network configuration was calculated in two steps (see p. 467). The first of these was concerned with the required transformation of the end-to-end vector distribution, and the second with the occurrence of groups of four ends of different chains in proper juxtaposition. Only the first step is involved here; the second step has no counterpart in the single molecule problem. The free energy of deformation we require is therefore $-kT \ln \Omega_1$, where $\ln \Omega_1$ is given by Eq. (XI-38) with $\nu = 1$, since the polymer molecule consists of only one elastic element or chain. Thus

$$\Delta F_{el} = kT[3(\alpha^2 - 1)/2 - \ln \alpha^3] \tag{8}$$

and

$$\partial \Delta F_{el}/\partial \alpha = 3kT(\alpha - 1/\alpha) \tag{9}$$

Substituting in Eq. (6) and equating to zero, we obtain

$$\alpha^5 - \alpha^3 = 2C_M\psi_1(1 - \Theta/T)M^{1/2} \tag{10}^2$$

This result is the analog of the swelling equation (XIII–39) for a network. It rests principally on the assumption that the intramolecular interactions of the segments with one another are the same as would obtain for a cloud of particles, not connected to each other, but having the same radial distribution as the average radial density distribution occurring for the molecule made up of a linear sequence of particles.

Equation (10) directs attention to a number of important characteristics of the molecular expansion factor α. In the first place, it is predicted to increase slowly with molecular weight (assuming $\psi_1(1 - \Theta/T) > 0$) and without limit even when the molecular weight becomes very large. Thus, the root-mean-square end-to-end distance of the molecule $\sqrt{\overline{r^2}}$ should increase more rapidly than in proportion to the square root of the molecular weight. This follows from the theory of random chain configuration according to which the *unperturbed* root-mean-square end-to-end distance $\sqrt{\overline{r_0^2}}$ is proportional to $M^{1/2}$ (Chap. X), whereas $\sqrt{\overline{r^2}} = \alpha\sqrt{\overline{r_0^2}}$.

Secondly, α depends on the intensity of the thermodynamic interaction as expressed by $\psi_1(1 - \Theta/T)$, which is equal to $\psi_1 - \kappa_1$ (see Chap. XII, p. 523). The larger this factor, the greater the value of α for a given M. As should have been expected, therefore, the better the solvent the greater the "swelling" of the molecule. Conversely, the poorer the solvent the smaller the molecule.   Ordinarily ψ_1 is positive, and in a poor solvent Θ will also be positive.   Hence $\alpha^5 - \alpha^3$, and therefore α, may be expected to decrease as the temperature is decreased in a poor solvent.   It is not to be overlooked in this connection that phase separation sets a practical limit on how poor a solvent may be and yet permit the existence of a stable homogeneous dilute solution. We have seen in Chapter XIII that Θ represents the lowest temperature for complete miscibility in the given poor solvent at the limit of infinite molecular weight.   (Θ, it will be recalled, is a constant which, like ψ_1 and κ_1 from which it is derived, is characteristic of a given solvent-polymer pair and independent of the molecular weight of the latter.)   For large finite molecular weights, the critical solution temperature T_c will be only slightly lower than Θ, and a measurable quantity of polymer can be maintained in the dilute phase down to temperatures only a few degrees below T_c (see Fig. 121 of Chap. XIII). Hence in practice, the quantity $\psi_1(1-\Theta/T)M^{1/2}$ will never be much less than zero, and α therefore will not be much less than unity. Although, as will be shown later (see Sec. 3c and Table XL, p. 621), in a good solvent the molecule may expand considerably beyond its unperturbed size ($\alpha = 1$), in a poor one it will join with other molecules to form a separate, more concentrated phase in preference to appreciable contraction ($\alpha < 1$) below the unperturbed size.

At the absolute temperature $T = \Theta$ in a poor solvent $\alpha^5 - \alpha^3 = 0$ according to Eq. (10), and α must equal unity irrespective of M; i.e., *at $T = \Theta$ the molecular dimensions are unperturbed by intramolecular interactions.*   This deduction is of foremost importance in the interpretation of polymer molecular characteristics in dilute solution.   From a physical point of view, it follows from the exact balance at this temperature between the effect of mutual volume exclusion of the segments, which tends to enlarge the molecule, and the effect of a positive energy (or standard state free energy) of mixing, which encourages first neighbor contacts between polymer segments and, hence, a more compact configuration for the molecule.   As was pointed out in the discussion of intermolecular interactions in Chapter XII (p. 532), the Θ-point corresponds to the Boyle point of an imperfect gas at which temperature PV is equal to RT over a considerable range in pressure. Thus, at $T = \Theta$, $\pi/c = RT/M$ up to concentrations of several percent, which was interpreted to mean that the net interaction between the segments of a pair of molecules is zero, for reasons exactly equivalent to those stated above.   The broad significance of the Θ-point is thus apparent.

One further matter relating to the nature of C_M as defined by Eq. (7)

deserves mention. This quantity depends inversely on the molar volume v_1 of the solvent. Extracting $1/v_1$ from C_M, we have in Eq. (10) the factor $\psi_1(1-\Theta/T)/v_1$, which according to Eqs. (XII-45) and (XII-46) is equal to $(1/2-\chi_1)/v_1$. Hence $\alpha^5-\alpha^3$ should be proportional to $1/2v_1-\chi_1/v_1$. Since according to Eq. (XII-21') χ_1 is proportional to the product of the interaction density and v_1, the second term of this expression can be considered constant for a series of solvents of increasing molar volume but otherwise thermodynamically equivalent to one another in their interactions with polymers of the given chemical type. For such a series of solvents α may be expected to decrease with v_1 owing to the decrease in the first term above. Consider now the polymer molecules present in an amorphous polymer in the absence of added solvent. Each individual molecule is dispersed in a "solvent" consisting of other polymers of the same kind. Necessarily χ_1 will be zero, and v_1 is very large. Hence α will be very nearly unity; i.e., molecules in an undiluted amorphous polymer should be substantially unperturbed. The statistical treatment of rubber elasticity is not subject to modification therefore by consideration of intramolecular interferences. The physical reason for this fortunate conclusion is easy to comprehend: although a molecule in the bulk state (or in a concentrated solution as well) interferes with itself, it has nothing to gain by expanding, for the decrease in interaction with itself is compensated by increased interference with its neighbors.

2. FRICTIONAL PROPERTIES OF THE POLYMER MOLECULE IN SOLUTION

2a. The Free-draining Molecule.[4,5]—The "pearl string" model represented in Fig. 138 is convenient for the purpose of discussing the hydrodynamic properties of the polymer chain. It consists of a sequence of beads, each of which offers hydrodynamic resistance to the flow of the surrounding medium, connected to one another by a string which does not. The force ξ_i exerted on the surrounding medium by the bead i of the chain owing to motion of the polymer molecule relative to the medium may be written

$$\xi_i = \zeta \left| u_i - u \right| \tag{11}$$

where u_i represents the vectorial velocity of the bead i and u is the velocity which the surrounding medium would have at the location of the center of the bead if this particular bead (only) were missing. The quantity $\left| u_i - u \right|$ representing the magnitude of the difference in the velocities thus defined will hereafter be written simply as Δu_i; ζ is the frictional coefficient for a bead, representing the ratio of the force to the velocity. If the bead is assigned a spherical shape, then according to Stokes' law

$$\zeta = 6\pi\eta_0 a \qquad (12)$$

where a is the radius of the bead and η_0 is the viscosity of the medium.

For the present we consider the case of very small frictional effects due to the beads; i.e., the Stokes' law radius a is small. We assume that the effects are so small that the motion of the surrounding medium is only very slightly disturbed by the movement of the polymer molecule relative to the medium. The frictional effects due to the polymer molecule are then comparatively easy to treat, for the velocity of the medium everywhere is approximately the same as though the polymer molecule were not present. The solvent streams through the molecule almost (but not entirely) unperturbed by it; hence the term free-draining is appropriate

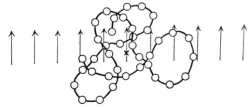

Fig. 138.—The free-draining molecule during translation through the solvent. Flow vectors of the solvent relative to the polymer chain are indicated.

for this case. The velocity difference Δu_i we require in Eq. (11) is simply defined by the motion of the molecule on the one hand and the unperturbed flow of the medium on the other.

If the motion of the molecule is one of translation, as it is during sedimentation in a centrifugal field, the velocity of every bead is the same, and in the free-draining case the difference in velocity Δu_i for each bead relative to the solvent is the same as the (relative) translational velocity u of the molecule as a whole. Fig. 138 is illustrative of this case. The total force on the molecule is then

$$\Xi = \sum \xi_i = \sum \zeta \Delta u_i = x\zeta u$$

where x is the number of beads. The frictional coefficient for the molecule as a whole is

$$f_0 = \Xi/u = x\zeta \qquad (13)$$

where the subscript zero specifies infinite dilution. The frictional coefficient f_0 is directly proportional to the chain length x of the free-draining molecule. The sedimentation constant s_0, defined by (Eq. VII-40), should therefore be independent of the molecular weight. The shape of the molecule, so long as the free-draining condition is fulfilled, is of no importance insofar as the frictional coefficient and the related sedimentation and diffusion constants are concerned.

The motion occurring during viscous flow consists of a shear in which different layers of the solution move with different velocities. The large polymer molecule finds it impossible to adjust its motions so as to coincide with the velocities of the different layers of the liquid through which it extends. Its situation is depicted in Fig. 139, where vectors representing the unperturbed velocity of the liquid relative to the position of the center of gravity of the molecule are shown. Let

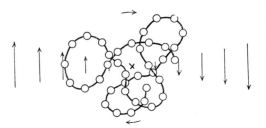

the velocity gradient be γ; that is, the velocity of the unperturbed liquid changes by γ centimeters per second for each centimeter displacement from left to right in the figure. Obviously, the liquid exerts a torque on the molecule tending to rotate it, as indicated by the curved arrows. When the molecule rotates, however, new frictional forces

FIG. 139.—The free-draining molecule during viscous flow of the solution. Flow of the solvent relative to the center of gravity (✕) of the molecule and the rotation of the molecule as a whole are indicated.

are brought into play as segments in the upper and lower portion of the figure move *across* the field. The condition of dynamic stability requiring that the sum of the torques acting on the molecule be zero leads to the conclusion that the molecule will strike a compromise by rotating with an angular velocity ω, in radians per second, equal to half the shear gradient; i.e., $\omega = \gamma/2$. It is easy to show that under these circumstances the velocity at any point in the thus rotating molecule relative to the flowing solvent is $s\gamma/2$, where s is the distance of the point from the center of the molecule.

In the free-draining case $s\gamma/2$ is also the relative velocity of the medium in the vicinity of a bead at a distance s from the center. Hence the frictional force acting on the bead is $\zeta s\gamma/2$, and the rate of energy dissipation by the action of the bead is the product of the force and the velocity, or $\zeta(s\gamma/2)^2$. The total energy dissipated per unit time by the molecule will be given by the sum of such terms for each bead, or

$$(\zeta\gamma^2/4) \sum s_j^2 = (\zeta\gamma^2/4)x\overline{s^2}$$

where $\overline{s^2}$ is the mean-square distance of a bead from the center of gravity. But the viscosity of the solution as a whole is equal to the rate of energy dissipation divided by the square of the rate of shear. Hence on the assumption that the solution is so dilute that there is no appreciable interaction between different polymer molecules, the polymeric solute makes a contribution to the viscosity which is given by $(cN/M)\zeta x\overline{s^2}/4$ where cN/M is the number of polymer molecules per cc., c being the concentration in grams per cc. The total viscosity is therefore

$$\eta = \eta_0 + (cN/M_0)\zeta\overline{s^2}/4$$

where $M_0 = M/x$ is the molecular weight of a bead. Hence the intrinsic viscosity $[\eta]$, which is defined as the limiting value of $(\eta - \eta_0)/\eta_0 c$ at infinite dilution, for a free-draining molecule should be given by

$$[\eta] = (\zeta N/\eta_0 M_0)\overline{s^2}/400 \qquad (14)$$

The factor of 100 has been introduced in the denominator in order to convert to the usual units (g./100 ml.)$^{-1}$ for the intrinsic viscosity. For a linear polymer molecule $\overline{s^2}$ is proportional to $\overline{r^2} = \alpha^2\overline{r_0^2}$. Hence, if it were legitimate to disregard the effects of intramolecular interactions on chain configuration, the intrinsic viscosity should then be proportional to the molecular weight.[4,5] Since α^2 normally increases slowly with M, the intrinsic viscosity of a free-draining linear molecule should increase a little more rapidly than with the first power of the molecular weight.

We may note in passing that the intrinsic viscosity of a fully extended rod molecule, for which $\overline{s^2}$ is proportional to the square of the length, should depend on the *square* of the molecular weight,[4,5] in the free-draining approximation. In a more accurate treatment[6] which avoids this approximation, the simple dependence on M^2 is moderated by a factor which depends on the effective thickness of the chain (or bead density along the chain) compared with the chain length.

2b. The Equivalent Sphere Model.—The free-draining approximation leads to the prediction that the frictional coefficient f_0 should increase with the first power of the molecular weight and that the intrinsic viscosity should increase with M raised to a power a little greater than unity. Experiments show that both quantities increase with a power of M less by about 0.4 to 0.5 than predicted. Closer inspection of the problem casts serious doubt on the free-draining approximation. Even with a frictional coefficient ζ less by a factor of ten than the value to be expected for a spherical bead comparable in size to a chain unit, the motion of the solvent must be markedly altered deep within a random-coiled molecule of high molecular weight.

Solvent toward the center of an actual polymer molecule will acquire a velocity more nearly that of the molecule than the velocity which the solvent medium would have at the same point in the absence of the polymer molecule. Evidence to be cited later leads to the conclusion that solvent situated in the interior of the molecule ordinarily moves almost in unison with the polymer beads as though the solvent were bound to the polymer (which to be sure it is not). This leads to the concept of an *equivalent hydrodynamic sphere*, impenetrable to solvent, which would display the same frictional coefficient f_0, or would enhance the viscosity equally, as the actual polymer molecule. Writing $\mathcal{R}_e$ for the radius of this equivalent sphere, we should then have according to Stokes' law

$$f_0 = 6\pi\eta_0\mathcal{R}_e \tag{15}$$

According to the Einstein viscosity relation

$$(\eta - \eta_0)/\eta_0 = 2.5(n_2/V)\,\mathcal{U}_e$$

where n_2/V is the number of molecules per unit volume and $\mathcal{U}_e = (4\pi/3)\mathcal{R}_e^3$. Or, writing $n_2/V = cN/100M$ where c is the concentration in grams per 100 ml., we find

$$[\eta] = 0.025N\mathcal{U}_e/M \tag{16}$$

This remarkably simple treatment suffers one serious deficiency: the value of $\mathcal{R}_e$ remains quantitatively undefined. More or less intuitively it has been suggested by various investigators that $\mathcal{R}_e$ should increase as the root-mean-square end-to-end distance $\sqrt{\overline{r^2}}$ for a linear chain, or, more generally, as the root-mean-square distance $\sqrt{\overline{s^2}}$ of the beads from the center of any polymer molecule, linear or branched. Accepting this postulate unquestioningly, we should then have f_0 proportional to $M^{1/2}\alpha$ and $[\eta]$ proportional to $M^{1/2}\alpha^3$. These conclusions happen to be correct, but the premises from which they have been deduced leave many questions unanswered. A more thoroughgoing examination of the hydrodynamic interaction is needed.

2c. Real Polymer Chains with Large Frictional Interactions.[7,8]—Let us first consider the simpler problem of linear translation of the molecule through the solvent. We shall, however, view the motion of the solvent with reference to the molecule. Near its center the solvent is very nearly stationary in this sense, but as we move outward its velocity increases.* The velocity will pass through a maximum before it finally, perhaps at some distance beyond the outer fringes of the molecule, reaches the unperturbed relative velocity of the surrounding medium. (The occurrence of this maximum is merely a consequence

---

* The discussion here follows the more quantitative treatment given by Debye and Bueche,[7] although we seek to avoid their model consisting of a sphere throughout which the segment density is uniform, and beyond which it is zero.

of retardation of some of the solvent by the molecule. This retardation must be compensated by an accelerated motion of other portions of the solvent. A corresponding acceleration occurs in the motion of a liquid medium past a rigid sphere.) The situation is illustrated in Fig. 140, without indicating the aforementioned maximum, however. Thus a transition region exists at some distance from the center of the molecule wherein the velocity of the solvent relative to the molecule increases from zero to a value approaching its external value, and in this region and beyond energy is dissipated owing to the difference in velocity between the polymer chain elements, or beads, and the solvent. The location of this transition region may be expected to depend on the number and spatial distribution of the beads and on the frictional coeffi-

cient ζ of a bead. Since the frictional coefficient ζ is proportional to η_0, a better variable to use the present connection is ζ/η_0, which is independent of η_0. Assuming laminar flow, the pattern of the flow, and therefore the depth of pene-tration of the molecule by the flow, will be independ-

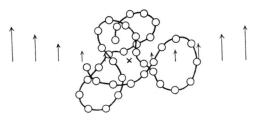

FIG. 140.—Translation (e.g., sedimentation) of a chain molecule with perturbation of the solvent flow relative to the molecule.

ent of the viscosity η_0 of the medium. Suppose the ratio ζ/η_0 is varied by altering the diameter of a bead, or possibly also its shape, while the number and spatial distribution of beads, or segments, remains fixed. If ζ/η_0 is progressively decreased, the solvent flow will penetrate closer to the center, leading eventually to the free-draining case. If ζ/η_0 is increased, the depth of penetration will decrease, and if it is made sufficiently large, only the outer fringes of the molecule will be permeated by the flow; the perturbation of the liquid flow by the molecule will extend considerably beyond the molecule, of course.

In view of the exponential decrease in segment density with the square of the distance from the molecular center, the depth of penetration of the molecule by the liquid flow may change only very slowly with further increase in ζ/η_0 once this ratio is great enough to exclude the flow lines from the inner (more densely populated) regions of the molecule. Hence a condition of comparative insensitivity of the flow pattern to ζ/η_0 should be approached. Moreover, with respect to the energy dissipation due to the molecule as a whole, any enlargement of the pattern of flow around (and through) the molecule which may result from an increase in ζ/η_0 will be compensated by a decrease in the number of beads which then resist the flow, again owing to the rapidly diminishing bead concentration with distance from the center. We are thus led to the important conclusion that *the frictional coefficient f_0 of the molecule as a whole (at fixed η_0) should become independent of the frictional coefficient ζ of a bead in the limit of sufficiently large ζ/η_0.*

The frictional coefficient for the molecule should be expected to depend in general on the following variables: the "size" of the molecule, which may be represented by the root-mean-square distance $\sqrt{\overline{s^2}}$ of a bead, or segment, from the center of gravity; the number x of beads; the nature of the distribution of beads; and ζ/η_0. It will be directly proportional to the viscosity of the medium; hence we shall be concerned with the functional dependence of f_0/η_0 on the variables listed. This quantity has the dimensions of a length, which immediately suggests direct proportionality to $\sqrt{\overline{s^2}}$. If attention is confined for the moment to molecules having equivalent spatial distributions (e.g., Gaussian), it is apparent that an increase in the number x of beads while the dimension $\sqrt{\overline{s^2}}$ of the molecule is held constant will have the same effect as an increase in the size of the existing beads, i.e., as an increase of ζ/η_0. Hence, in the limit of sufficient internal resistance to flow, the molecular frictional coefficient f_0 should be independent not only of ζ/η_0 but also of x, provided $\sqrt{\overline{s^2}}$ is maintained constant. On the basis of the foregoing considerations, the molecular frictional coefficient should be amenable to representation by a relationship of the form

$$f_0/\eta_0 = \sqrt{\overline{s^2}}\,\phi(\zeta/\eta_0,\, x,\, \sqrt{\overline{s^2}})$$

where ϕ is a dimensionless function of the variables indicated such that it approaches an upper asymptotic limit as ζ/η_0 and/or x increases while $\sqrt{\overline{s^2}}$ remains constant. In view of the equivalence in the effects of increases in ζ/η_0 and in x, their product $x\zeta/\eta_0$, having the dimensions of length, may be expected to represent them in the function ϕ; this product expresses the total internal flow-resisting capacity of the molecule. Since the function ϕ must be dimensionless, it may be assumed to depend on the dimensionless variable $x\zeta/\eta_0\sqrt{\overline{s^2}}$. Thus, on the basis of dimensional analysis, we arrive at the conclusion that the molecular frictional coefficient may be expressed as follows:

$$f_0/\eta_0 = \sqrt{\overline{s^2}}\,\phi(x\zeta/\eta_0\sqrt{\overline{s^2}}) \qquad (17)$$

where ϕ has the limiting characteristics already mentioned. It must also depend on the contour of the distribution of beads about the center of gravity. However, for nearly all polymers of interest this distribution will be approximately the same; i.e., for randomly coiled linear polymers it is approximately Gaussian, and even moderately branched polymers may present a similar spatial distribution of beads.

Of major importance is the fact that the specific character of polymer chains of a given type enters the relationship (17) only through the effective size of one of its beads as indicated by the ratio ζ/η_0. Even the effect of this factor vanishes when the total internal resistance to flow $x\zeta/\eta_0$ is sufficiently large. Hence, in this limit, which will include nearly all actual cases of interest (see Sec. 4), the molecular frictional coefficient *should depend only on the size $\sqrt{\overline{s^2}}$ and not otherwise on the nature of the polymer.* Accordingly, we choose to let

$$f_0/\eta_0 = P'\sqrt{\overline{s^2}} \tag{17'}$$

where P' represents the limiting value of the function ϕ appearing in Eq. (17). It should be *the same for all polymers in all solvents*, provided only that their segment spatial distributions are similar (i.e., approximately Gaussian). In dealing with linear polymers, the root-mean-square end-to-end distance $\sqrt{\overline{r^2}}$ usually has been chosen as a measure of the size. Hence we adopt the relation

$$f_0/\eta_0 = P\sqrt{\overline{r^2}} \tag{18}$$

in preference to Eq. (17'). According to Eq. (X-14) relating $\overline{s^2}$ to $\overline{r^2}$ for a Gaussian distribution, $P = P'/\sqrt{6}$.

The detailed hydrodynamic treatment of Kirkwood and Riseman[8] yields the following relationship

$$f_0/\eta_0 = \sqrt{\overline{r^2}}\,[2^{5/2}(3\pi)^{-3/2} + \eta_0\sqrt{\overline{r^2}}/x\zeta]^{-1} \tag{19}$$

which will be recognized to be of the form suggested by dimensional considerations leading to Eq. (17); $\sqrt{\overline{r^2}}$ appears in place of $\sqrt{\overline{s^2}}$ and the reciprocal of the bracketed expression corresponds to the function ϕ. According to Eq. (19), the reciprocal of the frictional coefficient should be a linear function of $(\overline{r^2})^{-1/2}$ with a slope given by $1/\eta_0 P$ where

$$P = (3\pi)^{3/2}/2^{5/2} = 5.11 \tag{20}$$

and an intercept equal to $1/x\zeta$. The importance to be attached to this intercept depends of course on the magnitude of $\eta_0\sqrt{\overline{r^2}}/x\zeta$ compared to $1/P$. Let us assume that Stokes' law may be used to estimate ζ, i.e., that $\zeta/\eta_0 = 6\pi a$ where a is the radius of a bead. A reasonable value for a would be 5×10^{-8} cm., and for $(\overline{r^2}/x)^{1/2}$ about 10^{-7} cm., where x is the number of beads of the above size in the chain of mean-square-length $\overline{r^2}$. Temporarily treating $\overline{r^2}/x$ as a constant for a given series of homologous polymers, we obtain from Eq. (19)

$$f_0/\eta_0 \cong \sqrt{\overline{r^2}}/[0.2 + 0.1/\sqrt{x}]$$

In the molecular weight range of interest x will exceed 10^2 and usually it will exceed 10^3; hence the second term in the bracketed expression in Eq. (19) should be very much smaller than the first, and Eq. (18) may be expected to serve as an acceptable approximation.*

The estimation of ζ from Stokes' law when the bead is similar in size to a solvent molecule represents a dubious application of a classical equation derived for a continuous medium to a molecular phenomenon. The value used for ζ above could be considerably in error. Hence the real test of whether or not it is justifiable to neglect the second term in Eq. (19) is to be sought in experiment. It should be remarked also that the Kirkwood-Riseman theory, including their theory of viscosity to be discussed below, has been developed on the assumption that the hydrodynamics of the molecule, like its thermodynamic interactions, are equivalent to those of a cloud distribution of independent beads. A better approximation to the actual molecule would consist of a cylinder of roughly uniform cross section bent irregularly into a random, tortuous configuration. The accuracy with which the cloud model represents the behavior of the real polymer chain can be decided at present only from analysis of experimental data.

The interpretation of the viscosity contribution of the partially permeated coil is similar to that given for the frictional coefficient, and detailed discussion will therefore be unnecessary. The molecule rotates with an angular velocity $\omega = \gamma/2$, assuming it to behave as if rigid. Solvent near the interior partakes of the velocity of the nearby chain segments, and permeation of the molecule by the flow lines of the surrounding liquid is less the greater ζ/η_0. The greater disturbance in the solvent caused by an increase in ζ/η_0 is again compensated by the decrease in the number of segments subjected to the counterflow of solvent. The Kirkwood-Riseman[8] theory leads in this case to

$$[\eta] = (\pi/6)^{3/2}(N/100)XF(X)(\overline{r^2})^{3/2}/M \tag{21}$$

where $[\eta]$ is expressed in the customary units of deciliters per gram;

$$X = (6\pi^3)^{-1/2}x\zeta/\eta_0(\overline{r^2})^{1/2} \tag{22}$$

and $F(X)$ is a function of X, values for which have been tabulated by Kirkwood and Riseman. Thus the intrinsic viscosity depends pri-

* The foregoing interpretation differs significantly from that given by Kirkwood and Riseman, who anticipated an important contribution from the second term in brackets in Eq. (19). On the basis of their treatment of intrinsic viscosity data, which erred in disregarding the molecular expansion factor α, they were led to adopt a value for the frictional coefficient ζ of a bead which is nearly two orders of magnitude smaller than we have used. Hence they postulated a correspondingly larger value for the term in question.

marily on the ratio $(\overline{r^2})^{3/2}/M$, consisting of a volume divided by the molecular weight. It depends secondarily on a function $XF(X)$ of the same variable entering into the function ϕ of Eq. (17) for the frictional coefficient f_0. In the limit of sufficiently large X, i.e., if $x\zeta/\eta_0$ is sufficiently large, $XF(X)$ approaches the asymptotic limit 1.588. Estimation of $x\zeta/\eta_0$ in the previous manner leads to the conclusion that for polymers having molecular weights in excess of about 10,000, $XF(X)$ should be at, or near, its limiting value. Then we may write, in analogy with Eq. (17)

$$[\eta] = \Phi(\overline{r^2})^{3/2}/M \qquad (23)$$

where

$$\Phi = 0.01588(\pi/6)^{3/2}N = 3.62 \times 10^{21} \qquad (24)^*$$

Like the analogous parameter P in the theory of the frictional coefficient, Φ is seen to be *independent of the characteristics of the given chain molecule* beyond the requirement that its spatial form be that characteristic of a randomly coiled chain molecule.

According to Eq. (18), the high polymer molecule should exhibit the frictional coefficient of an *equivalent sphere* (compare Eq. 15) having a radius *proportional* to the root-mean-square end-to-end distance $(\overline{r^2})^{1/2}$ (or to $(\overline{s^2})^{1/2}$). Similarly, according to Eq. (23) its contribution to the viscosity should be that of an equivalent sphere (compare Eq. 16) having a volume proportional to $(\overline{r^2})^{3/2}$. In analogy with Eq. (17'), we might write

$$[\eta] = \Phi'(\overline{s^2})^{3/2}/M \qquad (23')$$

which may be expected to hold approximately for many nonlinear molecules as well as for linear ones. Here $\Phi' = 6^{3/2}\Phi$ since for a randomly coiled linear chain $\overline{s^2} = \overline{r^2}/6$.

3. TREATMENT OF EXPERIMENTAL RESULTS: INTRINSIC VISCOSITIES OF NONIONIC POLYMERS[2]

According to the interpretation given above, the intrinsic viscosity is considered to be proportional to the ratio of the effective volume of the molecule in solution divided by its molecular weight. In particular (see Eq. 23), this effective volume is represented as being proportional to the cube of a linear dimension of the randomly coiled polymer chain,

---

* If the particles are asymmetric in shape, then as Kirkwood[9] has pointed out the theory leading to Eq. (21) requires amendment to include a contribution from rotatory diffusion. For spherically symmetric molecules, as we here assume, rotatory diffusion is, of course, inconsequential.

such as its root-mean-square end-to-end distance. To get at the underlying factors influencing the intrinsic viscosity, it is desirable to separate the quantity $(\overline{r^2})^{3/2}$ into its component factors $(\overline{r_0^2})^{3/2}$ and α^3. Eq. (23) may then be recast as follows:

$$\lfloor\eta\rfloor = \Phi(\overline{r_0^2}/M)^{3/2}M^{1/2}\alpha^3 \tag{25}$$

Since $\overline{r_0^2}/M$ is independent of M for a linear polymer of a given unit structure (see Chap. X), we may write

$$[\eta] = KM^{1/2}\alpha^3 \tag{26}$$

where

$$K = \Phi(\overline{r_0^2}/M)^{3/2} \tag{27}$$

If the preceding analysis of hydrodynamic effects of the polymer molecule is valid, K should be a constant independent both of the polymer molecular weight and of the solvent. It may, however, vary somewhat with the temperature inasmuch as the unperturbed molecular extension $\overline{r_0^2}/M$ may change with temperature, for it will be recalled that $\overline{r_0^2}$ is modified by hindrances to free rotation the effects of which will, in general, be temperature-dependent. Equations (26), (27), and (10) will be shown to suffice for the general treatment of intrinsic viscosities.

3a. The Intrinsic Viscosity at the Theta Point and the Evaluation of K.—Ordinarily the intrinsic viscosity should depend on the molecular weight not only owing to the factor $M^{1/2}$ occurring in Eq. (26), but also as a result of the dependence of the factor α^3 on M. The influence of this expansion resulting from intramolecular interactions may be eliminated by suitable choice of the solvent and temperature. Specifically, in an ideal solvent, or Θ-solvent, $\alpha=1$ and Eq. (26) reduces to

$$[\eta]_\Theta = KM^{1/2} \tag{26'}$$

If the contribution of a polymer molecule to the viscosity of the solution is in reality proportional to the cube of its linear dimension, the intrinsic viscosity in a Θ-solvent should be proportional to the square root of the molecular weight. The influence of intramolecular interactions on the configuration having been neutralized by this choice of solvent medium, it becomes possible to examine separately the hydrodynamic aspects of the problem.

The intrinsic viscosity usually changes rapidly with temperature in the vicinity of the Θ-point; hence it may be necessary to define this temperature experimentally with some accuracy—preferably within a degree or two. The Θ-point may be deduced from accurate osmotic or light-scattering measurements in the poor solvent over a range of temperatures. Θ may then be identified as the temperature at which the second virial coefficient is zero. A generally more satisfactory method involves determination of the consolute temperatures for a series of fractions covering an extended range in molecular weight. The Θ-point is obtained by extrapolating these critical temperatures to infinite molecular weight, using the method previously given in Chapter XIII. (See Eq. XIII-7 and Fig. 122, p. 548). It is essential that liquid-liquid phase preparation, rather than crystalline precipitation, shall prevail in the precipitation measurements (see Chap. XIII). If viscosity measurements are conducted in a mixture consisting of a solvent and a nonsolvent in suitable proportions, the required Θ-point, at which the second virial coefficient is zero and the molecule consequently assumes an unperturbed average configuration, corresponds to the critical point in the ternary system: nonsolvent-solvent-polymer of infinite molecular weight (see p. 552). This point may again be located by extrapolating appropriate precipitation data for a series of fractions of the polymer.[10,11]

The results of intrinsic viscosity measurements for four polymer-solvent systems made at the Θ-temperature of each are shown in Fig. 141. The four systems and their Θ-temperatures are polyisobutylene in benzene at 24°C, polystyrene in cyclohexane at 34°C, poly-(dimethylsiloxane) in methyl ethyl ketone at 20°C, and cellulose tricaprylate in γ-phenylpropyl alcohol at 48°C. In each case a series of poly-

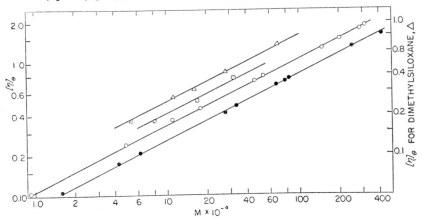

FIG. 141.—Double logarithmic plot of $[\eta]_\Theta$ against M for several polymer series: polyisobutylene in benzene at 24°C, ○; [12,13] polystyrene in cyclohexane at 34°C, ●;[13,14,15] cellulose tricaprylate in γ-phenylpropyl alcohol at 48°C, □;[10] and poly-(dimethylsiloxane) in methyl ethyl ketone at 20°C, △ (right-hand ordinate scale).[16] All lines are drawn with the theoretical slope of one-half.

mer fractions extending over the widest possible molecular weight range was used. Straight lines have been drawn through the log-log plots with the theoretical slope of one-half. In all cases the agreement is within experimental error. The results for polyisobutylene and polystyrene extend over very wide ranges in molecular weight from a few thousand to over a million. The agreement below $M = 10,000$ may be to some extent fortuitous; here the chains probably are not long enough to justify statistical procedures for obtaining their effective configurations. Furthermore, the postulated proportionality between the effective hydrodynamic volume and $(\overline{r^2})^{3/2}$ becomes increasingly suspect in this lower range. In terms of the Kirkwood-Riseman theory, at low molecular weights $XF(X)$ of Eq. (21) should be expected to fall below its asymptotic limit, and this should be manifested by a decrease in K at low M. These doubts in regard to theory at low M notwithstanding, it is an experimental fact, in the two cases mentioned at least, that $[\eta]_\Theta$ is quite accurately proportional to $M^{1/2}$ down to molecular weights of a few thousand; in other words K remains constant, substantially independent of M. The postulated proportionality between the effective hydrodynamic volume and the cube of the linear dimension $(\overline{r^2})^{1/2}$ is thus confirmed to an extent exceeding expectation.

Values are given in Table XXXVII for K's in different solvents. These have been obtained by measuring intrinsic viscosities of fractions of known molecular weights in the indicated solvents and solvent mixtures at their Θ-points (see Eq. 26′). The values given for K correspond, of course, to intercepts of lines such as are shown in Fig. 141. According to Eq. (27), they should be proportional to $(\overline{r_0^2}/M)^{3/2}$. Since this quantity depends only on chain bond dimensions and on the freedom of rotation about them, K should be independent of the solvent, in first approximation at least. The results for polystyrene, which are most extensive, tend to confirm this conclusion. Apart from a significant downward trend with increasing temperature, the values for K appear to be nearly independent of the specific character of the solvent medium. The unexplained small discrepancy between the K's obtained using two of the mixed solvents may, however, be somewhat greater than the experimental error arising mainly from uncertainty in the location of the Θ-point.* The trend with temperature

* It should be remarked that small differences between the average dimension $(\overline{r_0^2})^{1/2}$ of a polymer coil in different Θ-solvents at the same temperature may conceivably arise from a specific influence of the solvent on the average disposition of a chain bond relative to its predecessor in the chain. This would amount to an effect of the solvent on the restrictions to free rotation about the valence bonds of the polymer chain. Such effects have not been established. They may be presumed to be small.

may reasonably be ascribed to variation in $\bar{r^2}/M$ with T. Thus the effects of hindrances to free rotation, which invariably expand the chain to a larger configuration than would hold for completely free rotation, appear to be diminished by an increase in temperature. The close similarity between the values observed for K for polystyrene in different solvents was anticipated in earlier work[17] showing that the intrinsic viscosity near the precipitation point for different nonsolvent-solvent mixtures is nearly independent of the mixture used.

TABLE XXXVII.—VALUES OBTAINED FOR K IN DIFFERENT SOLVENTS

Polymer	Solvent	$\Theta(°K)$	$K \times 10^4$ (at $T = \Theta$)
Polystyrene[11]	Cyclohexane (0.869)-carbon tetrachloride (0.131)	288	8.45
Polystyrene[11]	Toluene (0.476)-n-heptane (0.524)	303	8.6
Polystyrene[11]	Methyl ethyl ketone (0.889)-methanol (0.111)	303	8.05
Polystyrene[13,15]	Cyclohexane	307	8.2
Polystyrene[a]	Methylcyclohexane	343.5	7.6
Polystyrene[11,15]	Ethylcyclohexane	343	7.5
Polyisobutylene[12,13]	Benzene	297	10.7
Polyisobutylene[12]	Phenetole	359	9.1 (± 0.5)
Polyisobutylene[12]	Anisole	378.5	9.1 (± 0.5)
Poly-(dimethylsiloxane)[16]	Methyl ethyl ketone	293	8.1
Poly-(dimethylsiloxane)[16]	Phenetole	356	7.9

[a] E. T. Dumitru and L. H. Cragg, private communication.

The results for polyisobutylene indicate a small but significant decrease of K with temperature. For the silicone K is the same, within experimental error, in two different Θ-solvents in spite of a 63° temperature difference.

The demonstration of proportionality of the intrinsic viscosity measured at the Θ-point to $M^{1/2}$ over wide ranges in molecular weight offers solid support for the basic premise of the simple interpretation presented in the preceding section, namely, that the intrinsic viscosity is proportional to the cube of a linear dimension such as $\sqrt{\bar{r^2}}$. This conclusion rests, of course, on the premise that $\bar{r^2} = \bar{r_0^2} \sim M$ at the Θ-point —a condition which seems inescapable on the basis of the most general considerations (see Sec. 1). Moreover, the factor of proportionality (i.e., Φ) between $(\bar{r^2})^{3/2}/M$ and $[\eta]$ appears to be independent of the solvent, as theory would predict. It remains to examine the values obtained for this factor for different polymer types.

3b. The Universal Parameter Φ; Unperturbed Dimensions of Linear Polymers.—If $\overline{r^2}$ is determined from dissymmetry measurements on the light scattered by dilute solutions of a polymer fraction, the results being extrapolated to infinite dilution as required (see Chap. VII), and if the intrinsic viscosity is determined in the *same solvent* at the *same temperature*, then it is possible to calculate Φ using Eq. (23) provided the molecular weight also is known. Results obtained in this manner are summarized in Table XXXVIII. Since $(\overline{r^2})^{1/2}$ can at best be determined with an accuracy of ± 5 percent, and this quantity must be cubed, inaccuracies in the values obtained for Φ are inevitable.

TABLE XXXVIII.—SUMMARY OF RESULTS LEADING TO VALUES FOR Φ

	Number of fractions measured	$M \times 10^{-5}$ (or range)	$\Phi \times 10^{21}$ (and mean deviation)
Polystyrene–methyl ethyl ketone[18]	9	2.3–17.6	2.30 (± 0.25)
Polystyrene-dichloroethane[18]	5	5.2–17.8	1.95 ($\pm$ .20)
Polystyrene-toluene[18]	2	16.2	2.6
Polystyrene-benzene[7]	1	10	2.0
Poly-(methyl methacrylate)-acetone[19]	4	7.5–14	2.2
Poly-(methyl methacrylate)-methyl ethyl ketone[19]	4	7.5–14	2.0
Poly-(methyl methacrylate)-chloroform[19]	4	7.5–14	2.2
Polyisobutylene-cyclohexane[20]	2	5.1–7.2	2.2
Polyisobutylene-ethyl-n-caprylate[20]	1	6.6	2.6
Poly-(acrylic acid)-dioxane[20]	4	8.4–14	2.2

Nevertheless, the Φ values for different systems derived from measurements carried out in several laboratories are strikingly consistent, and they lend support to the view that the value of Φ is the same in different systems, provided of course that the molecules conform to a random coil configuration. The best value for Φ appears at present to be 2.1 (± 0.2) $\times 10^{21}$, r being expressed in cm., M in units of molecular weight, and $[\eta]$ in deciliters per gram.

It is important to use well-fractionated samples for the purpose of determining the value of Φ in the manner set forth above. If the sample includes species covering a considerable range in molecular weight, then the nature of the averages of the measured quantities, $(\overline{r^2})^{3/2}$ and M, required by Eq. (23) assumes importance. It may be readily shown[20] that the number average of $(\overline{r^2})^{3/2}$ over the molecular weight distribution should be used in conjunction with $\overline{M}_n$ in this equation. If the sample is appreciably heterogeneous, the z-average of $\overline{r^2}$

which is obtained from the light-scattering dissymmetry method yields a value for $(\overline{r^2})^{3/2}$ which exceeds the required number average of $(\overline{r^2})^{3/2}$. The error thus introduced is only partially compensated by use of $\overline{M}_w$ instead of $\overline{M}_n$ in Eq. (23).   A rough estimate[20] of the magnitude of the error arising from residual heterogeneity of the fractions on which the data of Table XXXVIII were obtained leads to the conclusion that the true value of Φ probably is about 15 to 20 percent greater than the average value quoted above.   While the latter should be appropriate in applications to other fractions similarly prepared by conventional procedures, the true value probably is about 2.5×10^{21} in the previously given units.   This is somewhat smaller than the theoretical value. 3.6×10^{21}, obtained from Eq. (24) of the Kirkwood-Riseman theory. The origin of this discrepancy is not clear.   The likelihood that it represents an inaccuracy of the hydrodynamic theory is diminished by the excellent agreement of the theoretical and experimental values of the corresponding parameter P relating to the frictional coefficient (see p. 628).

If the universal constancy of Φ is accepted, it is then possible to calculate the average dimensions of polymer molecules in solution merely from knowledge of their intrinsic viscosities and molecular weights. More particularly, it is possible to calculate the natural, or unperturbed, dimensions of the polymer chain from the value of K.   Thus, using Eq. (27) one may obtain the quantity $\overline{r_0^2}/M$ which is characteristic of polymer chains of a given homologous series, i.e., consisting of a particular unit (or units in the case of a copolymer of specified composition).   This ratio, as already pointed out, depends on the bond dimensions and angles, and on hindrances to rotation; it also depends on the molecular weight M_0 per chain bond.   (See for example Eq. (X-24) for a simple chain made up exclusively of identical bonds of length l.   The number n of bonds in the chain may for present purposes be set equal to M/M_0, and $\overline{r^2}$ of Eq. (X-24) should be replaced with $\overline{r_0^2}$ in the present notation.)   The results of such calculations carried out using the value of Φ given above are shown in the fourth column of Table XXXIX.   The numbers appearing in this column correspond to $(\overline{r_0^2})^{1/2}$ expressed in Ångstrom units for a molecular weight of 10^6.   They are subject to an absolute error arising from the uncertainty in the value assigned to Φ.   If the constancy of Φ exceeds the experimental uncertainty in its determination (and this probably is true), then the relative values given for $(\overline{r_0^2}/M)^{1/2}$ are significant beyond their absolute accuracy.

Also shown for comparison in Table XXXIX are the values of $(\overline{r_0^2}/M)^{1/2}$ calculated assuming free rotation about each single bond of the chain (see Chap. X).   Appropriate bond lengths and angles have

TABLE XXXIX.—UNPERTURBED END-TO-END CHAIN LENGTHS CALCULATED
FROM K VALUES

Polymer	Temp. °C	K $\times 10^4$	$(\overline{r_0^2}/M)^{1/2}$ $\times 10^{11}$ calcd. from K	$(\overline{r_{0f}^2}/M)^{1/2}$ $\times 10^{11}$	$(\overline{r_0^2}/\overline{r_{0f}^2})^{1/2}$
Polyisobutylene[12]	24	10.6	795	412[a]	1.93
Polyisobutylene[12]	95	9.1	757	412	1.84
Polystyrene[11,14]	ca. 25	8.3	735	302[a]	2.44
Polystyrene[11,14]	70	7.5	710	302	2.35
Poly-(methyl methacrylate)[20]	30	6.5	680	310[a]	2.20
Polyacrylic acid[20]	30	7.6	710	363[a]	1.96
Natural rubber[21]	0–60	11.9	830	485[b]	1.71
Gutta-percha[21]	60	23.2	1,030	703[b]	1.46
Poly-(dimethylsiloxane)[16]	20	8.1	730	456[c]	1.60
Cellulose tributyrate[22]	30		2,000[d]	408[b]	4.9
Cellulose tributyrate[10]	90	12.7	845	408	2.07
Cellulose tributyrate[10]	130	8.2	730	408	1.80
Cellulose tricaprylate[10]	48	12.9	850	366	2.3
Cellulose tricaprylate[10]	140	11.3	810	366	2.2

[a] Calculated taking $l=1.54$Å and $\theta=109.5°$.

[b] Calculated from appropriate bond lengths and angles.

[c] Calculated from Eq. (X-25) with $l=1.65$ Å, $\theta_1=110°$, and $\theta_2=130°$.

[d] Value based on light-scattering dissymmetry for a fraction of molecular weight 220,000.[22]

been used in each case.   The ratio of the mean linear dimensions calculated from K and assuming free rotation are given in the last column. The observed end-to-end distances, unperturbed by long-range interactions, are consistently greater than would hold if bond rotation were completely unrestricted.   Steric effects of one sort or another probably are responsible.   The more compact arrangements of successive chain bonds are preferentially excluded because they tend to superpose neighboring substituents in sterically impossible ways.   Even in a polymethylene chain the *trans* (extended) arrangement is energetically favored over the *gauche* form obtained by a rotation of $\pm 120°$, because of repulsions between methylene groups (see Chap. X, p. 417).   In more highly substituted chains steric interactions are multiplied, but they do not always act in such a way as to render the fully extended form most stable.   In polyisobutylene, for example, with its pair of methyl substituents on alternate carbons of the chain, repulsions between hydrogens on 1,4 pairs of carbon atoms are severe in all configurations, but they are intolerably so in the fully extended planar zigzag form (see Chap. VI).   Hence sequences of successive units seek a compromise arrangement which is neither fully extended nor close-coiled.

Consistent with this interpretation is the fact that an extended unit such as the *trans* unit of isoprene

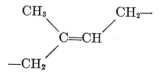

of gutta-percha yields a chain having an unperturbed length more nearly that which it would assume if all single bond rotations could be considered completely free. Somewhat crudely perhaps, we may consider that this unit is long relative to its breadth owing both to the presence of the double bond and to the absence of large substituents. Coiled configurations are therefore less hampered by steric interactions between near-neighboring units of the chain. Similarly, in the poly-(dimethylsiloxane) chain the greater length of the Si—O bond (and also the Si—C bond) and the wide valence angle at the oxygen atoms of the chain greatly reduce steric interactions between methyl groups, with the result that the chain is somewhat nearer its free rotation size than is true for its structural analog, polyisobutylene.

The effect of temperature in reducing $(\overline{r_0^2}/M)^{1/2}$ has already been considered. In most cases the change with temperature is small; it appears to be negligible for the silicone (see Table XXXVII). Of the polymers listed in Table XXXIX, cellulose tributyrate is exceptional in this respect. At high temperatures its $(\overline{r_0^2}/\overline{r_{0f}^2})^{1/2}$ ratio is more or less normal,* and compliance with the viscosity relations set forth above is indicated by constancy of the values obtained for K using fractions differing in molecular weight.[10] As the temperature is lowered, however, the intrinsic viscosity rises rapidly in an athermal solvent such as tributyrin; the intrinsic viscosity at a molecular weight of 220,000 increases over threefold between 130° and 0°C.[10] The change in molecular dimensions with temperature is indicated by the values given for $(\overline{r_0^2}/M)^{1/2}$ in Table XXXIX. Thus, at 30°C the value of $(\overline{r_0^2})^{1/2}$ for a molecular weight of 220,000 exceeds one-half of the value for maximum extension. A molecule of this form can no longer be represented as a spherically symmetric distribution of segments, much less as a random coil, and the foregoing treatment of viscosity

---

* The butyrate substituents are larger than those for other polymers listed, but they are not proportionately larger than the exceptional length of the structural unit of cellulose. An inspection of models shows that the three butyrate units can be accommodated about the cellulose ring with less obstruction than for the phenyl groups of polystyrene, for example.

consequently should not be expected to hold. In particular, the frictional resistance will be substantially less than that for a random coil of equal end-to-end dimension; hence the apparent value of Φ (and K) should be correspondingly lower. Newman[22] has found from light-scattering dissymmetry measurements that $[\eta]M/(\overline{r^2})^{3/2}$ is indeed less than the value of Φ (see Eq. 23) found for randomly coiled polymers.

Other cellulose derivatives, such as the nitrate and the acetate, give evidence of abnormally low chain flexibility also.[23] The cellulose trinitrate chain, for example, exceeds half of its fully extended length at molecular weights less than about 100,000 (degrees of polymerization less than 300). With the proportionately increased opportunities for chain bending in longer chains, the molecule tends to assume the character of a random coil, and above a molecular weight of 400,000 the value of $[\eta]M/(\overline{r^2})^{3/2}$ appears to approach its normal value, $\Phi\cong2\times10^{21}$, as an upper limit.[22,23]

Apparently, extended forms of the cellulosic chain are of substantially lower energies than others. It seems necessary to postulate some sort of specific interaction between successive units, including their substituents, in order to account for energies of the magnitude

$$\begin{matrix} & O & \\ & \diagup\ \diagdown & \\ C & & C \end{matrix}$$

required. These interactions may tend to lock interunit C C ether bonds in the preferred arrangement which gives rise to the extended form. In the case of cellulose tricaprylate, on the other hand, the temperature coefficient is small and $(\overline{r_0^2}/M)^{1/2}$ assumes a "normal" value (see Table XXXIX). The larger substituents appear to suppress the interaction otherwise responsible for the extended form.

3c. The Molecular Expansion Factor α and the General Dependence of the Intrinsic Viscosity on Molecular Weight.—Having established values for K from measurements of intrinsic viscosities in Θ-solvents using fractions of known molecular weight, and having shown also that K normally is independent of the solvent, though often dependent to some extent on the temperature, we are now in a position to deduce values for the volume expansion factor α^3 in good solvents from intrinsic viscosities measured in them. Eq. (26) may be used for this purpose. In effect we take $\alpha^3=[\eta]/[\eta]_\Theta$, where $[\eta]$ is the intrinsic viscosity in the given solvent and $[\eta]_\Theta$ is the intrinsic viscosity of the same polymer fraction (observed directly or calculated from knowledge of K) in an ideal solvent at the same temperature. Values of α^3 derived in this manner are listed in Table XL for several polyisobutylene fractions in cyclohexane and for polystyrene fractions in benzene. The intramolecular expansion factors α^3 exhibit unmistakable increases

TABLE XL.—DEPENDENCE OF α^3 ON MOLECULAR WEIGHT

$M \times 10^{-3}$	$[\eta]$	$\alpha^3 = [\eta]/[\eta]_\Theta$	$[(\alpha^5 - \alpha^3)/M^{1/2}] \times 10^2$
Polyisobutylene in Cyclohexane at 30°C[12,13]			
9.5	0.145	1.39	3.5
50.2	.47	1.96	4.9
558	2.48	3.10	4.7
2,720	7.9	4.46	4.6
Polystyrene in Benzene at 20°C[15]			
44.5	0.268	1.55	2.5
65.5	.356	1.70	2.9
262	1.07	2.54	4.3
694	2.07	3.03	4.0
2,550	5.54	4.22	4.3
6,270	11.75	5.73	5.0

with the molecular weight. This is a most important observation, for it shows directly that the intramolecular interactions alter the molecular dimensions in a manner which increases with M without apparent limit, in conformity with the theory given earlier in this chapter.

According to Eq. (10), the quantity $(\alpha^5 - \alpha^3)/M^{1/2}$, given in the last column of Table XL, should be independent of molecular weight. An appreciable variation over the wide range covered by these data

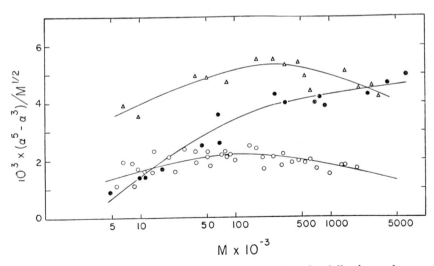

FIG. 142.—$(\alpha^5 - \alpha^3)/M^{1/2}$ plotted against log M for the following polymer-solvent systems: polyisobutylene-cyclohexane ($\triangle$), polystyrene-benzene ($\bullet$), and polyisobutylene-diisobutylene ($\bigcirc$).[12,13]

is evident. These and similar results are shown in Fig. 142.[13] The deviations from constancy at low molecular weights may be dismissed on the grounds that such molecules contain too few segments to justify the adoption of the cloud distribution assumed in the derivation of Eq. (10). The change at higher molecular weights appears to be real and indicates that the function $\alpha^5 - \alpha^3$ occurring in Eq. (10) is only approximately correct. In defense of this relationship it should be pointed out that the logarithmic scale used in Fig. 142 encompasses an abnormally large molecular weight range. Over the molecular weight range usually of interest—from 5×10^4 to 10^6—Eq. (10) offers a useful approximation. Actually, intrinsic viscosities may be calculated with considerable accuracy over a wide range in molecular weight and temperature through the use of Eqs. (26) and (10).[12,24]

The intrinsic viscosity of a linear polymer may usually be approximated by the empirical relation (see Chap. VII)

$$[\eta] = K'M^a \tag{28}$$

This observation is not at variance with the present interpretation, for it may be shown that the dependence of α on M prescribed by Eq. (10) can be represented quite satisfactorily over considerable ranges in M by a power dependence $\alpha^3 \sim M^{a'}$. The exponent in Eq. (28) is given by $a = 1/2 + a'$. The better the solvent the larger will be the right hand member of Eq. (10) and the larger the exponent a' required. If the right hand member in Eq. (10) were sufficiently large, α^3 could be neglected compared with α^5 and we should have $\alpha \sim M^{1/10}$. Then a' would equal 0.30 and a would be 0.80. This is an upper limit, according to the theory of randomly coiled linear polymers, for the exponent a in the empirical equation given above. At the Θ point $\alpha = 1$, $a' = 0$ and $a = 0.50$. This is a virtual lower limit, inasmuch as appreciably poorer solvents (i.e., at appreciably lower temperatures) will not dissolve the polymer. Experimentally observed intrinsic viscosities display good agreement with these predictions concerning the dependence of the exponent a in the empirical equation on the solvent.[12]

3d. Intrinsic Viscosity—Temperature Relations and Thermodynamic Parameters.—A change in temperature may affect the intrinsic viscosity through alteration of both K and α^3. Except for cellulose derivatives which display an unusually rapid alteration in average configuration with temperature, the change of the latter of these factors occurring in Eq. (26) is likely to be dominant. In a very good solvent where the heat (κ_1) of dilution is negative and ψ_1 is normal (i.e., positive), Θ will be negative, and according to Eq. (10) α should decrease with increase in temperature. In an athermal solvent ($\kappa_1 = 0$,

$\Theta = 0$) it should be independent of temperature, and in a poor solvent ($\Theta > 0$) it should increase with T. As is evident from inspection of Eq. (10), this increase should be most rapid in the immediate vicinity of $T = \Theta$; at temperatures well removed from Θ the temperature coefficient of α should be relatively small. Finally, α should change more rapidly with temperature the higher the molecular weight. These deductions from Eq. (10) are well borne out by numerous observations on the temperature coefficients of intrinsic viscosities.[12] The intrinsic viscosities for a polyisobutylene fraction of high molecular weight in several solvents are plotted against the temperature in Fig. 143. In the poor solvents, benzene and toluene, the intrinsic viscosity rises rapidly but at a decelerating rate as the temperature departs from Θ (24° and −13°C, respectively). In cyclohexane the intrinsic viscosity happens to be very nearly independent of T over the range covered; the influence of a small positive Θ (and κ_1) is almost exactly compensated by the decrease in K with temperature. Diisobutylene is clearly a poorer solvent than cyclohexane, but its temperature coefficient indicates a lower value for Θ (and κ_1). The quality of a

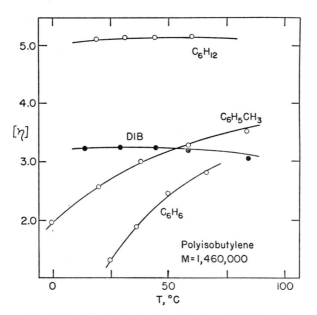

FIG. 143.—The intrinsic viscosity of a polyisobutylene fraction of high molecular weight plotted against temperature in four solvents: cyclohexane, diisobutylene (DIB), toluene and benzene. The lines shown have been calculated according to theory. (Fox and Flory.[12])

solvent evidently is not determined entirely by the energy of inter-action; the entropy parameter ψ_1 enters as an important variable also (see below).

In order to achieve a quantitative separation of the factors respon-sible for the temperature coefficient of the intrinsic viscosity,[2] K should first be established as a function of temperature by carrying out measurements in Θ-solvents having Θ's covering the temperature range of interest. The expansion factor α^3 may then be obtained from the intrinsic viscosity measured at the temperature T in the given solvent. If C_M occurring in Eq. (10) were independent of the temper-ature, $(\alpha^5 - \alpha^3)/M^{1/2}$ should then plot linearly with $1/T$. However, according to its definition given in Eq. (7), C_M is proportional to $(\overline{r_0^2}/M)^{-3/2}$, and hence should change with temperature inversely as K changes. To circumvent this complication it is required merely to plot the quantity $(K_T/K_0)(\alpha^5 - \alpha^3)/M^{1/2}$ against $1/T$, where K_T is the value of K at the temperature T and K_0 is its value at some refer-

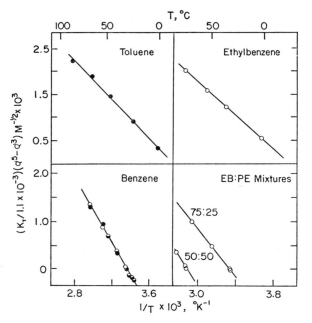

Fig. 144.—The treatment of expansion factor–tem-perature data obtained from intrinsic viscosities of poly-isobutylene fractions in three pure solvents and in ethyl-benzene-diphenyl ether mixtures. Data for fractions having molecular weights $\times 10^{-6}$ of 1.88, 1.46, and 0.180 are represented by O, ●, and ◑, respectively. (Fox and Flory.[12])

ence temperature.[12] Such plots for polyisobutylene in several solvents are shown in Fig. 144. The slopes give $2C_{M,0}\psi_1\Theta$, where $C_{M,0}$ represents C_M at the reference temperature (0°C in Fig. 144); the abscissa intercepts occur at $T=\Theta$. Analysis in this manner of intrinsic viscosity-temperature measurements yields values for $C_M\psi_1$ and Θ for the solvent-polymer pair. The results shown in Fig. 144 for several fractions of polyisobutylene in benzene fall on the same line, which is an indication of the adequacy of the $(\alpha^5-\alpha^3)/M^{1/2}$ function over the tenfold molecular weight range covered.

If we now calculate C_M from Eq. (7), the results of the foregoing analysis yield numerical values for the entropy of dilution parameters ψ_1 in the various solvents. From the Θ's obtained simultaneously, the heat of dilution parameter $\kappa_1=\Theta\psi_1/T$ may be computed. To recapitulate, the value of α^3, in conjunction with $M^{1/2}$, gives at once $C_M\psi_1(1-\Theta/T)$. Acceptance of the value of C_M given by Eq. (7) as numerically correct makes possible the evaluation of the total thermodynamic interaction $\psi_1(1-\Theta/T)$, which is equal to $(\psi_1-\kappa_1)$. If the temperature coefficient is known, this quantity may be resolved into its entropy and energy components.

Thermodynamic parameters deduced as described above are shown in Table XLI for polyisobutylene[12] and for polystyrene.[15,24] It will be recalled that these primary parameters are obtained only with consider-

TABLE XLI.—THERMODYNAMIC PARAMETERS CALCULATED FROM INTRINSIC
VISCOSITIES AND THEIR TEMPERATURE COEFFICIENTS

Solvent	Θ in °K	ψ_1	κ_1 at 25°C
Polyisobutylene[12]			
Benzene	297	0.15	0.15
Toluene	261	.14	.12
Ethylbenzene	251	.14	.117
Cyclohexane	126	.14	.059
Diisobutylene	84	.056	.016
n-Hexadecane	175	.094	.055
n-Heptane	0	.035	0
2,2,3-Trimethylbutane	0	.047	0
Polystyrene[15,24]			
Cyclohexane	307	0.13	0.134
Benzene	100	.09	.03
Toluene	160	.11	.06
Dioxane	198	.10	.07
Ethyl acetate	222	.03	.02
Methyl ethyl ketone	0	.006	0
Methyl n-amyl ketone	210	.05	.04

able difficulty in an unequivocal manner from conventional thermodynamic measurements on dilute solutions, i.e., by osmotic or light-
scattering measurements (see Chap. XII). Where comparisons are
possible with the values obtained by the latter methods,[25] the results
of the viscosity method appear to be lower, perhaps by a factor of about
two. It should be borne in mind in this connection that the absolute
significance of values such as are given in Table XLI rests on the assumption that the theoretically calculated C_M is correct. While the *form* of
the theory given is well confirmed by the various experiments cited,
by no means does it follow that the absolute magnitude calculated for
the parameter C_M is also reliable. If assumptions involved in the derivation of Eq. (10) distort the value of C_M, then relative significance
only should be ascribed to thermodynamic parameters found by the
viscosity method. Within this limitation, the viscosity method provides a remarkably simple means for evaluating the thermodynamic
interaction between polymer and solvent.

It will be observed that entropies of dilution (as indicated by ψ_1)
are highly variable from one system to another. This is contrary to
the theory developed from consideration of lattice arrangements,
according to which ψ_1 should be approximately 1/2 and nearly independent of the system. For polystyrene in methyl ethyl ketone, the
entropy of dilution is nearly zero; i.e., this solvent is a poor one not
because of an adverse energy of interaction but because of the low
entropy.[15] First neighbor interactions apparently contribute to the
entropy as well as to the energy, a point which was emphasized in
Chapter XII. It will be noted also that cyclic solvents almost without
exception are more favorable from the standpoint of the entropy than
acyclic ones.

4. TREATMENT OF EXPERIMENTAL RESULTS: FRICTIONAL COEFFICIENTS

Theory presented earlier in this chapter led to the expectation that
the frictional coefficient f_0 for a polymer molecule at infinite dilution
should be proportional to its linear dimension. This result, embodied
in Eq. (18) where P is regarded as a universal parameter which is the
analog of Φ of the viscosity treatment, is reminiscent of Stokes' law for
spheres. Recasting this equation by analogy with the formulation of
Eqs. (26) and (27) for the intrinsic viscosity, we obtain[26]

$$f_0/\eta_0 = K_f M^{1/2}\alpha \tag{29}$$

where

$$K_f = P(\overline{r_0^2}/M)^{1/2} \tag{30}$$

Equations (29), (30), and (10) might be applied to the elucidation of the frictional coefficient in a manner paralleling the procedure applied to the intrinsic viscosity. One should then determine f_0 (from sedimentation or from diffusion measurements extrapolated to infinite dilution) in a Θ-solvent in order to find the value of K_f, and so forth. Instead of following this procedure, one may compare observed frictional coefficients with intrinsic viscosities, advantage being taken of the relationships already established for the viscosity. Eliminating $\overline{r^2}$ from Eqs. (18) and (23) we obtain[26]

$$f_0/\eta_0 = P\Phi^{-1/3}(M[\eta])^{1/3} \tag{31}$$

which says in effect merely that the effective hydrodynamic radius for translation (i.e., for the frictional coefficient) is *proportional* to the cube root of the effective volume in shear, this latter quantity being proportional to $M[\eta]$; it does *not* imply that the effective spheres are of the same size.* In the light of observations on the intrinsic viscosity, confirmation of Eq. (31) is all that is required to substantiate the extension of the treatment to the frictional coefficient, and in particular to justify the basic equation (18).

The sedimentation constant s_0 at infinite dilution (see Eq. VII-40) is given by

$$s_0 = M(1 - \bar{v}\rho)/Nf_0 \tag{32}$$

where $\bar{v}$ is the partial specific volume of the polymer and ρ is the density of the solvent. From Eq. (31)

$$s_0[\eta]^{1/3}M^{-2/3} = \Phi^{1/3}P^{-1}(1 - \bar{v}\rho)/\eta_0 N \tag{33}$$

Thus the quantity on the left evaluated for a series of polymer fractions differing only in chain length should be independent of M. Results shown in Table XLII for fractions of poly-(methyl methacrylate)[27] and of polyisobutylene[28] covering unusually wide ranges confirm this prediction within experimental error. It is borne out also by less extensive results of sedimentation measurements on several other systems.[26] Introduction of the values of $\bar{v}$, ρ, and η_0 enables

---

* If the frictional effects of the molecule in both translation and in shear were equivalent to those which would be exhibited by the same hard sphere of radius $\mathcal{R}_e$, we should then be justified in eliminating $\mathcal{R}_e$ between Eqs. (15) and (16), with the result

$$f_0/\eta_0 = 0.474 \times 10^{-6}(M[\eta])^{1/3}$$

i.e., $\Phi^{1/3}P^{-1} = 2.11 \times 10^6$, which is somewhat smaller than the observed value (cf. seq.). Evidently, the same equivalent sphere cannot be adopted for both f_0 and $[\eta]$, but such a coincidence is not required by the treatment given here.

TABLE XLII.—SEDIMENTATION CONSTANTS FOR TWO SERIES OF POLYMER FRACTIONS

$M \times 10^{-4}$	$[\eta]_{20°}$	$(s_0 \times 10^{13})_{20°}$ in sec.	$s_0[\eta]^{1/3}M^{-2/3} \times 10^{17}$
		Poly-(methyl methacrylate) in acetone[27],[*]	
744	5.95	107	50.5
321	3.19	69	46.3
142	1.85	48.5	46.7
61.1	0.90	36.5	48.8
30.6	.62	25.2	47.4
14.8	.348	18.8	47.0
7.72	.188	14.1	44.3
		Polyisobutylene in cyclohexane[28]	
142	4.89	4.45 ±0.20	6.0±0.3
67.2	2.87	3.33 ± .11	6.2± .2
17.2	1.12	1.94 ± .04	6.5± .2
8.67	0.706	1.49 ± .02	6.8± .1
3.09	.342	0.925± .01	6.6± .2

* Data given in Tables XLII and XLIII for poly-(methyl methacrylate) are based on the results of Meyerhoff and Schulz[27a] as interpreted by Fox and Mandelkern.[27b]

TABLE XLIII.—VALUES OF $\Phi^{1/3}P^{-1}$ FOR SEVERAL POLYMER-SOLVENT SYSTEMS[26],[27],[28]

Polymer	Solvent	$\Phi^{1/3}P^{-1} \times 10^{-6}$
Polystyrene	Methyl ethyl ketone	2.6
Polystyrene	Toluene	2.3
Cellulose acetate	Acetone	2.7
Polysarcosine	Water	2.3
Polyisobutylene	Cyclohexane	2.5
Poly-(methyl methacrylate)	Acetone	2.6
	Ave.	2.5±0.1

$\Phi^{1/3}P^{-1}$ to be calculated according to Eq. (33) from the mean of $s_0[\eta]^{1/3}M^{-2/3}$. Results for polymer-solvent systems for which suitable data are available are summarized in Table XLIII. The agreement for these diverse polymers, which include a synthetic polypeptide and a cellulose derivative, leaves little basis for doubt as to the essential validity of the simple theory employed. Taking 2.5×10^6 as the best value for $\Phi^{1/3}P^{-1}$ and 2.1×10^{21} for Φ, the experimental P is 5.1, in perfect agreement with the value (see Eq. 20) calculated from the theory of Kirkwood and Riseman.

Let us examine briefly some of the more important consequences of this treatment of the frictional coefficient of flexible chain molecules. It is evident from Eq. (29) and the previous discussion of the manner in which α changes with M according to Eq. (10) that the frictional coefficient f_0 should vary with a power of M equal to one-half in a Θ-solvent, and slightly greater than one-half in a good solvent. In no case should the power be greater than 0.60, provided the primary equation (18) applies. If a is the exponent in the empirical intrinsic viscosity relation, Eq. (28), then

$$f_0 \sim M^{1/2+a''} \qquad (34)$$

where $a'' = (a - 1/2)/3$, as may easily be seen by comparing Eqs. (29) and (26). The sedimentation constant (Eq. 32) will increase with a power of M generally less than one-half; i.e.

$$s_0 \sim M^{1/2-a''} \qquad (35)$$

Experimental results are consistent with this relation,[27,28] but inaccuracies in sedimentation constants preclude precise evaluation of the empirical exponent. Similarly, the diffusion constant at infinite dilution, given by

$$D_0 = kT/f_0 \qquad (36)$$

should vary *inversely* as a power of M generally exceeding one-half, but less than six-tenths.

Diffusion and sedimentation measurements on dilute solutions of flexible chain molecules could be used to determine the molecular extension $(\overline{r^2})^{1/2}$ or the expansion factor α. However, the same information may be obtained with greater precision and with far less labor from viscosity measurements alone. For anisometric particles such as are common among proteins, on the other hand, sedimentation velocity measurements used in conjunction with the intrinsic viscosity may yield important information on the effective particle size and shape.[29]

5. LINEAR POLYELECTROLYTES

5a. General Characteristics.[30]—By incorporating ionic substituents in the polymer chain, substances are obtained which combine the properties of electrolytes and of polymers. Their solutions show large conductances, and thermodynamic measurements give further evidence of ionic dissociation. The configuration of the polyelectrolyte molecule in dilute solution may be greatly expanded by the electrostatic repulsion between its charged groups, and this effect on the configuration manifests itself in a very large intrinsic viscosity. Fuoss and co-workers[30] have carried out extensive investigations on cationic polyelectrolytes prepared by quarternizing poly-(4-vinylpyridine) with butyl bromide. The product

etc., —CH$_2$—CH—CH$_2$—CH—, etc.

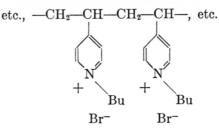

is an analog of polystyrene with a substituent and positive charge at the *para* position in each aromatic ring. It is a strong electrolyte by virtue of the fact that the quarternary ammonium ion claims the bromide ion only by electrostatic forces; the monomeric analog, butyl pyridinium bromide, is completely dissociated in media of high dielectric constant. Sodium polyacrylate

etc., —CH$_2$—CH—CH$_2$—CH—, etc.
            | |
          COO⁻   COO⁻
          Na⁺   Na⁺

is an example of an anionic polyelectrolyte.

The polyelectrolyte molecule in an infinitely dilute solution can be considered to be the microscopic analog of a cross-linked space-network polyelectrolyte gel of the type discussed in Chapter XIII. Thus the lone polyelectrolyte molecule in solution bears a relation to a swollen polyelectrolyte similar to that existing between an uncharged polymer molecule in solution and a swollen macroscopic cross-linked network of similar chemical constitution. The latter analogy was pointed out earlier in this chapter. For the purposes of a qualitative account, we may again replace the restraining influence of the polymer chain, which connects the units together, by an imaginary elastic membrane; the polymer units will be confined within this membrane, and some of them, at least, are presumed to be charged.

For definiteness, let us consider a cationic polyelectrolyte; the polymer units will then be positively charged. The associated counter-anions, i.e., the mobile ions belonging to the polyelectrolyte, may diffuse through the hypothetical membrane into the outer solution but in doing so they leave a residual *net* positive charge within the domain of the molecule. As a result of this net charge, the potential within the molecule is increased relative to the surroundings and further loss of anions is consequently discouraged. A state of equilibrium eventually is reached at which the potential is just enough to support the difference in the concentrations of ions inside and outside the molec-

ular domain, i.e., across the hypothetical membrane.* If we assume
tentatively that the condition of neutrality may be applied to the single molecule in solution, i.e., that the stoichiometric excess of positive
charge within the molecule is negligible compared to its total (fixed)
charge, the problem reduces to one of Donnan equilibrium across the
hypothetical membrane.   The mobile counter-ions of the polyelectrolyte exert an osmotic pressure on the membrane which may be
very large, in accordance with the findings for polyelectrolyte gels.†
The polymer molecule should therefore expand enormously if a substantial portion of its units are ionized.   Addition of a salt to the solution tends to equalize the concentrations of ions inside and outside the
molecule, and thus to decrease the osmotic expansive force.

We may point out parenthetically that it is usually customary to
attribute the expansion to electrostatic repulsions between the net
(positive) charges on the polymer molecule which are uncompensated
due to loss of counter-ions to the outer solution.[30]   It may be shown
that the osmotic force owing to the excess of mobile ions within the
molecule must be equal to the force of electrostatic repulsion when the
molecule is in equilibrium with its surroundings.   Hence either point
of view is equally satisfactory in principle.   The two are, of course,
mutually related; no net charge would develop in the molecule were
it not for the mobile counter-ions, and no excess of mobile ions would
be retained to exert an osmotic pressure if it were not for the charges
on them.

While the condition of stoichiometric neutrality invariably must
hold for a macroscopic system such as a space-network polyelectrolyte gel, its application to the polyelectrolyte molecule in an infinitely
dilute solution may justifiably be questioned.   In a polyelectrolyte gel
of macroscopic size the minute excess charge is considered to occur in
the surface layer (the gel being conductive), which is consistent with
the assumption that the potential changes abruptly at the surface.
This change is never truly abrupt, for it must take place throughout a
layer extending to a depth which is of the order of magnitude of the

---

* With reference to an actual polymer molecule we should, of course, speak
of the potential at a point within the molecule, since the potential will decrease
radially from its center in the manner dictated by the spatial distribution of the
fixed charges (which like the segment density, may often be approximated by
a Gaussian distribution) and that of the counter-ions.   For the purpose of the
present qualitative discussion, however, we refer merely to the potential "inside"
the molecule.

† The units of the polymer chain also may contribute to this osmotic pressure.   This contribution, corresponding to that for the uncharged polymer,
usually is much smaller than the osmotic pressure exerted by the mobile ions.

thickness $1/\kappa$ of the Debye-Hückel ionic atmosphere, where κ (not to be confused with κ_1) is a quantity which increases as the square root of the ionic strength in the surrounding solution. Within this layer there will be a deficiency of the counter-ions (i.e., anions in the case of a cationic polyelectrolyte). Since $1/\kappa$ is always extremely small compared with the dimensions of an ordinary space-network gel, the deviation from electroneutrality in the surface layer does not alter appreciably the ion concentration in the gel as a whole. In the case of a polyelectrolyte molecule this condition may not invariably hold. In water containing an electrolyte at an ionic strength of 10^{-4} molar, for example, $1/\kappa \cong 300$ Å, which may well be an appreciable fraction of the "size" (i.e., $\sqrt{\bar{r^2}}$) of the polymer molecule, even when allowance is made for a considerable expansion of the molecule. Within this surface zone, an appreciable fraction of the mobile ions will have been lost to the outer solution. The condition of stoichiometric neutrality for the molecule as a whole will be acceptable in general only if $1/\kappa$ is small compared to $\sqrt{\bar{r^2}}$. This condition will not hold in the absence of added simple electrolytes at very low concentrations of polymer. On the other hand, it should apply in the presence of even a small quantity of added electrolyte, which reduces the value of $1/\kappa$, and also tends to equalize the difference in concentration of mobile ions inside and outside the molecule (see discussion of polyelectrolyte gels, p. 585).

Let us now consider the process of dilution of a more concentrated solution comprising polyelectrolyte and solvent without added simple electrolyte (i.e., salt). At higher concentrations the polyelectrolyte molecules overlap one another and no incentive is offered for counter-ions to leave the domain of a given molecule. As the solution is diluted regions tend to appear which are not occupied by polymer molecules. The greater the dilution, the greater will be the volume of these regions compared with those occupied by polymer molecules, and more of the mobile counter-ions will diffuse from the molecular domains into the intervening regions of pure solvent. At the same time the osmotic forces of the counter-ions expand the polymer chains so that the unoccupied volume is greatly reduced on this account; or we may with equal validity attribute the molecular expansion to the development of a net charge due to loss of counter-ions. The dilution may eventually be increased until it is quite impossible for the molecules to pervade the entire volume. Continued dilution removes more counter-ions from the domains of the molecules. As the net charge is thereby increased, further loss of counter-ions becomes progressively more difficult. Extreme dilutions may be required before most of them are removed. Calculations indicate[31] that a rather small net charge—less than one electronic charge per ten units—should be sufficient to open the molecule to a configuration approaching full extension. Hence, if a large fraction of the units bear ionic groups, the molecule may be expected to approach full extension long before it is deprived of most of its shield of counter-ions.

The results of osmotic pressure measurements[32] are shown in Fig. 145 for poly-(4-vinyl-N-butylpyridinium bromide) in alcohol, and for the parent uncharged polymer, poly-(4-vinylpyridine), likewise in alcohol. The value of π/c for the former (note difference in scales) is much larger than for the latter, and it *increases* with dilution. The

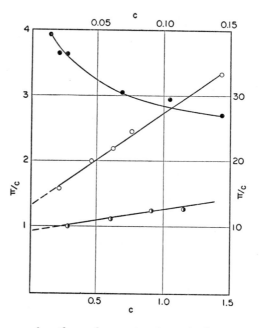

Fig. 145.—Osmotic pressure–concentration ratios (π in g./cm² and c in g./100 ml.) for poly-(4-vinylpyridine) in alcohol, O, coordinates left and below; poly-(N-butyl-4-vinylpyridinium bromide) in alcohol, ● coordinates right and above; and the same polymer in alcoholic 0.61 N lithium bromide, ◑ coordinates left and below.[30,32]

bromide ions effectively given up by the polymer to the solution act like separate osmotic units, in the same way as do the dissociated ions of any strong electrolyte. They therefore contribute to the osmotic pressure. Those "which are electrostatically associated with the polycation are, from the point of view of the osmometer, merely indistinguishable constituent parts of the macromolecule and not independent kinetic units."* As the concentration is decreased, a larger fraction of them escape the influence of the electrostatic field of the polymer chain; hence π/c increases. Over the concentration range covered by the results shown in Fig. 145, from 10 to 20 percent of the anions

* Quotation from Fuoss.[30]

act as though they were free.   The differentiation should, of course, be interpreted as one of degree rather than kind; that is to say, an individual ion is neither completely free nor completely bound, but its momentary situation may vary between these extremes.

If 0.6 N lithium bromide is added to the solution of the polyelectrolyte and also to the solvent on the opposite side of the osmometer membrane, the lowermost set of points in Fig. 145 (lower and left scales) is observed.[32]   The anion concentration inside and outside the coil is now so similar that there is little tendency for the bromide ions belonging to the polymer to migrate outside the coil.   Hence the osmotic pressure behaves normally in the sense that each polyelectrolyte molecule contributes essentially only one osmotic unit.   The π/c intercept is lower than that for the parent poly-(vinylpyridine) owing to the increase in molecular weight through addition of a molecule of butyl bromide to each unit.

Conductance experiments shed further light on the nature of polyelectrolytes in solution.   Transference studies by Wall and co-workers[33] on polyacrylic acid in water showed that nearly all of the current was carried by the few hydrogen ions released by the polyacid.   The high mobility of the former and the small charge on the latter in relation to its bulk (i.e., its large frictional coefficient) accounts for this result.   Partial neutralization of the acid with sodium hydroxide brought about a large increase in the proportion of the current carried by the polymeric ion.[34]   Furthermore, a portion of the sodium ion, approximating half of the total at higher degrees of neutralization, moved with the polymeric anion toward the anode compartment.   It was possible on this basis to differentiate "free" sodium ions which migrate toward the cathode in normal fashion from "bound" sodium ions which were carried toward the anode.   This distinction doubtless is apparent rather than real, inasmuch as the migration tendencies of the sodium ions must vary in a continuous manner depending on their locations with respect to the polyanion, which presumably was rather highly extended under the experimental conditions adopted by Wall and co-workers.[34]   The rather arbitrary distinction is significant nevertheless, for it provides direct evidence that some of the mobile ions effectively escape the influence of the electric field of the polymeric ion while others appear to be retained by it.   At degrees of neutralization exceeding one-fourth, roughly half the current was carried by the polyanion-sodium ion aggregate, having a net charge equal to the difference between the number of ionized groups and the number of retained sodium ions.[34]   The thus-defined net charge is large at higher degrees of neutralization (one-fourth or greater); hence the force exerted on the polyion by the externally applied electric field is large. The polymeric ion consequently migrates, in spite of its size, at a rate comparable to that for a simple ion.

Cathers and Fuoss[35] have shown that the conductance of poly-(4-vinyl-N-butylpyridinium bromide) increases with the dielectric constant of the medium. The energy of removal of a mobile bromide ion from the electrostatic field of the molecule decreases as the dielectric constant is increased; hence the number of "free" ions and the net charge on the polymeric ion should increase. Both contribute toward increasing the conductance.[30]

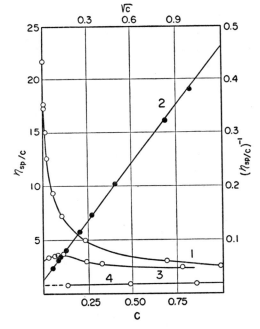

Fig. 146.—Reduced viscosity η_{sp}/c of poly-(N-butyl-4-vinylpyridinium bromide) in aqueous solutions. Curve 1, in pure water; 3, in 0.001 N KBr; curve 4, in 0.0335 N KBr. Curve 2 is a test of Eq. (37) using the data of curve 1. Upper and right-hand coordinates apply to curve 2 only. (Fuoss.[30,36])

5b. Intrinsic Viscosities of Polyelectrolytes.—The viscosities of dilute solutions of polyelectrolytes display a unique dependence on concentration which distinguishes them from nonionic polymers. For the latter the reduced viscosity, η_{sp}/c, increases only gradually with increasing concentration, whereas for the polyelectrolyte it undergoes a marked increase with dilution, reaching values at high dilutions which are many times the intrinsic viscosity which would be expected for the polymer in the absence of charges. Curve 1 of Fig. 146 is representative of results which have been reported for various polyelectrolytes, e.g., sodium pectinate[37] in water, sodium salts of polymethacrylic acid in water,[38] and polyamides[39] in formic acid (which deposits protons on some of the amide units).   At concentrations in the neighborhood of 1 percent, the molecules are in close contact and partially overlap one another.   They are not appreciably expanded, and the specific viscosity is approximately normal.   As the solution is diluted, the polymer molecules no longer fill all of the space and intervening regions extract some of the mobile ions.   Net charges develop in the domains of the polymer molecules causing them to expand.   As this process continues with further dilution, the expansive force increases.   As already mentioned, it may easily be great enough at very high dilutions, where the polymer molecule will have lost many of its mobile ions, to extend it virtually to its maximum length.   The observed strong increase in η_{sp}/c with dilution is in harmony with these considerations.   The curve seems almost to be asymptotic to the ordinate axis, although this cannot actually be true.

Theoretical treatment of the viscosity-concentration relationship for polyelectrolyte solutions would involve both the cumbersome statistics of highly elongated chains beyond the range of usefulness of the Gaussian approximation and the even more difficult problem of their electrostatic interactions when highly charged.   There appears to be little hope for a satisfactory solution of this problem from theory. Fuoss has shown, however, that experimental data may be handled satisfactorily through the use of the empirical relation[36]

$$\eta_{sp}/c = A/(1 + Bc^{1/2}) \tag{37}$$

where A and B are constants.   In practice, straight lines are obtained when $(\eta_{sp}/c)^{-1}$ is plotted against $c^{1/2}$, as in curve 2 of Fig. 146.   Whether or not deviations from this formula would occur at lower concentrations than those at which measurements may be made is of course a matter of conjecture.   The good agreement at higher concentrations suggests that extrapolation to $c=0$ probably is justified.   If this is assumed to be the case, A represents the intrinsic viscosity, $(\eta_{sp}/c)_{c\to0}$.   Values obtained for A in several instances increase with a power of the molecular weight approaching the square,[30,38,39 40] as should be expected for the intrinsic viscosity when the molecules are fully extended.

Addition of potassium bromide (curves 3 and 4 of Fig. 146) suppresses the loss of mobile ions; hence the rise in η_{sp}/c at low concentrations may be eliminated in this manner.[30]   A fixed *low* concentration (curve 3) of potassium bromide has little effect at higher polymer concentrations where the concentration of bromide ions in the polymer greatly exceeds that of the added salt.   As the polymer concentration is diminished, the added salt becomes effective in suppressing further loss of mobile ions from the polymer and η_{sp}/c ceases to rise. The maximum in the curve is explained in this way. It may be eliminated by maintaining a constant total concentration of electrolyte, i.e., by reducing the added salt content as the polyelectrolyte concentration is increased.[41]

Although the polyelectrolyte molecule in the highly extended state occurring in solutions of very low total mobile ion concentration has evaded theoretical treatment, a plausible theory applicable to the less extended polymer molecule in the presence of added electrolyte has been given.[42,43] This theory is merely an extension of the theory of intramolecular interactions for uncharged molecules developed earlier in this chapter. Each element of volume in the domain of the polymer molecule is considered to be in Donnan equilibrium with the outer solution containing electrolyte at an ionic strength S^*, and the osmotic effects of the mobile ions are taken into account in a manner corresponding to the treatment of polyelectrolyte gels (pp. 586 ff.). Eq. (5) may again be used, but to the expression for the chemical potential there must be added a term for the osmotic effects of the ions, whose concentrations are determined by the condition of equilibrium with the outer solution. In this manner, the following relationship may be derived:[43]

$$\alpha^5 - \alpha^3 = 2C_M\psi_1(1 - \Theta/T)M^{1/2} + 2C_I i^2 M^{1/2}/S^* + \cdots \qquad (38)$$

where

$$C_I = (3^3 \cdot 10^3/2^{9/2}\pi^{3/2})(\overline{r_0^2}/M)^{-3/2}N^{-1}M_0^{-} \qquad (39)$$

and i is the number of electronic charges per polymer unit of molecular weight M_0. The first term in Eq. (38) corresponds to the previous equation (10). Its effect usually is small compared with that of the ionic term. The latter predicts a linear dependence of $\alpha^5 - \alpha^3$ on $1/S^*$. Experiments[44] seem to confirm the form of the result. However, the effect observed is much smaller than that predicted by the equations when the degree i of charge is large (i.e., near unity). It appears likely that the ions within the polymer are considerably restrained by the large charge density on the polymer chain, with the result that their full osmotic effect is by no means realized within the molecule.

REFERENCES

1. P. J. Flory, *J. Chem. Phys.*, **17**, 303 (1949).
2. P. J. Flory and T. G. Fox, Jr., *J. Am. Chem. Soc.*, **73**, 1904 (1951); *J. Polymer Sci.*, **5**, 745 (1950).
3. P. Debye and F. Bueche, *J. Chem. Phys.*, **20**, 1337 (1952).
4. M. L. Huggins, *J. Phys. Chem.*, **42**, 911 (1938); **43**, 439 (1939).

5. P. Debye, *J. Chem. Phys.*, **14**, 636 (1946); J. J. Hermans, *Physica*, **10**, 777 (1943); H. A. Kramers, *J. Chem. Phys.*, **14**, 415 (1946).

6. J. G. Kirkwood and P. L. Auer, *J. Chem. Phys.*, **19**, 281 (1951).

7. P. Debye and A. M. Bueche, *J. Chem. Phys.*, **16**, 573 (1948); H. C. Brinkman, *Applied Sci. Res.*, **A1**, 27 (1947).

8. J. G. Kirkwood and J. Riseman, *J. Chem. Phys.*, **16**, 565 (1948).

9. J. G. Kirkwood, *Rec. trav. chim.*, **68**, 649 (1949).

10. L. Mandelkern and P. J. Flory, *J. Am. Chem. Soc.*, **74**, 2517 (1952).

11. A. R. Shultz, thesis, Cornell University, 1953.

12. T. G. Fox, Jr., and P. J. Flory, *J. Am. Chem. Soc.*, **73**, 1909 (1951); *J. Phys. Colloid Chem.*, **53**, 197 (1949).

13. W. R. Krigbaum and P. J. Flory, *J. Polymer Sci.*, **11**, 37 (1953).

14. W. R. Krigbaum, L. Mandelkern, and P. J. Flory, *J. Polymer Sci.*, **9**, 381 (1952).

15. T. G. Fox, Jr., and P. J. Flory, *J. Am. Chem. Soc.*, **73**, 1915 (1951).

16. P. J. Flory, L. Mandelkern, J. Kinsinger, and W. B. Shultz, *J. Am. Chem. Soc.*, **74**, 3364 (1952).

17. T. Alfrey, A. Bartovics, and H. Mark, *J. Am. Chem. Soc.*, **64**, 1557 (1942); L. H. Cragg and T. M. Rogers, *Can. J. Research*, **B26**, 230 (1948).

18. P. Outer, C. I. Carr, and B. H. Zimm, *J. Chem. Phys.*, **18**, 830 (1950).

19. J. Bischoff and V. Desreux, *J. Polymer Sci.*, **10**, 437 (1953); *Bull. soc. chim. Belg.*, **61**, 10 (1952).

20. S. Newman, W. R. Krigbaum, C. F. Laugier, and P. J. Flory, to be published.

21. H. L. Wagner and P. J. Flory, *J. Am. Chem. Soc.*, **74**, 195 (1952).

22. S. Newman, private communication.

23. S. Newman and P. J. Flory, *J. Polymer Sci.*, **10**, 121 (1953); P. M. Doty, N. S. Schneider, and A. Holtzer, *J. Am. Chem. Soc.*, **75**, 754 (1953); R. M. Badger and R. H. Blaker, *J. Phys. Chem.*, **53**, 1051 (1949); R. S. Stein and P. M. Doty, *J. Am. Chem. Soc.*, **68**, 159 (1946).

24. L. H. Cragg, E. T. Dumitru, and J. E. Simkins, *J. Am. Chem. Soc.*, **74**, 1977 (1952).

25. W. R. Krigbaum and P. J. Flory, to be published.

26. L. Mandelkern and P. J. Flory, *J. Chem. Phys.*, **20**, 212 (1952).

27. (a) G. Meyerhoff and G. V. Schulz, *Makromol. Chem.*, **7**, 294 (1952); G. V. Schulz, H.-J. Cantow, and G. Meyerhoff, *J. Polymer Sci.*, **10**, 79 (1953); (b) T. G. Fox and L. Mandelkern, *J. Chem. Phys.*, **21**, 187 (1953).

28. L. Mandelkern, W. R. Krigbaum, H. A. Scheraga, and P. J. Flory, *J. Chem. Phys.*, **20**, 1392 (1952).

29. H. A. Scheraga and L. Mandelkern, *J. Am. Chem. Soc.*, **75**, 179 (1953).

30. R. M. Fuoss, *Faraday Society Discussions*, **11**, 125 (1951); see also, *J. Polymer Sci.*, **12**, January (1954).

31. A. Katchalsky, O. Künzle, and W. Kuhn, *J. Polymer Sci.*, **5**, 283 (1950).

32. U. P. Strauss and R. M. Fuoss, *J. Polymer Sci.*, **4**, 457 (1949).

33. F. T. Wall, G. S. Stent, and J. J. Ondrejcin, *J. Phys. Colloid Chem.*, **54,** 979 (1950).

34. J. R. Huizenga, P. F. Greiger, and F. T. Wall, *J. Am. Chem. Soc.*, **72,** 2636 (1950).

35. G. I. Cathers and R. M. Fuoss, *J. Polymer Sci.*, **4,** 121 (1949).

36. R. M. Fuoss and U. P. Strauss, *Annals N.Y. Acad. Sci.*, **51,** 836 (1949); *J. Polymer Sci.*, **3,** 246 (1948).

37. D. T. F. Pals and J. J. Hermans, *J. Polymer Sci.*, **3,** 897 (1948).

38. A. Oth and P. M. Doty, *J. Phys. Chem.*, **56,** 43 (1952).

39. J. R. Schaefgen and C. F. Trivisonno, *J. Am. Chem. Soc.*, **73,** 4580 (1951); **74,** 2715 (1952).

40. A. Katchalsky and H. Eisenberg, *J. Polymer Sci.*, **6,** 145 (1951).

41. J. J. Hermans and D. T. F. Pals, *J. Polymer Sci.*, **5,** 733 (1950).

42. J. J. Hermans and J. Th. G. Overbeek, *Rec. trav. chim.*, **67,** 761 (1948).

43. P. J. Flory, *J. Chem. Phys.*, **21,** 162 (1953).

44. P. J. Flory, W. R. Krigbaum, and W. B. Shultz, *J. Chem. Phys.*, **21,** 164 (1953).

Glossary of Principal Symbols

a Exponent in the empirical relationship between intrinsic viscosity and molecular weight (Chaps. VII and XIV).

a Distance between the centers of gravity of a pair of interacting polymer molecules (Chap. XII).

a_1, a_2 Activities of solvent and solute, respectively.

A, B   Functional groups. Also units of a polymer or copolymer.

A Work function.

A   Ångström unit (10^{-8} cm.).

A_p, A_t, etc.   Frequency factors in the Arrhenius equation as applied to chain propagation, termination, etc. (Chaps. IV and V).

A_1, A_2, A_3 Coefficients in the virial expansion of the osmotic pressure as a power series in the concentration c (Chap. XII $et\ seq.$).

B Cohesive energy (or free energy) density parameter (Chaps. XII and XIII).

c Concentration, usually in g./cc. but sometimes (especially when used in relation to viscosity measurements) in g./100 cc.; concentration of functional groups in equivalents per unit volume in Chap. III.

c_i Molar concentration of polymer units (Chap. XII).

640

c_s, $c_s{}^*$	Molar salt concentration within a polyelectrolyte and in the surrounding medium, respectively (Chap. XIII).
$\tilde{c}$	Velocity of light in vacuum.
C_M, C_S, C_P	Chain transfer constants (k_{tr}/k_p) for monomer, solvent, and polymer (Chaps. IV, V, and IX).
C_M	Parameter occurring in thermodynamic relations for dilute polymer solutions (Chaps. XII and XIV).
D, D_0	Diffusion constant at finite and at zero concentration, respectively.
e	Base of natural logarithms.
e_1, e_2	Polarity constants in the "Q,e" scheme for copolymerization reactivity ratios (Chap. V).
E	Internal energy (Chap. XI).
E_p, E_t, etc.	Activation energies for chain propagation, termination, etc.
f	Fraction of primary radicals, released by an initiator, which initiate polymer chains (Chaps. IV and V).
f	Functionality of a structural unit (Chaps. VIII and IX).
f	Elastic force associated with a given deformation—usually an elongation (Chap. XI).
f, f_0	Frictional coefficient for a polymer molecule at a finite concentration and at infinite dilution (Chaps. VII and XIV).
f_1, f_2	Mole fractions of monomers 1 and 2, respectively, in the monomer mixture during copolymerization (Chap. V).
f_x, f_x'	Fractions of x-mer species remaining in dilute and in concentrated phases, respectively, formed in the process of fractionation; $f_x' = 1 - f_x$ (Chap. XIII).

f_i Expectancy that a given lattice site is filled when a neighboring site is known to be vacant.

$\bar{f}_i$ Average expectancy that a lattice site is occupied (Chap. XII).

$F, \Delta F$ Free energy (Gibbs) and free energy change.

ΔF_{el} Free energy change for elastic deformation.

ΔF_M Free energy change on mixing.

ΔF_u Free energy of fusion per mole of polymer units.

F_1, F_2 Mole fractions of monomer units 1 and 2 in the increment of polymer formed from monomer mixture having the composition f_1 (Chap. V).

g Factor relating the third virial coefficient Γ_3 to Γ_2^2 (Chaps. VII and XII).

g Ratio of mean-square size of a branched polymer molecule to that of a linear one of the same molecular weight (Chap. X).

H Heat content, or enthalpy.

ΔH_M Heat of mixing.

$\Delta \bar{H}_1$ Partial molar heat of dilution.

$\Delta H_f, \Delta H_p$ Heats of formation and polymerization, respectively.

ΔH_u Heat of fusion per mole of polymer units.

H Parameter relating the turbidity-concentration ratio to the molecular weight (Chap. VII).

i Number of electronic charges per polymer unit in a polyelectrolyte (Chaps. XIII and XIV).

i_θ Intensity of light scattered at the angle θ by a solution containing N particles in a volume V (Chap. VII).

i_θ^0 Same corrected for dissymmetry.

I_0 Intensity of incident light.

$I, [I]$ Initiator and its concentration (Chaps. IV and V).

I_{abs}	Intensity of light absorbed in photochemical polymerization.
J	See Eq. (XII-55).
k	Boltzmann's constant. (Where confusion with a rate constant may arise, $\boldsymbol{k}$ is used.)
k, k'	Reaction rate constants. Units of moles (or equivalents), liters, seconds.
$k_d, k_i, k_p, k_{tr}, k_t, k_z$	Reaction rate constants for initiator decomposition, thermal initiation, chain propagation, chain transfer, chain termination, and inhibition, respectively.
k_{pP}	Rate constant for addition of a diene polymer unit to a chain radical (Chap. IX).
k_{11}, k_{12}, etc.	Copolymerization propagation constants for a radical of the type indicated by the first subscript with a monomer indicated by the second (Chap. V).
K	Constant in the theoretical intrinsic viscosity relationship $[\eta] = KM^{1/2}\alpha^3$.
K_f	Corresponding constant relating to the frictional coefficient f_0 (Chap. XIV).
K'	Constant in the empirical intrinsic viscosity relationship $[\eta] = K'M^a$.
K^*	Parameter relating the Rayleigh scattering ratio to molecular weight (Chap. VII).
l	Number of bifunctional units in a nonlinear polymer containing n polyfunctional units (Chap. IX).
l	Bond length.
ln, log	Natural and decadic logarithms, respectively.
L, L_0	Length of an elastically deformed specimen, and its relaxed length (Chap. XI).
m	Mass of a particle or polymer molecule.

M, $[M]$	Monomer and its concentration
M_1, M_2, $[M_1]$, $[M_2]$	Monomers 1 and 2 in copolymerization and their concentrations (Chap. V).
M_x, M_y	Inactive polymers of x or y units.
$M_x\cdot$, $M\cdot$	Chain radical of size x, and a chain radical of any size.
$[M\cdot]$	Total chain radical concentration.
M, M_i	Molecular weight and molecular weight of species i.
$\overline{M}_n$, $\overline{M}_w$, $\overline{M}_v$	Number, weight, and viscosity average molecular weights, respectively.
M_0	Molecular weight of a polymer chain unit.
M_c	Molecular weight per cross-linked unit. Hence also the molecular weight per chain of a perfect network.
n	Number of polyfunctional units in a nonlinear polymer molecule (Chap. IX).
n	Number of actual bonds (of length l) in a polymer chain (Chap. X).
$\tilde{n}$, $\tilde{n}_0$	Refractive indexes of solution and of pure solvent.
n_1	Number of solvent molecules in a solution (Chaps. XII, XIII, and XIV).
n_2	Number of solute (polymer) molecules in a solution (Chaps. XII and XIII).
n_i	Number of polymer molecules of species i in a solution.
N	Number of particles (Chaps. V and VII). Total number of polymer molecules of all sizes (Chaps. VIII and IX). Number (or number of moles) of primary molecules (Chap. XI).
N_0	Total number of structural units in the system.
N_A, N_B	Number (or number of moles) of A and of B functional groups in the system (Chaps. III, VIII, and IX).
N_x, N_i	Number (or number of moles) of x-mers and of polymer species i.

N_1, N_2, N_i	Mole fractions of solvent, of polymer (total), or of polymer species i in a solution (Chaps. VII and XII).
N_x	Mole fraction of x-mers in the polymer distribution (Chaps. VIII and IX).
N	Avogadro's number.
p	Extent of reaction (condensation polymers), or probability of continuation of the chain (condensation or addition polymers).
p_c	Critical extent of reaction for incipient gelation.
p_A, p_B	Extents of reaction of A and of B groups.
P	Pressure.
P	Parameter relating the frictional coefficient f_0 to the molecular dimension $\sqrt{\overline{r^2}}$ (Chap. XIV).
$P(\theta)$	Factor expressing the reduction in scattered intensity at the angle θ owing to intraparticle interference (Chap. VII).
q	Swelling ratio by volume.
q_m	Same at equilibrium with pure solvent, i.e., maximum swelling (Chap. XIII).
r	Ratio (N_A/N_B) of A to B groups in the system (Chaps. III, VIII, and IX).
r	Ratio of dark to light intervals in photopolymerization with sectored radiation (Chap. IV).
$r, \sqrt{\overline{r^2}}, \sqrt{\overline{r_0^2}}$	Distance between the ends of a polymer chain (i.e., the displacement length), its root-mean-square averaged over all configurations, and the same for unperturbed molecules (Chaps. X to XIV).
r_m	Maximum extension of the polymer chain (i.e., the contour length).
$\mathbf{r}$	Vector connecting the ends of a chain.
r_1, r_2	Monomer reactivity ratios in copolymerization (Chap. V).

R The gas constant.

R_i, R_p, R_t Rates of initiation, propagation, and termination reactions for addition polymerization (in moles per liter per second).

R_θ, R_θ^0 Rayleigh scattering ratio at the angle θ, and the same corrected for dissymmetry (Chap. VII).

$\mathcal{R}$ Ratio of volumes of two liquid phases in equilibrium (Chap. XIII).

$\mathcal{R}_e$ Equivalent hydrodynamic radius of a polymer molecule (Chap. XIV).

$s, \overline{s^2}$ Distance of a specified unit from the center of gravity of the polymer chain, and the square of this distance averaged over all units (Chap. X *et seq.*).

s, s_0 Sedimentation constants at a finite concentration and at infinite dilution (Chaps. VII and XIV).

S Entropy.
ΔS_M Entropy of mixing.
ΔS_M^* Entropy of mixing calculated from external configurational considerations and neglecting first neighbor interactions (Chap. XII).

$\Delta \overline{S}_1, \Delta \overline{S}_1^*$ Corresponding partial molar entropies of dilution.

S^* Ionic strength (molar).

$[S]$ Concentration of solvent or transfer agent.

t Time.

T Absolute temperature (°K).
T_g Glass transition temperature.
T_m, T_m^0 Melting point, melting point of pure polymer.
T_p, T_c Precipitation temperature (liquid phase separation), and critical solution temperature, respectively.

u Excluded volume (Chap. XII).

u Translational velocity of a polymer molecule in sedimentation (Chap. XIV).

U_M, U_A, U_P Resonance stabilization energies for monomer, attacking radical and product radical, respectively (Chap. V).

$\bar{v}$ Specific volume of polymer.

$v_1, v_2, v_x, v_i, v_{2m}$ Volume fractions of solvent, solute (polymer, including all species), x-mer, and species i, and volume fraction of polymer in swollen network in equilibrium with pure solvent, respectively.

V Total volume of the system, or volume of the deformed (swollen) polymer network.

V_0 Volume of the undeformed (unswollen) polymer network (Chaps. XI and XIII).

$V_1, V_2, V_i, V_x, V_s, V_u$ *Molecular* volumes of solvent, polymer, polymer species i, x-mer, a polymer segment, and a repeating unit, respectively.

v_1, v_2, etc.   Corresponding molar volumes.

$\mathcal{U}_{\bullet}$ Equivalent hydrodynamic volume of a polymer molecule (Chap. XIV).

w_x, w_i Weight fractions of x-mer and of species i.

w_s, w_g Weight fractions of sol and of gel.

w Valency factor, z_+z_- (Chap. XIII).

Δw_{ij} Interaction free energy change associated with the formation of a contact between a pair of segments of molecules i and j.

$W(x, y, z), W(r)$ Density and radial distribution functions for the end-to-end coordinates of a polymer chain (usually Gaussian functions).

x Number of structural units in the given polymer molecule (Chaps. I through XII), or its number of segments (Chap. XII *et seq.*).

x_i Number of segments in species i (Chap. XII *et seq.*)

$\bar{x}_n, \bar{x}_w$ Number and weight average number of units, or segments.

X   A monomer substituent.

$y, \bar{y}_n, \bar{y}_w$ Number of units in a given primary molecule, and the weight and number averages for the distribution as a whole.

z Number of primary molecules combined in a given non-linear molecule (Chap. IX).

z Lattice coordination number (Chap. XII).

z_+, z_- Valencies of cation and anion (Chap. XIII).

z_β Dissymmetry coefficient, i.e., ratio of scattered intensities at $(90-\beta)^0$ to that at $(90+\beta)^0$ [Chap. VII].

Z, Z·   Inhibitor or retarder and its radical.

$[Z], [Z\cdot]$ Concentrations of same.

α Optical polarizability (Chap. VII).

α Branching probability (Chap. IX).

α_c Critical value of α for incipient network formation.

α Factor expressing the linear deformation of a polymer molecule owing to solvent-polymer interaction, i.e., $\alpha = (\overline{r^2/r_0^2})^{1/2}$ [Chaps. X, XII, and XIV].

α Deformation of a network structure by elongation, i.e., $\alpha = L/L_0$ (Chaps. XI and XIII).

$\alpha_x, \alpha_y, \alpha_z$ Corresponding factors characterizing a general homogeneous deformation with reference to Cartesian coordinates.

α_s Factor expressing the linear deformation of a network structure by isotropic swelling, i.e., $\alpha_s = (V/V_0)^{1/3} = 1/v_2^{1/3}$ (Chap. XIII).

α_T Thermal expansion coefficient (Chap. XII).

β Characteristic function of the branching probability α (Eq. IX-20) or of the cross-linking index γ (Eq. IX-48).

β Parameter characterizing the Gaussian distribution of the end-to-end (displacement) length of a polymer chain (Chap. X *et seq.*).

β_0 The same parameter for the unperturbed state.

β', β_0' Corresponding parameters for the distribution of chain segments about the center of gravity of the molecule, in the Gaussian approximation.

γ Cross-linking index, i.e., number of cross-linked units per primary molecule in the system as a whole (Chap. IX).

Γ_2, Γ_3 Coefficients in the alternative virial expansion of the osmotic pressure (see Eqs. VII-13 and XII-76).

$\epsilon, \epsilon_0, \Delta\epsilon$ Optical dielectric constant of the medium, of the solvent, and their difference (Chap. VII).

ϵ Expected number of cross-linked units in a primary molecule (Chap. IX).

ζ Frictional coefficient for an element, or bead, of the polymer chain (Chap. XIV).

η, η_0 Viscosity, and the viscosity of the pure solvent.

η_r Relative viscosity, η/η_0.

η_{sp} Specific viscosity, $\eta_r - 1$.

$[\eta]$ Intrinsic viscosity (expressed in deciliters per gram).

θ Angle between transmitted and scattered beam (Chap. VII).

θ Degree of conversion of monomer to polymer in addition polymerization (Chap. IX).

θ Valence angle between successive bonds in the polymer chain (Chaps. X and XIV).

Θ "Ideal" temperature at which van't Hoff's law is obeyed for a given poor solvent-polymer system; $\Theta = \kappa_1 T / \psi_1$.

κ Coefficient of compressibility (Chap. XI).

κ_1 Parameter expressing the energy, divided by kT, of interaction between a solvent molecule and polymer (Chap. XII $et~seq.$).

λ, λ' Wavelength in vacuum and in the medium of refractive index $\tilde{n}$ or $\tilde{n}_0$.

μ_1, μ_2, μ_x, μ .   Chemical potential (i.e., partial molar free energy) of solvent, polymer (all species collectively), x-mer, and a polymer unit, respectively.

μ_1^0, etc.   Chemical potentials in standard states (pure liquids in all cases).

$(\mu_1 - \mu_1^0)_E$ Excess (i.e., nonideal) chemical potential of the solvent.

ν Kinetic chain length (Chaps. IV and V).
ν Number of monomers combined per growing center in a system yielding the Poisson distribution (Chap. VIII).

ν Number (or number of moles) of cross-linked or branched units (Chaps. IX and XI). Hence, also the number of chains in a perfect network structure (Chap. XI).
ν_e Effective number (or number of moles) of chains in a real network (Chaps. XI and XIII).

ν_i Number of ways of arranging the i-th polymer molecule on a lattice to which $i-1$ polymer molecules have been added previously (Chap. XII).

ν_+, ν_-, ν Number of cations, anions, and total ions $(\nu = \nu_+ + \nu_-)$ into which the given electrolyte dissociates (Chap. XIII).

ξ Number of bonds in the repeating unit of a condensation polymer (Chap. VIII).

ξ Variable depending on thermodynamic parameters (Chap. XII; see Eq. XII-56).

π — Osmotic pressure.

$\prod$ — Product sign.

ρ — Rate of radical generation in the aqueous phase during emulsion polymerization (Chap. V).

ρ — Density of solvent medium. (Used especially in sedimentation velocity equation, Chaps. VII and XIV.)

ρ — Degree, or density, of cross-linking (or branching), i.e., fraction of units cross-linked in the entire system (Chaps. IX and XI).

ρ', ρ'' — Same for sol and gel components, respectively (Chaps. IX and XI).

ρ — Segment density, expressed in number per unit volume, within a given volume element δV.

ρ_k, ρ_l — Same for polymer molecules k and l (Chap. XII).

σ — Parameter governing the partitioning of polymer species between phases in fractionation (Eq. XIII-23).

$\sum$ — Summation sign.

τ, τ_s — Lifetime of a free radical, and the same in the steady state (Chap. IV).

τ, τ^0 — Turbidity as determined by light-scattering measurements, and the same quantity corrected for dissymmetry (Chap. VII).

τ — Equilibrium retractive force, referred to unit undeformed cross section, for elongated rubber (Chap. XI).

τ_α — Same quantity specifically for the elongation α.

ϕ — Angle of rotation about one of the single bonds of the polymer chain (Chap. X).

Φ — Parameter relating the intrinsic viscosity to the molecular dimension $\sqrt{\bar{r}^2}$ (Chaps. VII and XIV).

χ_1 — Parameter expressing the first neighbor interaction free energy, divided by kT, for solvent with polymer (Chap. XII *et seq.*).

χ_{ij} — Corresponding parameter for the interaction of component i with j in a polycomponent system.

ψ_1 — Parameter characterizing the entropy of dilution of polymer with solvent.

ω — Angular velocity (Chaps. VII and XIV).

ω_z, ω_z' — Combinatory factors occurring in size distribution functions for nonlinear condensation polymers (see Eqs. IX-18 and -34).

ω_i — Probability density for a chain displacement vector $\mathbf{r}_i$ (Chap. XI).

Ω — Total number of configurations.

Author Index

Roman-type numbers refer to discussion of work of the author cited. Italic numbers, with reference numbers superscripted, refer to chapter bibliographies.

658 AUTHOR INDEX

Subject Index

Activation energies, *see* Free radical polymerization, *etc.*

Addition polymers and polymerization:
of cyclic compounds, 39, 57–61
definition of, 37–39
of unsaturated monomers, 51–55, 106 ff. (*see also* Anionic polymerization, Cationic polymerization, Free radical polymerization, *and* Vinyl polymers

Allylic polymerization, 172–174

Anionic polymerization, 61, 224–226

Association theory of polymers, 7–11, 17, 20–21

Asymmetry in structural units, 37, 55–56, 237–239

Autoacceleration, 124–129, 160

Binary systems:
critical conditions in, 543–545
crystallization in, 568–576
experimental results on liquid, 546–548
liquid-liquid, 542–548
phase diagrams for, 545–547, 574–576
of two polymers, 554–555

Bond angle restrictions, 410–413 (*see also* Hindrance to free rotation)

Branched polymers, 30, 32–35, 256–260 (*see also* Branching, Cross-linked polymers, Polyfunctional condensation, *etc.*)
molecular configuration in, 422–423, 611
molecular distribution in, 365–377

Branching:
kinetics of, 385
short chain, 259–260
special case of, 361 ff.
in vinyl polymers, 34, 256–260, 384–386, 390

Branching coefficient, 350–353

critical value of, 352–353

Cage effect in chain initiation, 119–121

Cationic polymerization:
catalysts for, 217
kinetics of, 222–224
mechanism of, 219–222

Cellulose:
early investigations on, 10–11, 22
kinetics of hydrolytic degradation of, 83–85

Center of gravity, mean-square distance from, *see* Radius of gyration

Chain dimensions, *see* Configuration of polymer chains *and* Molecular dimensions

Chain displacement length distribution:
for freely jointed chain, 402–410, 426–428
Gaussian approximation for, 404–410, 426–427
at high extensions, 409–410, 427–428
for hindered chains, 410–413
intramolecular interactions and, 424–426, 599–600
in rubber elasticity theory, 465–467

Chain initiation (*see also* Initiators):
absolute rate of, 117–121, 155
efficiency of, 113–114, 117–121, 141
by inhibitors, 163, 168–169
mechanism of, 107–110, 119–121
thermal, 116, 130–132, 202–203

Chain length (*see also* Degree of polymerization):
kinetic, 132–135, 157

Chain propagation:
activation energies and steric factors for, 158–160, 191–194, 232, 236–237
effects of substituents on, 159–160, 189–194, 231–232, 238, 240